Nano-electrocatalyst
for
Oxygen Reduction Reaction
Fundamentals to Field Applications

Editors

Omar Solorza Feria
Departamento de Química
Centro de Investigación y de Estudios Avanzados (Cinvestav IPN)
Ciudad de México, México

Sathish-Kumar Kamaraj
IPN-Centro de Investigación en Ciencia Aplicada y Tecnología Avanzada
Unidad Altamira, (CICATA Altamira)
Tamps., Mexico

CRC CRC Press
Taylor & Francis Group
Boca Raton London New York

CRC Press is an imprint of the
Taylor & Francis Group, an **informa** business
A SCIENCE PUBLISHERS BOOK

Cover credit: The images reproduced by kind courtesy of Dr. Daniel Bahena Uribe and Dr. Omar Solorza Feria.

First edition published 2024
by CRC Press
2385 NW Executive Center Drive, Suite 320, Boca Raton FL 33431

and by CRC Press
4 Park Square, Milton Park, Abingdon, Oxon, OX14 4RN

© 2024 Omar Solorza Feria and Sathish-Kumar Kamaraj

CRC Press is an imprint of Taylor & Francis Group, LLC

Library of Congress Cataloging-in-Publication Data (applied for)

ISBN: 978-1-032-36973-0 (hbk)
ISBN: 978-1-032-37018-7 (pbk)
ISBN: 978-1-003-33490-3 (ebk)

DOI: 10.1201/9781003334903

Typeset in Times New Roman
by Shubham Creation

Dedication

I'm dedicating this book to my mentor and founder of the
Mexican Hydrogen Society, **Dr Omar Solorza Feria**.
Thank you for your wisdom, your wit, and your friendship. I am truly grateful.

Preface

The environmental impact of global warming combined with the global demand for energy forces us to completely transition away from our reliance on fossil fuels towards greener and more sustainable forms of renewable energy. In the field of renewable energy, nanoelectrocatalysts with higher efficiency and more promising properties have garnered a lot of attention recently. This is because of the intention to use them. Hydrogen fuel cell/polymer electrolyte membrane (PEM) vehicles are the primary environmental-friendly electromobility vehicles that possess a higher energy density with fast refuelling technology, which triggers the interest towards the automobile industry to launch various types of PEM fuel-cell vehicles in the global market. PEM fuel cells face a number of significant technical challenges, the most significant of which are their high cost and low reliability and durability. Despite this, the commercialization of PEM fuel cells has not been successful due to a number of factors. Fuel cell catalysts, such as those based on platinum (Pt), as well as their associated catalyst layers are the primary contributors to these difficulties. Exploring new catalysts, improving catalyst activity and stability/ durability, and lowering catalyst cost are currently the primary approaches being taken in fuel cell technology and commercialization. These steps are being taken in an effort to overcome the challenges. The oxygen reduction reaction, also known as ORR, is primarily of greater interest in fuel cells and metal-air batteries. Non-noble catalysts, the next generation of ORR-based energy generation catalysts slated to replace expensive Pt, have been recognized as the sustainable solution for the commercialization of PEM fuel cells. This is in relation to the cost reduction of Pt-based catalysts. In more recent years, the global community of fuel cell researchers and developers has increased their efforts in this area of research and development. Additional worries about the slow kinetic reaction of ORR, which is responsible for the rate-limiting reaction at the cathodic system of the energy generation systems, will, as a result, cause the systems to have a lower energy efficiency. Despite this, optimizing ORR for market expansion in a way that is both cost-effective and efficient based on nanoelectrocatalysts continues to be a challenge and an ongoing task. As a result, the various fundamental issues pertaining to the ORR reaction kinetics theories, measurement tools, techniques and nanoelectrocatalysts of noble, non-noble metals, and nano-carbon materials were focused. To support readers in comprehending the scientific and technological aspects of nano-electrocatalysis, this book presents a number of significant and representative figures, tables, photographs, and a comprehensive list of reference papers. In our opinion, this book will be of great use to scientists and engineers working in the energy or fuel cell industries. We expect that the reader will be able to quickly find the most up-to-date information on the basics and applications of catalyzing the oxygen reduction reaction in next-generation energy-generation systems as a direct result of reading this book. Those studying energy, electrochemistry science/technology, fuel cells, electrocatalysis, and metal air batteries at the undergraduate and graduate levels, as well as scientists and engineers in these fields, will find this book useful.

Now, we would like to explicitly thank all of the contributors who worked so diligently on their respective chapters.

Omar Solorza Feria
Sathish-Kumar Kamaraj

Acknowledgements

In the first place, we want to give thanks to God for blessing us with good health and the ability to edit this book. Our deepest gratitude goes to the series editor and advisory board for believing in our work and accepting our book for publication. Thanks to everyone who helped make this book a reality—the authors, the reviewers, and everyone in between. We are grateful to the many publishers and authors who gave us permission to use their work, especially the figures and tables.

Kamaraj, Sathish-Kumar would like to express his gratitude to Director General of Instituto Politécnico Nacional (IPN) and Director of Centro de Investigación en Ciencia Aplicada y Tecnología Avanzada, Unidad Altamira (CICATA Altamira) for their constant support and facilities to enable promotion of research activities; to the Secretaria de Invesigacion y Posgrado (SIP) for the project number 20231443; and for further extensions to the funding agency of the National Council for Science and Technology (CONACyT, México) and Secretary of Public Education (SEP, México). He extended his gratitude to Mrs Kamaraj Mounika and Bbg Aarudhraa for their affection and family support.

Omar Solorza Feria
Sathish-Kumar Kamaraj

Contents

Chapter **1**

Surface Active Nano-islands Bounded PGM and non-PGM-based Nano-electrocatalysts-driven Efficient ORR Activity for DMFCs and AAEMFCs

P. Anandha Ganesh[1]* and D. Jeyakumar[2,3]

[1]State Key Laboratory for Mechanical Behavior of Materials, School of Materials, Science and Engineering, Xi'an Jiaotong University, No. 28, Xianning West Road, Xi'an, 710049 Shaanxi, China.

[2]Fuel Cell Catalysis and Nano-materials Group, Functional Materials Division, CSIR, Central Electrochemical Research Institute, Karaikudi–630006, Tamil Nadu, India.

[3]V.B. Ceramics Research Centre, Chennai–600041, Tamil Nadu, India.

1.1 INTRODUCTION

Energy is vital in our life and most of our global energy needs are being satisfied through non-renewable and environmentally hazardous conventional resources, like fossil fuels, nuclear energy, and thermoelectric powers. It has been predicted by researchers that coal would be the only fossil fuel resource

*For Correspondence: Email: nanoganesh@xjtu.edu.cn

after the year 2042 and it will replenish before the year 2110. Burning of these fossil resources accounts for over 75% of global emissions (greenhouse gases like CO_2, CH_4, N_2O, etc.). In addition to these environmental issues, rapidly increasing global population (to reach 8 billion by November 2022), worldwide energy crisis coupled with infrastructure expansion and increasing utilization of dwindling fossil resources urges the research and development of alternative and/or sustainable clean energy devices. It was well known that the United Nations Climate Change Conference that took place in Paris on 12th December 2015, was signed to decrease the global temperature by 1.5 degrees Celsius. Hence, research on the development of sustainable and renewable energy sources, like solar, wind, hydroelectric, tidal are being pursued as alternative sources of energy for reduced or zero hazardous emissions. But these renewable energy resources are intermittent and their energy deliverances are seasonal and cannot be relied upon as a regular source of energy throughout the year. In this regard, energy sources including fuel cells, super-capacitors, batteries, etc. are considered as alternatives in which batteries and super-capacitors need regular charging (Hoogers, 2003). But, fuel cells don't need to be charged and continuously deliver energy as long as the fuel is fed into the system. Further benefits of fuel cells are noiseless operation, less hazardous and exhibit high volumetric energy densities compared to batteries and super-capacitors. Presently, the main target for fuel cell researchers is to improve the power density, overall efficiency and durability at reduced cost. Among numerous kinds of fuel cells, polymer electrolyte membrane fuel cell (PEMFC) is one of the best suitable and clean energy conversion type for vehicles or stationary power applications (Hoogers, 2003). PEMFC uses H_2 and O_2 as anode and cathode side fuels and Nafion® ($C_7HF_{13}O_5S \cdot C_2F_4$), a perfluoro sulfonic acid-based polymer membrane as electrolyte which conducts H^+ ion (3.63×10^{-3} cm^2/V.s for H^+ (H_3O^+)) through it and hence it is called as proton exchange or polymer electrolyte membrane fuel cell (Fuller, 2008). It covers 70–80% of small-scale fuel cell market, owing to its low operation temperature (25–80°C), short start up time (at ambient temperature), moderate energy density for H_2 and power densities along with water and heat as the only by-products (Hoogers, 2003). But the major challenges in PEMFC commercialization are the relatively expensive and low stability issues of the Pt-based PEMFC electrocatalyst. On the other hand, direct methanol fuel cell (DMFC), a PEMFC variant, involves reduction of oxygen at the cathode and oxidation of methanol at anode with the aid of electrocatalysts present at the respective electrodes. Methanol holds many advantages, like high room temperature, volumetric energy density value at 15.6 MJ/L, ease of handling-cum-storage and no complex C-C cleavage compared to other small organic molecules (Fuller, 2008) and is more advantageous than H_2 (volumetric energy density of 0.01 MJ/L at ambient conditions). However, the DMFC efficiency was not completely realized because of three major reasons: (i) methanol cross-over, (ii) slow oxygen reduction reaction (ORR) kinetics, and (iii) CO poisoning of catalytic Pt centres at the anode side. Conventional DMFCs use Pt-Ru/C catalyst for methanol oxidation reaction (MOR) and carbon supported Pt catalyst for ORR. It is clearly established that, Pt-Ru/C anode catalyst performs well (MOR) at the anode side and minimizes the CO poisoning owing to its bimetallic (electronic effect) nature. But unfortunately, the Pt/C ORR catalyst oxidizes the methanol that cross over from anode to cathode during ORR. This leads to mixed cathode potential generation due to the simultaneous MOR and ORR at cathode side, thereby reducing the DMFC performance. Hence, an efficient methanol-tolerant platinum group metal (PGM)-based electrocatalyst is needed for PEMFC or DMFC (owing to its high acidic flux) (Fuller, 2008) because a non-PGM-based electrocatalyst exhibits low stability and mass transport for ORR process in acidic medium. Alternatively, hydrogen-oxygen served alkaline exchange membrane fuel cells (AEMFCs) offer several advantages, like reduced fuel cross-over from anode to cathode, OH$^-$ environment, low-cost non-PGM-based ORR electrocatalysts and decreased water flooding issues compared to acidic PEMFCs. (Ganesh et al., 2021). But, slow ORR kinetics and carbonation issues caused by the alkaline electrolyte and carbon-supported non-PGM electrocatalysts decrease the AAEMFCs performance (Ganesh et al., 2021). However, several non-PGM-based electrocatalysts (transition metal oxides, M-N-C catalysts, heteroatom doped porous/graphitic carbon materials, Ag-based catalysts, etc.) (Ganesh and Jeyakumar, 2018a; Balaji et al., 2020; Kaipannan et al. 2020; Ganesh et al., 2021) were shown to have better ORR performance and stability. In the past decades, huge efforts have been made in developing novel nanostructured Pt (active component) based efficient bimetallic ORR catalyst, Pt-monolayer catalysts, Pt-skin catalysts with tuneable size, surface morphologies and compositions at low Pt content[7–17]. A combined geometric and electronic effect in these nanostructured Pt-based ORR electrocatalysts that involve either Pd or Au or transition elements (Fe, Co, Ni, etc.), as the other component enhance the sluggish ORR kinetics and tolerance to cross-over methanol (Hoogers, 2003). These bimetallic catalysts in the form of core shell nanoalloys (Ilayaraja et al., 2013), dendrite-like structures, porous surfaced and Pt monolayer/adorned nanostructures not only improve the ORR kinetics through synergistic and electronic coupling effects, but also help to maximize the Pt utilization and reduce the Pt content in PEMFC and DMFC catalyst layers, respectively (Cho et al., 2017; Góral-Kurbiel et al., 2016; Yang et al., 2016). Considering the requirements for DMFC, the ORR electrocatalyst should also be tolerant to methanol. Thus it is clear that there is a great need for the development of a stable, size- and surface-textured efficient methanol-tolerant nanostructured ORR catalyst with low Pt content for next generation DMFCs. This chapter begins with a brief introduction on PGM and non-PGM-based nanostructured DMFC ORR electrocatalysts. Next, the wet chemical reduction process that has been used for the synthesis of nano-porous surface active PGM and non-PGM ORR catalysts will be demonstrated. This is followed by a discussion on significant characterization techniques used for corroborating the size, shape and surface texture of electrocatalysts with their ORR activity. Finally, it summarizes the electrochemical ORR activity, fuel cell performance and stability of the surface active PGM-based nano-porous ORR electrocatalysts and non-PGM-based ORR electrocatalysts with that of benchmark HiSPEC Pt/C ORR catalysts.

1.2   A HISTORICAL VIEW OF ORR ELECTROCATALYSIS

1.2.1   Pioneering Classical Works on ORR

Currently, a huge number of researchers are working on ORR electrocatalytic materials with the aim to improve their kinetics in electrochemical fuel cell or metal/redox flow air batteries. But, the idea on ORR came long back around 1940s. H.A. Laitinen and I.M. Kolthoff in 1940, first noticed the reduction of oxygen to hydrogen peroxide during their voltammetric studies using stationary platinum micro-electrodes. They used rotating platinum micro-electrodes and found a tremendous increment in the diffusion current (61 mA) compared to that obtained from using stationary platinum micro-electrodes (3 mA) for ORR (Laitinen and Kolthoff, 1940). In 1949, Krasil'schikov carried out ORR studies on smooth Ag, Au and Pt polycrystalline cathodes (Breiter, 1969). B.E. Conway et al. in 1950, corroborated the ORR activity response as observed by Krasil'schikov, using polycrystalline Pt and Hg electrodes (Azzam et al., 1950). They also postulated the major rate-controlling processes in ORR *ca.* oxygen diffusion to electrode surface, oxygen adsorption on electrode surface and the reduction of adsorbed oxygen involving and not involving hydrogen atoms and observed the formation of H_2O_2 intermediate during ORR process on Hg electrodes. In 1961, Sawyer evaluated the reduction of oxygen on Pt, Pd, Au, Ag, Ni, Cu, Ta, W and Pb-based polycrystalline electrodes using cyclic voltammetry and chrono-potentiometric studies (Sawyer and Interrrante, 1961). Roger Parsons in 1964 explained about the kinetics of electrode reactions and revisited the oxygen and hydrogen adsorption on platinum wires (Parsons, 1964). Charles C. Liang and Andre L. Juliard investigated the issue of H_2O_2 intermediate generation during ORR process on Pt electrode in sulphuric acid electrolyte (Liang and Juliard, 1965) and measured the ORR threshold rate and over-potential at the rate of surface oxides removal on Pt electrode.

In 1960s, A. Damjanovic and his group carried out pioneering works on ORR to understand its behavior on polycrystalline electrode surface. In 1966, A. Damjanovic et al. used bare, oxide-covered Pt electrode surfaces to measure the rate constants of oxygen reduction and found better ORR activity on bare Pt surfaces (Damjanovic and Brusic, 1966a). They also used pre-treated Pt electrodes for ORR in both acid and alkaline medium (Damjanovic and Brusic, 1966b). In 1967, A. Damjanovic and V. Brusic measured the ORR kinetics like Tafel behavior and exchange current densities using Pt, Pd and Au alloys in acidic medium (Damjanovic and Bockris, 1966c) and the ORR activity was ascribed to the heat change of adsorbed reaction intermediates and alloy compositions (Damjanovic et al., 1966). The effect of thermal, chemical and electrochemical treatment on Pt-Au and Pd-Au alloy electrodes and oxide-free Pt surfaces for ORR studies has also been carried out by this group (Damjanovic et al., 1966). Genshaw along with A. Damjanovic in 1967, used rotating ring disk electrode (RRDE) measurements reported the two-electron pathway for ORR and H_2O_2 intermediate formation on gold electrodes (Genshaw et al., 1967). H. Wroblowa, studied the O_2 adsorption and its kinetics on oxide covered Pt surfaces in relation to the oxygen partial pressure and rest potential effect in $1N$ H_2SO_4 (Wroblowa et al., 1967). The Temkin and Langmuir adsorption isotherms of oxygen adsorption were proposed to determine the ORR kinetics on oxide-free Pt surfaces in acid and alkaline conditions (Wroblowa et al., 1967) and proposed a simplified scheme for ORR mechanism, based on two-electron and four-electron pathways (Wroblowa et al., 1976). The rest potential on oxide-free Pt electrodes in electrolyte with very less impurity was emphasized by A.J. Appleby (Appleby, 1972). In collaboration with A. Damjanovic, many researchers tried to understand the ORR kinetics and mechanism on Pt electrodes in the late 1970s[18–20]. In 1982, J.P. Hoare from General Motors research laboratories studied ORR on both oxide-covered and oxide-free polycrystalline Pt electrodes in acidic electrolytes and ascribed the improved ORR activity to the carbon clusters adsorbed on Pt (Hoare, 1982). Steininger et al. in 1982 evaluated the oxygen adsorption behavior on low index Pt (111) plane electrodes using temperature programmed thermal desorption spectroscopy (TPTDS), high-resolution electron energy loss spectroscopy and low energy electron diffraction studies (Steininger et al., 1982). In 1983, Srinivasan et al. carried out RRDE studies to understand the ORR kinetics and intermediates generation on Pt-based electrodes in H_2SO_4 electrolyte (Hsueh et al., 1983a). They also performed RRDE studies to evaluate the anion effect adsorption on Pt surface oxide formation using different electrolytes (Hsueh et al., 1983b). In 1983, E. Yeager used various metals and carbon-based electrocatalysts and proposed the mechanisms of ORR through direct four e^- pathway, two e^- peroxide pathway and one e^- superoxide pathway in acidic and alkaline solutions with thermodynamic reversible potentials for those pathways (Yeager, 1984). A. Damjanovic in 1986 described the symmetric factor and transfer coefficient values in acidic and alkaline medium for ORR as 0.5 and 1 respectively for Pd- and Pt-based electrodes (Sepa, et al., 1986a). Based on activation enthalpy calculations, they concluded that the ORR rate determining step is a slow first e^- transfer process. The effect of activation enthalpies and pH in determining the kinetics and mechanisms of ORR was also evaluated (Sepa et al., 1986b). In addition, they proposed two ORR Tafel values (60 and $120\,mV/dec$) attributed to the O_2 adsorption conditions and reaction intermediates (Sepa, et al., 1986c). ORR studies were carried out on faceted single crystal based Pt surfaces including Pt (111), (110) and (100) in various acidic electrolytes (Kadiri et al., 1991). They observed a very low anion adsorption and increased reaction rates for ORR using $HClO_4$ compared to H_2SO_4, H_3PO_4, etc. In 1992, Trasatti et al. proposed different possible methodologies for the electrochemical surface area (ECSA) calculations through *in-situ* and *ex-situ* processes applicable to most of the electrode types (Trasatti and Petrii, 1993) along with the conditions and limitations in each method towards real surface area measurements. In 1993, A.J. Appleby proposed kinetic models for ORR for Pt electrodes with H^+ transfer step as the slow ORR rate-determining step in both acidic and alkaline aqueous media (Appleby, 1993). In 1993, Sanjeev Mukerjee et al. used Pt and Pt-M (M = Ni, Cr, Co) bulk alloys as ORR electrocatalysts for PEMFCs (Mukerjee and Srinivasan, 1993). They noticed an enhanced ORR activity in alloy catalysts compared to bare Pt catalysts and in contrast, a reduced ORR activity with increasing particle size (decreasing surface area) for all catalysts.

ORR was studied on low index Pt single crystal planes using rotating disk electrode (RDE) in $HClO_4$ electrolyte (Markovic et al., 1994). They observed an increase in half-wave potentials for ORR through four electron pathways in well-ordered planes and the Tafel values of 60 and 120 mV/decade was ascribed to the oxygen containing intermediate adsorption. Conway et al. in 1995 described the mechanism of electrochemical oxide film formation on polycrystalline Au, Pt and Rh surfaces using voltammetry and surface techniques (Conway, 1995). ORR kinetics was studied on thin film Pt electrode with varying thickness (1 to 50 nm) using RDE measurements (Tammeveski et al., 1997) and the ORR performance is endorsed to the presence of small sized Pt in the thin film catalyst. Hall et al. used RDE and cyclic voltammetry to evaluate the H_2O_2 oxidation on Pt as a function of potential, pH and temperature (Hall et al., 1998). Maruyama et al. in 1998 evaluated the ORR activity and H_2O_2 generation on Nafion-solution-coated Au electrodes using RRDE measurements (Maruyama et al., 1998). Gonzalez et al. discussed the particle size effect in thin and porous carbon-coated Pt electrodes and faceted crystallites on ORR using RDE technique in both acidic and alkaline media (Perez et al., 1998).

1.2.2   Nanostructured Materials for ORR

Nanostructures hold increase surface area to volume ratio owing to their reduced particle size in the nano dimension range (1–100 nm) compared to their bulk counterparts. This enhanced surface area forms one of the essential criteria for efficient electrochemical reactions, since electrochemical reactions are surface dependent. As a result, nanostructures exhibit distinct electrochemical properties. In addition, nanostructures with porous moiety or surface roughening (i.e. nano-porous materials) offer further enhanced surface area to volume ratio and efficiently catalyze through facile adsorption of reaction intermediates. As a result nano-porous surface-textured materials act as better catalysts for various electrochemical reactions (Kesavan et al., 2018) including fuel cell electrocatalysis (Hoogers, 2003; Ganesh and Jeyakumar, 2017a). Facile aqueous synthesis of nano-porous structures with controllable size, shape, surface roughness with kinks, steps and catalytic edges at atomic levels are inevitable to determine its efficacy in electrocatalysis (Ganesh and Jeyakumar, 2014). In addition to these factors, composition, facets, choice-cum-ratio of capping and reducing agents and support-dependent catalytic performance of porous nanostructures have also been especially investigated for fuel cell ORR catalysis. Conventionally, to improve the sluggish ORR kinetics, huge amount of expensive and less abundant Pt is needed. Hence, recent researches focus on effective utilization of minimal Pt, by confining it to the surface in the form of decorated shell or monolayer Pt on Au or Pd core (Fuller, 2008). In this regard, Pt clusters having steps and kink sites exhibit enhanced O_2 adsorption. In addition, Au-Pt or Pd-Pt bimetallic catalysts (Ganesh and Jeyakumar, 2018b) with various topographic profiles show superior ORR activity attributed to their geometry, bond distance between Pt-Pt atoms and lowered Pt *d*-band centre (Fuller, 2008). Antoine et al. in 2001, used commercial

Vulcan XC-72 carbon-supported Pt nanoparticles to study the O_2 pressure effect and H^+ activity on the ORR kinetics and mechanism in acidic medium (Antoine et al., 2001). Effect of Pt nanoparticle size towards ORR was also studied and the highest ORR mass activity was observed in 3 nm-sized Pt particles. Before the report of Olivier Antoine group, the ORR catalysis was mostly studied on bulk single and polycrystalline electrodes or thin film porous electrodes. Inspiringly, Olivier Antoine et al. used Pt-C nanostructures to enhance the ORR and hence this report was considered as one of the milestones in the field of ORR catalysis.

Schmidt and Gasteiger et al. determined the amount of H_2O_2 formed during ORR on Pt-C nanostructures using RRDE measurements to evaluate the ratio of two electron to four electron pathway (Paulus et al., 2001). They also described the sulphate ion adsorption effect on ORR activity by using H_2SO_4 and $HClO_4$ electrolytes. In 2002, Markovic et al. extensively studied the ORR kinetics using thin film Pt single crystal plane electrodes, Pt@Pd bimetallic electrodes and Pt alloy surfaces in acidic and alkaline electrolytes (Markovic and Ross Jr, 2002). Stamenkovic along with this group prepared high surface area Pt-M-based bimetallic and tri-metallic alloy catalysts (M = Ni, Co, Fe, Rh) at various ratios for ORR studies in $HClO_4$ electrolyte (Paulus et al., 2002). Compared to commercial Pt-C catalyst, an improved ORR performance in alloyed catalysts was revealed from RRDE studies. Anderson et al. understood the ORR kinetics on Pt dual site based on activation enthalpies used density functional theory (Sidik and Anderson, 2002). He concluded that the slow first e^- transfer step is the ORR rate determining step owing to its highest activation barrier. This report corroborates with the experimental observation of Damjanovic in 1986 (Sepa et al., 1986(a)). Tao Li and Perla B. Balbuena in 2003 used DFT to study the ORR on small Pt_5 clusters and noticed a steep decrement in the activation barrier for oxygen adsorption (Li and Balbuena, 2003). Stamenkovic et al. prepared sputtered thin film Pt_3Ni polycrystalline alloy, Pt mono-layered thin films and Pt films in ultra-high vacuum (UHV) for ORR catalysis (Stamenković et al., 2003). The RRDE measurements showed four electron pathways for ORR on surface-reconstructed high surface area PtNi alloys and pure Pt electrodes in $HClO_4$ electrolyte. The maximum ORR activity was observed in Pt monolayer catalyst than Pt alloys and Pt catalysts. Enhanced ORR kinetics was observed on stepped Pt single crystal surfaces compared to symmetric and terraces in both H_2SO_4 and $HClO_4$ electrolytes using RRDE measurements (Maciá et al., 2004). They claimed a lower ORR activity on Pt electrodes owing to the strongly adsorbed bisulphate anions from H_2SO_4 electrolyte. Savadogo et al. used non-Pt based Pd alloys for the first time for ORR and showed better performance than bare Pt electrodes in acidic medium (Savadogo et al., 2004). In 2004, Zhong et al., investigated an enhanced ORR activity using Au-Pt alloy nanoparticles owing to the synergism and composition effect of Au and Pt (Maye et al., 2004). Zhdanov et al. in 2006 proposed the 'associative and dissociative' mechanisms for atomic oxygen adsorption on Pt using DFT studies (Zhdanov and Kasemo, 2006). In 2007, Stamenkovic et al. reported surface active sites enriched Pt_3Ni (111) single crystal surface with several fold enhancements in ORR activity than Pt (111) and commercial Pt-C catalysts owing to lowered *d*-band centre of Pt (Stamenkovic et al., 2007). Various monolayer deposited Pt-M nanoparticles (M = Pd, Ir, Ru, Au, Os, Re, Rh) on carbon supported Pd (111) single crystal surfaces was reported to have enhanced Pt utilization for ORR (Vukmirovic et al., 2007). M-N-C catalysts (M = Co, Fe, Ni, etc.) with lower ORR performance and operational stability compared to commercial Pt-C catalyst was reported (Bezerra et al., 2008). Mayrhofer et al. used various Pt-based high surface area and single crystal electrocatalysts and compared the particle size effect on ORR mass and specific activities using thin film RRDE measurements (Mayrhofer et al., 2008). Rossmeisl et al. in 2010 proposed that the ORR over potential on Pt is determined by the proton transfer step from water and OH removal from Pt surface (Tripković et al., 2010). They also proposed the concept of 'volcano plot' that describes the ORR performance vs. O_2 binding energies for various catalysts. A DFT model was developed to screen Pt based alloy ORR catalysts based on the shift in Pt adsorption site *d*-band centre and Pt-OH site after alloying with secondary metals (Xin et al., 2012). Both experimental and computational studies was carried out to validate the volcano behaviour of Pd-M-PdPt-C (M = Ni, Co, Fe, Cr and Pt) electro-catalysts (Trinh et al., 2012). While these ORR catalysts performed well as predicted from the volcano trend, they showed insufficient ORR activity after methanol addition due to their strong methanol oxidation characteristics. Pt monolayer based Ir and Re ORR electro-catalysts supported on carbon was developed to enhance the Pt utilization for ORR (Karan et al., 2012). They found enhanced mass activities for their catalysts compared to that of Pt-C and 2015 USA Department of Energy goal along with enhanced stability after 30,000 accelerated durability test cycles in aqueous 0.1 M perchlorate electrolyte. In 2014, $Pt_3Fe@Pt-FeO_x$ nanostructures prepared through acid leaching by Bao et al. were found to have enhanced ORR performance than Pt-C ORR catalysts owing to the synergistic activity of sub-surface and surface Fe species (Li et al., 2014). *In-situ* transmission electron microscopy (TEM) studies was carried out using thermally annealed octahedral Pt-Ni ORR nano-catalysts to observe the changes in the morphology and surface composition (Gan et al., 2009). The ORR performance was ascribed to the presence of active Pt (111) rich and Ni (111) sub-surface rich corners and edges on PtNi catalyst. Zidong Wei recently reported a review on Pt- and non-Pt-based ORR catalysts along with their efficacies and possibilities for future fuel cells and metal air batteries applications (Nie et al., 2015). Recently, many PdPt-based ORR catalysts are projected for PEMFCs (Venarusso et al., 2016; Ghosh and Raj, 2015). Shao et al. reported an extensive review of recent ORR catalysts based on clean Pt surfaces, Pt and Pd core-shell/alloy structures, transition metal-based oxides, oxy-nitrides, nitrides and carbo-nitrides, chalcogenides, noble, non-noble metal and metal-free carbon-based nanostructures (Shao et al., 2016). They also discussed the ORR performance of these catalysts depending on their size and shape through half-cell and H_2-O_2 full cell studies. Finally, they concluded that more enhancements in the activity and robustness of non-PGM electrocatalysts was needed along with minimal production cost to bypass the usage of commercial Pt-C fuel cell catalysts. Baletto et al. used DFT slab model approach to understand the interface effect on ORR activity at MgO (100) supported 1.5 nm sized PtNi core shell structures (Asara et al., 2016). Multi-layered AuCuPt and PtNi nano-alloys was reported with higher ORR activity attributed to the local strain effect and Au-Pt ligand effect from AuCu alloy core (Cleve et al., 2016). Morais et al. recently projected Pd-M alloys (M = Pd, Cu and Pt) with enhanced ORR activities in alkaline medium ascribed to their electronic structure (Castegnaro et al., 2017). Very recently, Lee et al. described tuneable ORR activity in core shell Pd-Pt nano-cubes by controlling the thickness of Pt shell in sub-nm range (Yang et al., 2017). Pt- and Pd-based bimetallic nano-catalysts with various surface morphologies and compositions in the form of nano-alloys, porous shell/core and Pt monolayer or Pt decorated structures were projected as better ORR catalysts (Cho et al., 2017; Góral-Kurbiel et al., 2016; Yang et al., 2016). Several Pt-based nano-porous core shell-structured electrocatalysts were extensively used for ORR due to their enhanced surface area and activity (Liu et al., 2012a; Ye et al., 2012; Lima et al., 2007; Zhang et al., 2012; Hong et al., 2012; Chen et al., 2013; Guo et al., 2007; Chen et al., 2008; Bing et al., 2010 and Wen et al., 2008). α-PtO$_2$ catalysts were reported as better ORR catalysts owing to the presence of steps and kink sites for favorable O$_2$ adsorption (Aricò et al., 2001 and Gao et al., 2012). The ORR catalysts discussed till now were particularly focused on improving the ORR performance in H$_2$ based PEMFCs. For DMFC ORR catalysis, the electrocatalysts must not only have efficient ORR activity, but should also be tolerant to methanol.

1.2.3 Methanol Tolerant ORR Catalysis and Carbon Corrosion Effect

ReRuS and MoRuS based transition metal sulphides was first reported as methanol tolerant ORR catalysts (Reeve et al., 1998). But, the reason for the methanol tolerant property in these catalysts was not discussed. Shukla et al. in 2003 categorized various methanol resistant ORR catalysts into four classes (i) Pt-based alloy catalysts, (ii) macrocyclic derivatives of transition metal compounds, (iii) metallic oxides, (iv) metal chalcogenides, etc. (Shukla and Raman, 2003). They concluded RuSe-C as one of the effective methanol resistant ORR catalysts for DMFC applications. In 2011, Xing et al. reported a review on DMFC ORR catalysts based on Pt-, Pd- and non-Pt-based structures along with their performance and stability for DMFC applications (Zhao et al., 2011). Many Pd-based catalysts were claimed as methanol-tolerant ORR catalysts, but they oxidize methanol under hydrodynamic situations (Choi et al., 2015; Wang et al., 2010; Yang et al., 2010a; Yang et al., 2010b and Ghosh et al., 2012). Pt particle dissolution and aggregation due to its growth on carbon supported PGM- and non-PGM-based electrocatalysts were also discussed in relation to long-term performance in DMFCs (Tuaev et al., 2016). Another important phenomenon called 'carbon corrosion' was discussed as a major cause for catalyst-support degradation (Janthon et al., 2017). The occurrence of Pt accelerates the oxidation of carbon and leads to corrosion at potentials around 1.1 V vs. RHE (Tuaev et al., 2016). It may also happen during fuel cell start-stop cycles, fuel starvation, etc. Hence, there is a great need for developing corrosion–resistant free carbon support or non-carbon support (which is conductive) for fuel cell catalysts (Tuaev, 2016). Extensive researches on carbon-free fuel cell electrocatalysts have been investigated to avoid the carbon corrosion-triggered H_2–O_2 fuel cell performance loss. Considering the DMFC cathode catalysis, the ORR catalyst should exhibit ideal methanol tolerant ORR activity and durability in acidic medium without support degradation at reduced Pt content (Ganesh and Jeyakumar, 2018b).

1.2.4 Surface Active Nano-islands

Surface active nano-islands can be defined as the surface-active sites which are composed of small groups of catalytically active atoms bound at the nano-electrocatalyst surface (i.e. kinks, edges and step sites). Further to emphasize the active sites principle, a factor called Taylor Ratio was defined as the amount of active sites present among total number of catalyst surface sites. Previous studies on single crystals and poly-crystals dispersions of Pt nanoparticles (Pan et al., 2018) suggest that every surface Pt atoms behave as surface active sites, irrespective of its atomic planes and/or sites. Similarly, various other metal/metal oxide surfaces have also been reported for many surface-sensitive hydrocarbon conversion reactions. For example, Fe (111) plane was reported as active site in the surface-sensitive ammonia synthesis using Fe catalyst (Pan et al., 2018). It was revealed that while the size reaches nanoscale dimension, its surface structure sensitivity alters the electronic band structure of transition metals and results in distinct electrocatalytic properties (Pan et al., 2018). Hu et al. reported FeO islands on Pt (111) surface as interfacial surface active sites for gas phase oxidation of benzyl alcohol (Pan et al., 2018) whereas Bandarenka et al. showed that Pd islands on Au (111) surface act as active HER catalyst (Pfisterer et al., 2017). Likewise, Stimming et al. proposed Pt nano-islands on Au (111) surface and Pd nano-islands on Au (111) surface for proton reduction and hydrogen reactions (Pfisterer et al., 2017). Huang *et al.* reported FeO (111) islands at Pt-FeO interface for low temperature CO oxidation (Lin et al., 2022) while Siyu Ye proposed a uniform distribution of co-adsorbed Sn islands on Pt (111) as a prerequisite for better CO oxidation (Zhang, 2008). Wieckowski and co-workers suggested that Ru islands on Pt increased the rate of methanol oxidation reaction activity compared to PtRu alloy (Zhang, 2008). Similarly, Watanabe and Motoo showed that Ru islands on Ru/Pt (111), Rh islands on Pt (111) and Pt islands on Ru particle exhibit facile CO_{ads} electro-oxidation behavior (Shao, 2013). SnO_2 nanoparticle cores adorned with Pt, Ir and Rh based nano-islands and SnO_x nano-islands on Pt (111) was prepared via seeded-growth method for ethanol electro-oxidation reaction catalysis (Shao, 2013). While these reports focused on the application of surface nano-islands for electro-oxidation reactions, Ganesh et al. successfully demonstrated the efficient role of sub-nm-sized surface active Pt islands bound nano-porous AuPt (< 10 nm) electrocatalyst for ORR prepared via facile wet chemical reduction process in water that resulted in enhanced DMFC performance at low Pt content (Ganesh and Jeyakumar, 2014). The improved DMFC activity might be ascribed to the favorable O_2 adsorption and methanol-tolerant ORR activity of nano-porous AuPt electrocatalyst triggered by the surface Pt (IV) states of Pt nano-islands. This activity enhancement might also be due to the sub-nm sized surface active Pt islands (surface active sites) and nano-porous nature (enhanced electrochemical active surface area (ECSA)) of AuPt ORR electrocatalyst (Ganesh and Jeyakumar, 2014). It will be revealed that surface active nano-islands-bound nanostructured electrocatalyst can perform as efficient methanol tolerant ORR catalyst for DMFC.

1.3   PGM AND NON-PGM-BASED ORR ELECTROCATALYSTS

1.3.1   Wet Chemical Reduction Process

A chemical reaction that involves the reduction of metallic ions into their zero state with the help of a reducing agent to form metallic nanoparticles in aqueous or non-aqueous is called as wet chemical reduction process (Turkevich et al., 1951; Xia et al., 2003). The size, morphology and uniformity of nanoparticles prepared by this method depend on the choice of solvents, capping agent, reducing agent and reaction conditions (Panigrahi et al., 2004). Nanoparticle formation involves mainly three steps (Turkevich et al., 1951) *ca.* nucleation, growth and Ostwald ripening as follows: 'nucleation' involves the formation of nucleates like initial particles (1 to 2 nm sized or even less); 'growth' is a thermodynamically controlled step that involves the increase in particle size to decrease their surface energy. This step should be controlled by using suitable capping agents to prevent agglomeration of nanoparticles (Turkevich et al., 1951). A capping agent should have a great attraction for the initially formed nuclei and binds to surface atoms (*Https://En.Wikipedia.Org/Wiki/Colloidal_gold*, 2000). This process helps in stabilizing the surface energy of those juvenile nuclei and prevents binding to other nuclei. Lack of control leads to a process called Ostwald ripening that involves the dissolving and redeposit of smaller nanoparticles on to larger-sized particles to form agglomerates (*Https://En.Wikipedia.Org/Wiki/Colloidal_gold*, 2000). Hence, this step is also called coagulation process. The stability of nanoparticles can be achieved either through electrostatic or steric stabilization or the mixture of two (electro-steric stabilization) (*Https://En.Wikipedia.Org/Wiki/Colloidal_gold*, 2000). Turkevich et al. in 1951 first proposed the wet chemical reduction of chloroaurate to gold nanoparticles using sodium citrate as both capping and reducing agent (Turkevich et al., 1951). Stabilizing agents like phosphine, poly vinyl pyrrolidone and sodium dodecyl sulphate can also be used as capping agents in wet chemical reduction method (Panigrahi et al., 2004; Younan Xia et al., 2003). Brust et al. prepared 5–6 nm gold nanoparticles from chloroaurate using tetra octylammonium bromide (TOAB) in toluene and sodium borohydride as stabilizing agent and reducing agents respectively (*Https://En.Wikipedia.Org/Wiki/Colloidal_gold*, 2000). Based on these reports, sodium citrate and sodium borohydride were used as capping and reducing agents with water as the solvent for the synthesis of nano-alloys. In addition, galvanic replacement reaction (GRR) method was employed to prepare nano-porous structures from nano-alloys.

1.3.2   Galvanic Replacement Reaction (GRR)

GRR is a one-step method which involves the dissolution of a less noble metallic element and deposition of a higher noble metallic element (Bansal et al., 2008). The metal with lower standard reduction potential will get replaced by the one with higher standard reduction potential (Bansal et al., 2008). Hence, the motivating power for the GRR is the favorable free energy change that arises from the variance in standard electrochemical reduction potential values of the elements that participate in the reaction. GRR usually results in the formation of nano-porous or hollow nanostructures (Ganesh and Jeyakumar, 2018a) and is more advantageous than the chemical etching or other de-alloying process (Bansal et al., 2008). It involves the conversion of solid structure into hollow interiors by etching through pinhole sites (Bansal et al., 2008). It is a consequence of differences in diffusing rates of two different species involved in the replacement process (Yin et al., 2004). Thus the variance in the standard electrochemical reduction potential values is the main criterion for the GRR and some values of standard reduction potential related to the present work is given as follows:

Table 1.1   Standard electrochemical reduction potential of elements involved in the work (da Silva and Ângelo, 2010 and Vanysek, 2000)

Components involved	Standard reduction potential values (vs. RHE)
AgCl/Ag	0.22 V
$[AuCl_4]^-$/Au	1.002 V
$[PdCl_4]^{2-}$/Pd	0.591 V
Mn^{2+}/Mn	−1.18 V
$[PtCl_4]^{2-}$/Pt	0.75 V
Pt^{4+}/Pt	1.3 V

From Table 1.1 it is clear that Mn and Ag have lower standard reduction potential compared to Pt, Pd and Au. Hence, it is envisaged that Ag and Mn could be galvanically replaced by Pt, Pd or Au. For example, if $[PtCl_4]^{2-}$ is added to Au-Ag nano-alloys, the Ag from the nano-alloy will get replaced and substituted by Pt. This GRR is thermodynamically feasible owing to the lesser electrochemical standard reduction potential of AgCl/Ag (0.22 V) compared to $[PtCl_4]^{2-}$/Pt (0.75 V). As a result, for each Pt atom, two Ag atoms will get replaced and to compensate this loss, nano-void will form resulting in nano-porous Au-Pt structures (AuPt NPoS) (Ganesh and Jeyakumar, 2014).

1.3.3 PGM-based Nano-porous ORR Electrocatalysts

1.3.3.1 Synthesis of AuPt NPoS

AuAg nano-alloy was prepared using wet chemical reduction method (Ganesh and Jeyakumar, 2014). Typically, required amounts of chloroaurate aqueous solution and sodium nitrate were taken in water and stirred continuously. The slightly turbid mixture was heated to solubilize AgCl. To this solution, 1% trisodium citrate decahydrate was added to stabilize the newly forming nanostructures. Finally, 1 ml of 0.15% sodium borohydride was added and the solution color turned brownish yellow, intimating $Au_{100-x}Ag_{2x}$ nano-alloy formation (Ganesh and Jeyakumar, 2014). Nano-porous $Au_{100-x}Pt_x$ NPs were formed through GRR from AuAg nano-alloys in water at room temperature (Ganesh and Jeyakumar, 2014). Typically, AuAg nano-alloy solution was treated with required amount of K_2PtCl_4 solution resulting in nano-porous Au-Pt formation (Ganesh and Jeyakumar, 2014). The resultant $Au_{85}Pt_{15}$ suspension was centrifuged to remove AgCl and the clear solution was used for further analysis. $Au_{85}Pt_{15}/C$ materials were prepared by adding commercially activated porous carbon to the suspensions. As a result, the nanoparticles get adsorbed on the carbon surface, leaving a nearly transparent solution of water and results in the formation of $Au_{85}Pt_{15}/C$ electrocatalyst materials. These carbon supported samples were used for XPS, voltammetry and DMFC studies, etc.

1.3.3.2 Synthesis of PdPt NPoS

$Pd_{100-x}Mn_{2x}$ nano-alloy was synthesized through a facile wet chemical reduction process with their respective precursor ratio in aqueous medium (Ganesh and Jeyakumar, 2017b). In a typical case of (x = 15) nano-alloy synthesis, required amounts of K_2PdCl_4 and Mn $(CH_3CO_2)_2$ were added with 1% trisodium citrate decahydrate (1 ml) and co-reduced by adding 0.15% $NaBH_4$ (1 ml) under stirring condition. As a result, reaction mixture turns from pale yellow to orangish brown, revealing the $Pd_{100-x}Mn_{2x}$ nano-alloy formation (Ganesh and Jeyakumar, 2017b). GRR of $Pd_{100-x}Mn_{2x}$ nano-alloys using $H_2PtCl_6.H_2O$ at ambient temperature results in $Pd_{100-x}Pt_x$ NPoS formation. Typically, $Pd_{100-x}Mn_{2x}$ (x = 15) suspension was added with required amounts of $H_2PtCl_6.H_2O$. The solution color turns to blackish brown from orangish brown, indicating $Pd_{85}Pt_{15}$ NPoS formation (Ganesh and Jeyakumar, 2017b). This $Pd_{85}Pt_{15}$ NPoS suspension were centrifuged and used for further analysis. Carbon supported $Pd_{85}Pt_{15}$ NPoS (i.e. $Pd_{85}Pt_{15}/C$ electrocatalyst materials) were prepared following the similar procedure used for the preparation of $Au_{85}Pt_{15}/C$ materials as discussed in Section 1.3.3.1. These carbon supported samples were used for XPS, electrochemical and DMFC studies, etc. Furthermore, a carbon support free $Pd_{85}Pt_{15}$ NPoS (C-free $Pd_{85}Pt_{15}$ NPoS) (Ganesh and Jeyakumar, 2018b) was also prepared using a 'self-settlement approach' (a little modified approach of Schmidt et al. strategy (Henning et al., 2016)) for further analysis.

1.3.4 Non-PGM-based Nano-porous ORR Electrocatalysts

1.3.4.1 Synthesis of Shaped Silver Nanostructures (AgNs)

AgNs were synthesized through a facile wet chemical reduction method in water at ambient conditions via a new 'hierarchical shape tuning approach' (Ganesh et al., 2021). Various shapes, such as vermiform (worm-like) Ag nanostructures (V-AgNs), spherical Ag nanostructures (S-AgNs), sphere-in worm/worm-in-sphere were obtained (Ganesh et al., 2021). Typically, $AgNO_3$ (8 mmol) in 100 ml of water and trisodium citrate decahydrate (0.75 mmol) was stirred continuously for 10 minutes. To this reaction mixture required amounts of $NaBH_4$ and ascorbic acid were simultaneously added and stirred for 30 minutes. Likewise, precisely controlling the dual reducing agent ratios to trisodium citrate decahydrate using same $AgNO_3$ amount under similar reaction condition resulted in various shaped Ag nanostructures. 'Self-settlement approach' (Schmidt et al. strategy (Henning et al., 2016)) (Ganesh et al., 2021) was employed to obtain V-AgNs and S-AgNs and used for morphological, electrochemical and AAEMFCs studies.

1.4 MORPHOLOGICAL CHARACTERIZATION OF PGM AND NON-PGM ELECTROCATALYSTS

1.4.1 High Resolution-scanning Transmission Electron Microscopy (HR-STEM) and High Angle Annular Dark Field (HAADF)-STEM Studies

The HR-STEM image of sub-10 nm sized $Au_{85}Pt_{15}$ NPoS was presented in Figure 1.1A. It reveals its spherical shape along with lattice fringes and porous moiety. Additionally, the HAADF-STEM elemental mapping image of Au and Pt (Figs. 1.1B and 1C) reveal core confinement of core Au and surface confined scattered Pt atoms respectively. The composite elemental mapping (Fig. 1.1D) clearly indicates the presence of core Au with surface Pt nano-islands and peripheral white blotches over the surface, indicative of porous moiety. This kind of results have been previously noticed for several nano-porous materials with pores-like feature in de-alloyed $PtCo_3$ materials owing to the pore formation (Ganesh and Jeyakumar, 2014; Liu et al., 2012b). Hence, the dominant white patches observed in HAADF-STEM images confirm the nano-porous moiety in $Au_{85}Pt_{15}$ NPoS (Ganesh and Jeyakumar, 2014).

The HR-STEM analysis of $Pd_{85}Pt_{15}$ NPoS[114] (Fig. 1.1E) shows its hexagonal shape with lattice fringes and porous nature as noticed in $Au_{85}Pt_{15}$ NPoS (Ganesh and Jeyakumar, 2014). In addition, the NPoS shows d-spacing values consistent with that of standard values of Pd and Pt, thereby confirming its elemental presence as shown in Fig. 1.2D. A closer look (Fig. 1.2D) confirms the existence of kinks and edges along the surface. The HAADF-STEM elemental color profile of Pd and Pt (Figs. 1.1F and 1.1G) indicates the presence of core concentrated Pd atoms and surface concentrated Pt. Its composite mapping (Fig. 1.1H) reveals

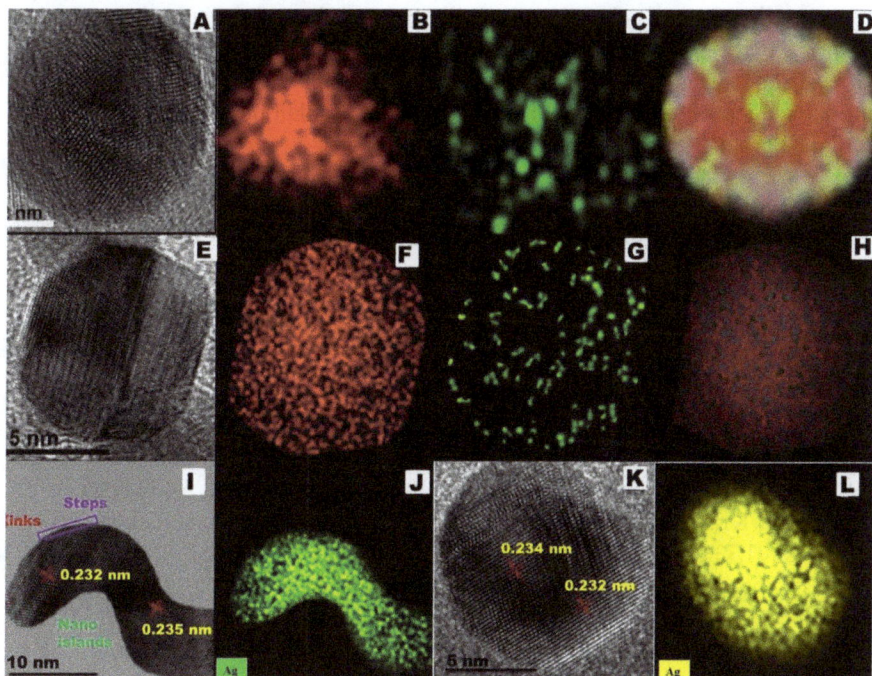

Figure 1.1 HR-STEM image of Au85Pt15 NPoS (A), Pd85Pt15 NPoS (E), V-AgNs (I), S-AgNs (K). HAADF-STEM elemental mapping pictures of Au85Pt15 NPoS (B and C), Pd85Pt15 NPoS (F and G) with their respective composite elemental mapping images (D and H) and HAADF-STEM elemental mapping images of V-AgNs (J), S-AgNs (L) (Reproduced with permission from ref. (Ganesh and Jeyakumar, 2018b; Ganesh and Jeyakumar, 2014; Ganesh et al., 2021)).

the porous nature of $Pd_{85}Pt_{15}$ NPoS resembled by channelled white parts (voids) and Pd-Pt islands throughout the surface. The prevailing white blotches owe to the nano-porous moiety in $Pd_{85}Pt_{15}$ NPoS as reported previously (Ganesh and Jeyakumar, 2014). These features depict Pd and Pt surface nano-islands occurrence and support the XPS characteristics of $Pd_{85}Pt_{15}$NPoS (i.e. Pd (II) and Pt (IV) states) as revealed in Figs. 1.2B and 1.2C. Hence, the surface Pd-Pt nano-islands occurrence in $Pd_{85}Pt_{15}$ NPoS confirmed from the HAADF-STEM studies will be further corroborated from XPS analysis (Ganesh and Jeyakumar, 2018b). The HR-STEM image of a worm shaped V-AgNs (Fig. 1.1I) displays the atomic fringes (planes) and nano-islands-like features (kinks and edge sites) and *d*-spacing values corresponding to Ag (Ganesh et al., 2021). Likewise, the HAADF-STEM elemental mapping image of V-AgNs (Fig. 1.1J) shows occurrence and well-dispersed Ag. In contrast, the HR-STEM image of a spherical shaped S-AgNs (Fig. 1.1K) indicates the lattice fringes with *d*-spacing values consistent with that of standard values of Ag. The HAADF-STEM elemental mapping image of S-AgNs (Fig. 1.1L) shows the occurrence of Ag over the structure (Ganesh et al., 2021). In view of growth mechanism of V-AgNs, primarily the low citrate concentration capped silver ions and the aging effect of sodium citrate lead to chaining of silver nuclei. Then, the silver nuclei undergo a simplistic anisotropic growth and trigger to form a worm-like structure. Finally, the surface nano-islands formation happens with the slow and meticulous addition of ascorbic acid and sodium borohydride (Ganesh et al., 2021).

1.5 SURFACE CHARACTERIZATION OF PGM ELECTROCATALYSTS

1.5.1 X-ray Photoelectron Spectroscopic (XPS) Studies

De-convoluted XPS spectra of AuPt and PdPt NPoS (Ilayaraja et al., 2013) are presented in Fig. 1.2. XPS spectra of $Au_{85}Pt_{15}$/C NPoS (Fig. 1.2A) show peaks. Au center gives rise to dual set of peaks at 84.43 eV and 86.05 eV and 88.07 eV and 89.50 eV for Au $4f_{7/2}$ and $4f_{5/2}$ energy levels relating to Au^0 and Au-O respectively (Naumkin, Kraut-Vass and Gaarenstroom, 2012). In contrast, the Pt shows tri sets of peaks (71.33 eV and 74.73 eV), (72.71 eV and 76.11 eV), (74.51 eV and 77.91 eV) for $4f_{7/2}$ and $4f_{5/2}$ (Naumkin, Kraut-Vass and Gaarenstroom, 2012) energy levels consigned for Pt^0, Pt (II) and Pt (IV) states (Ganesh and Jeyakumar, 2014). It is well known that Pt^0 and Pt (II) will be noticed prevalently with minimal Pt (IV) levels. Furthermore, surface nano-island like Pt are expected to show more Pt (IV) states owing to the steps and kinks sites that result in favorable O_2 adsorption (Aricò et al., 2001). Hence, it is clear that Pt in $Au_{85}Pt_{15}$ NPoS could be availed as surface nano-islands on nano-porous Au. In addition, the trace amount of Au-O states detected attributes to Pt (IV) solid interface with nano-porous Au core. Hence, XPS results reveal the presence of surface Pt nano-islands and nano-porous Au surface in $Au_{85}Pt_{15}$ NPoS as corroborated from the HR-STEM and HAADF STEM elemental mapping studies (Ganesh and Jeyakumar, 2014).

Likewise, the de-convoluted XPS spectra of $Pd_{85}Pt_{15}$/C NPoS (Fig. 1.2B) show dual pair of XPS peaks intended for Pd $3d_{5/2}$ and Pd $3d_{3/2}$ energy points (335.83 eV and 341.14 eV and 337.85 eV and 343.30 eV) consigned to Pd (0) and Pd (II) (Naumkin, Kraut-Vass and Gaarenstroom, 2012) respectively. On contrary, the Pt center (Fig. 1.2C) shows tri sets of XPS peaks (71.10 eV and 73.54 eV), (72.40 eV and 76.03 eV), (74.82 eV and 78.55 eV) for Pt $4f_{7/2}$ and Pt $4f_{5/2}$ energy points correspondingly consigned to Pt (0), Pt (II) and Pt (IV) (Naumkin, Kraut-Vass and Gaarenstroom, 2012; Ganesh and Jeyakumar, 2018b). Based on XPS outcomes, it was clear that Pd was present as Pd (0) and Pd (II) while Pt center shows more Pt (IV) in addition to Pt (0) and Pt (II). This XPS feature attributes to the robust interaction of Pt (IV) with porous Pd moiety (Ganesh and Jeyakumar, 2018b). These results reveal the existence of significant levels of surface Pd and Pt islands on nano-porous Pd surface. Interestingly, $Pd_{85}Pt_{15}$ NPoS with surface Pd and Pt nano-islands exhibits enhanced steps and kinks sites (Fig. 1.2D) that could improvise better ORR activity than $Au_{85}Pt_{15}$ NPoS (Ganesh and Jeyakumar, 2014).

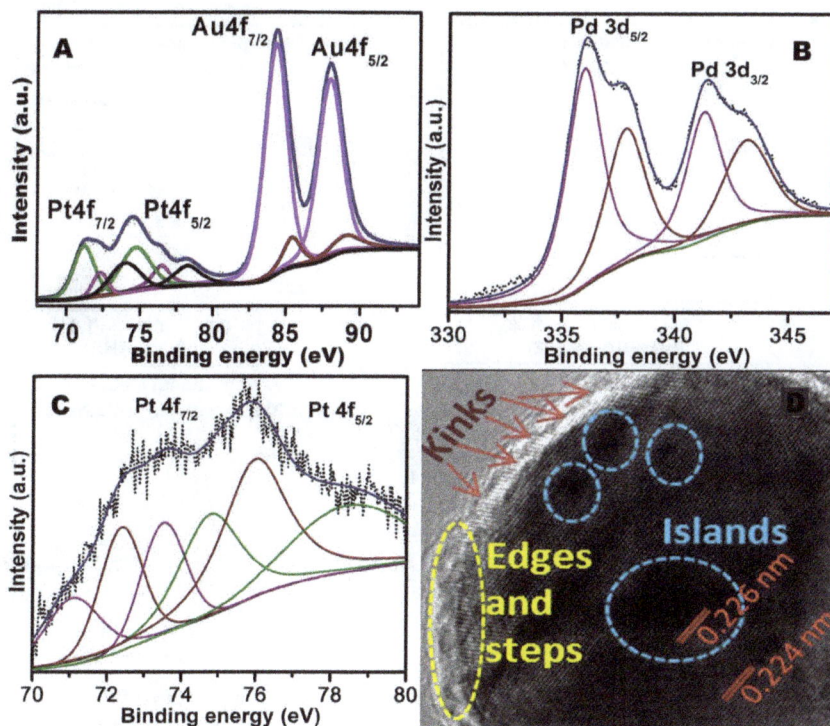

Figure 1.2 XPS spectra of Au$_{85}$Pt$_{15}$ NPoS (A), Pd$_{85}$Pt$_{15}$ NPoS (B and C) and HRSTEM image (closer view) of Pd$_{85}$Pt$_{15}$ NPoS (D) clearly show the presence of surface active sites. (Reproduced with permission from ref. (Ganesh and Jeyakumar, 2014; Ganesh and Jeyakumar, 2018b)).

1.6 ELECTROCHEMICAL CHARACTERIZATION OF PGM ELECTROCATALYSTS

1.6.1 Linear Sweep Voltammetry—RDE Studies

1.6.1.1 Methanol Tolerant ORR Studies

It is well established that an efficient DMFC cathode catalyst should not only have a better ORR activity, it should also have enhanced tolerance to methanol to circumvent the mixed cathode potential formation owing to concurrent ORR and MOR (methanol crossover) at cathode (Ganesh and Jeyakumar, 2014). Hence, the methanol tolerant ORR behavior of Au$_{85}$Pt$_{15}$/C NPoS was studied using LSV-RDE studies under O$_2$ saturated conditions in the presence of methanol (0.5 M) at 1600 rpm (Fig. 1.3A). After methanol addition, there is no shift in the onset potential of ORR graphs. More interestingly there occurs no methanol oxidation peak in ORR graphs of nano-porous materials, but negligible decrement in the ORR current after methanol addition, manifesting the resistance of Au$_{85}$Pt$_{15}$/C NPoS to MOR (Ganesh and Jeyakumar, 2014).

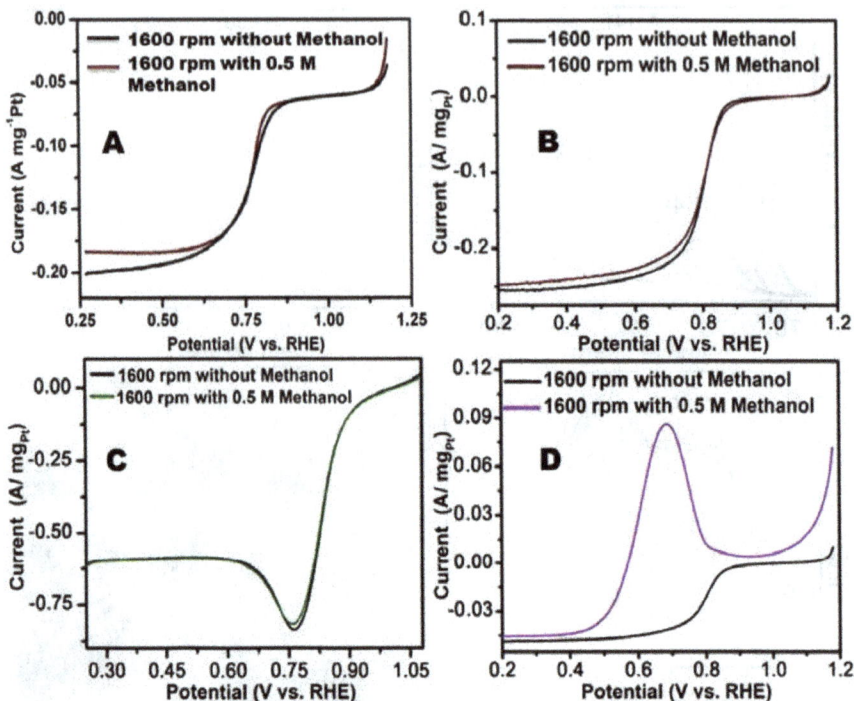

Figure 1.3 LSV-RDE methanol-tolerant ORR results of $Au_{85}Pt_{15}$ NPoS (A), $Pd_{85}Pt_{15}$ NPoS (B), C-free $Pd_{85}Pt_{15}$ NPoS (C) and commercial HiSPEC Pt/C (D) ORR electrocatalysts with and without methanol addition (Reproduced with permission from ref. (Ganesh and Jeyakumar, 2014; Ganesh and Jeyakumar, 2018b)).

Likewise, the methanol tolerant ORR activity of $Pd_{85}Pt_{15}$/C NPoS was studied using LSV-DE at 1600 rpm with 0.5 M methanol addition under O_2 saturated conditions. As exposed in Fig. 1.3B, $Pd_{85}Pt_{15}$/C NPoS lacks methanol oxidation peak at hydrodynamic condition after methanol addition. Further, the ORR $E_{1/2}$ and onset potential values of $Pd_{85}Pt_{15}$/C NPoS did not shift, but only exhibit a negligible decrement in ORR current value (Ganesh and Jeyakumar, 2018b). These observations clearly show that $Pd_{85}Pt_{15}$/C NPoS is highly tolerant to methanol and depicts its efficacy towards methanol tolerant ORR activity due to the presence of Pd (vital role of Pd for MOR tolerance). Hence, it can be concluded that the efficient methanol tolerant ORR catalysis could be ascribed to the Pt surface nano-island (sub-nm size) as illustrated through HR-STEM studies. This could be correlated with Pt island size of around 0.5 nm (nearly four atoms of Pt) as reported in the case of $Au_{85}Pt_{15}$/C ORR catalyst (Ganesh and Jeyakumar, 2014). Furthermore, the LSV-RDE studies of a completely carbon support free (i.e. C-free $Pd_{85}Pt_{15}$ NPoS) (Ganesh and Jeyakumar, 2018b) at 1600 rpm under O_2 saturated conditions with 0.5 M methanol addition was shown in Fig. 1.3C. More interestingly, there is an absence of MOR peak and no alteration in ORR $E_{1/2}$ potential and current values. The enhanced methanol tolerant ORR activity observed in C-free $Pd_{85}Pt_{15}$ NPoS could be due to the maximal utilization of catalyst surface for ORR and absence of carbon support corrosion (Ganesh and Jeyakumar, 2018b), thereby increasing it activity and stability several folds higher than observed in $Au_{85}Pt_{15}/C$ NPoS and $Pd_{85}Pt_{15}/C$ NPoS. The ORR performance of benchmark HiSPEC Pt/C ORR electrocatalyst was studied under similar electrochemical environments to compare its methanol-tolerant ORR activity with NPoS. As a result, HiSPEC Pt/C (Fig. 1.3D) shows a strong peak for methanol oxidation with high current magnitude at around 0.7 V and displays a prominent alteration in the ORR $E_{1/2}$ and onset potential value (Ganesh and Jeyakumar, 2018b).

Based on the surface Pt ratio and LSV-RDE ORR steady state currents, the figure of merit (FOM) value for $Au_{85}Pt_{15}/C$ NPoS (Ganesh and Jeyakumar, 2014), $Pd_{85}Pt_{15}/C$ NPoS (Ganesh and Jeyakumar, 2018b), C-free $Pd_{85}Pt_{15}$ NPoS (Ganesh and Jeyakumar, 2018b) and HiSPEC Pt/C (Ganesh and Jeyakumar, 2018b) catalysts were calculated in order to understand their surface Pt atom utilization efficiency (Yang et al., 2010a). Accordingly, before methanol addition, the FOM values of $Au_{85}Pt_{15}/C$ NPoS, $Pd_{85}Pt_{15}/C$ NPoS, C-free $Pd_{85}Pt_{15}$ NPoS and HiSPEC Pt/C catalysts catalyst was 7.2×10^{-23} A per surface Pt atom, 6.8×10^{-22} A per surface Pt atom, 4.3×10^{-22} A per surface Pt atom and 3.8×10^{-24} A per surface Pt atom, respectively. On the contrary, the FOM for $Au_{85}Pt_{15}/C$ NPoS, $Pd_{85}Pt_{15}/C$ NPoS, C-free $Pd_{85}Pt_{15}$ NPoS and HiSPEC Pt/C catalysts catalyst was 6.8×10^{-23} A per surface Pt atom, 5.2×10^{-22} A per surface Pt atom, 3.9×10^{-22} A per surface Pt atom and 4.2×10^{-25} A per surface Pt atom respectively after methanol addition. Based on FOM analysis, it is clear that $Au_{85}Pt_{15}/C$ NPoS, $Pd_{85}Pt_{15}/C$ NPoS, C-free $Pd_{85}Pt_{15}$ NPoS catalysts did not suffer drastically. But, the FOM in commercial HiSPEC Pt/C ORR catalyst fell by approximately one order. Thus, NPoS catalysts show higher retention of FOM as compared to HiSPEC Pt/C ORR catalyst after methanol addition. However, a further meticulous examination shows that the C-free $Pd_{85}Pt_{15}$ NPoS exhibit enhanced FOM values compared to $Pd_{85}Pt_{15}/C$ NPoS and $Au_{85}Pt_{15}/C$ NPoS catalysts (Ganesh and Jeyakumar, 2014). This behavior discloses the maximal catalyst consumption, porous moiety and surface active Pd and Pt surface nano-islands of C-free $Pd_{85}Pt_{15}$ NPoS (Ganesh and Jeyakumar, 2018b) but also determines its methanol resistant ORR features grander to NPoS and commercial HiSPEC Pt/C ORR catalysts. Based on FOM results, DMFC performance studies were carried out for C-free $Pd_{85}Pt_{15}$ NPoS, $Au_{85}Pt_{15}/C$ NPoS catalysts and compared with commercial HiSPEC Pt/C catalyst.

1.7 DMFC SINGLE-CELL PERFORMANCE STUDIES OF PGM ELECTROCATALYSTS

1.7.1 Current-voltage (i-V) Curve and Cross-sectional Studies of PGM Electrocatalysts-based Membrane Electrode Assemblies (MEAs)

DMFC studies of MEA fabricated using $Au_{85}Pt_{15}/C$ NPoS (1.0 mg/cm^2) and commercial HiSPEC Pt-Ru/C (0.5 mg/cm^2) as cathode and anode-based catalyst materials respectively with Nafion 117 membrane electrolyte (Ganesh and Jeyakumar, 2014) displays the polarization curves and FESEM images in Figure 1.4A and 1.4B. The i-V curves were recorded by serving 2M methanol (2.0 ml/minute) and oxygen (300 ml/minute) at anode and cathode sides respectively at different temperatures. Fig. 1.4A shows the DMFC performance of $Au_{85}Pt_{15}$/C NPoS ORR catalyst with negligible drop in the potential value confirming its methanol resistant ORR feature as anticipated from LSV-RDE ORR studies (Ganesh and Jeyakumar, 2014). Furthermore, the power density increases with increase in temperature. For comparison, an MEA with HiSPEC Pt/C (1.0 mg/cm^2) cathode material and Pt-Ru/C (0.5 mg/cm^2) anode material were used under similar fuel cell operating conditions (Ganesh and Jeyakumar, 2014).

More interestingly, the power density of $Au_{85}Pt_{15}$/C NPoS ORR catalyst (Ganesh and Jeyakumar, 2014) based MEA was 38 mW/cm^2 and is greater than benchmark HiSPEC Pt/C ORR electrocatalyst based MEA (16 mW/cm^2) at 60°C. Antonino et al. testified a performance of 34 mW/cm^2 (60°C) for Pt-Pd$_3$Co cathode material (Aricò et al., 2014). It is well known that the key tasks in enhancing the DMFC performance are to increase the methanol-tolerant ORR performance and stability of catalysts (by maximizing its effective use), ORR mass conveyance of fuels and to circumvent electrocatalyst-support instability issue. Hence, a carbon-free cathode catalyst (C-free $Pd_{85}Pt_{15}$ NPoS) (Ganesh and Jeyakumar, 2018b) was used to realize a stable and efficient DMFC performance at minimal Pt loadings and to get rid of aforementioned issues. The polarization i-V curves of all the MEAs were recorded by serving 2.0 M methanol (2.0 ml/minute) and oxygen (300 ml/minute) at anode and cathode sides respectively at various functioning temperatures. In general, the fuel cell performance consistently increases with increasing temperature. To increase the DMFC performance and to decrease the voltage fall, MEA fabricated using support less Pt NPs (0.2 mg$_{Pt}$/cm^2) anode and support less $Pd_{85}Pt_{15}$ NPoS (0.4 mg$_{Pt}$/cm^2) cathode (Ganesh and Jeyakumar, 2018b) catalysts were employed and their DMFC polarization curve and cross-sectional FESEM view was shown in Figs. 1.4C and 1.4D respectively. It is important to recall that carbon-free $Pd_{85}Pt_{15}$ NPoS displayed higher methanol-resistant ORR activity after methanol (0.5 M) addition and no alterations in the ORR $E_{1/2}$ and onset potential values compared to benchmark HiSPEC Pt/C and other NPoS ORR electrocatalysts in LSV-RDE studies. As a result, carbon support free DMFC MEA displays a insignificant initial fall in the activation kinetic area with better performance values (0.75 V and 51 mW/cm^2) at 70°C (Fig. 1.4C). This DMFC performance is higher than that for benchmark HiSPEC Pt/C (1 mg$_{Pt}$/cm^2) cathode material of 30 mW/cm^2 and literature report (Baglio et al., 2006). This enhancement is endorsed to the better methanol resistant ORR activity of C-free $Pd_{85}Pt_{15}$ NPoS cathode catalyst and also attributed to the reduced anode and cathode catalyst layer thickness (Fig. 1.4B), enhanced mass transport at both anode and cathode compartments and 'zero carbon environments' compared to $Au_{85}Pt_{15}$/C NPoS based MEA (Fig. 1.4D) (Ganesh and Jeyakumar, 2018b). The increased catalyst layer thickness and catalyst degradation problems in carbon supported NPoS lead to severe ECSA loss. But, the enhanced ORR and efficient methanol resistant activity of C-free $Pd_{85}Pt_{15}$ NPoS ORR catalyst with distinctive surface bound Pd-Pt islands reflected in DMFC studies make it a promising ORR catalyst for DMFC (Ganesh and Jeyakumar, 2018b).

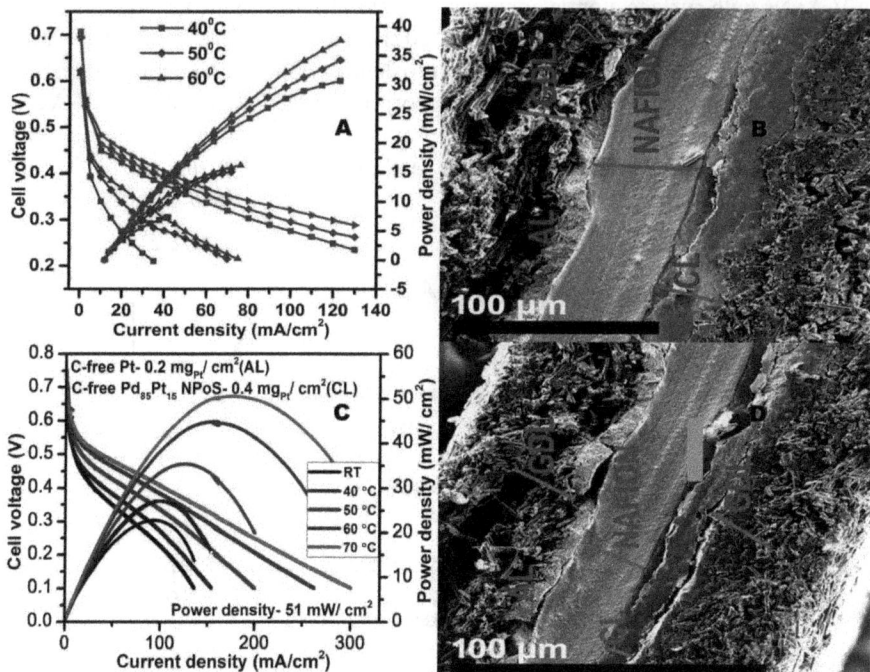

Figure 1.4 DMFC performance results of $Au_{85}Pt_{15}$ NPoS and C-free $Pd_{85}Pt_{15}$ NPoS ORR electrocatalysts (A and C) compared with benchmark HiSPEC Pt/C ORR electrocatalyst at various temperatures. FESEM cross-sectional view of $Au_{85}Pt_{15}$ NPoS and C-free $Pd_{85}Pt_{15}$ NPoS ORR electrocatalysts based MEAs (B and D) (Reproduced with permission from ref. (Ganesh and Jeyakumar, 2014; Ganesh and Jeyakumar, 2018b)).

1.8 SURFACE, ELECTROCHEMICAL AND FUEL CELL STUDIES OF NON-PGM ELECTROCATALYSTS

1.8.1 XPS, LSV- RDE-accelerated Endurability Test (AET) and AAEMFCs Studies

De-convoluted XPS spectra (Ilayaraja et al., 2013) of S-AgNs and V-AgNs electrocatalysts (Ganesh et al., 2021) spectra were presented (Fig. 1.5A). S-AgNs displays peaks (374.2 eV and 368.2 eV) for Ag $3d_{3/2}$ and Ag $3d_{5/2}$ energy levels respectively relating to Ag(0) states (Naumkin, Kraut-Vass and Gaarenstroom, 2012). V-AgNs show (374.2 eV and 373.9 eV) for Ag $3d_{3/2}$ energy level w.r.t to

Ag(0) and Ag(I) surface states (Naumkin, Kraut-Vass and Gaarenstroom, 2012). Similarly, one set of peak (368.2 eV and 367.9 eV) was observed for Ag $3d_{5/2}$ energy level in accordance with Ag(0) and Ag(I) states (Naumkin, Kraut-Vass and Gaarenstroom, 2012). Thus, the XPS outcomes of S-AgNs reveal the absence of Ag(I) and presence of Ag(0). In contrast, Ag(I) and Ag(0) surface states were seen in V-AgNs with surface atomic composition ratio of 1.25 : 98.75, i.e. Ag(I) is lower than Ag(0) surface states (Ganesh et al., 2021). This negligible Ag(I) states may perhaps be validated with Ag surface nano-islands feature that employ as surface active kinks and steps for simplistic O_2 adsorption (Aricò et al., 2001) on V-AgNs surface (Fig. 1.1I) as discovered in our previous work (Ganesh and Jeyakumar, 2014). The surface nano-island like moiety and exclusive worm-like morphology in

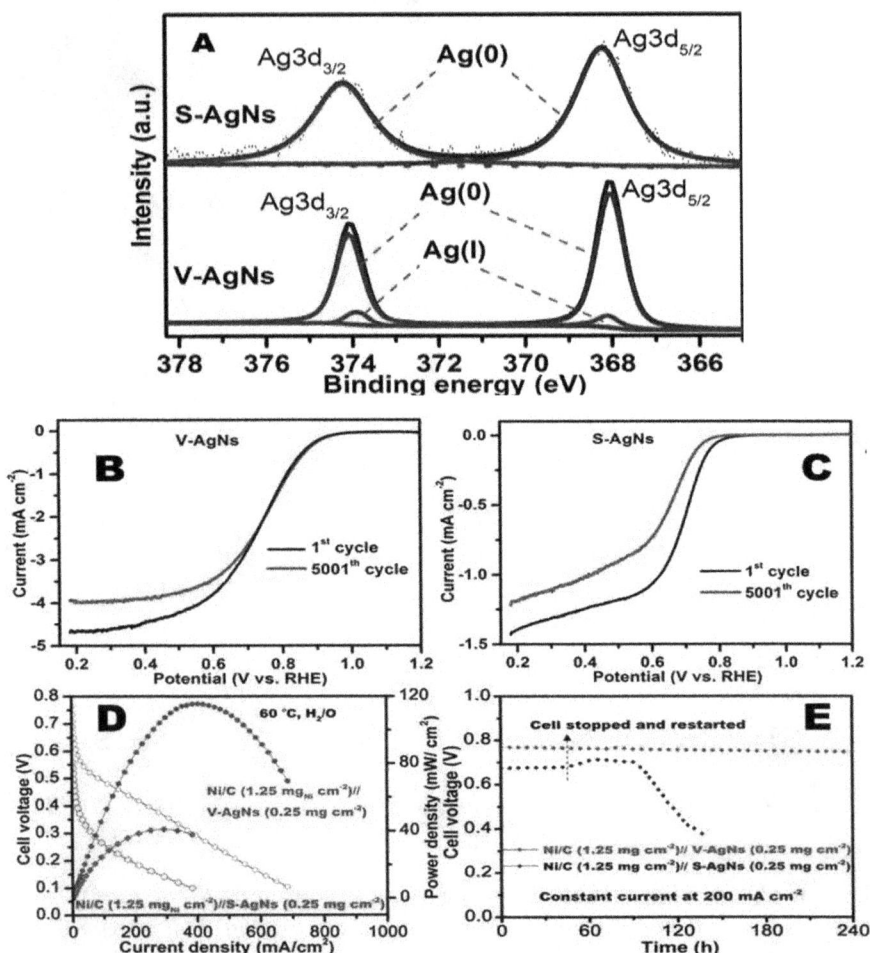

Figure 1.5 XPS spectra (A), LSV-RDE-AET-short-term durability studies (B and C). Complete n-PAAEMFC performance results (D) and their respective durability results (E) of AgNs based MEAs (Reproduced with permission from ref. (Ganesh et al., 2021)).

V-AgNs might stimulate its surface towards efficient electrocatalytic ORR activity. AET studies in 0.5 M KOH (O_2 saturated) and ambient operating environments were performed to understand the ORR steadiness of S-AgNs and V-AgNs electrocatalysts (Ganesh et al., 2021) as presented in Figs. 1.5B and 1.5C. It is valuable to note from Fig. 1.5B, that 88% of ORR activity was observed after AET with little alteration in the onset and $E_{1/2}$ potential for V-AgNs electrocatalyst (Ganesh et al., 2021). On the other hand, 85% of ORR activity was observed after AET for S-AgNs electrocatalyst (Ganesh et al., 2021) (Fig. 1.5C). This enhanced short-term durability of V-AgNs (Ganesh et al., 2021) (above 5000 AET cycles) was greater than Ag nano-plates (50 AET cycles) (Garlyyev et al., 2017) Ag_3Sn (2000 AET cycles) (Wang et al., 2017) and Ag nano-flower (3000 AET cycles) (Narayanamoorthy et al., 2016) ORR catalysts. Based on volcano plot (Trinh et al., 2012), the ORR performance of Ag was predicted to increase with decreasing size. In addition, the improved ORR mass transport, enhanced electrocatalyst utilization at nanoscale regime (10 nm), shaped morphology and surface active enrichment increase the O_2 binding energy for each surface active electrocatalytic site for efficient ORR performance and durability in V-AgNs related to S-AgNs and other similar reports in literature (Ganesh et al., 2021). Considering these vital advantages, a V-AgNs/S-AgNs cathode-based complete n-PAAEMFC was made up with commercial Ni/C anode and alkaline exchange electrolyte (Ganesh et al., 2021).Fully n-PAAEMFC performance of V-AgNs cathode (Ganesh et al., 2021) demonstrates an OCV of 0.77 V at 60°C and a performance value of 115.6 mW cm^2 (Fig. 1.5D). On the contrary, a fully n-PAAEMFC using S-AgNs cathode (Ganesh et al., 2021) shows an OCV of 0.58 V at 60°C resulting in maximum power density of 41.3 mW cm^2 (Fig. 1.5D). The AET of fully n-PAAEMFC (Ganesh et al., 2021) employed using V-AgNs/S-AgNs cathode materials at a continual current density (200 mA cm^2) was done to monitor the fully n-PAAEMFC voltage oscillations w.r.t time. It is worthy to mention that for V-AgNs cathode at 60°C the durability (Fig. 1.5E) was greater and displays insignificant drop for about 240 h (Ganesh et al., 2021). On the other hand, the fully n-PAAEMFC fabricated using S-AgNs cathode (Fig. 1.5E) shows reduced stability for 136 h. Fascinatingly, V-AgNs cathode based complete n-PAAEMFC maintain its durability after 80 h of AET (stopping and reviving the cell). On the other hand, the fully n-PAAEMFC mode using S-AgNs cathode illustrates a unfavorable drop in durability after resuming the n-PAAEMFC function with similar conditions (Ganesh et al., 2021). Furthermore, AET was seldom shown for fully n-PAAEMFC based previous literature reports (Alesker et al., 2016; Gu et al., 2013; Hu et al., 2013; Lu et al., 2008). Hence, the AET results demonstrated in the current case would deliver appreciable information that might help in deducing the durability and activity of a complete n-PAAEMFC.

1.9   CONCLUSION

Among NPoS based ORR electrocatalysts, $Au_{85}Pt_{15}$/C NPoS, $Pd_{85}Pt_{15}$/C NPoS, C-free $Pd_{85}Pt_{15}$ NPoS proved to be effectual ORR electrocatalysts with outstanding characteristics, such as better ORR onset and half-wave potential values, methanol tolerant ORR activity and FOM than benchmark Pt/C ORR electrocatalyst. The higher ORR activity in NPoS electrocatalysts principally arise owing to the surface bound Pt/Pd nano-islands on nano-porous Au/Pd surface, i.e. enriched Pt (IV) or Pd (II) surface states. LSV-RDE analysis undoubtedly demonstrates the methanol tolerant ORR behavior of NPoS electrocatalysts that arise from nano-porous Au/Pd structure. In addition, ORR-RDE studies in the presence of methanol evidently show negligible shift in ORR onset and half-wave potential values. These intriguing methanol tolerant ORR characteristics of NPoS help in reducing the mixed cathode potential generation and accomplish the conditions of a proper DMFC ORR electrocatalyst. More importantly, the facile synthesis method employed for NPoS electrocatalysts will guide to prepare effective DMFC ORR electrocatalysts using ultra-low Pt content. Although Pt ORR catalysts exhibit enhanced ORR activity, their ORR activity is unfavorably affected in the presence of methanol. However, $Pd_{85}Pt_{15}$ NPoS with surface-bound Pd-Pt islands supported on carbon were shown to exhibit excellent methanol tolerant ORR activity with higher retention of FOM, ORR mass-specific activity values than benchmark Pt/C ORR electrocatalysts and surpass state-of-art DMFC ORR electrocatalysts. 'Post-GRR self-settlement' strategy and a facile hierarchical tactic of decreasing thickness of catalyst layers are described for C-free $Pd_{85}Pt_{15}$ NPoS to alleviate the mass transfer issues during DMFC operation. Notably, a fully carbon free MEA fabricated using support less $Pd_{85}Pt_{15}$ NPoS cathode and support less bare Pt NPs anode display an improved DMFC performance endorsed to the carbonless situation, better electrocatalyst use and mass transport at anode and cathode. Thus, two innovative approaches were demonstrated (i) carbonless and distinctive $Pd_{85}Pt_{15}$ NPoS (surface confined Pd-Pt islands) with effective methanol resistant ORR activity, (ii) a fully carbon less MEA made using $Pd_{85}Pt_{15}$ NPoS cathode and support less Pt NPs anode at low Pt levels to improve the DMFC activity. The unique carbon free $Pd_{85}Pt_{15}$ NPoS with enhanced DMFC performance at a lower Pt amount outperforms the previous similar DMFC results and hopefully opens up a new avenue in the area of DMFC energy research. A novel hierarchical shaped tuning method was performed to synthesize various shaped AgNs. V-AgNs displayed higher ORR $E_{1/2}$ and onset potential values, mass-specific activity and enhanced endurability for about 5,000 potential cycles than S-AgNs and other similar reports. The effective and stable ORR activity of V-AgNs electrocatalyst was endorsed to their nanoscale dimension, 'unique worm-like shape with surface active nano-islands' and 'support free design' prompting improved ORR mass transport and boosting utilization of catalyst. More interestingly, a complete n-PAAEMFC with maximum fuel cell performance revealed vital understanding towards generating high performance n-PAAEMFCs with various other non-PGM elements as ORR electrocatalyst materials. In addition it might promote the surface active V-AgNs as a sustainable and low-cost ORR electrocatalyst material not only for n-PAAEMFCs but also for redox-flow air or metal–air batteries. The carbon support free $Pd_{85}Pt_{15}$ ORR catalysts showed very high DMFC activity at lowest Pt loading. It also helped to mitigate the carbon corrosion and improve mass transfer due to reduced catalyst layer thickness. This carbon free ORR catalyst at low Pt content loading would help as a reliable model for developing advanced DMFC cathode electrocatalysts and alternative to conventional Pt/C ORR electrocatalysts. In future, attempts could be made in tuning the Pt loading levels and scale up the preparation of carbon free $Pd_{85}P_{15}$ NPoS, to evaluate their performance in DMFC single-cell studies. Since these NPoS exhibit better ORR activities; they may be used as cathode catalysts to enhance the performance of H_2–O_2 fuel cells. Further attempts can be made to realize the fascinating part of Pd and Pt islands on nano-porous surface, the mechanism of ORR on NPoS electrocatalysts through *in-situ* spectroscopic and *in-situ* TEM studies. Electrocatalysts with ultra-low PGM (Pd, Pt, Rh, Ir, Ru and Os) or non-PGM (Ag, Cu, Fe, Co, Mn, Ni or carbon-based materials, Mxenes, etc.) will be prepared using facile aqueous method and their electrocatalytic ORR performances will be evaluated.

ACKNOWLEDGMENTS

Authors acknowledge Director, CSIR, Central Electro-Chemical Research Institute, Karaikudi for organizational help, DST-INSPIRE fellowship for funding, CSIR-CIF, JNU and IITM for the core characterization amenities used for the current research work.

REFERENCES

Alesker, M., M. Page, M. Shviro, Y. Paska, G. Gershinsky, D.R. Dekel and Zitoun, D. 2016. Palladium/nickel bifunctional electrocatalyst for hydrogen oxidation reaction in alkaline membrane fuel cell. J. Power Sources. 304: 332–339. https://doi.org/10.1016/j.jpowsour.2015.11.026

Antoine, O., Y. Bultel and R. Durand. 2001. Oxygen reduction reaction kinetics and mechanism on platinum nanoparticles inside Nafion®. J. Electroanal. Chem. 499: 85–94. https://doi.org/10.1016/S0022-0728(00)00492-7

Appleby, A.J. 1972. Rest potentials of oxide-free platinum electrodes. J. Electroanal. Chem. 35: 193–207.

Appleby, A.J. 1993. Electrocatalysis of aqueous dioxygen reduction. J. Electroanal. Chem. 357: 117–179. https://doi.org/10.1016/0022-0728(93)80378-U

Aricò, A.S., A.K. Shukla, H. Kim, S. Park, M. Min and V. Antonucci. 2001. An XPS study on oxidation states of Pt and its alloys with Co and Cr and its relevance to electroreduction of oxygen. Appl. Surf. Sci. 172: 33–40. https://doi.org/10.1016/S0169-4332(00)00831-X

Aricò, A.S., A. Stassi, C. D'Urso, D. Sebastián and V. Baglio. 2014. Synthesis of Pd3Co1@ Pt/C core-shell catalysts for methanol-tolerant cathodes of direct methanol fuel cells. Chem. Eur. J. 20: 1–7. https://doi.org/10.1002/chem.201402062.

Asara, G.G., L.O. Paz-Borbón and F. Baletto. 2016. 'Get in touch and keep in contact': Interface effect on the oxygen reduction reaction (ORR) activity for supported PtNi nanoparticles. ACS Catal. 6: 4388–4393. https://doi.org/10.1021/acscatal.6b00259

Azzam, A., J.O'M. Bockris, B.E. Conway and H. Rosenberg. 1950. Some aspects of the measurement of hydrogen overpotential. Trans. Faraday Soc. 46: 918. https://doi.org/10.1039/tf9504600918

Baglio, V., A. Di Blasi, E. Modica, P. Creti, V. Antonucci and A.S. Aricò. 2006. Electrochemical analysis of direct methanol fuel cells for low temperature operation. J. Electrochem. Soc. 1: 71–79.

Balaji, S.S., P.A. Ganesh, M. Moorthy and M. Sathish. 2020. Efficient electrocatalytic activity for oxygen reduction reaction by phosphorus-doped graphene using supercritical fluid processing. Bull. Mater. Sci. 43: 151. https://doi.org/10.1007/s12034-020-02142-2

Bansal, V., H. Jani, J.D. Plessis, P.J. Coloe and S.K. Bhargava. 2008. Galvanic replacement reaction on metal films: a one-step approach to create nanoporous surfaces for catalysis. Adv. Mater. 20: 717–723. https://doi.org/10.1002/adma.200701297

Bezerra, Cicero W.B., L. Zhang, K. Lee, H. Liu, A.L.B. Marques, E.P. Marques, et al., 2008. A review of Fe-N/C and Co-N/C catalysts for the oxygen reduction reaction. Electrochim. Acta. 53: 4937–4951. https://doi.org/10.1016/j.electacta.2008.02.012

Bing, Y., H. Liu, L. Zhang, D. Ghosh and J. Zhang. 2010. Nanostructured Pt-alloy electrocatalysts for PEM fuel cell oxygen reduction reaction. Chem. Soc. Rev. 39: 2184–2202. https://doi.org/10.1039/b912552c

Breiter, M.W. 1969. Electrochemical Processes in Fuel Cells. Springer, US.

Castegnaro, M.V., W.J. Paschoalino, M.R. Fernandes, B. Balke, M.d.C.M. Alves, E.A. Ticianelli, et al.,2017. Pd-M/C (M = Pd, Cu, Pt) electrocatalysts for oxygen reduction reaction in alkaline medium: Correlating the electronic structure with activity. Langmuir. 160: 155–163. https://doi.org/10.1021/acs.langmuir.7b00098

Chen, S., P.J Ferreira, W. Sheng, N. Yabuuchi, L.F. Allard and Y. Shao-Horn. 2008. Enhanced activity for oxygen reduction reaction on 'Pt$_3$Co' nanoparticles: direct evidence of percolated and sandwich-segregation structures. JACS 130: 13818–13819. https://doi.org/10.1021/ja802513y

Chen, L., L. Kuai and B. Geng. 2013. Shell structure-enhanced electrocatalytic performance of Au–Pt core–shell catalyst. CrystEngComm 15: 2133. https://doi.org/10.1039/c3ce 27058k

Cho, K.Y., Y.S. Yeom, H.Y. Seo, P. Kumar, A.S. Lee, K.-Y. Baek and H.G. Yoon. 2017. Molybdenum-doped PdPt@Pt core–shell octahedra supported by ionic block copolymer-functionalized graphene as a highly active and durable oxygen reduction electrocatalyst. AACS Appl. Mater. Interfaces. 9: 1524–1535. https://doi.org/10.1021/acsami.6b13299

Choi, B., W.-H. Nam, D.Y. Chung, I.-S. Park, S.J. Yoo, J.C. Song and Y.-E.Sung. 2015. Enhanced methanol tolerance of highly Pd rich Pd-Pt cathode electrocatalysts in direct methanol fuel cells. Electrochim. Acta 164: 235–242. https://doi.org/10.1016/j.electacta.2015.02.203

Cleve, T.V., S. Moniri, G. Belok, K.L. More and S. Linic. 2016. Nanoscale engineering of efficient oxygen reduction electrocatalysts by tailoring local chemical environment of Pt surface sites. ACS Catal. 4: 200–215. https://doi.org/10.1021/acscatal.6b01565

Conway, B.E. 1995. Electrochemical oxide film formation at noble metals as a surface-chemical process, Prog. Surf. Sci. 49: 331–452. https://doi.org/Doi 10.1016/0079-6816 (95)00040-6

da Silva, M.R. and A.C.D. Ângelo. 2010. Synthesis and characterization of ordered intermetallic nanostructured PtSn/C and PtSb/C and Evaluation as Electrodes for Alcohol Oxidation. Electrocatal 1: 95–103. https://doi.org/10.1007/s12678-010-0010-5

Damjanovic A. and V. Brusic. 1966a. Electrode kinetics of oxygen reduction on oxide-free platinum electrodes. Electrochim. Acta. 12: 615–628.

Damjanovic, A. and V. Brusic. 1966b. Pretreatment of Pt-Au and Pd-Au alloy electrodes in the study of oxygen reduction. J. Electroanal. Chem. 15: 29–33.

Damjanovic, A. and J.O'M. Bockris. 1966c. The rate constants for oxygen dissolution on bare and oxide-covered platinum. Electrochim. Acta. 11: 376–377. https://doi.org/10.1016/0013-4686(66)87048-2

Damjanovic, A., A. Dey and J.O'M. Bockris. 1966. Kinetics of oxygen evolution and dissolution on platinum electrodes. Electrochim. Acta. 11: 791–814. https://doi.org/10.1016/0013-4686(66)87056-1

Fuller, T. 2008. Proton Exchange Membrane Fuel Cells. ECS transactions.

Gan, L., M. Heggen, C. Cui and P. Strasser. 2009. Thermal facet healing of concave octahedral Pt-Ni nanoparticles imaged *in-situ* at the atomic scale: Implications for the rational synthesis of durable high performance ORR electrocatalysts. ACS Catal. 6: 692–695. https://doi.org/10.1021/acscatal.5b02620

Ganesh, P.A. and D. Jeyakumar. 2014. One pot aqueous synthesis of nanoporous $Au_{85} Pt_{15}$ material with surface bound Pt islands: An efficient methanol tolerant ORR catalyst. Nanoscale. 6: 13012–13021. https://doi.org/10.1039/c4nr04712e

Ganesh, P.A. and D. Jeyakumar. 2017a. Intriguing catalytic activity of surface active gold-platinum islands on nano-porous Au in determining efficient direct formic acid oxidation pathway. ChemistrySelect. 2: 3562–3571. https://doi.org/10.1002/slct.201700670

Ganesh, P.A. and D. Jeyakumar. 2017b. Nano-porous electrocatalyst with textured surface active pd-pt islands for efficient methanol tolerant oxygen reduction reaction. ChemistrySelect. 2: 7544–7552. https://doi.org/10.1002/slct.201700992

Ganesh, P.A. and D. Jeyakumar. 2018a. A facile strategy for the synthesis of flower-like structures with enhanced nano-porous surfaces. J. Adv. Microsc. Res. 13: 48–53. https://doi.org/10.1166/jamr.2018.1356

Ganesh, P.A. and D. Jeyakumar. 2018b. Hierarchical approach of mitigating carbon influence in nano-porous electrocatalyst with unique surface islands for efficient methanol resistive oxygen reduction. Electrochim. Acta. 262: 306–318. https://doi.org/10.1016/j.electacta.2018.01.002

Ganesh, P.A., A.N. Prakrthi, S. Selva Chandrasekaran and D. Jeyakumar. 2021. Shape-tuned, surface-active and support-free silver oxygen reduction electrocatalyst enabled high performance fully non-PGM alkaline fuel cell. RSC Adv. 11: 24872–24882. https://doi.org/10.1039/D1RA02718B

Gao, Min-Rui, Z.-Y. Lin, J. Jiang, C.-H. Cui, Y.-R. Zheng and S.-h. Yu. 2012. Completely green synthesis of colloid adams catalyst α-PtO_2 nanocrystals and derivative Pt nanocrystals with high activity and stability for oxygen reduction. Chem. Eur. J., 18: 8423-8429. https://doi.org/10.1002/chem.201200353

Garlyyev, B., Y., Liang, F.K. Butt and A.S. Bandarenka. 2017. Engineering of highly active silver nanoparticles for oxygen electroreduction via simultaneous control over their shape and size. Adv. Sustainable Syst. 1(12): 1–7. https://doi.org/10.1002/adsu.201700117

Genshaw, M.A., A. Damjanovic and J.O'M. Bockris. 1967. Hydrogen peroxide formation in oxygen reduction at gold electrodes: II. Alkaline solution. J. Electroanal. Chem. Interfacial Electrochem. 15: 163–172. https://doi.org/10.1016/0022-0728(67)85021-6

Ghosh, S., R.K. Sahu and C.R. Raj. 2012. Pt-Pd alloy nanoparticle-decorated carbon nanotubes: a durable and methanol tolerant oxygen reduction electrocatalyst. Nanotechnology. 23: 385602. https://doi.org/10.1088/0957-4484/23/38/385602

Ghosh, S. and C.R. Raj. 2015. Pt-Pd nanoelectrocatalyst of ultralow Pt content for the oxidation of formic acid: Towards tuning the reaction pathway. J. Chem. Sci. 127: 949–957. https://doi.org/10.1007/s12039-015-0854-6

Góral-Kurbiel, M., R. Kosydar, J. Gurgul, B. Dembińska, P.J. Kulesza and A. Drelinkiewicz. 2016. Carbon supported PdxPty nanoparticles for oxygen reduction. The effect of Pd:Pt ratio. Electrochim. Acta. 222: 1220–1233. https://doi.org/10.1016/j.electacta.2016.11.096

Gu, S., W. Sheng, R. Cai, S.M. Alia, S. Song, K.O. Jensen and Y. Yan. 2013. An efficient Ag–ionomer interface for hydroxide exchange membrane fuel cells. Chem. Comm. 49(2): 131–133. https://doi.org/10.1039/c2cc34862d

Guo, S., Y. Fang, S. Dong and E. Wang. 2007. High-efficiency and low-cost hybrid nanomaterial as enhancing electrocatalyst: spongelike Au/Pt core/shell nanomaterial with hollow cavity. J. Phys. Chem. C. 111: 17104–17109. https://doi.org/10.1021/jp075625z

Hall, S.B., E.A. Khudaish and A.L. Hart. 1998. Electrochemical oxidation of hydrogen peroxide at platinum electrodes. Part II: Effect of potential. Electrochim. Acta. 43: 2015–2024. https://doi.org/10.1016/S0013-4686(97)10116-5

Henning, S., L. Kühn, J. Herranz, J. Durst, T. Binninger, M. Nachtegaal, et al., 2016. Pt-Ni aerogels as unsupported electrocatalysts for the oxygen reduction reaction. J. Electrochem. Soc. 163(9): F998–F1003. https://doi.org/10.1149/2.0251609jes

Hoare, J.P. 1982. On the interaction of oxygen with platinum. Electrochim. Acta. 27: 1751–1761.

Hong, J.W., S.W. Kang, B.-S. Choi, D. Kim, S.B. Lee and S.W. Han. 2012. Controlled synthesis of Pd-Pt alloy hollow nanostructures with enhanced catalytic activities for oxygen reduction. ACS Nano. 6: 2410–2419. https://doi.org/10.1021/nn2046828

Hoogers, Gregor. 2003. Fuel Cell Technology Handbook. CRC Press.

Hsueh, K-L., E.R. Gonzalez and S. Srinivasan. 1983a. Electrolyte effects on oxygen reduction kinetics at platinum: a rotating ring-disc electrode analysis. Electrochim. Acta. 28: 691–697.

Hsueh, K.-L., D.-T. Chin and S. Srinivasan. 1983b. Electrode kinetics of oxygen reduction. J. Electroanal. Chem. 153: 79–95.

Hu, Q., G. Li, J. Pan, L. Tan, J. Lu and L. Zhuang. 2013. Alkaline polymer electrolyte fuel cell with Ni-based anode and Co-based cathode. Int. J. Hydrogen Energy. 38(36): 16264–16268. https://doi.org/10.1016/j.ijhydene.2013.09.125

Ilayaraja, N., N. Prabu, N. Lakshminarasimhan, P. Murugan and D. Jeyakumar. 2013. Au–Pt graded nano-alloy formation and its manifestation in small organics oxidation reaction. J. Mater. Chem. A. 1: 4048–4056. https://doi.org/10.1039/c3ta01451g

Janthon, P., F. Viñes, J. Sirijaraensre, J. Limtrakul and F. Illas. 2017. Carbon dissolution and segregation in platinum. Catal. Sci. Technol. 7: 807–816. https://doi.org/10.1039/C6CY02253G

Kadiri, F.E., R. Faure and R. Durand. 1991. Electrochemical reduction of molecular oxygen on platinum single crystals. J. Electroanal. Chem. 301: 177–188.

Kaipannan, S., P.A. Ganesh, K. Manickavasakam, S.K. Sundaramoorthy, K. Govindarajan, S. Mayavan, et al., 2020. Waste engine oil derived porous carbon/ZnS nanocomposite as bi-functional electrocatalyst for supercapacitor and oxygen reduction. J. Energy Storage. 32: 101774. https://doi.org/10.1016/j.est.2020.101774

Karan, H.I., K. Sasaki, K. Kuttiyiel, C.A. Farberow, M. Mavrikakis and R.R. Adzic. 2012. Catalytic activity of platinum mono layer on iridium and rhenium alloy nanoparticles for the oxygen reduction reaction. ACS Catal. 2: 817–824. https://doi.org/10.1021/cs200592x

Kesavan, M., A. Arulraj, K. Rajendran, P. Anbarasu, P. Anandha Ganesh, D. Jeyakumar and M. Ramesh. 2018. Performance of dye-sensitized solar cells employing polymer gel as an electrolyte and the influence of nano-porous materials as fillers. Mater. Res. Express. 5: 115305. https://doi.org/10.1088/2053-1591/aade2a

Laitinen, H.A. and I.M. Kolthoff. 1940. Voltammetry with stationary microelectrodes of platinum wire. J. Phys. Chem. 18: 1061–1079.

Li, T. and P.B. Balbuena. 2003. Oxygen reduction on a platinum cluster. Chem. Phys. Lett. 367: 439–447. https://doi.org/10.1016/S0009-2614(02)01755-4

Li, J., G. Wang, J. Wang, S. Miao, M. Wei, F. Yang, et al., 2014. Architecture of PtFe/C catalyst with high activity and durability for oxygen reduction reaction. Nano Res. 7: 1519–1527. https://doi.org/10.1007/s12274-014-0513-0

Liang, C.C. and A.L. Juliard. 1965. The overpotential of oxygen reduction at platinum electrodes. J. Electroanal. Chem. 9: 390–394.

Lima, F.H.B., J. Zhang, M.H. Shao, K. Sasaki, M.B. Vukmirovic, E.A. Ticianelli and R.R. Adzic. 2007. Catalytic activity-d-band center correlation for the O_2 reduction reaction on platinum in alkaline solutions. J. Phys. Chem. C. 111: 404–410. https://doi.org/10.1021/jp065181r

Lin, G., H. Li and K. Xie. 2022. Identifying and engineering active sites at the surface of porous single-crystalline oxide monoliths to enhance catalytic activity and stability. CCS Chem. 4: 1441–1451. https://doi.org/10.31635/CCSCHEM.021.202000740

Liu, C-W, Y.-C. Wei, C.-C. Liu and K.-W. Wang. 2012a. Pt–Au core/shell nanorods: Preparation and applications as electrocatalysts for fuel cells. J. Mater. Chem. 22: 4641. https://doi.org/10.1039/c2jm16407h

Liu, Z., H. Xin, Z. Yu, Y. Zhu, J. Zhang, J.A. Mundy et al., 2012b. Atomic-scale compositional mapping and 3-dimensional electron microscopy of dealloyed $PtCo_3$ catalyst nanoparticles with spongy multi-core/shell structures. J. Electrochem. Soc. 159: F554–F559. https://doi.org/10.1149/2.051209jes

Lu, S., J. Pan, A. Huang, L. Zhuang and J. Lu. 2008. Alkaline polymer electrolyte fuel cells completely free from noble metal catalysts. PNAS. 105(52): 20611–20614.

Maciá, M.D., J.M. Campiña, E. Herrero and J.M. Feliu. 2004. On the kinetics of oxygen reduction on platinum stepped surfaces in acidic media. J. Electroanal. Chem. 564: 141–150. https://doi.org/10.1016/j.jelechem.2003.09.035

Markovic, N.M., R.R. Adzic, B.D.Cahan and E.B. Yeager. 1994. Structural effects in electrocatalysis: oxygen reduction on platinum low index single-crystal surfaces in perchloric acid solutions. J. Electroanal. Chem. 377: 249–259. https://doi.org/10.1016/0022-0728(94)03467-2

Markovic, N.M. and P.N. Ross Jr. 2002. Surface science studies of model fuel cell electrocatalysts. Surf. Sci. Rep. 45: 117–229.

Maruyama J., M. Inaba and Z. Ogumi. 1998. Rotating ring-disk electrode study on the cathodic oxygen reduction at Nafion®-coated gold electrodes. J. Electroanal. Chem. 458: 175–182. https://doi.org/10.1016/S0022-0728(98)00362-3

Maye, M.M., N.N. Kariuki, J. Luo, L. Han, P. Njoki, L. Wang et al., 2004. Electrocatalytic reduction of oxygen: gold and gold-platinum nanoparticle catalysts prepared by two-phase protocol. Gold Bull. 37: 217–223. https://doi.org/10.1007/BF03215216

Mayrhofer, K.J.J., D. Strmcnik, B.B. Blizanac, V.Stamenkovic, M. Arenz and N.M. Markovic. 2008. Measurement of oxygen reduction activities via the rotating disc electrode method: From Pt model surfaces to carbon-supported high surface area catalysts. Electrochim. Acta. 53: 3181–3188. https://doi.org/10.1016/j.electacta.2007.11.057

Mukerjee, S. and S. Srinivasan. 1993. Enhanced electrocatalysis of oxygen reduction on platinum alloys in proton exchange membrane fuel cells. J. Electroanal. Chem. 357: 201–224. https://doi.org/10.1016/0022-0728(93)80380-z

Narayanamoorthy, B., N. Panneerselvam, C. Sita, S. Pasupathi, S. Balaji and I.S. Moon. 2016. Enhanced stabilities of Ag electrocatalyst as self-standing and multiwalled carbon nanotube supported nanostructures for oxygen reduction in alkaline medium. J. Electrochem. Soc. 163(5): H313–H320. https://doi.org/10.1149/2.0931605jes

Naumkin, A.V., K.-V. Anna and C.J.P. Stephen W. Gaarenstroom. 2012. NIST X-ray Photoelectron Spectroscopy (XPS) Database, Version 3.5, U.S. Secretary of Commerce. https://srdata.nist.gov/xps/

Nie, Y., L. Li and Z. Wei. 2015. Recent advancements in Pt and Pt-free catalysts for oxygen reduction reaction. Chem. Soc. Rev. 44(8) https://doi.org/10.1039/c4cs00484a

Pan, yanbo, X. Shen, L. Yao, A. Bentalib and Z. Peng. 2018. Active sites in heterogeneous catalytic reaction on metal and metal oxide: theory and practice. In: Catal. 8(10). https://doi.org/10.3390/catal8100478

Panigrahi, S., S. Kundu, S. Ghosh, S. Nath and T. Pal. 2004. General method of synthesis for metal nanoparticles. J. Nanopart. Res. 6: 411–414. https://doi.org/10.1007/s11051-004-6575-2

Parsons, R. 1964. The kinetics of electrode reactions and the electrode material. Surf. Sci. 2: 418–435.

Paulus, U.A., T.J. Schmidt, H.A. Gasteiger and R.J. Behm. 2001. Oxygen reduction on a high-surface area Pt/Vulcan carbon catalyst: A thin-film rotating ring-disk electrode study. J. Electroanal. Chem. 495: 134–145. https://doi.org/10.1016/S0022-0728(00)00407-1

Paulus, U.A., A. Wokaun, G.G. Scherer, T.J. Schmidt, V. Stamenkovic, N.M. Markovic, et al., 2002. Oxygen reduction on high surface area Pt-based alloy catalysts in comparison to well-defined smooth bulk alloy electrodes. Electrochim. Acta. 47: 3787–3798. https://doi.org/10.1016/S0013-4686(02)00349-3

Perez, J., E.R. Gonzalez and E.A. Ticianelli. 1998. Oxygen electrocatalysis on thin porous coating rotating platinum electrodes. Electrochim. Acta. 44: 1329–1339. https://doi.org/10.1016/S0013-4686(98)00255-2

Pfisterer, J.H.K., Y. Liang, O. Schneider and A.S. Bandarenka. 2017. Direct instrumental identification of catalytically active surface sites. Nature. 549: 74–77. https://doi.org/10.1038/nature23661

Reeve, R.W., P.A. Christensen, A. Hamnett, A. Haydock and S.C. Roy. 1998. Methanol tolerant oxygen reduction catalysts based on transition metal sulfides. J. Electrochem. Soc. 145: 3463–3471.

Savadogo, O., K. Lee, K. Oishi, S. Mitsushima, N. Kamiya and K.-I. Ota. 2004. New palladium alloys catalyst for the oxygen reduction reaction in an acid medium. Electrochem. Commun. 6: 105–109. https://doi.org/10.1016/j.elecom.2003.10.020

Sawyer, D.T. and L.V. Interrante. 1961. Electrochemistry of dissolved gases II reduction of oxygen at platinum, palladium, nickel and other metal electrodes. J. Electroanal. Chem. 2: 310–327. https://doi.org/10.1016/0022-0728(61)85004-3

Sepa, D.B., M.V. Vojnovic, Lj.M. Vracar and A. Damjanovic. 1986a. Apparent enthalpies of activation of electrodic oxygen reduction at platinum in different current density regions-I. Acid solution. Electrochim. Acta. 31: 91–96.

Sepa, D.B., M.V. Vojnovic and A. Damjanovic. 1986b. Different views regarding the kinetics and mechanisms of oxygen reduction at Pt and Pd electrodes. Electrochem. Commun. 32: 129–134.

Sepa, D.B., Lj.M. Vracar, M.V. Vojnovic and A. Damjanovic. 1986c. Symmetry factor and transfer coefficient in analysis of enthalpies of activation and mechanisms of oxygen reduction at platinum electrodes. Electrochim. Acta. 31: 1401–1402.

Shao, M. (ed.) 2013. Electrocatalysis in Fuel Cells: A Non- and Low- Platinum Approach. Springer, London. https://doi.org/10.1007/978-1-4471-4911-8/COVER

Shao, M., Q. Chang, J.-P. Dodelet and R. Chenitz. 2016. Recent advances in electrocatalysts for oxygen reduction reaction. Chem. Rev. 116. https://doi.org/10.1021/acs.chemrev.5b00462

Shukla, A.K. and R.K. Raman. 2003. Methanol-resistant oxygen-reduction catalysts for direct methanol fuel cells. Annu. Rev. Mater. Res. 33: 155–168. https://doi.org/10.1146/annurev.immunol.21.120601.141107

Sidik R.A. and A.B. Anderson. 2002. Density functional theory study of O_2 electroreduction when bonded to a Pt dual site. J. Electroanal. Chem. 528: 69–76. https://doi.org/10.1016/S0022-0728(02)00851-3

Stamenković, V., T.J. Schmidt, P.N. Ross and N.M. Marković. 2003. Surface segregation effects in electrocatalysis: Kinetics of oxygen reduction reaction on polycrystalline Pt_3Ni alloy surfaces. J. Electroanal. Chem. 554–555: 191–199. https://doi.org/10.1016/S0022-0728(03)00177-3

Stamenkovic, V.R., B. Fowler, B.S. Mun, G. Wang, P.N. Ross, C.A. Lucas, et al., 2007. Improved oxygen reduction activity on $Pt_3Ni(111)$ via increased surface site availability. Science. 315: 493–497. https://doi.org/10.1126/science.1135941

Steininger, H., S. Lehwald and H. Ibach. 1982. Adsorption of oxygen on Pt(111). Surf. Sci. 123: 1–17.

Tammeveski, K., T. Tenno, J. Claret and C. Ferrater. 1997. Electrochemical reduction of oxygen on thin-film Pt electrodes in 0.1 M KOH. Electrochim. Acta. 42: 893–897. https://doi.org/10.1016/S0013-4686(96)00325-8

Trasatti, S. and O.A. Petrii. 1993. Real surface area measurements in electrochemistry. J. Electroanal. Chem. 321: 353–376. https://doi.org/10.1016/0022-0728(92)80162-W

Trinh, Q.T., J. Yang, J.Y. Lee and M. Saeys. 2012. Computational and experimental study of the Volcano behavior of the oxygen reduction activity of PdM@PdPt/C (M = Pt, Ni, Co, Fe, and Cr) core–shell electrocatalysts. J. Catal. 291: 26–35. https://doi.org/10.1016/j.jcat.2012.04.001

Tripković, V., E. Skúlason, S. Siahrostami, J.K. Nørskov and J. Rossmeisl. 2010. The oxygen reduction reaction mechanism on Pt(111) from density functional theory calculations. Electrochim. Acta. 55: 7975–7981. https://doi.org/10.1016/j.electacta.2010.02.056

Tuaev, X., S. Rudi and P. Strasser. 2016. The impact of the morphology of the carbon support on the activity and stability of nanoparticle fuel cell catalysts. Catal. Sci. Technol. 6: 8276–8288. https://doi.org/10.1039/C6CY01679K

Turkevich, J., P.C. Stevenson and J. Hillier. 1951. A study of the nucleation and growth processes in the synthesis of colloidal gold. Discuss. Faraday Soc. 11: 55. https://doi.org/10.1039/df9511100055

Vanysek, P. 2000. Standard reduction potential of electrochemical series. pp. 110–160. *In*: Handbook of Chemistry and Physics. CRC Press LLC.

Venarusso, L.B., J. Bettini and G. Maia. 2016. Superior catalysts for oxygen reduction reaction based on porous nanostars of aPt, Pd, or Pt–Pd alloy shell supported on a gold core. Chem. Electro. Chem. 3: 749–756. https://doi.org/10.1002/celc.201600046

Vukmirovic, M.B., J. Zhang, K. Sasaki, A.U. Nilekar, F. Uribe, M. Mavrikakis, et al., 2007. Platinum monolayer electrocatalysts for oxygen reduction. Electrochim. Acta. 52: 2257–2263. https://doi.org/10.1016/j.electacta.2006.05.062

Wang, D.L., H.L. Xin, Y.C. Yu, H.S. Wang, E. Rus, D.A. Muller, et al., 2010. Pt-Decorated PdCo@Pd/C core-shell nanoparticles with enhanced stability and electrocatalytic activity for the oxygen reduction reaction. J. Am. Chem. Soc. 132: 17664–17666. https://doi.org/10.1021/ja107874u

Wang, Q., F. Chen, Y. Liu, N. Zhang, L. An and R.L. Johnston. 2017. Bifunctional electrocatalysts for oxygen reduction and borohydride oxidation reactions using Ag_3Sn nanointermetallic for the ensemble effect. AACS Appl. Mater. Interfaces. 9(41): 35701–35711. https://doi.org/10.1021/acsami.7b05186

Wen, Z., J. Liu and J. Li. 2008. Core/Shell Pt/C nanoparticles embedded in mesoporous carbon as a methanol-tolerant cathode catalyst in direct methanol fuel cells. Adv. Mater. 20: 743–747. https://doi.org/10.1002/adma.200701578

Wroblowa, H., M.L.B. Rao, A. Damjanovic and J.O'M. Bockris. 1967. Adsorption and kinetics at platinum electrodes in the presence of oxygen at zero net current. J. Electroanal. Chem. 15: 139–150.

Wroblowa, H.S., Yen-Chi-Pan and G. Razumney 1976. Electroreduction of oxygen. J. Electroanal. Chem. 69: 195–201.

Xia, Y., P. Yang, Y. Sun, Y. Wu, B. Mayers, B. Gates, Y. Yin, et al., 2003. One-dimensional nanostructures: synthesis, characterization, and applications. Adv. Mater. 15: 353–389. https://doi.org/10.1002/adma.200390087

Xin, H., A. Holewinski and S. Linic. 2012. Predictive structure-reactivity models for rapid screening of Pt-based multimetallic electrocatalysts for the oxygen reduction reaction. ACS Catal. 2: 12–16. https://doi.org/10.1021/cs200462f

Yang, J., C.H. Cheng, W. Zhou, J.Y. Lee and Z. Liu. 2010a. Methanol-tolerant heterogeneous PdCo@PdPt/C electrocatalyst for the oxygen reduction reaction. Fuel Cells. 10: 907–913. https://doi.org/10.1002/fuce.200900205

Yang, J., W. Zhou, C.H. Cheng, J.Y. Lee and Z. Liu. 2010b. Pt-decorated Pdfe nanoparticles as methanol-tolerant oxygen reduction electrocatalyst. ACS Appl. Mater. Interfaces. 2: 119–126. https://doi.org/10.1021/am900623e

Yang, Y., J.-J. Du, L.-M. Luo, R.-H. Zhang, Z.-X. Dai and X.-W. Zhou. 2016. Facile aqueous-phase synthesis and electrochemical properties of novel PtPd hollow nanocatalysts. Electrochim. Acta. 212: 966–972. https://doi.org/10.1016/j.electacta.2016.07.085

Yang, C.-C., Z.-T. Liu, Y.-P. Lyu and C.-L. Lee. 2017. Shell-thickness-controlled synthesis ofcore-shell Pd@Pt nanocubes and tuning of their oxygen reduction activities. J. Electrochem. Soc. 164: 112–118. https://doi.org/10.1149/2.1411702jes

Ye, F., H. Liu, W. Hu, J. Zhong, Y. Chen, H. Cao, et al., 2012. Heterogeneous Au-Pt nanostructures with enhanced catalytic activity toward oxygen reduction. Dalton Trans. 41: 2898–2903. https://doi.org/10.1039/c2dt11960a

Yeager E. 1984. Electrocatalysts for O_2 reduction. Electrochim. Acta. 29: 1527–1537.

Yin, Y., R.M. Rioux, C.K. Erdonmez, S. Hughes, G.A. Somorjai and A.P. Alivisatos. 2004. Formation of hollow nanocrystals through the nanoscale kirkendall effect. Science. 304: 711–714. https://doi.org/10.1126/science.1096566

Zhang, G.-R., D. Zhao, Y.-Y. Feng, B. Zhang, D.S. Su, G. Liu, et al., 2012. Catalytic Pt-on-Au nanostructures: Why Pt becomes more active on smaller Au particles. ACS Nano. 6: 2226–2236. https://doi.org/10.1021/nn204378t

Zhang, J. 2008. PEM Fuel Cell Electrocatalysts and Catalyst Layers: Fundamentals and Applications. Springer, London. https://doi.org/10.1007/978-1-84800-936-3/COVER

Zhao, X., M. Yin, L. Ma, L. Liang, C. Liu, J. Liao, et al., 2011. Recent advances in catalysts for direct methanol fuel cells. Energy Environ. Sci. 4: 2736. https://doi.org/10.1039/c1ee01307f

Zhdanov, V.P. and B. Kasemo. 2006. Kinetics of electrochemical O_2 reduction on Pt. Electrochem. Commun. 8: 1132–1136. https://doi.org/10.1016/j.elecom.2006.05.003

Chapter **2**

Recent Developments of Transition Metal Oxide Nanoparticles on Oxygen Reduction Reaction

Vaithiyanathan Sankar Devi, Mattath Athika,
Sekar Sandhiya and Perumal Elumalai*

Electrochemical Energy Storage Lab, Department of Green Energy Technology
Madanjeet School of Green Energy Technologies, Pondicherry University,
Puducherry-605014, India. Tel: +61-413-2654867

2.1 INTRODUCTION

In recent times, there has been a spike in the growth of consumer electronics devices such as mobile phones, tablets, etc. that has led to tremendous interest on the development of energy storage technologies. In addition, the population growth, new machineries in the production sectors and the utilization of novel technologies in the agricultural fields has elevated significant energy consumption. The nature and the mode of the energy usage have been dramatically changing almost year-by-year in the past two decades. Such an enormous amount of the energy use has become indispensable for modern civilizations, leading to comfortable living, including smart homes, smart cities etc. (Cheng and Chen,

*For Correspondence: Email: drperumalelumalai@pondiuni.ac.in

2012; Romm, 2006). Currently, most of the energy requirements are met by non-renewable energy sources that have resulted in the release of harmful ozone-layer-depleting gases into the environmental eco-system (Goodenough, 2014). On the other side, the lack of continuous availability of the renewable energy sources like solar, wind, and others necessitates the design and development of efficient energy storage systems to store energy for continuous use. In this view, the development of a reliable, long-lasting, cost-effective, and efficient energy storage/conversion systems is rising (Yang et al., 2010). There are countable energy storage/conversion systems that include fuel cells, solar cells, supercapacitors and rechargeable batteries including metal-air batteries. Over the years, petrol and the diesel prices are being deregulated and increasing day-by-day, forcing the integration of battery electric vehicles (BEVs) for passenger transportation, as the electric/hybrid vehicles can lead to significant reduction of toxic emissions and lower greenhouse effect (Sadek, 2012). In this regard also, the development of reliable, energy storage systems is considered important. The fuel cell and metal-air batteries have many advantages, such as less ecological impact, low operating temperature, high energy density, etc. The oxygen reduction reaction is an important and inevitable reaction that takes place in the fuel cells and metal-air batteries. Thus, the most important cathode reaction in the fuel cell or the metal-air batteries is the oxygen reduction reaction (ORR) – a challenging reaction in the field of catalysis and electrochemistry. The ORR is a multi-electron transfer process that follows an electrocatalytic inner-sphere mechanism. The reaction is highly dependent on the nature of the electrode and its features, such as surface area, electronic structure, redox feature, ability to adsorb and diffusion O_2, etc. The oxygen evolution reaction (OER) is also important in the field of the metal-air batteries as it occurs during charging of the battery. Both ORR and OER can be made to occur in acidic and alkaline media. In this chapter, detailed mechanism of the ORR and OER, fundamental parameters, design and the role of electrocatalysts pertaining to the nanostructured transition metal oxide materials are extensively discussed.

2.2 BASICS OF OXYGEN REDUCTION REACTION

The oxygen reduction reaction (ORR) is the most critical reaction in the field of electrochemistry, which involves multistep electron transfer process by adsorbing molecular oxygen (O_2) species on the surface of the catalyst. Depending on the mode of oxygen adsorption, the ORR can occur in two different pathways: (i) The $4e^-$ transfer pathway and (ii) $2e^-$ transfer pathway. The direct four-electron pathway involves simultaneous co-ordination of both the oxygen atoms of the O_2 molecule on the catalyst surface. It is a complete reduction pathway whereas the two electron pathway adopts one oxygen atom co-ordination on the catalyst surface and is a partial reduction path. The adsorbed oxygen on the surface of the electrode is then converted into H_2O and OH^- based on the electrolytes (acidic or alkaline). The direct conversion of the O_2 to H_2O involves a dissociative mechanism where the first step is adsorption of O_2 on the catalyst surface followed by breaking of the oxygen–oxygen bond to give adsorbed oxygen atoms. In the $2e^-$ transfer pathway, an undesirable peroxide (HO_2^-) is formed as an intermediate which leads to sluggish kinetics as the formed HO_2^- must further undergo reduction to form H_2O/OH^-. The four-electron transfer process is highly desirable rather than two e^- transfer process because the reactivity of peroxide is higher than the stability of the H_2O formed.

The various reaction steps involved in the direct four-electron transfer pathway from O_2 to H_2O/OH^- and the two-electron transfer pathway from O_2 to H_2O_2/HO_2^- occurring in the aqueous solutions are shown in the following reactions:

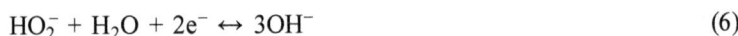

$$O_2 + 4H^+ + 4e^- \leftrightarrow 2H_2O \qquad \text{(Acidic aqueous)} \qquad (1)$$

$$O_2 + 2H^+ + 2e^- \leftrightarrow H_2O_2 \qquad\qquad\qquad\qquad\qquad (2)$$

$$H_2O_2 + 2H^+ + 2e^- \leftrightarrow 2H_2O \qquad\qquad\qquad\qquad\quad (3)$$

$$O_2 + 2H_2O + 4e^- \leftrightarrow 4OH^- \qquad \text{(Alkaline aqueous)} \quad (4)$$

$$O_2 + H_2O + 2e^- \leftrightarrow HO_2^- + OH^- \qquad\qquad\qquad\quad (5)$$

$$HO_2^- + H_2O + 2e^- \leftrightarrow 3OH^- \qquad\qquad\qquad\qquad\quad (6)$$

The standard reduction potential for the direct four-electron pathway O_2 to OH^- is 1.227 V vs RHE (or 0.401 V *vs* SHE), while the standard reduction potential for the two e^- pathway is much lower to 0.6 V [6]. This leads to lower resultant fuel cell operating voltage or metal-air battery voltage if $2e^-$ transfer pathway is followed. Thus, desirable high cell voltage can result, involving $4e^+$ transfer ORR for better practical devices.

2.3 MORE ON ORR KINETICS

Experimentally, the kinetics of ORR on any electrocatalyst is mainly determined in a three-electrode cell system by means of linear sweep voltammetry (LSV) technique, where the current density is measured as a function of the applied voltage on the working electrode. The ORR working potential window of an electrode in O_2 saturated solution is usually 0–1 V vs RHE. Schematics of ORR LSV curve is shown in Fig. 2.1. The LSV curve is categorized into three regions, where each region corresponds to a particular electro-catalytic process (Athika, 2022). The high-potential region is referred as the surface controlled-reaction zone because the process is bound by the nature of the electrode surface that leads to adsorption of molecular O_2. The low-potential region is a diffusion-controlled region, where the current density remains nearly in a steady state as the potential increases. It is due to the rapid reaction rate of the dissolved oxygen on the electrode and dissolution and diffusion of oxygen become the limiting factors. The middle region is the mixed diffusion-kinetic controlled region, where surface reaction and mass diffusion have major roles. The onset potential (E_{onset}) and the half-wave potential ($E_{1/2}$) along with the limiting current density of the LSV curves are used to investigate the catalyst performances quantitatively. The more positive potential with high limiting current density indicate that the catalyst is more active towards ORR (Ge et al., 2015).

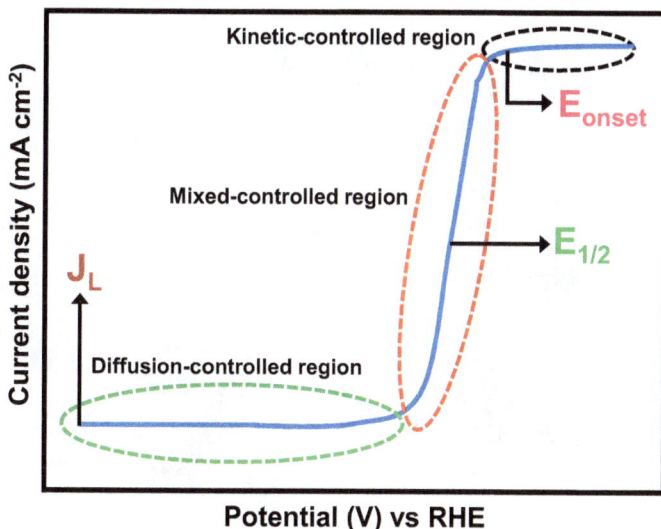

Figure 2.1 Schematics of the ORR LSV curve, where the E_{onset} is the onset potential, $E_{1/2}$ is half-wave potential and J_L refers to the diffusion-limiting current density.

The O=O bond has an extremely high bond energy of 498 kJ mol^{-1} and is difficult to break electrochemically. So, efficient electrocatalyst should be used to accelerate the O=O bond breaking by providing a lower energy pathway. In order to understand the kinetics of the ORR on any catalytic material, the rotating-disk electrode (RDE) or rotating ring-disk electrode (RRDE) approach is the most commonly used technique. As shown in Fig. 2.2, the RDE is a convective electrode system comprising of an electrode disc and a rotating shaft. Typically, the RRDE technique is used to determine the reaction pathways, Eqs. (1) to (6) involved on any given electro-catalyst. In addition to the oxygen reduction currents monitored in RDE, the RRDE includes a coaxial ring electrode to monitor the intermediate products that form on the disc electrode (*see* Fig. 2.2). Consider the following general reactions occurring on the RRDE with the amount of disc reaction products or intermediates reaching the ring electrode:

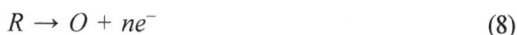

$$O + ne^- \rightarrow R \tag{7}$$

$$R \rightarrow O + ne^- \tag{8}$$

As a result of reaction (Eq. 7) on the disc electrode producing R, a percentage of R (*xR*) will be pushed on to the ring electrode, where reaction (Eq. 8) can occur at a limited mass-transport rate as long as the ring electrode is held at a sufficiently positive potential which can be measured as ring current. The straight four-electron transfer pathway (Eq. 1–4) is preferred even in the metal-air batteries as the formed HO_2^- intermediate reduces the amount of electrons transferred per

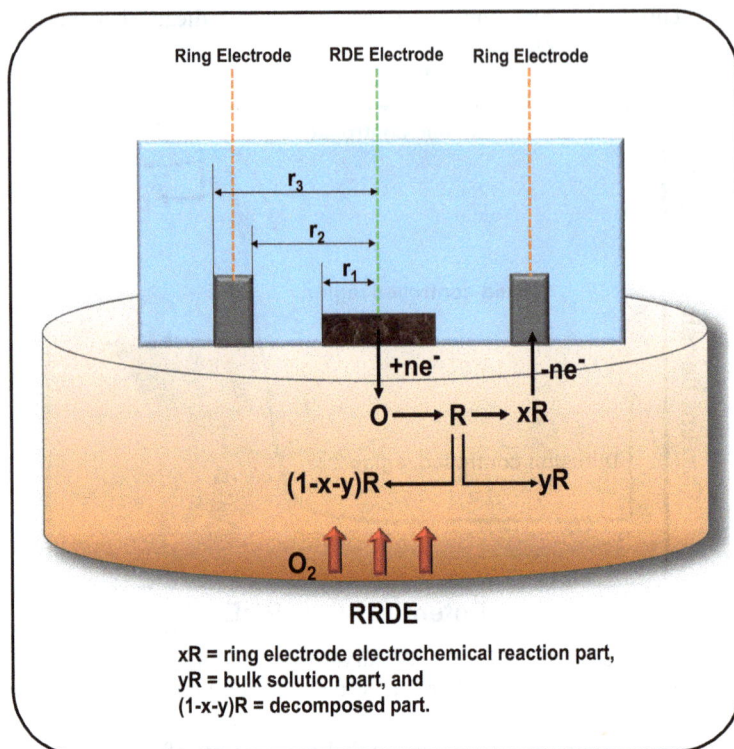

Figure 2.2 Schematics of a typical rotating-ring disc electrode (RRDE), where r_1 is the radius of the disk, r_2 is the inner radius and r_3 is the outer radius.

oxygen molecule reacted, leading to sluggish rate for O_2 reaction. During the experiment, the RRDE (or RDE) system is rotated. Thus, the spinning of the RRDE draws the electrolyte solution to its surface and flings it outwards in a radial path from the center due to centrifugal force. As a result, the diffusion layer that forms near the disc/ring electrode surface due to the electrochemical reaction becomes thinner as the electrode rotation rate increases. Thus, the reaction rate can be controlled by changing the electrode rotation speed and can be quantified by measuring ring and disc currents (Zhu et al., 2016).

In the molecular oxygen, two unpaired electrons are present in a doubly degenerate π^* antibonding orbitals and the bond order of the O_2 molecule is two. When O_2 is reduced, the added electron occupies in the anti-bonding π^* orbitals, thereby decreasing the bond order to less than two. This results in an increase in the O–O bond length while simultaneously decreasing the bond energy to easily break at lower energy. The ORR is a heterogeneous reaction as it involves solid, liquid and gas interactions at the triple phase. Fig. 2.3 demonstrates the probable (1:1) and (2:1) metal-dioxygen interactions. The adsorption of O_2 on the catalyst surface in the (1:1) metal-dioxygen complex structures involves two distinct bonding patterns: (i) O_2 side-on metal interaction (ii) O_2 end-on metal interaction. Side-on interaction as a result of two bonds: a σ-bond formed by the overlap of a π orbital of O_2 and a d_z^2 (or s) orbital on the metal, and a π back bonding between the metal $d\pi$ orbitals and the partially filled π^* antibonding orbital of oxygen. In the case of end-on interactions, it is able to form σ-bonds between O_2 and the metal ions by donating electrons from the σ-rich orbital of dioxygen to the d_z^2 orbital of the metal ions. While the π-backbond donation is created by charge transmission from the metal ions (d_{xz}, d_{yz}) to the O_2 molecule of the (π^* orbital). However, in the metal-dioxygen (2:1) complex, the bonding is caused by the interaction of the metal d-orbital with π^* and π orbitals combinations of the O_2 molecule (Taube, 1965; Zhu et al., 2016; Zhang, 2008). The ORR electrocatalyst must be chemically stable, not oxidized or corroded by the O_2, and electrically conductive for efficient electron transfer reaction to occur. The steps occurring in the ORR in an aqueous electrolyte are as follows:

(a) Diffusion and adsorption of O_2 molecule on the surface of electrocatalysts.
(b) Electron transfer to the adsorbed O_2 molecules.
(c) Weakening and breaking of the O=O bond to form OH^-/H_2O.
(d) Removal of the produced OH^- ions/H_2O to the electrolyte.

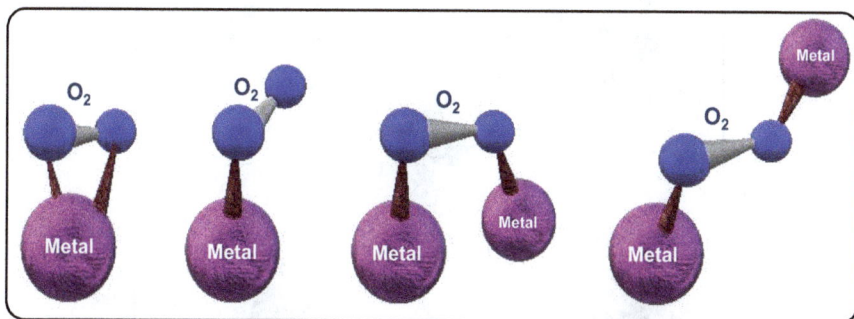

Figure 2.3 Schematics of different possible interactions of molecular O_2 with metal catalyst.

In spite of several decades of researches and accomplishments in the ORR, the slow kinetics and obscure mechanisms of the ORR still remain unclear due to difficulty in precisely identifying the active sites. In addition, the complex process of the proton/electron transfer to the oxygen molecule as well as the scissoring of the O=O bond remains unknown (Griffith, 1956). Nevertheless, a few reports have concluded the significant advancements on the ORR mechanism occurring on the transition metal oxide-based electro-catalyst, such as perovskite oxide (Sun et al., 2020; Liang et al., 2021; Liu et al., 2021).

2.4 OXYGEN EVOLUTION REACTION

Oxygen evolution reaction (OER) is a counter reaction to the ORR. It is also a vital reaction in energy devices, particularly in the metal-air batteries. The kinetics evaluation for the OER on any electrocatalysts is almost identical to that for ORR, with the exception that the working potential of the OER is typically in the range of 1–2 V vs RHE. The electrolyte should be saturated with Ar (or N_2) gas during the measurement. The current density and the onset potential of OER LSV curves is recorded on the RDE/RRDE. It can be used to evaluate the OER catalyst performance. Several decades of research have gone into understanding and analyzing the OER mechanisms. The commonly accepted OER mechanism consists of four-electron/proton transfer steps in acidic and alkaline electrolytes. The OER mechanism is the reverse of the ORR mechanism in terms of reaction steps. The reversible reaction of Eq. (3) illustrates the OER mechanism in alkaline medium as shown below:

$$4OH^- \leftrightarrow O_2 + 2H_2O + 4e^- \tag{9}$$

The OER involving $4e^-$ transfer is complex and involves high energy loss during the reaction. In alkaline condition, the following reactions (Eqs. 10–14) in the OER on a typical catalyst surface and its schematic illustration are shown in Fig. 2.4:

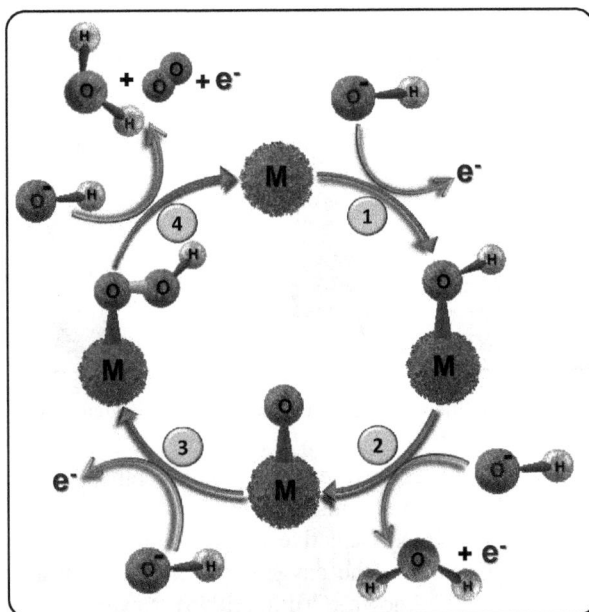

Figure 2.4 Flowchart of the OER reaction occurring in alkaline medium, where M is the metal reaction center and O is oxygen and R is product.

$$M + OH^- \rightarrow MOH + e^- \tag{10}$$

$$MOH + OH^- \rightarrow MO + H_2O + e^- \tag{11}$$

$$2MO \rightarrow 2M + O_2 \tag{12}$$

$$MO + OH^- \rightarrow MOOH + e^- \tag{13}$$

$$MOOH + OH^- \rightarrow M + O_2 + H_2O + e^- \tag{14}$$

where M is the surface metal active site. Intermediates are also formed in the OER electron transfer process, and understanding of the same is also critical for improving the OER performances. Herein, the direct dissociation of the MO (Eq. 12) and the generation as well as decomposition of the MOOH intermediate are the two ways to form the oxygen molecule. Thus, the strength of the M–O bond can be employed as a measure of the catalytic activity. A strong M–O bond restricts product desorption, while a weak M–O bond prevents intermediate binding on the catalyst surface. Hence, a catalyst with optimal M–O bond strength is required for strong catalytic activity towards OER. Accordingly, material scientists and chemists focused their efforts on developing more stable, highly active, and earth-abundant electro-catalysts to lower the kinetic barriers for the OER catalysis (Liang et al., 2021).

2.4.1 Basic Parameters of the ORR/OER Kinetics

2.4.1.1 Tafel Slope, Overpotential and Current Density

The kinetics and mechanism of both OER and ORR are frequently studied using Tafel analysis. Normally, it gives two important physical parameters, such as Tafel slope and the exchange current density. The typical Tafel equation for the ORR/OER is represented as shown below:

$$\eta = a + b \log (i) \tag{15}$$

where η denotes the difference between the applied and equilibrium potentials, i.e. overpotential (mV), a indicates a constant related to the magnitude of forward and reverse current at equilibrium, b denotes the Tafel slope represented in mV/decade current, i is the current density. The Tafel plots are logarithmic graphical representation of the current (i) produced in an electrochemical reaction and its overpotential (η) (Kapalka et al., 2008). Typically, these plots are obtained from the electrochemical experiments conducted under controlled atmospheric conditions. The slope of the Tafel plot is referred as Tafel slope. The typical Tafel slope should be as low as possible and less than 120 mV dec^{-1}. Lower overpotential implies more active the catalyst is (Holewinski and Linic, 2012).

2.4.1.2 Koutecky–Levich Plot, Number of Electrons and Peroxide

Diffusion and kinetics-limited ORR performance of a catalyst can be analyzed using Koutecky–Levich (K-L) plot. The K-L plot specifically describes the RDE/RRDE current density variation upon the electrode rotation rate that can be represented using the following Eq.:

$$\frac{1}{j} = \frac{1}{j_K} + \frac{1}{j_L} = \frac{1}{B}\omega^{1/2} + \frac{1}{j_K} \tag{16}$$

where j, j_K, and j_L are the current densities measured, kinetic-limited and mass transfer-limited currents, respectively. At a given potential, j_K is assumed to be a constant, and j_L is proportional to the angular velocity of the electrode (ω). The proportionality constant B can be expressed by the following Eq.:

$$B = 0.62nFC^*D^{2/3}v^{-1/6} \tag{17}$$

where n denotes the number of electrons transferred, F indicates the Faraday constant, C^* denotes the bulk oxygen concentration in the electrolyte, D represents the oxygen diffusion coefficient, and v is the viscosity of the electrolyte. The slope of the linear plot of j^{-1} versus $\omega^{-1/2}$ is known as K-L plot that gives the transferred electron number (n) in the ORR. The RRDE method is a more precise approach for determining the value of n. The rate of formation of peroxide ions (Eq. 4) and hydroxyl ions (Eq. 3) can also be used to calculate the number of electrons involved in the ORR on the catalyst surface. The selectivity of a catalyst towards the formation of hydrogen peroxide ions is (HO_2^-) determined using the following Eq.:

$$\%HO_2^- = \frac{200\dfrac{i_r}{N}}{i_d + \dfrac{i_r}{N}} \tag{18}$$

$$n = \frac{4i_d}{i_d + \dfrac{i_r}{N}} \tag{19}$$

$$N = \frac{-i_{r(N)}}{i_{d(N)}} \tag{20}$$

where N is the collection efficiency of the RRDE, $i_{d(N)}$ denotes the disc current, and $i_{r(N)}$ denotes the ring current in the standard reactions of ferricyanide/ferrocyanide solution at a constant electrode rotation of RRDE (Paulus et al., 2001; Tsai et al., 2014; Faulkner and Bard, 2002).

2.4.1.3 Stability and Cycle Life

Cycle life or stability of the electrocatalyst is an important parameter; it refers to how long the catalyst can be used before losing 80% of its original activity. The percentage loss of electrocatalytic activity can be calculated using the following eq.:

$$\Delta i\% = \frac{i_{BOL} - i_{EOL}}{i_{BOL}} \times 100 \tag{21}$$

where the $\Delta i\%$ is the percentage loss of electrocatalytic limiting current density, i_{BOL} is the limiting current density at the beginning of cycle-life test (cycle 1), and i_{EOL} is the current density at the end of the cycle life test (last cycle tested) (Zhu et al., 2016).

2.4.2   Electrochemical Techniques to Follow ORR/OER Kinetics

2.4.2.1   Cyclic Voltammetry

Cyclic voltammetry (CV) is a powerful tool for analyzing the electrochemical reduction of oxygen molecules in a three-electrode cell setup consisting of a working electrode, a reference electrode, and a counter electrode. The CV is typically performed in an O_2 or Ar saturated KOH electrolyte in the oxygen reduction potential between 0 and 1 V vs RHE. Cyclic voltammogram, which is a plot of potentio-dynamic current response recorded at the vicinity of electrode against the applied voltage at a specific scan rate, can provide information about the redox reaction that occurs. The high current density of the reduction peak in the O_2 atmosphere is a direct indication of the catalytic activity of the catalyst toward ORR.

2.4.2.2   Linear Sweep Voltammetry

Linear sweep voltammetry (LSV) is a fundamental potentio-static sweep technique. It is analogous to a single-segment cyclic voltammetry experiment. Working electrode potential is moved linearly from one point of voltage to another while current is measured. The LSV is a fast and reliable characterization tool that gives both qualitative and quantitative information on the electrode material. This enables the study of the electrochemical species formed at the electrode surface. The onset potential and limiting current density along with the half-wave potential of an LSV curve recorded in RDE or RRDE are used to diagnose the ORR performance as a catalyst.

2.5   DESIRED REQUIREMENTS FOR ORR/OER ELECTROCATALYST

When designing an ORR/OER electrocatalyst, there are numerous considerations that must be taken into account. High catalytic activity, electronic conductivity and porosity are the three critical features for a superior catalyst. Since ORR is a sluggish reaction, the performance of the fuel cell/metal-air battery is also limited. The triple-phase boundary (TPB) at which the solid electrode is synchronously linked with the liquid electrolyte and the gaseous O_2 from the atmosphere, is where the reaction occurs predominantly. In real devices generally an air electrode consists of three layers. They are gas diffusion layer (GDL), a conductive current collector, and an oxygen electrocatalyst layer. It is constructed with the catalyst layer facing the internal liquid electrolyte and the GDL facing the external atmosphere. All the components of the air electrode have distinct and equal roles calling for modification of the air electrode's overall structure to create a robust structural design that can provide available active sites for both ORR and OER (Lu et al., 2013).

More importantly, due to the difficulty in determining the electrode architecture that is responsible for the device performances, some major modification strategies of the air electrode are reported (Wang et al., 2018; Pan et al., 2018; Wang and Xu, 2019; Wu et al., 2020; Tomboc et al., 2020). Doping with heteroatoms by means of improved oxygen chemisorption and electron transfer and defect structures in electro-catalysts aids in charge transferability. It increase the number of low co-ordinated sites that can act as active sites (Lima et al., 2007; Cheng et al., 2009; Bardenhagen et al., 2015; Pan et al., 2017; Kang et al., 2020). The following requirements should be met by an ORR electrocatalyst:

- Extraordinary catalytic activity.
- Extraordinary electrical conductivity.
- High electrolyte wettability.
- Ability to accelerate both the ORR and OER
- Extraordinary electrochemical stability (not oxidizing at high electrode potentials).
- Extraordinary chemical stability (being insoluble in both an acidic or basic aqueous solution and not being oxidized by protons and oxygen).
- Positive structure optimum composition, favourable morphology, high specific surface area (SSA), high porosity, small particle size, and uniform distribution of catalyst particles on the support.
- Strong bonding between the support surface and the catalyst particle.

An electrocatalyst cannot simultaneously satisfy each of these conditions. Depending on the type of applications and the conditions of the applications, there are some trade-off and tolerance. It is necessary to design and apply the most appropriate catalyst in order to enhance the catalytic efficiency towards the ORR and the OER. Various catalyst materials with high bifunctional catalytic activity are discussed below.

2.5.1 Classification of Electrocatalysts for ORR/OER

The electro-catalysts are broadly classified into: (i) carbonaceous materials, (ii) noble metal/metal oxides, and (iii) transition metal oxides. Among them, transition metals and their oxides, including transition metal carbides, nitrides, carbonitrides, oxynitrides, and chalcogenides, have been widely investigated as electro-catalysts as an alternative to the benchmark catalyst. Such catalysts can minimize electrode overpotential, leading to better performance of the metal-air battery/fuel cell. Researchers have attributed the improved performances to the surface properties of metallic oxides, such as distinctive shape and changeable oxygen stoichiometry, and a significant fraction of active surface defect sites (Li et al., 2012). Cobalt, nickel, and manganese-based oxides are the most effective catalysts used in the earlier researches as electrocatalysts for ORR/OER (Oloniyo et al., 2012; Jung et al., 2013). Transition metal oxides have gained a lot of interest as extremely effective ORR electrocatalysts. They are capable of forming several cationic oxidation states; their variety of oxides makes it possible to adjust them with high tolerance for better ORR performance. They cost less than the traditional noble metal catalyst, which makes them more viable for use in real world applications. The oxides of the metal are in different structures, such as rock salt, spinel, perovskites, etc. The transition metals of Group VII and VIII elements (e.g. Mn, Fe, Co, and Ni) possess multiple valences, resulting in a variety of oxides and chalcogenides/carbides. In addition to manganese oxides, the performance of other transition metal oxides NiO, CuO, Co_3O_4, Fe_2O_3, V_2O_5, MoO_3, Y_2O_3, $Co_3O_4@Ni$, $ZnMn_2O_4$, $MnCo_2O_4$, were also broadly investigated. These materials exhibit enhanced ORR/OER compared to the noble metal catalysts. When they are applied to energy storage/conversion applications, such as in metal-air batteries and fuel cell, they exhibit better cell performances (Thapa et al., 2010; Anandan et al., 2011; Zhao et al., 2013: Barlie and Gewirth, 2013; Park et al., 2013; Zhang, 2016; Maiyalagan and Elumalai, 2019). Mixed metal oxides such as spinel structures have aroused great interest due to their good catalytic performance originating from the synergistic effect among the metal ions (Cuma and Koroglu, 2015; Liu et al., 2017). Lots of progress have been recently made on the spinel structure of mixed/metal oxides as electro-catalysts. In the following sections, the commonly explored electrocatalysts are discussed.

2.6   METAL-BASED ELECTROCATALYSTS

Several researchers have reported that metal oxides, sulfides, hydroxides and carbides are the most promising candidates for the ORR/OER activities with satisfactory stability even in alkaline electrolyte. In recent years, various non-precious transition metal-based electro-catalysts, such as Co, Ni, Fe and Mn have been investigated for the ORR/OER activity and also used to enhance the performances for energy conversion and storage devices. In addition, the precursors of these catalysts are abundant and inexpensive, making them practical, unlike the precious metal electro-catalysts (Zhang et al., 2016). However, these metals possess limited surface area and unsatisfactory stability due to limited site activity as well as site population, which depend on the intrinsic properties, nature of the metal, crystallite size, crystallinity, morphology, composition, including doping levels, etc. (Ibrahim et al., 2019) and extrinsic properties, such as synthesis methods, temperature, type of electrolyte, and concentration of electrolyte, etc. Such limited activity has been enhanced by introducing carbon support materials. Table 2.1 shows the various metal-based electro-catalysts that were explored for ORR/OER.

Su et al., synthesized a hybrid composite of Co nanoparticles embedded in nitrogen-doped carbon (Co/N-C-800) via a solvothermal carbonization strategy (Su et al., 2014). All electrochemical tests were done in RDE at room temperature in 0.1 M KOH aqueous solution. The Co NPs are uniformly embedded in the granular-like carbon matrix and are non-agglomerated, indicating a good dispersion in the carbon matrix leading to effective prevention being excessively oxidized. This resulted in fast electron transport between the carbon matrix and the Co NPs. The high specific surface area, low charge transfer resistance and good synergistic interaction led to better bifunctional catalytic activity of Co/N-C-800 than other Co/N-C catalyst. The electrocatalyst Co/N-C-800 showed outstanding activity for both ORR and OER with the onset potential of 0.74 and 1.59 V, Tafel slope of 61 and 61.4 mV dec^{-1}, and OER with an overpotential of 371 mV at the current density of 10 mA cm^{-2} and ΔE value of 0.859 V, respectively.

Liu et al. synthesized transition metal nanoparticles (Fe, Co, and Ni) encapsulated in nitrogen-doped carbon nanotube hybrids by a one-step solid-state reaction of cyanamide and transition metal chloride (Liu et al., 2016c). RDE was used to conduct electrochemical tests in 0.1 M KOH at room temperature. The most effective bifunctional catalytic activity and good stability towards both ORR and OER were showed by the optimized Co/N-CNT hybrid. The fast electron transport among the carbon matrix and the cobalt nanoparticles, leading to effective electrical conductivity is due to the intermingling of small-sized cobalt nanoparticles dispersed in the carbon matrix. The electrocatalytic activity of Co/N-CNTs, Fe/N-CNTs and Ni/N-CNTs for ORR was with the onset potential of 0.94, ~0.96, 0.91 V and half wave potential of 0.84, 0.81, 0.73 V exhibiting OER with the half wave potential of 1.62 V, 1.75 V, 1.82 V, the Tafel slope of 50, 63, 59 mV dec^{-1} and ΔE value of 0.78, 0.94, 1.09 V, respectively. The transferred electron number 'n' for the Fe/NCNT and the Co/N-CNT were 3.85 and 3.90, respectively. Universal strategy to directly synthesize single-layer graphene encapsulating uniform earth-abundant 3d transition-metal nanoparticles, such as Fe, Co, Ni and their alloys in a confined channel of mesoporous silica was developed by Cui et al. (Cui et al., 2016). OER performance in 1 M NaOH solution was assessed using an electrochemical cell. The electron transfer from the metals that were encapsulated to the graphene surface was greatly facilitated by the single atomic thickness of the graphene shell, which effectively optimized the electronic structure of the graphene surface and consequently triggered the OER activity of the inactive graphene surface. The FeNi@NC electrocatalyst showed outstanding electrochemical performance towards OER with the onset potential of 1.44 V, the Tafel slope of 70 mV dec^{-1} and with the overpotential of 280 mV at a current density of 10 mA cm^{-2}. The carbon-encapsulated NiFe$_2$ alloy nanocrystals (NiFe@NCx) by means of a unique two-stage encapsulation strategy method displayed the best ORR and OER activities in the alkaline medium, as reported by Zhu et al. (Zhu et al., 2016). The electrocatalytic performances of the prepared materials were evaluated systematically in 0.1 M KOH using RRDE technique. The superb electrocatalytic performance originated from the modulation of the electronic structure of the outer carbon layers by electron penetration from the NiFe core. Reducing the size of the encapsulated nanoalloy could significantly increase the active sites density and the electron density in the graphene shells, which enhanced the ORR and OER activities. The mesoporous feature provided plenty of catalytic active sites and facilitated electron and mass transport during the electrocatalysis process. The η and i exhibited by the catalyst were found to be 0.78 V and 50 mA cm^{-2}, respectively. A 4-electron pathway was found to dominate the ORR that was established using K–L plots (n = 4.1) and RRDE

Table 2.1 List of metal-based electrocatalysts examined for the ORR and OER activities

S. No.	Catalyst	η (mV) at J = 10 mA cm⁻¹	E_{OER} (V) at J = 10 mA cm⁻¹	E_{ORR} (V) at J = -3 mA cm⁻¹	Oxygen electrode (OER-ORR) E(V)	Tafel slope (mV dec⁻¹)	Electron transfer number (n)	References
1	Co/N-C	371	1.599	0.74	0.859	61	3.91–4.03	Su et al., 2014
2	Co/N-CNT	390	1.62	0.84	0.78	50	3.85–3.90	Liu et al., 2016c
3	FeNi@NC	280	1.44	–	–	70	–	Cui et al., 2016
4	NiFe@NCx	320	1.53	1.03	0.50	60	3.96	Zhu et al., 2016

Table 2.2 List of Ni-based electro-catalysts examined for the ORR and OER activities

S. No.	Catalysts	E_{ORR} (V) at I = -3 mA cm⁻²	E_{OER} (V) at I = 10 mA cm⁻²	Oxygen electrode (OER-ORR) E(V)	$Tafel_{OER}$ (mV dec⁻¹)	References
1	N-NiO	0.90	1.73	0.83	83	Qian et al., 2019
2	Ni/NiO	0.87	1.48	0.61	62	Liu et al., 2020
3	NiCo₂O₄/G	0.54	1.67	1.13	164	Lee et al., 2013
4	NiFe₂O₄/MWCNT	0.23	1.63	1.40	93	Li et al., 2015a
5	Ni/NiFe₂O₄/C	0.60	1.65	1.05	75	Athika et al., 2020

experiments (n = 3.96). The catalyst exhibited high stability and activity with an onset potential of 1.03 V for ORR and an overpotential of only 0.23 V at 10 mA cm^{-2} for the OER.

2.7   METAL OXIDE-BASED ELECTROCATALYSTS

2.7.1   Ni-based Electrocatalysts

Table 2.2 shows the list of nickel-based electrocatalysts that were reported for ORR/OER. Qian et al., synthesized nitrogen-doped NiO nanosheets via two-step hydrothermal method that exhibited good bifunctional electrocatalytic performances for both OER and ORR (Qian et al., 2019). A standard three-electrode cell system was used to measure the performances in 1 M KOH. The electrocatalytic ability was attributed to the conductivity and effective active surface area caused mainly by doping of the N atoms and the porous structure. There was lower onset potential (0.90 V), the ideal half-wave potential ($E_{1/2}$, 0.69 V), and ΔE of 0.83 V for ORR. The small overpotential of 270 mV was at a current density of 10 mA/cm^2 and Tafel slope of 83 mV dec^{-1} for the OER. The electron transfer number nearly 4e$^-$ (~3.8) were resulted for the N-NiO electrocatalyst.

Recently, highly efficient electrocatalyst based on mesoporous Ni/NiO nanosheets were generated by means of scalable process of hydrothermal growth and post-acid etching by Liu et al. (Liu et al., 2020). The electrochemical tests were done in 0.1 M KOH using RRDE. The generated NiO nanosheets consisting of Ni nanoparticles were purposely embedded into the mesoporous nanosheets that had well-established pore channels which led to charge transport at the interface between the nanosheets. The oxygen deficiencies at the pore edges (giving rise to enhanced electrical conductivity) and an overall mesoporous structure led to a higher surface area and therefore, effective exposure of the active sites. The nanosheets showed outstanding electrochemical performance and catalytic activity for ORR with the onset potential of 0.87 V. The number of electrons transferred per O$_2$ (n) was as high as 3.81 for the porous Ni/NiO, confirming that the porous Ni/NiO nanosheets favored a nearly four-electron pathway at a relatively low overpotential.

Mesoporous NiCo$_2$O$_4$ nanoplatelets@graphene sheets (NiCo$_2$O$_4$-G) were developed by Lee et al. via a one-pot precipitation reaction and hydrothermal process that exhibited excellent electrochemical activity (Lee et al., 2013). The hybrid effect of NiCo$_2$O$_4$–G on ORR and OER activities was confirmed by RDE and CV measurements. The increased electrical conductivity and the creation of new active sites with the incorporation of Ni atoms into the octahedral site of the spinel crystal structure led to fast charge transfer facilitated by the graphene sheets as a support material. The presence of mesopores enhanced the electrocatalytic activity by increasing the surface exposure of the active sites. There was onset potential of –0.12 V, the half wave potential of –0.27 V, the electron transfer number of 3.9 for ORR, the onset potential of 0.55 V and Tafel slope of 799 mV per dec^{-1} for OER, respectively.

Li et al. synthesized $NiFe_2O_4$ nanoparticles crosslinked with the outer walls of multi-walled carbon nanotubes (MWCNTs) via a simple and scalable hydrothermal method (Li et al., 2015a). Using a traditional three-electrode cell system through 0.1 M KOH aqueous solution, the electrochemical activities of the materials were assessed through RDE, RRDE, and CV for ORR. The composite showed a distinct nano-network structure that provided more active sites and made it simple to transfer electrolyte ions and oxygen during the catalytic reaction. The strong coupling and synergistic impact between the $NiFe_2O_4$ nanoparticles and the MWCNT matrix, as well as the nano network supplied by the MWCNTs, were credited with superior catalytic activity and stabilities. The $NiFe_2O_4$/MWCNT nanohybrid exhibited Tafel slope of 93 mV dec^{-1} and the range of electron number transferred (n) is 3.61–3.68 for ORR and onset potential of 0.42 V and Tafel slope of 93 mV dec^{-1} for OER. Athika et al. synthesized $Ni/NiFe_2O_4$@carbon (NNFOC) nanocomposite by a simple solution-combustion method and its bifunctional electro-catalytic activity towards ORR/OER was examined (Athika et al., 2020). The surface had a cauliflower-like morphology and was made up of grains and pores. The specific surface area of NNFOC nanocomposite was 50 m^{2}g^{-1}. The fact that the pore size ranged from 2 to 50 nm suggests that the NNFOC catalyst is highly enriched in mesopores. The NNFOC composites mesoporous characteristics produced more active oxygen adsorption/desorption sites, which enhanced the ORR/OER kinetics. When compared to the benchmark catalyst, Pt/C, the produced catalyst, was shown to be a capable non-precious bifunctional catalyst. In oxygen-saturated 0.1 M KOH at various electrode rotations, the ORR activity of NNFOC was investigated on RRDE in the potential range of 0–0.9 V *vs.* RHE at a scan rate of 10 mVs^{-1}. In an Ar-saturated 0.1 M KOH solution, the OER activity of NNFOC was tested on RRDE using the LSV technique at a scan rate of 5 mVs^{-1} and 1600 rpm electrode rotation. The NNFOC composite on RRDE exhibited good ORR and OER activities. Extraordinarily, the NNFOC nanocomposite exhibited lesser onset potential of 0.60 V for ORR, 1.65 V for OER, 1.05 V for ΔE and a Tafel slope of 75 mV dec^{-1} and lower overpotential during ORR and OER. Fig. 2.5 shows the CV curves, RRDE LSV plots, K-L plots, and the no. of e$^-$ transferred and percentage of peroxide ion formed on the NNFOC electrode. Fast electron transfer kinetics, a sizable electrochemical active surface area, and strong chemical-electronic coupling were the causes of NNFOC nanocomposites increased catalytic activity. Increased ORR and OER activities were the result of synergistic interaction between Ni, nickel ferrite, and the porous carbon network, which produced more reaction sites, high conductivity, and quick diffusion paths.

2.7.2 Co-based Electrocatalysts

Table 2.3 shows the list of Co-based electrocatalysts explored for ORR/OER. Liang et al., synthesized a hybrid material consisting of Co_3O_4 nanocrystals and reduced graphene-oxide (rGO) by controlled nucleation and subsequent hydrothermal treatment to report a high-performance bi-functional catalyst (Liang et al., 2011).

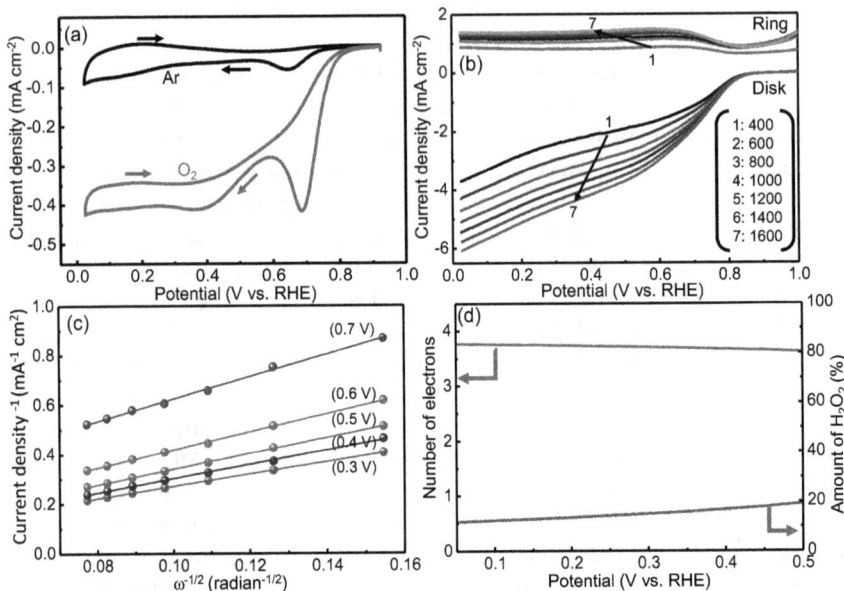

Figure 2.5 (a) CV curves recorded for the NNFOC electrode at a scan rate of 10 mV s^{-1}, (b) LSV profiles recorded for the NNFOC catalyst at a scan rate of 10 mV s^{-1} at different rotations using RRDE at a constant potential of 0.6 V on the ring electrode, (c) corresponding K-L plots and (d) the average number of e^{-} (n) transferred and percentage of peroxide ion formed on NNFOC composite on RRDE at 1600 rpm. (Reproduced with permission from Ref. [Athika et al., 2020]).

The catalytic pathways of the hybrid materials were verified by performing RRDE. Although Co_3O_4 or graphene oxide alone has little catalytic activity, their hybrid exhibited high ORR activity that was further enhanced by nitrogen doping in the graphene. The catalyst exhibited a half wave potential for ORR at 0.83 V, the Tafel slope of ~37 mV dec^{-1} and n value of ~4.0. The Co_3O_4/N-doped Ketjenblack composite by a facile strategy was synthesized by Liu et al. and used as a high performance catalyst for ORR (Liu et al., 2016a). The CV measurements were carried out in an O_2 saturated or Ar saturated 0.1 M KOH solution with a scan rate of 10 mV s^{-1}. The synergistic effect between Co_3O_4 and N-KB enabled the Co_3O_4/N-KB composite much higher cathodic currents, more positive half-wave potential and high electron transfer number in comparison with the Co_3O_4 or N-KB catalysts. The catalyst exhibited a half wave potential for ORR at 0.79 V, the Tafel slope of ~74.7 mV dec^{-1} and n value of ~4.0.

The Co_3O_4 nanocrystals embedded in N-doped mesoporous graphitic carbon layer/multi-walled carbon nanotubes (MWCNTs) hybrids by a facile carbonization and subsequent oxidation process of the MWCNTs-based metal-organic framework (MOF) were synthesized by Li et al. (Li et al., 2015b). All the electrochemical measurements were carried out in 0.1 M KOH aqueous electrolyte at room temperature using RRDE. The highly dispersed Co_3O_4 NPs were embedded inside the graphitic carbon layer and MOF successfully combined with MWCNTs,

Table 2.3 List of single metal Co-based electrocatalysts examined for the ORR and OER activities

S. No.	Catalyst	E_{ORR} (V) at I = –3 mA cm^{-2}	E_{OER} (V) at I = 10 mA cm^{-2}	$E_{1/2}$(V)	Electron transfer number (n)	Tafel slope (mV dec^{-1})	Over potential (mV)	References
1	Co$_3$O$_4$/N-rmGO	0.92	1.50	0.83	~4.0	~37	310	Liang et al., 2011
2	Co$_3$O$_4$/N-KB	0.83	–	0.79	~4.0	74.7	–	Liu et al., 2016a
3	Co$_3$O$_4$@C-MWCNTs	0.85	1.50	–	3.8–3.9	62	320	Li et al., 2015b
4	Co$_3$O$_4$ /NPGC	0.97	1.52	0.842	~4.0	–	450	Li et al., 2016

Table 2.4 List of poly metal-based electrocatalysts examined for the ORR and OER activities

S. No.	Catalysts	E_{ORR} (V) at I = –3 mA cm^{-2}	E_{OER} (V) at I = 10 mA cm^{-2}	Oxygen electrode (OER-ORR) E (V)	Tafel$_{OER}$ (mV dec^{-1})	References
1	Fe$_3$O$_4$@CoO	0.79	1.67	0.88	89	Zhou et al., 2019
2	(Co$_3$O$_4$–CeO$_2$/C)	0.12	0.23	0.11	176	Goswami et al., 2021
3	CoFe$_2$O$_4$/CNF	0.82	1.85	1.03	82	Liu et al., 2016b
4	ZnCo$_2$O$_4$@rGO	0.69	1.65	0.96	102	Chakrabarty et al., 2019
5	MnCo$_2$O$_4$/N-doped carbon nanofiber	0.82	1.76	0.94	85	Xu et al., 2014
6	Co/CoFe$_2$O$_4$/C	0.83	1.75	0.92	106	Athika et al., 2020

resulting in high surface area that was the key for enhancing the transport of oxygen and the electrolyte on to the catalyst surface. The catalyst exhibited overpotential of 320 mV and onset potential of 1.50 V.

The Co_3O_4 nanocrystals embedded nitrogen-doped partially graphitized carbon framework (Co_3O_4/NPGC) for ORR was prepared by Li et al. (Li et al., 2016). The electro-catalytic activity was investigated by LSV in O_2 and N_2 saturated in 0.1 M KOH, using a three-electrode cell system. A pomegranate-like composite architecture was found which provided low dimension of highly active Co_3O_4 nanocrystal seeds possessing active sites for the electrochemical reactions and the pomegranate-like structure efficiently prevented the metal oxide from self-accumulation and provided the mass transfer pathways. Further graphitized carbon shell and the framework were not only highly conductive but chemically stable and highly robust, enhancing the durability of the catalyst. The catalyst exhibited onset potential of 0.97 V, half wave potential of 0.842 V for ORR and an overpotential of 450 mV.

2.7.3   Poly Metal-based Electrocatalysts

Table 2.4 shows the list of poly metal-based electrocatalysts that were reported for ORR/OER activities. Zhou et al. prepared the Fe_3O_4@CoO nanocomposites by a seed-mediated growth approach (Zhou et al., 2019). The RRDE tests were done in 0.1 M KOH electrolyte under O_2 saturated condition. The particular interactions between the core and shell components gave rise to the electrocatalysts high catalytic activity. The Fe_3O_4@CoO NCs inherited the spherical morphology of the Fe_3O_4 seeds. The catalytic activity of Fe_3O_4@CoO NCs was significantly influenced by the CoO shell thickness. With regard to ORR and OER in particular, the Fe_3O_4@CoO NCs with two monolayers of CoO displayed a relatively greater catalytic performance. The ORR and OER onset potentials were 0.953 and 1.51 V, and an overpotential of 390 mV at a current density of 10 mA cm^{-2}, respectively and had a Tafel slope of 89.3 mV dec^{-1} for the OER. The electron transfer number was greater than 3.8 with ΔE of 0.794 V.

The mixed metal oxide hybrid consisting of nanostructured Co_3O_4 and CeO_2 supported on carbon (Co_3O_4–CeO_2/C) is synthesised by a facile and cost-effective hydrothermal method followed by calcination and carbonization which was prepared by Gowsami et al. (Goswami et al., 2021). The ORR and OER activities were systematically investigated in N_2 and O_2 saturated 0.1 M KOH solution using a standard three-electrode analyzer by CV, LSV, and chronoamperometry methods. The Co_3O_4–CeO_2 particles assume a spherical morphology and are evenly dispersed over the carbon matrix of a relatively larger size. The process has adequate oxygen vacancies and strong oxide/oxide and oxide/carbon heterointerfaces. The remarkable electrocatalytic performance of the resultant hybrid was directly correlated to its high electrochemical active surface area, surface oxygen vacancies, and synergistic effects between the Co_3O_4 and the CeO_2 phases. The Co_3O_4–CeO_2/C hybrid was highly active towards ORR and OER with the onset potential of -0.12 V and 0.23 V, and Tafel slope of 69 mV dec^{-1}

and 176 mV dec^{-1}. ORR has ΔE value of 1.04 V, a low overpotential of 520 mV and half wave potential of -0.27 V.

Liu et al. synthesized the CoFe$_2$O$_4$ nanoparticles supported on carbon nanofibers (CFO/CNF) derived from bacterial cellulose fabricated via an electrostatic assembly method, owing to the covalent coupling between CFO and CNF that resulted in high electronic conductivity and high surface area, leading to improved mass transport of O$_2$ and electrolyte (Liu et al., 2016b). Fig. 2.6 shows the schematics of synthesis of the Co/CoFe$_2$O$_4$@carbon electro-catalyst. The electro-catalytic activity for ORR of the samples was studied using the RRDE technique. The mesopores along with the 3D network of CFO/CNF benefited mass transport during the catalytic process. ORR has half-wave potential of only 72 mV and the onset potential of -0.09 V. The diffusion limiting current density reached -5.41 mA cm^{-2} for the catalyst. The OER onset potential was 0.58 V and the current density of CFO/CNF at 1.0 V reached 23.02 mA cm^{-2}. A flower-like porous ZnCo$_2$O$_4$ microstructure and the composite of rGO-ZnCo$_2$O$_4$ by the one-step solvothermal method were successfully synthesized by Chakrabarty et al. (Chakrabarty et al., 2019). The electrochemical property was measured by a three-electrode cell in 0.1 M aqueous KOH solution using RDE. The mixed oxidation state of Co ion ($+2$ and $+3$) in the ZnCo$_2$O$_4$ was reported to be the origin of the bifunctional catalytic activity, whereas the porous ZnCo$_2$O$_4$ on to the rGO layer increased the catalytic surface area. Simultaneously high conducting rGO layer improved their charge transport mechanism. The uniform distribution of ZnCo$_2$O$_4$ microspheres composed of three-dimensional interconnected porous networks which endowed the high surface area facilitated the electrochemical performance,

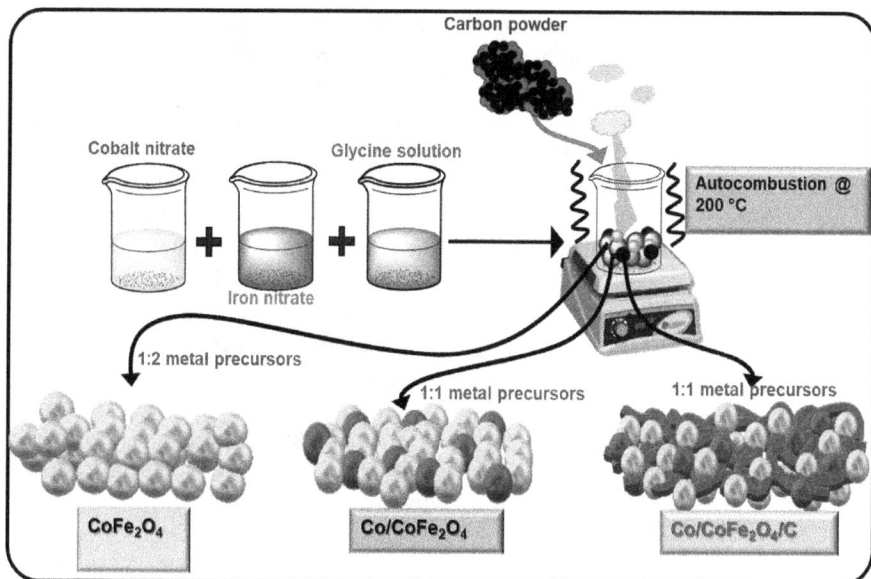

Figure 2.6 Schematic of various steps adopted in the generation of the CFO, CCFO and the CCFOC samples (Reproduced with permission from Ref. [Athika et al., 2020]).

thus improving the electrolyte penetration capabilities. ORR has the rGO-ZnCo$_2$O$_4$ and ZnCo$_2$O$_4$ with the onset potential of 0.95 and 0.81 V, the Tafel slope of 59.2 and 101.7 mV dec^{-1}. The n values of ~3.95 and ~3.4 and ΔE value of 0.679 for the rGO-ZnCo$_2$O$_4$ were observed. The OER revealed the maximum current density of 145 mA/cm^2 at 1.9 V for RGO-ZnCo$_2$O$_4$ with the overpotential of 0.30 V to reach the benchmark current density of 10 mA cm^{-2}. Xu et al. prepared a MnCo$_2$O$_4$ nanoparticles on nitrogen-enriched carbon nanofibers (NCFs) by means of solvothermal method (Xu et al., 2014). The ORR and OER kinetics were measured by RDE in O$_2$ saturated 0.1 M KOH solution and also by CV. The improved electrochemical performance was correlated to the abundance of active graphitic-N sites in the NCFs. The effective integration of MnCo$_2$O$_4$ nanoparticles and nitrogen-enriched carbon nanofibers led to effective charge transfer and post-charge transfer electron conduction. The MnCo$_2$O$_4$–NCF composite exhibited onset potential of –0.08 V and 0.6 V for ORR and OER respectively, and the half-wave potential of –0.21 V. The electron transfer numbers (n) was ~3.96.

The Co/CoFe$_2$O$_4$@Carbon (CCFOC) nanocomposite was generated by means of a facile one-pot solution combustion method by Athika et al. (Athika et al., 2020). Using a linear sweeping voltammetry (LSV) investigation on the RRDE electrode, the ORR activity of the electrocatalysts was thoroughly examined. When compared to the pristine CoFe$_2$O$_4$ (CFO) and Co/CoFe$_2$O$_4$ (CCFO) electro-catalysts, the CCFOC electrocatalysts showed a spectacular ORR onset potential, a lower OER overpotential, and an improved catalytic limiting current density. Both the Co/CoFe$_2$O$_4$ nanoparticles and the porous carbon support nanonetwork contributed catalytically in a synergistic manner to the CCFOC electro-catalysts remarkable catalytic activity when compared to the other two electrocatalysts. Increased oxygen adsorption sites and quicker electron transfer kinetics through a low-activation energy pathway were made possible by the CCFOC materials high surface area, structural stability, reduced band gap, and superior conductivity. Due to the additional carbon, it was discovered that the CCFOC sample had larger pores than the CFO and CCFO samples. These pores developed during autocombustion as some of the injected carbon broke down into CO or CO$_2$. At a scan rate of 10 mVs^{-1} in O$_2$ saturated 0.1 M KOH, the LSV plots of all three electrocatalysts were acquired. This was done by rotating the electrocatalysts at 1600 rpm. Fig. 2.7 shows the Tafel plots and LSV stability curves along with the combined ORR-OER LSV curves recorded for the electrocatalysts. The CCFOC electrocatalysts onset potential (E$_{onset}$) was 0.94 V, which was higher than that of the CFO (0.89 V) and the CCFO (0.92 V) catalysts. Additionally, compared to the CFO (E$_{1/2}$= 0.74 V) and CCFO (E$_{1/2}$= 0.81 V) catalysts, the CCFOC electrocatalyst demonstrates a superb half-wave potential (E$_{1/2}$) of 0.85 V. Over the investigated potential range of 0.1 - 0.6 V, the average n value for the ORR on the CCFOC electrode was 3.98, which was marginally higher than that of the other two catalysts (CFO = 3.94 and CCFO = 3.97). For the CCFOC electrocatalysts, the equivalent H$_2$O$_2$ percentage was as low as 1%. Since the ORR followed the preferred 4e$^-$ transfer pathway with the CCFOC catalyst, the main byproduct of the reduction was H$_2$O/OH$^-$. These findings demonstrated that adding porous carbon networks and metallic Co nanoparticles to $CoFe_2O_4$ improved O_2 adsorption and further it is reduction.

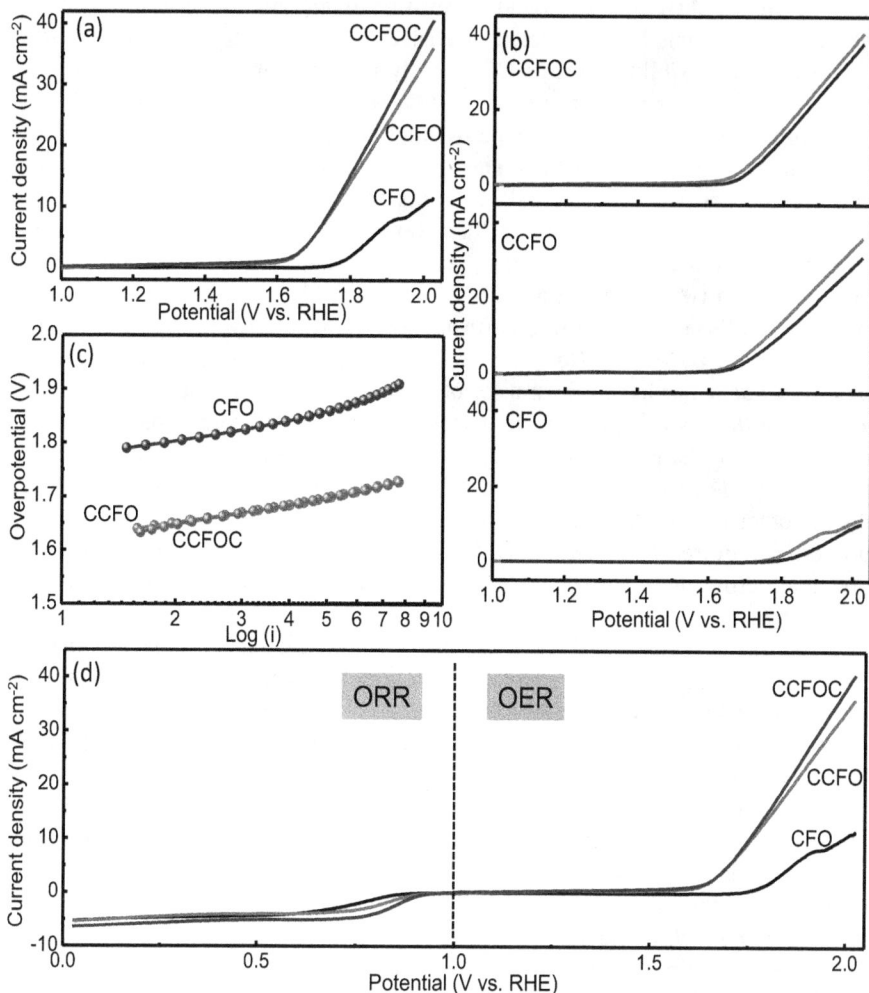

Figure 2.7 (a) Polarization curves, (b) cycle life test and (c) Tafel plots recorded for the CFO, CCFO and the CCFOC electrocatalysts at 1600 rpm at a scan rate of 10 mVs^{-1} in 0.1 M KOH in Ar atmosphere, and the overall ORR/OER LSV plots (Reproduced with permission from Ref. [Athika et al., 2020]).

2.7.4 Perovskite-based Electrocatalysts

Perovskites have general formula of ABO_3, where A is a typical lanthanide, alkaline or alkaline-earth metal (e.g. La, Sr or Ca) and B is a transition metal such as Mn, Co, Fe, Ni, Cr or Ti. The physical and chemical flexibility of the perovskite oxides are related to the A- and B-site metal ions present (Wang et al., 2019). Therefore, A- and B-site metal ions of the perovskite oxides are partially or completely substituted to form new-doped perovskite oxides with attractive properties (Chilvery et al., 2016). Lanthanum manganese oxide ($LaMnO_3$) with a structure of ABO_3 is a typical perovskite oxide. In the last few decades, perovskite materials have attracted extensive interest due to their compositional and structural flexibility towards ORR catalysts in alkaline solution. Besides, the perovskites can be synthesized easily at relatively low-cost (Jin et al., 2013). The physico-chemical properties of the perovskites can be widely altered, which results in enhanced catalytic activities. The great flexibility in both the composition and crystal structures of perovskites leads to the tunable electronic structure of perovskite oxides, which make them possess diverse physical and chemical properties and thus ideal candidates as electrocatalysts for ORR and OER in alkaline media (Malkhandi et al., 2012; Lee et al., 2015a; Xu et al., 2018). Earlier studies established that the rare-earth metals ORR activity in ABO_3 perovskite oxides follows La > Pr > Nd > Sm > Gd > Y > Dy > Yb. La-based perovskite oxides in particular have drawn a lot of attention because of their substantial La abundance, tunable shape, tunable defects, and composite formation with carbon or other materials. For instance, in each of $LaMnO_3$, $LaCoO_3$, $LaFeO_3$, $LaNiO_3$, anion/cation defects were introduced depending on the method of synthesis or by doping other metal ions, which made them possess diverse physical and chemical properties required for ORR and OER application (Ashok et al., 2018). The list of potential perovskite materials used as electrocatalysts for ORR/OER is presented in Table 2.5.

Currently, Sankar Devi et al. synthesized vacancy-induced nanostructured $LaMnO_3$ electrocatalyst by means of hydrothermal method and tested its ORR and OER performances (Sankar Devi et al., 2022). The $LaMnO_3$ electrocatalyst on the RRDE study showed excellent ORR and OER activities in 0.1 M KOH electrolyte. The LSV curves recorded on the RRDE-based $LaMnO_3$ catalyst at rotation speed of 1600 rpm and the benchmark catalyst, Pt/C catalyst at a scan rate of 10 mV s^{-1} is shown in Fig. 2.8. $LaMnO_3$ electrocatalyst demonstrated a superior onset potential, a decreased OER overpotential, and an enhanced catalytic limiting current density The ORR and OER had the onset potential of 0.749 V and 1.839 V and ΔE value of 1.09 V. The $LaMnO_3$ electrodes Tafel plot was essentially linear, with a slope value of 84 mV. The optimum $4e^-$ transfer ORR is quite close to this Tafel slope value. The Pt/C catalysts Tafel slope on the other hand, was 90 mV. In the first cycle and the 250th cycle, the onset potential and the limiting current density are almost identical. Even in the 250th cycle, the drop in the limiting current density was just 3%. A comparable result suggested that $LaMnO_3$ is a successful electrocatalyst for ORR. We acquired and reported an outstanding onset potential, a decreased OER overpotential, and an enhanced catalytic limiting current density. As roughly depicted in Fig. 2.9, the cationic and oxygen vacancies increased O_2 adsorption and furthered its effective decrease. The other perovskites were also investigated for the ORR/OER applications and layered $PrBaMn_2O_{5+d}$ (H-PBM) was simply prepared by annealing pristine $Pr_{0.5}Ba_{0.5}MnO_{3-\delta}$ in H_2 prepared by Chen et al. (Chen et al., 2016). The ORR and OER stability were evaluated in 0.1 M KOH for over 1000 LSV cycles using RDE. By using H-PBM, the oxygen reduction/evolution reaction

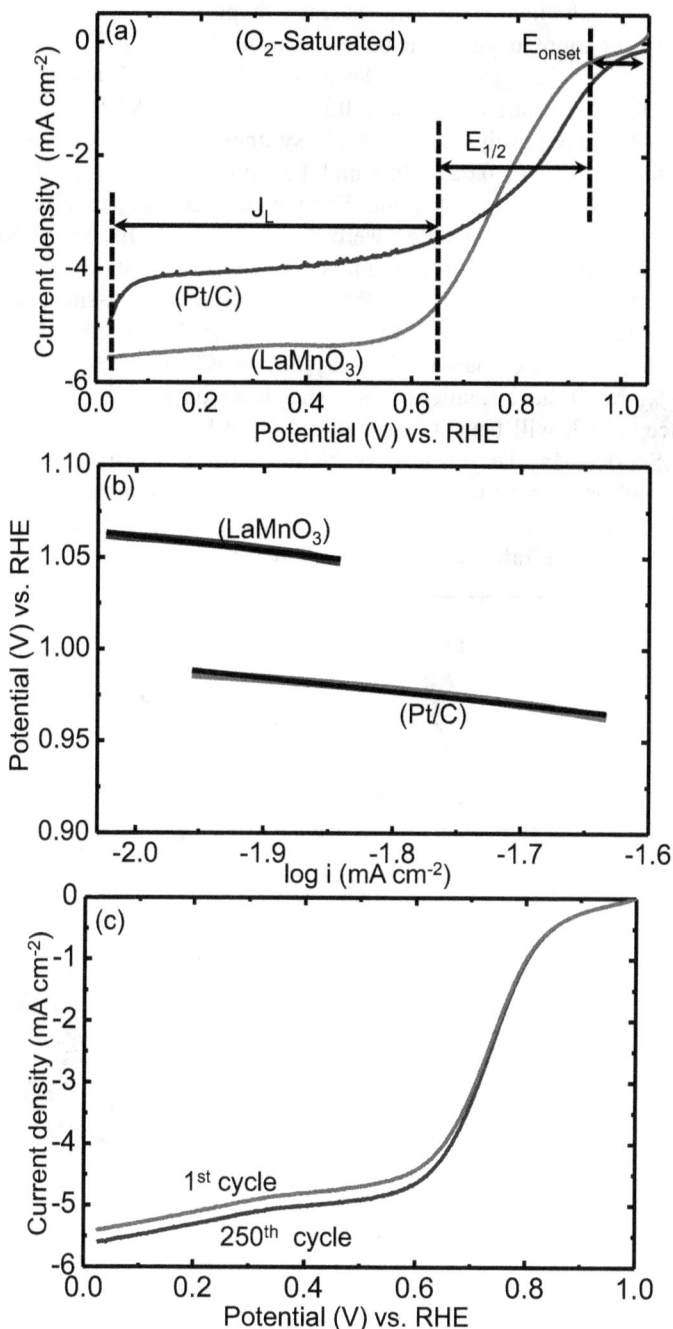

Figure 2.8 (a) Polarization curves recorded for the LaMnO$_3$ electrode and Pt/C electrode at 1600 rpm at a scan rate of 5 mVs^{-1} in 0.1 M KOH in the OER region, (b) Tafel plots and (c) representative cycle life tests (LSV) curves recorded on the LaMnO$_3$ electrode(Reproduced with permission from Ref. [Sankar Devi et al., 2022]).

activities noticeably improved. The increase in performance was attributed to the insertion of more oxygen vacancies, improved Mn ion e_g filling, and the simple integration of oxygen into the layered H-PBM. PBM/C and H-PBM/C had ORR and OER potentials of 0.68 and 0.74 V, 1.07 and 0.89 V, and n values of ~2.9 and ~4, respectively. Yuan et al. synthesized A-site cation deficient $(La_{0.8}Sr_{0.2})_{1-x}MnO_3(x=0, 0.02, 0.05)$ and Fe doped $(La_{0.8}Sr_{0.2})_{0.95}Mn_{0.5}Fe_{0.5}O_3$ perovskites by sol-gel process (Hua Yuan et al., 2019). The electrochemical catalytic activities of the samples were measured by RDE 0.1 M KOH. The catalyst exhibited irregular shapes and non-uniform distribution. The particle size decreased ($LSM_1 > LSM_2 > LSM_3$) with increase in A-site deficiency. In comparison with LSM_3, the B-site iron doped LSMF sample had a slightly larger particle size. Compared with $La_{0.8}Sr_{0.2}MnO_3$, $(La_{0.8}Sr_{0.2})_{0.98}MnO_3$ and $(La_{0.8}Sr_{0.2})_{0.95}MnO_3$ have smaller particle size, more oxygen vacancies and proper Mn valence, which will benefit both ORR and OER. The experiments revealed that $(La_{0.8}Sr_{0.2})_{0.95}Mn_{0.5}Fe_{0.5}O_3$ has the highest current density of 4.5 mA cm^{-2}. The ORR had onset potential of −0.124 V had a Tafel slope of −137 mV dec^{-1}. The LSM_1 exhibited onset potential of 0.2 and 0.7 V for ORR and OER; the ΔE value of 0.5 V and a Tafel slope of 167 mV dec^{-1}.

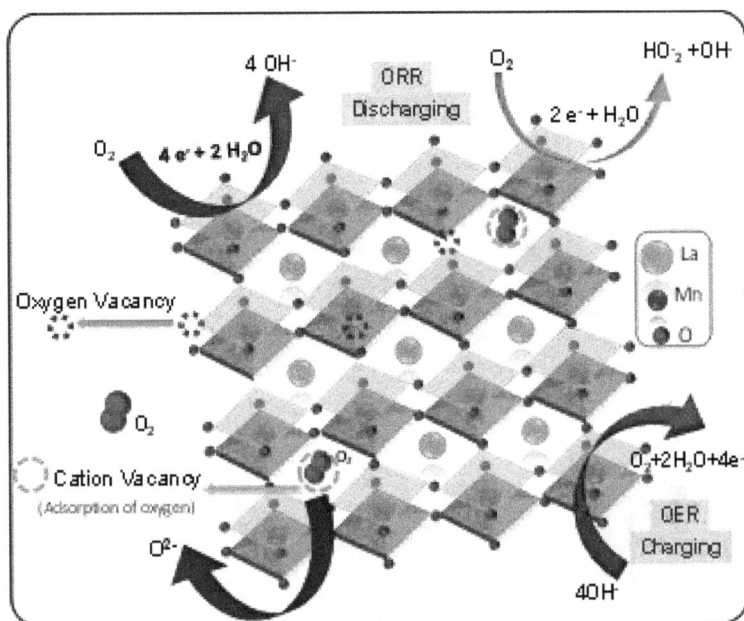

Figure 2.9   Schematic of oxygen adsorption, ORR and OER on the vacancy-induced $LaMnO_3$ (Reproduced with permission from Ref. [Sankar Devi et al., 2022]).

Hu et al. synthesized $La_{(1-x)}Ca_xMnO_3$-graphene composite by a sol-gel method (Hu et al., 2014). The ORR activity of the sample was studied with RRDE technique. The perovskite phase adhered on the surface of the graphene sheets, and adding graphene significantly improved the electrochemical performance of $LaMnO_3$. The prepared graphene had a gauze-shaped wrinkles and folds structure,

Table 2.5 List of perovskite-based electrocatalysts examined for the ORR and OER activities

S. No.	Catalysts	E_{ORR} (V) at I = -3 mA cm^{-2}	E_{OER} (V) at I = 10 mA cm^{-2}	Oxygen electrode (OER-ORR) E (V)	Tafel$_{OER}$ (mV dec^{-1})	References
1	$LaMnO_3$	0.749	1.839	1.09	109	Sankar Devi et al., 2022
2	$PrBaMn_2O_{5+d}$ (H-PBM)	0.74	0.89	0.15	–	Chen et al., 2016
3	$(La_{0.8}Sr_{0.2})_{1-x}MnO_3$	0.2	0.7	0.5	167	Hua yuan et al., 2019
4	$LaMn_{0.7}Co_{0.3}O_{3-x}$	0.73	1.62	0.8	–	Lee et al., 2015b

Table 2.6 List of composite electro-catalysts examined for the ORR and OER activities

S. No.	Catalyst	E_{ORR} (V) at I = -3 mA cm^{-2}	E_{OER} (V) at I = 10 mA cm^{-2}	Half-wave potential (V)	Electron transfer number (n)	Tafel slope (mV dec^{-1})	Over potential (mV)	References
1	Co_3O_4/MnO_2	-0.10	–	-0.19	3.92–3.95	–	–	Cui et al., 2020
2	$Co-CeO_2/C$ aerogels	0.92	1.45	0.75	3.66–3.78	99	380	Liu et al., 2022
3	FCNC 900	1.01	1.45	0.868	–	–	360	Jose et al., 2019
4	$Ni_xCo_yMn_zO_4$-300	–	1.54	–	–	85	400	Priamushko et al., 2020
5	$CoO@Co_3O_4/$NSG-650	0.9	1.5	0,79	–	63	–	Huang et al., 2018
6	$Fe_{0.3}Ni_{0.7}O_X/$MWCNT	0.84	1.50	–	0.5–0.6	-	342	Morales et al., 2020

which may be caused by oxygenic functional group and the resultant defects during the preparation of graphene oxide, which exhibited porous structure that led to increased three-phase region, thereby improving the mass transfer process. The electron transfer number of $La_{0.6}Ca_{0.4}MnO_3$ graphene was 3.6. Cobalt-doped lanthanum manganese oxide nanoparticles combined with nitrogen-doped carbon nanotubes (LMCO/NCNT) were successfully synthesized by Lee et al. by means of solvothermal followed by injection chemical vapour deposition (CVD) (Lee et al., 2015b). The electro-catalytic activity was evaluated using RDE. When $LaMnO_3$ was combined with cobalt and nitrogen-doped carbon nanotubes (LMCO/NCNT), high efficient and durable ORR characteristics were seen. It was subjected to half-cell evaluation, which delivered significantly improved ORR onset and half-wave potentials of −0.11 and −0.24 V, the OER current density of 27 mA cm^{-2} and n value of 3.9.

2.7.5   Composite Electrocatalysts

There are numerous combinations, such as metal oxide, hydroxide and sulfide, functional carbon material, metal, and their composites (Yaseen et al., 2021). The combination of these metals/metal oxides resulted in noticeable synergistic effects, increasing the bifunctional activity towards the redox reactions of interest. On the other hand, less studies were dedicated to the investigation of quaternary metals/ metal oxides (MMOs) despite their potential regarding the catalytic activity of OER and/or ORR. The various possible combinations of mixed oxides composites are discussed here (Mathumba et al., 2020). Table 2.6 shows the list of mixed metal oxides that were examined as electrocatalysts for ORR/OER.

Cui et al. synthesized a highly active and durable Co_3O_4/MnO_2 electrocatalyst which was synthesized through a two-step hydrothermal method and subsequent heat treatment for ORR (Cui et al., 2020). Using a potentiostat/galvanostat with three electrodes and an RDE, the electro-catalytic activity of the catalysts towards the ORR (LSV) was assessed. The surface of MnO_2 nanorods was covered with evenly scattered Co_3O_4 nanoparticles. When compared to its counterparts, such as MnO_2 nanorods or Co_3O_4 nanoparticles, the high electro-catalytic performance of Co_3O_4/MnO_2 revealed a synergetic connection between Co and Mn. This is one of the explanations to explain the improved catalytic performance and superior durability of Co_3O_4/MnO_2. The electron transfer numbers were in the range of 3.92–3.95 and exhibited onset potential of −0.10 V and half wave potential of −0.19 V. Liu et al. synthesized a hybrid composed of CeO_2-decorated Co nanoparticles supported on 3D porous carbon aerogels (Co-CeO$_2$/C aerogels) by a facile and scalable $CeCl_3/K_3Co (CN)_6$-chitosan hydrogel-derived approach, and demonstrated that the incorporation of CeO_2 into metallic Co and 3D network structure was beneficial to the bifunctional oxygen electrochemical performance (Liu et al., 2022). The reversible conversion between Ce^{2+} and Ce^{4+} in CeO_2 provided numerous oxygen vacancies, and the synergistic effects of CeO_2 and Co optimized the surface electronic structure of the catalyst. Moreover, the 3D porous network structure and the introduction of N enhanced the electronic conductivity and accelerated the mass transport, as well as prevented the aggregation and dissolution of the active components. The prepared $Co\text{-}CeO_2/C$ aerogels exhibited a satisfactory bifunctional electro-catalytic activity as well as outstanding durability with merely 0.38 V at 10 mA cm^{-2} for OER and a high half-wave potential of 0.75 V and onset potential of 0.92 V for ORR, n value in the range of 3.66–3.78, the Tafel slope of 99 mV dec^{-1} and a overpotential 380 mV. Jose et al. synthesized bifunctional ORR and OER electro-catalysts by means of thermal treatment of core shell Zeolitic imidazolate framework (ZIF) base units with Fe doping and N enrichment (Jose et al., 2019). Electro-catalytic activity of the hybrid ZIF derived materials was analyzed by a simple cyclic voltammogram (CV) conducted in N_2-saturated and O_2-staurated solutions. This step of Fe doping and N enrichment resulted in boosting the electrocatalytic activity of the material. Promising ORR performance of the FCNC 900 was judged from $E_{1/2}$ of 0.868 V and E_{onset} of 1.01 V while OER overpotential for same catalyst was 360 mV, much smaller than others.

Moreover, nanocasting or hard-templating is a versatile method to produce ordered mesoporous mixed transition metal oxides (MTMOs) synthesized by Priamushko et al. as a promising potential for both ORR and OER (Priamushko et al., 2020). The catalyst was subjected to cyclic voltammetry measurements in Ar-saturated 1 M KOH solution followed by LSV measurements. The results indicated that the calcination temperature greatly affects the porosity, crystalline structure, phase composition, and the activity of the catalysts toward OER. The best sample, $Ni_xCo_yMn_zO_4$ calcined at 300°C, provided a reasonable current density of 25 mA/cm^2 at 1.7 V and an overpotential of 400 mV at 10 mA/cm^2, and demonstrated increased current density of above 200 mA/cm^2 at 1.7 V once loaded into a Ni foam compared to bare foam. It exhibited an overpotential of 400 mV and Tafel slope of 85 mV dec^{-1}. The 3D nitrogen, sulfur-codoped carbon nanomaterial-supported cobalt oxides with polyhedron-like particles grafted on to graphene layers was reported by X. Huang et al. as highly active catalysts (Huang et al., 2018). Electrochemical performance was evaluated using RDE and RRDE in O_2 and H_2 saturated in 0.1 M KOH. Here, a cobalt, nitrogen and sulfur codoped nanomaterial with 3D hierarchical porous architectures is produced using property-flexible ZIF-67 and sulfur-functionalized graphene oxide, thanks to their abundant dopant species and high conductivity. The mass transport and charge delivery operations during the oxygen-evolving reactions are sped up by the crosslinked structures of polyhedron particles across the whole carbon framework, the best-performing $CoO@Co_3O_4/NSG\text{-}650$ with onset potential of 0.79 V for ORR and a Tafel slope of 63 mV dec^{-1}. Besides, the metric between ORR and OER difference to evaluate its overall electrocatalytic activity is 0.90 V. A trimetallic Mn-Fe-Ni oxide nanoparticles supported on multi-walled carbon nanotubes was synthesized by Morales et al. as high-performance bifunctional ORR/OER electrocatalyst in alkaline media (Morales et al., 2020). The ORR/OER activity and ORR selectivity of the samples were evaluated by RDE and RRDE voltammetry. The $Fe_{0.3}Ni_{0.7}O_X$ supported on oxygen-functionalized multi-walled carbon nanotubes is substantially activated into a bifunctional ORR/OER catalyst by means of additional incorporation of MnO_X. The carbon nanotube-supported trimetallic (Mn-Ni-Fe) oxide catalyst achieves remarkably low ORR and OER overpotentials with a low reversible ORR/OER, overvoltage of only 0.73 V, as well as selective reduction of O_2 predominantly to OH^- and onset potential of 0.84 V and n value ranges from 0.5 to 0.6.

2.8 APPLICATION OF ORR/OER IN LITHIUM-AIR BATTERY

One of the applications of the ORR/OER is metal-air battery. A typical metal-air battery combines a reactive metal as anode, such as Li, Na with atmospheric oxygen as cathode and with a suitable electrolyte. For example, a rechargeable $Li-O_2$ battery can be constructed using Li metal as anode and air-cathode having an electrocatalyst in Li^+ conducting electrolyte in non-aqueous solvent. The following reactions are occurring during charging-discharging of the $Li-O_2$ battery:

Charging reactions:

Type I:

$$Li_2O_2 \rightarrow LiO_2 + Li^+ + e^- \tag{22}$$

$$LiO_2 \rightarrow O_2 + Li^+ + e^- \tag{23}$$

Type II:

$$Li_2O_2 \rightarrow O_2 + 2Li^+ + 2e^- \tag{24}$$

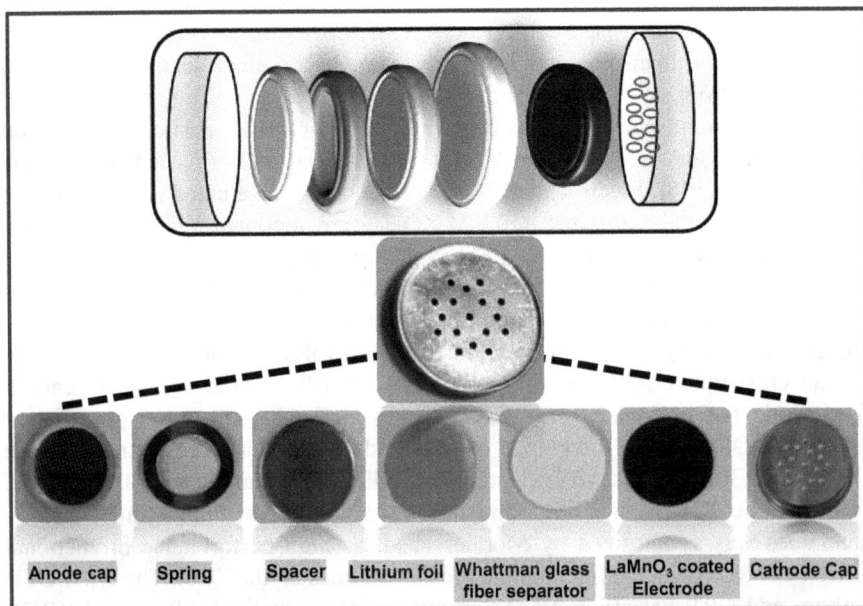

Figure 2.10 Schematic of air-coin-cell components and photographs of air-coin cell components and assembled air-coin cell (Reproduced with permission from Ref. [Sankar Devi et al., 2022]).

Discharging reactions:

Type I:

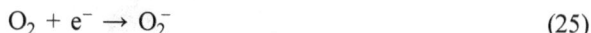

$$O_2 + e^- \rightarrow O_2^- \tag{25}$$

$$2O_2^- \leftrightarrow O_2 + O_2^{2-} \tag{26}$$

$$O_2^{2-} + 2Li^+ \rightarrow Li_2O_2 \tag{27}$$

Type II:

$$O_2 + e^- \rightarrow 2O_2^- \tag{28}$$

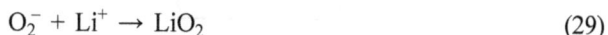

$$O_2^- + Li^+ \rightarrow LiO_2 \tag{29}$$

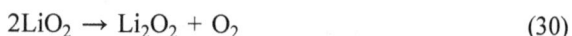

$$2LiO_2 \rightarrow Li_2O_2 + O_2 \tag{30}$$

Lithium metal undergoes oxidation, producing Li^+ and e^-. The oxygen from the atmospheric air diffuses into the electrolyte through the air-cathode and undergoes reduction to form Li_2O, Li_2O_2 or LiO_2 during discharge of the battery. During charging of the battery, Li is plated at anode side and the OER occurs at the cathode. Sankar Devi et al. conducted practical prototype Li-Air battery by fabricating one-side meshed CR-2032 air-coin cell (Li et al., 2016; Lv et al., 2019). The schematics and different parts of the one-side meshed CR-2032 air-coin cells are presented in Fig. 2.10. N-methyl-2-pyrrolidone (NMP) solvent was used to combine the $LaMnO_3$ nanocatalyst, super P carbon, and poly-(vinylidene fluoride) (PVDF) binder in the amounts of 70, 20, and 10 weight per cent, respectively to prepare the cathode ink. The carbon cloth gas diffusion layer-cum-current collector was then coated with the catalyst ink, which was then dried for 12 hours in an oven at 80°C. The catalyst-coated carbon fabric was then cut into discs and utilized as an electrode for air breathing (cathode). Lithium disc was cut from the lithium foil and utilized as an anode. The electrolyte was 1 M $LiCF_3SO_3$ (lithium trifluromethane sulfonate) in tetraethylene glycol dimethyl ether, and the separator was Whatman glass micro-fiber filter paper. The one-side meshed air-coin cell was assembled in the glovebox, which was maintained at <1 ppm O_2 and H_2O. The open circuit voltage (OCV) of the fabricated air coin-cell was initially recorded and shown in Fig. 2.11(a). Over a period of five days, the cell showed a constant OCV of roughly 3.3 V, which is about the theoretical cell voltage. The OCV of the coin cell was non-zero and as high as 3.0 V even on the sixth day of self-decay. This proved that the $LaMnO_3$ cathode functions as an electrode that can breathe air in a real-world Li-Air battery. The acquired galvanostatic charge-discharge patterns captured at various current rates are shown in Fig. 2.11(b). At a current density of 50 mAg^{-1}, it can be noted that the coin-type Li-Air battery displayed a discharge capacity of up to 3100 $mAhg^{-1}$ (catalyst). The discharge capacity decreases as the current density rises. The once-charged laboratory prototype Li-Air complete cell could power the LED bulb for around 24 hours constantly, according to Sankar Devi et al. practical demonstration. Inset of Fig. 2.11 also displays digital images of the manufactured air coin cell that supplied power to commercial LED bulb. The working LED bulb that was driven by the CR-2032 coin-type Li-Air battery and used $LaMnO_3$ as the air-breathing cathode is depicted in Fig. 2.12 in typical timeline images.

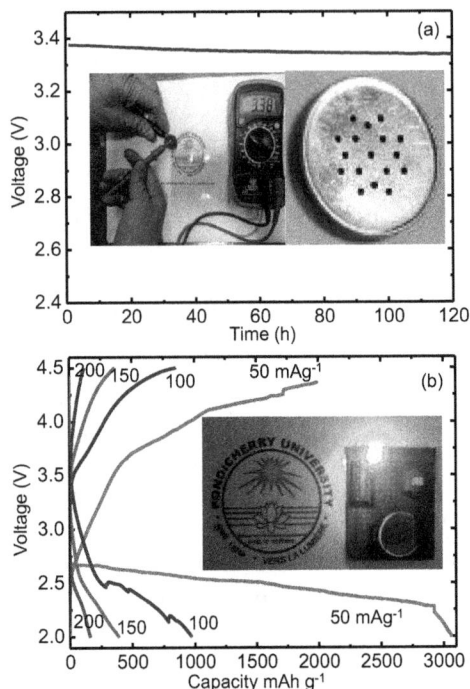

Figure 2.11 (a) OCV decay recorded for the CR-2032 coin-type lithium-air battery consisting of the LaMnO$_3$ air cathode Inset: Photograph of the fabricated Li-Air coin cell showing OCV and (b) charge-discharge curves at different current densities recorded for the coin cell containing the LaMnO$_3$ air-cathode and lithium metal anode. Inset: Photograph of Li-Air coin cell powering LED bulb (Reproduced with permission from Ref. [Sankar Devi et al., 2022]).

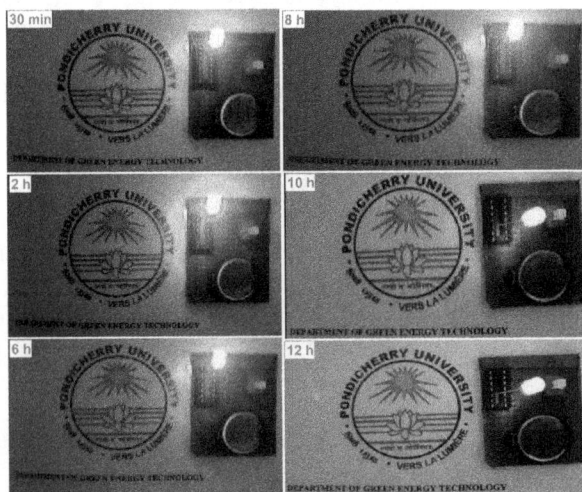

Figure 2.12 Representative time-line photographs of the fabricated CR-2032-type Li-Air coin cell consisting of the LaMnO$_3$ air-breathing cathode powering green LED bulb (Reproduced with permission from Ref. [Sankar Devi et al., 2022]).

2.9 SUMMARY AND RECOMMENDATIONS

As the demand for developing energy conversion/storage devices is increasing, the ORR/OER occurring on the devices is to be improved with effective electrocatalyst. The rate of oxygen adsorption and subsequent diffusion is expected to be significantly increased. Additionally, the electron transfer process reaction sites need to be enhanced. As much as possible, the complete 4e$^-$ transfer reaction pathway is to be progressed on any catalyst. The past research and developments on various catalysts discussed above reveal that significant progress has been made pertaining to transition metal oxides. As a result the following vital aspects are to be considered: large surface area, high porosity, tunable morphology, ability of catalyst to adsorb molecular O_2, the electronic structure of the metal or metal oxide, the large contact area with electrolyte provided by the support layer/matrix, etc.

Instead of pristine metal or metal oxide, use of the composite along with the support matrix led to enhanced catalytic activities. The composite electrocatalyst had high reaction site, better conductivity and O_2 adsorption which led to favorable onset potential, overpotential, limiting current density and nearly 4e$^-$ transfer for ORR.

There are countable metal/metal oxides that have exhibited favorable electrocatalytic activity towards ORR/OER. Transition metals like Ni, Co, Mn and perovskite oxide, such as $LaMnO_3$ are such typical catalysts that can find practical use as ORR/OER catalyst as they exhibit onset pontential in the range of –0.10 to 1.03 V with limiting current density in the range of –3 mA cm^{-2} for ORR, with nearly 4e$^-$ transfer reaction pathway as well as low peroxide percentage. Different synthetic strategies can be employed to tune the nanostructure morphology and composition of the catalyst layer to further enhance the ORR/OER activities.

ACKNOWLEDGEMENTS

PE thanks the DST-IISc Energy Storage Platform on Supercapacitor and Power Dense Devices through MECSP-2K17 Program Under Grant DST/TMD/MECSP/2K17/20. VSD Thanks DST for JRF fellowship.

REFERENCES

Anandan, V., R. Kudla, J. Adams, M. Karulkar and A. Drews. 2011. Mixed Metal Oxide Catalysts for Rechargeable Lithium-Air Batteries. ECS Meet. Abst. MA2011-02: 1330–1330.

Ashok, A., A. Kumar, R.R. Bhosale, F. Almomani, S.S. Malik, S. Suslov, et al., 2018. Combustion synthesis of bifunctional $LaMO_3$ (M = Cr, Mn, Fe, Co, Ni) perovskites for oxygen reduction and oxygen evolution reaction in alkaline media. J. Electroanal. Chem. 809: 22–30.

Athika, M. and P. Elumalai. 2020. Porous carbon networks decorated with cobalt on CoFe₂O₄ as an air-breathing electrode for high-capacity rechargeable lithium-air batteries: role of metallic cobalt nanoparticles. Chem. Electro. Chem. 7: 4188–4200.

Athika, M., V.S. Devi and P. Elumalai. 2020. Cauliflower-like hierarchical porous nickel/nickel ferrite/carbon composite as superior bifunctional catalyst for lithium-air battery. Chemistry Select. 5: 3529–3538.

Athika, M. 2022. Metal oxide-carbon composite as air-breathing electrode for rechargeable lithium-air battery: role of metal nanoparticles. Ph.D. Thesis, Pondicherry University, Puducherry, India.

Bardenhagen, I., M. Fenske, D. Fenske, A. Wittstock and M. Bäumer. 2015. Distribution of discharge products inside of the lithium/oxygen battery cathode. J. Power Sources. 299: 162–169.

Barlie, C.J. and A.A. Gewirth. 2013. Investigating the Li-O₂ Battery in an ether-based electrolyte using differential electrochemical mass spectrometry. J. Electrochem. Soc. 160: A549–A552.

Chakrabarty, S., A. Mukherjee, W.N. Su and S. Basu. 2019. Improved bi-functional ORR and OER catalytic activity of reduced graphene oxide supported ZnCo₂O₄ microsphere. Int. J. Hydrogen Energy. 44: 1565–1578.

Chen, D., J. Wang, Z. Zhang, Z. Shao and F. Ciucci. 2016. Boosting oxygen reduction/evolution reaction activities with layered perovskite catalysts. Chem. Commun. 52: 10739–10742.

Cheng, F., J. Shen, W. Ji, Z. Tao and J. Chen. 2009. Selective synthesis of manganese oxide nanostructures for electrocatalytic oxygen reduction. ACS Appl. Mater. Interfaces. 1: 60–466.

Cheng, F. and J. Chen. 2012. Metal–air batteries: From oxygen reduction electrochemistry to cathode catalysts. Chem. Soc. Rev. 41: 2172–2192.

Chilvery, A., S. Das, P. Guggilla, C. Brantley and A. Sunda-Meya. 2016. A perspective on the recent progress in solution-processed methods for highly efficient perovskite solar cells. Sci. Technol. Adv. Mater. 17: 650–658.

Cui, X., P. Ren, D. Deng, J. Deng and X. Bao. 2016. Single layer graphene encapsulating non-precious metals as high-performance electrocatalysts for water oxidation. Energy Environ. Sci. 9: 123–129.

Cui, C., G., Du, K. Zhang, T. An, B. Li, X. Liu, et al., 2020. Co₃O₄ nanoparticles anchored in MnO₂ nanorods as efficient oxygen reduction reaction catalyst for metal-air batteries. J. Alloys Compd. 814: 152239.

Cuma, M.U. and T. Koroglu. 2015. A comprehensive review on estimation strategies used in hybrid and battery electric vehicles. Renew. Sustain. Energy Rev. 42: 517–531.

Faulkner, L.R. and A.J. Bard. 2002. Electrochemical Methods: Fundamentals and Applications. John Wiley and Sons.

Ge, X., A. Sumboja, D. Wuu, T. An, B. Li, F.W.T. Goh, et al., 2015. Oxygen reduction in alkaline media: from mechanisms to recent advances of catalysts. ACS Catal. 5: 4643–4667.

Goodenough, J.B. 2014. Electrochemical energy storage in a sustainable modern society. Energy Environ. Sci. 7: 14–18.

Goswami, C., K.K. Hazarika, Y. Yamada and P. Bharali. 2021. Nonprecious hybrid metal oxide for bifunctional oxygen electrodes: endorsing the role of interfaces in electrocatalytic enhancement. Energy and Fuels. 35: 13370–13381.

Griffith, J.S. 1956. On the magnetic properties of some haemoglobin complexes. Proc. R. Soc. London. Ser. A. Math. Phys. Sci. 235: 23–36.

Holewinski, A. and S. Linic. 2012. Elementary mechanisms in electrocatalysis: revisiting the ORR tafel slope. J. Electrochem. Soc. 159: H864–H870.

Hu, J., L. Wang, L. Shi and H. Huang. 2014. Preparation of $La_{1-x}Ca_xMnO_3$ perovskite-graphene composites as oxygen reduction reaction electrocatalyst in alkaline medium. J. Power Sources. 269: 144–151.

Hua Yuan, R., Y. He, W. He, M. Ni and M.K.H. Leung. 2019. Bifunctional electro-catalytic activity of $La_{0.8}Sr_{0.2}MnO_3$-based perovskite with the A-site deficiency for oxygen reduction and evolution reactions in alkaline media. Appl. Energy. 251: 113406.

Huang, X., J. Wang, H. Bao, X. Zhang and Y. Huang. 2018. 3D Nitrogen, sulfur-codoped carbon nanomaterial-supported cobalt oxides with polyhedron-like particles grafted onto graphene layers as highly active bicatalysts for oxygen-evolving reactions. ACS Appl. Mater. Interfaces. 10: 7180–7190.

Ibrahim, K.B., M.C. Tsai, S.A. Chala, M.K. Berihun, A.W. Kahsay, T.A. Berhe, et al., 2019. A review of transition metal-based bifunctional oxygen electro-catalysts. J. Chinese Chem. Soc. 66: 829–865.

Jin, C., X. Cao, L. Zhang, C. Zhang and R. Yang. 2013. Preparation and electrochemical properties of urchin-like $La_{0.8}Sr_{0.2}MnO_3$ perovskite oxide as a bifunctional catalyst for oxygen reduction and oxygen evolution reaction. J. Power Sources. 241: 225–230.

Jose, V., A. Jayakumar and J.M. Lee. 2019. Bimetal/Metal oxide encapsulated in graphitic nitrogen doped mesoporous carbon networks for enhanced oxygen electro-catalysis. ChemElectroChem. 6: 1485–1491.

Jung, K.N., A. Riaz, S.B. Lee, T.H. Lim, S.J. Park, R.H. Song, et al., 2013. Urchin-like α-MnO_2 decorated with Au and Pd as a bi-functional catalyst for rechargeable lithium-oxygen batteries. J. Power Sources. 244: 328–335.

Kang, J.H., A. Riaz, S.B. Lee, T.H. Lim, S.J. Park, R.H. Song, et al., 2020. Lithium-air batteries: Air-breathing challenges and perspective. ACS Nano. 14: 14549–14578.

Kapałka, A., G. Fóti and C. Comninellis. 2008. Determination of the Tafel slope for oxygen evolution on boron-doped diamond electrodes. Electrochem. Commun., 10: 607–610.

Lee, D.U., B.J. Kim and Z. Chen. 2013. One-pot synthesis of a mesoporous $NiCo_2O_4$ nanoplatelet and graphene hybrid and its oxygen reduction and evolution activities as an efficient bi-functional electro-catalyst. J. Mater. Chem. A. 1: 4754–4762.

Lee, D.U., H.W. Park, M.G. Park, V. Ismayilov and Z. Chen. 2015a. Synergistic bifunctional catalyst design based on perovskite oxide nanoparticles and intertwined carbon nanotubes for rechargeable zinc-air battery applications. ACS Appl. Mater. Interfaces. 7: 902–910.

Lee, D.U., M.G. Park, H.W. Park, M.H. Seo, V. Ismayilov, R. Ahmed, et al., 2015b. Highly active Co-doped $LaMnO_3$ perovskite oxide and N-doped carbon nanotube hybrid bi-functional catalyst for rechargeable zinc-air batteries. Electrochem. commun. 60: 38–41.

Li, Y., J. Wang, X. Li, D. Geng, M.N. Banis, R. Li, et al., 2012. Nitrogen-doped graphene nanosheets as cathode materials with excellent electro-catalytic activity for high capacity lithium-oxygen batteries. Electrochem. Commun. 18: 12–15.

Li, P., R. Ma, Y. Zhou, Y. Chen, Q. Liu, G. Peng, et al., 2015a. Spinel nickel ferrite nanoparticles strongly cross-linked with multiwalled carbon nanotubes as a bi-efficient electrocatalyst for oxygen reduction and oxygen evolution. RSC Adv. 5: 73834–73841.

Li, X., Y. Fang, X. Lin, M. Tian, X. An, Y. Fu, et al., 2015b. MOF-derived Co_3O_4 nanoparticles embedded in N-doped mesoporous carbon layer/MWCNT hybrids:

Extraordinary bi-functional electrocatalysts for OER and ORR. J. Mater. Chem. A. 3: 17392–17402.

Li, G., X. Wang, J. Fu, J. Li, M.G. Park, Y. Zhang, et al., 2016. Pomegranate-inspired design of highly active and durable bifunctional electrocatalysts for rechargeable metal-air batteries. Angew. Chemie–Int. Ed. 55: 4977–4982.

Li, C., Z. Yu, H. Liu and K. Chen. 2018. High surface area LaMnO$_3$ nanoparticles enhancing electro-chemical catalytic activity for rechargeable lithium-air batteries. J. Phys. Chem. Solids. 113: 151–156.

Liang, Y., Y. Li, H. Wang, J. Zhou, J. Wang, T. Regier, et al., 2011. Co$_3$O$_4$ nanocrystals on graphene as a synergistic catalyst for oxygen reduction reaction. Nat. Mater. 10: 780–786.

Liang, Y., D. Ye, N. Han, P. Liang, J. Wang, G. Yang, et al., 2021. Nano-porous silver-modified LaCoO$_3$-δ perovskite for oxygen reduction reaction. Electrochim. Acta. 391: 138908.

Liang, Q., G. Brocks and A. Bieberle-Hütter. 2021. Oxygen evolution reaction (OER) mechanism under alkaline and acidic conditions. J. Phys. Energy. 3(2): 026001.

Lima, F.H.B., M.L. Calegaro and E.A. Ticianelli. 2007. Electrocatalytic activity of manganese oxides prepared by thermal decomposition for oxygen reduction. Electrochim. Acta. 52: 3732–3738.

Liu, K., Z. Zhou, H. Wang, X. Huang, J. Xu, Y. Tang, et al., 2016a. N-doped carbon supported Co$_3$O$_4$ nanoparticles as an advanced electrocatalyst for the oxygen reduction reaction in Al-air batteries. RSC Adv. 6: 55552–55559.

Liu, S., W. Yan, X. Cao, Z. Zhou and R. Yang. 2016b. Bacterial-cellulose-derived carbon nanofiber-supported CoFe$_2$O$_4$ as efficient electrocatalyst for oxygen reduction and evolution reactions. Int. J. Hydrogen Energy. 41: 5351–5360.

Liu, Y., H. Jiang, Y. Zhu, X. Yang and C. Li. 2016c. Transition metals (Fe, Co, and Ni) encapsulated in nitrogen-doped carbon nanotubes as bi-functional catalysts for oxygen electrode reactions. J. Mater. Chem. A. 4: 1694–1701.

Liu, Y., L. Wang, L. Cao, C. Shang, Z. Wang, H. Wang, et al., 2017. Understanding and suppressing side reactions in Li-Air batteries. Mater. Chem. Front. 1: 2495–2510.

Liu, P., J. Ran, B. Xia, S. Xi, D. Gao and J. Wang. 2020. Bifunctional oxygen electrocatalyst of mesoporous Ni/NiO nanosheets for flexible rechargeable Zn–Air batteries. Nano-Micro Lett. 12: 1–12.

Liu, Y., H. Huang, L. Xue, J. Sun, X. Wan, P. Xiong, et al., 2021. Recent advances in the heteroatom doping of perovskite oxides for efficient electro-catalytic reactions. Nanoscale. 13: 19840.

Liu, Z., J. Wan, M. Li, Z. Shi, J. Liu and Y. Tang. 2022. Synthesis of Co/CeO$_2$hetero-particles with abundant oxygen-vacancies supported by carbon aerogels for ORR and OER. Nanoscale. 14: 1997–2003.

Lu, Y.C., B.M. Gallant, D.G. Kwabi, J.R. Harding, R.R. Mitchell, M.S. Whittingham, et al., 2013. Lithium-oxygen batteries: Bridging mechanistic understanding and battery performance. Energy Environ. Sci. 6: 750–768.

Lv, Y., Z. Li, Y. Yu, J. Yin, K. Song, B. Yang, et al., 2019. Copper/cobalt-doped LaMnO$_3$ perovskite oxide as a bifunctional catalyst for rechargeable Li-O$_2$ batteries. J. Alloys Compd. 801: 19–26.

Maiyalagan, T. and P. Elumalai. 2021. Rechargeable Lithium-Ion Batteries. CRC Press, India.

Malkhandi, S., B. Yang, A.K. Manohar, A. Manivannan, G.K.S. Prakash and S.R. Narayanan. 2012. Electrocatalytic properties of nanocrystalline calcium-doped lanthanum cobalt oxide for bifunctional oxygen electrodes. J. Phys. Chem. Lett. 3: 967–972.

Mathumba, P., D.M. Fernandes, R. Matos, E.I. Iwuoha and C. Freire. 2020. Supplementary materials: metal oxide impregnation (Co_3O_4 and Mn_3O_4) into S, N-doped graphene for oxygen reduction reaction (ORR). Materials. 13(7): 1562.

Morales, D.M., M.A. Kazakova, S. Dieckhöfer, A.G. Selyutin, G.V. Golubtsov, W. Schuhmann, et al., 2020. Trimetallic Mn-Fe-Ni oxide nanoparticles supported on multi-walled carbon nanotubes as high-performance bifunctional ORR/OER electrocatalyst in alkaline media. Adv. Funct. Mater. 30: 1–12.

Oloniyo, O., S. Kumar and K. Scott. 2012. Performance of MnO_2 crystallographic phases in rechargeable lithium-air oxygen cathode. J. Electron. Mater. 41: 921–927.

Pan, W.X., J.B. Yong and M. Wang. 2017. Optimizing discharge capacity of $Li-O_2$ batteries by design of air-electrode porous structure: multifidelity modeling and optimization. J. Electrochem. Soc. 164: E3499–E3511.

Pan, J., Y.Y. Xu, H. Yang, Z. Dong, H. Liu and B.Y. Xia. 2018. Advanced architectures and relatives of air electrodes in Zn–Air batteries. Adv Sci (Weinh). 5(4): 1700691.

Park, C.S., K.S. Kim and Y.J. Park. 2013. Carbon-sphere/Co_3O_4 nanocomposite catalysts for effective air electrode in Li/air batteries. J. Power Sources. 244: 72–79.

Paulus, U.A., T.J. Schmidt, H.A. Gasteiger and R.J. Behm. 2014. Oxygen reduction on a high-surface area Pt/Vulcan carbon catalyst: A thin-film rotating ring-disk electrode study oxygen reduction on a high-surface area Pt/Vulcan carbon catalyst: A thin-film rotating ring-disk electrode study. Res. Gate. 495: 134–145.

Priamushko, T., R. Guillet-Nicolas, M. Yu, M. Doyle, C. Weidenthaler, H. Tuÿsüz, et al., 2020. Nanocast mixed Ni-Co-Mn oxides with controlled surface and pore structure for electrochemical oxygen evolution reaction. ACS Appl. Energy Mater. 3: 5597–5609.

Qian, J., X. Bai, S. Xi, W. Xiao, D. Gao and J. Wang. 2019. Bifunctional electrocatalytic activity of nitrogen-doped NiO nanosheets for rechargeable zinc-air batteries. ACS Appl. Mater. Interfaces. 11: 30865–30871.

Romm, R. 2006. The car and fuel of the future. Energy Policy. 34: 2609–2614.

Sadek, N. 2012. Urban electric vehicles: a contemporary business case. Eur. Transp. Res. Rev. 4: 27–37.

Sankar Devi, V., M. Athika and P. Elumalai. 2022. Vacancy-induced $LaMnO_3$ perovskite as bifunctional air-breathing electrode for rechargeable lithium-air battery. ChemistrySelect. 7: 1–12.

Sun, C., J.A. Alonso and J. Bian. 2020. Recent advances in perovskite-type oxides for energy conversion and storage applications. Adv. Energy Mater. 11: 2000459.

Su, Y., Y. Zhu, H. Jiang, J. Shen, X. Yang, W. Zou, et al., 2014. Cobalt nanoparticles embedded in N-doped carbon as an efficient bifunctional electrocatalyst for oxygen reduction and evolution reactions. Nanoscale. 6(24): 15080–15089.

Taube, H. 1965. Mechanisms of oxidation with oxygen. J. Gen. Physiol. 49(1): 29–50.

Thapa, A.K., K. Saimen and T. Ishihara. 2010. Pd/MnO_2 air electrode catalyst for rechargeable lithium/air battery. Electrochem. Solid-State Lett. 13: 165–167.

Tomboc, G.M., P. Yu, T. Kwon, K. Lee and J. Li. 2020. Ideal design of air electrode – A step closer toward robust rechargeable Zn-air battery. APL Mater. 8(5): 050905.

Tsai, Y.L., K.L. Huang, C.C. Yang, J.S. Ye, L.S. Pan and C.L. Lee. 2014. Preparation and cyclic voltammetric dissolution of core-shell-shell Ag-Pt-Ag nanocubes and their comparison in oxygen reduction reaction in alkaline media. Int. J. Hydrogen Energy. 39: 5528–5536.

Wang, Y., J. Wu, S. Yang, H. Li and X. Li. 2018. Electrode modification and optimization in air-cathode single-chamber microbial fuel cells. Int. J. Environ. Res. Public Health. 15(7): 1349.

Wang, H.F. and Q. Xu. 2019. Materials design for rechargeable metal-air batteries. Matter. 1: 565–595.

Wang, H., M. Zhou, P. Choudhury and H. Luo. 2019. Perovskite oxides as bifunctional oxygen electrocatalysts for oxygen evolution/reduction reactions—a mini review. Appl. Mater. Today. 16: 56–71.

Wu, S., Q. Zhang, J. Ma, D. Sun, Y. Tang and H. Wang. 2020. Interfacial design of Al electrode for efficient aluminum-air batteries: issues and advances. Mater. Today Energy. 18: 100499.

Xu, C., M. Lu, Y. Zhan and J.Y. Lee. 2014. A bifunctional oxygen electrocatalyst from monodisperse $MnCo_2O_4$ nanoparticles on nitrogen enriched carbon nanofibers. RSC Adv. 4: 25089–25092.

Xu, X., W. Wang, W. Zhou and Z. Shao. 2018. Recent advances in novel nanostructuring methods of perovskite electrocatalysts for energy-related applications. Small Methods. 2: 1–35.

Yang, Z., J. Liu, S. Baskaran, C.H. Imhoff and J.D. Holladay. 2010. Enabling renewable energy- and the future grid-with advanced electricity storage. JOM. 62: 14–23.

Yaseen, W., N. Ullah, M. Xie, B.A. Yusuf, Y. Xu, C. Tong, et al., 2021. Ni-Fe-Co based mixed metal/metal-oxides nanoparticles encapsulated in ultrathin carbon nanosheets: a bifunctional electrocatalyst for overall water splitting. Surfaces and Interfaces. 26: 101361.

Zhang, J. (ed.). 2008. PEM Fuel Cell Electrocatalysts and Catalyst Layers. Fundamentals and Applications. Springer, London.

Zhang, B., Y. Niu, J. Xu, X. Pan, C.-M. Chen, W. Shi, et al., 2016. Tuning the surface structure of supported $PtNi_x$ bimetallic electrocatalysts for the methanol electro-oxidation reaction. Chem. Commun. 52: 3927–3930.

Zhao, G., Z. Xu and K. Sun. 2013. Hierarchical porous Co_3O_4 films as cathode catalysts of rechargeable $Li-O_2$ batteries. J. Mater. Chem. A. 1: 2862–12867.

Zhou, L., B. Deng, Z. Jiang and Z.J. Jiang. 2019. Shell thickness controlled core-shell $Fe_3O_4@CoO$ nanocrystals as efficient bifunctional catalysts for the oxygen reduction and evolution reactions, Chem. Commun. 55: 525–528.

Zhu, J., M. Xiao, Y. Zhang, Z. Jin, Z. Peng, C. Liu, et al., 2016. Metal-organic framework-induced synthesis of ultrasmall encased NiFe nanoparticles coupling with graphene as an efficient oxygen electrode for a rechargeable Zn-Air battery. ACS Catal. 6: 6335–6342.

Chapter 3

Electrochemical Characterization of Oxygen Reduction Reaction Catalysts: A Step-by-Step Guide

José Luis Reyes-Rodríguez[1]*, Adrián Velázquez-Osorio[2],
Elvia Terán-Salgado[3] and Daniel Bahena-Uribe[4]

[1]Escuela Superior de Ingeniería Química e Industrias Extractivas (ESIQIE),
Instituto Politécnico Nacional (IPN). Av. Luis Enrique Erro S/N,
Unidad Profesional Adolfo López Mateos, Zacatenco,
Alcaldía Gustavo A. Madero, C.P. 07738, Ciudad de México, México.
Email: jlreyes@ipn.mx; Tel. +52 (55) 3084 9664.

[2]Departamento de Química, Centro de Investigación y de Estudios Avanzados
(Cinvestav IPN), Av. IPN 2508, Col. San Pedro Zacatenco,
Alcaldía Gustavo A. Madero, C.P. 07360, Ciudad de México, México .
Email: aovelazquez@gmail.com; Tel. +52 (55) 5747 3715.

[3]Centro de Investigación en Ingeniería y Ciencias Aplicadas-(IICBA),
Universidad Autónoma Del Estado de Morelos, Av. Universidad 1001,
C.P, 62209, Cuernavaca, Morelos, México .
Email: elvia.teran@uaem.mx; Tel. +52 (777) 225 1890.

[4]Laboratorio Avanzado de Nanoscopía Electrónica (LANE),
Centro de Investigación y de Estudios Avanzados (Cinvestav IPN),
Av. IPN 2508, Col. San Pedro Zacatenco, Alcaldía Gustavo A. Madero, C.P.07360,
Ciudad de México, México.
Email: dbahenau@cinvestav.mx; Tel. +52 (55) 5747 3800 Exts. 1740, 1742.

*For Correspondence: Email: jlreyes@ipn.mx

3.1   INTRODUCTION

Fuel cells, in combination with water electrolyzers and other renewable energy technologies promise to be an attractive near-future alternative to generate electricity in a direct, sustainable, and environmentally-friendly way.

To understand the energy potential of fuel cells, it is important to first review key characteristics of hydrogen (H_2) as an energy vector. One important feature is hydrogen's high-energy density by mass ratio. When compared to traditional fossil fuels, such as gasoline, natural gas, diesel, and even coal, it is striking that burning 1 kg of H_2 provides an impressive 143 MJ kg^{-1} of energy. This is approximately 3x the energy generated by burning the same amount of any other liquid hydrocarbon fossil fuel (Mazloomi and Gomes, 2012). Another characteristic of H_2, a gas under normal conditions, is that it has a very low-energy density by volume ratio compared to liquid fuels. This property underscores an existing challenge in the development of hydrogen technology—it is easier to store a liquid fuel than a gaseous one. To put this in perspective, burning one liter of gaseous H_2 stored at a pressure of 690 atm provides 5.6 MJ l^{-1} of energy. This amount corresponds to one-sixth of the energy released by burning one liter of liquid gasoline, which has a volume energy density of 34.2 MJ l^{-1} (Mazloomi and Gomes, 2012).

Hydrogen's energy potential is best highlighted by the large amount of energy that it can release in combustion processes and by being a zero-emission fuel.

According to recent data from the International Energy Agency (IEA)[1], the automotive and transportation sector continues to be the most energy-demanding industry, accounting for 29.1% of the total annual global energy consumption, and contributing a considerable portion of harmful carbon emissions. While considerable efforts are being made to integrate hydrogen technologies in the transportation sector with the deployment of combined fuel cell and battery-based vehicles, challenges remain, particularly, as related to low-cost mass commercialization. From a technical perspective, it has been difficult to develop safe hydrogen storage materials and more durable batteries to support the vehicle's operation. In the case of fuel cells, one of the primary technical challenges lies in developing catalytic materials with high activity, durability, stability, with little or no noble metals.

This chapter addresses the electrochemical processes that underlie the characterization of catalytic materials for use in proton exchange membrane fuel cells (PEMFCs). This type of fuel cell is distinguished by having a polymer membrane with proton exchange capability placed between anode and cathode. This feature makes PEMFCs suitable for use in transportation due to their small size and portability. Fuel cells are electrochemical devices that produce electricity, heat, and water from spontaneous oxidation-reduction reactions occurring at its electrodes. Two electrochemical reactions drive the fuel cell, the hydrogen oxidation reaction (HOR) at the anode and the oxygen reduction reaction (ORR) at the cathode.

---

[1]Data can be consulted from the International Energy Agency website: https://www.iea.org/

The reactions that occur in a fuel cell are accelerated by using catalytic materials generally based on noble metals, such as platinum supported on carbon (Pt/C). In the presence of Pt/C catalysts, the HOR is kinetically fast. This has been observed by Döbereiner as early as 1823 (Sandstede et al., 2010), when he noted the immediate combustion of hydrogen in the presence of powder platinum at room temperature (Bard et al., 2008; Sandstede et al., 2010; Lee et al., 2013). The HOR, carried out at the anode, is responsible for generating the electron flow in a fuel cell and its reaction consists of two steps: the adsorption of hydrogen molecules on the surface of the Pt catalyst and the dissociation of these molecules by releasing electrons and H^+ ions (Eq. 1 and 2) (Carrette et al., 2001).

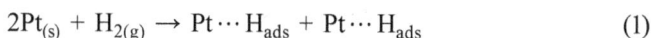

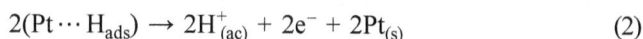

$$2Pt_{(s)} + H_{2(g)} \rightarrow Pt \cdots H_{ads} + Pt \cdots H_{ads} \tag{1}$$

$$2(Pt \cdots H_{ads}) \rightarrow 2H^+_{(ac)} + 2e^- + 2Pt_{(s)} \tag{2}$$

From the kinetic point of view, ORR is the reaction of greatest interest because it limits the overall performance of the fuel cell due to its slower kinetics (at least three orders of magnitude lower) compared to the anodic HOR. The reason for its slow kinetics lies in the fact that more energy is required to dissociate the O–O bond from the oxygen molecule once it is absorbed on the surface of the catalyst. Electrochemically, this dissociation requires a high overpotential to achieve.

It is widely accepted that ORR is a complex multi-electron process that involves several fundamental steps that give rise to different reaction intermediates. One of the most accepted models used to explain the ORR reaction is the mechanism proposed by Wroblowa et al. in 1976, shown below (Wroblowa and Razumney, 1976) (Eq. 3).

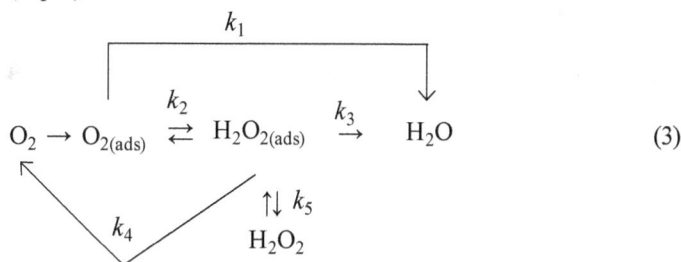

$$
\begin{array}{c}
\xrightarrow{\hspace{3cm}} \quad k_1 \quad \xrightarrow{\hspace{3cm}} \\[4pt]
O_2 \rightarrow O_{2(ads)} \;\overset{k_2}{\underset{}{\rightleftarrows}}\; H_2O_{2(ads)} \;\xrightarrow{k_3}\; H_2O \\[6pt]
\nwarrow \qquad \nearrow \; \updownarrow k_5 \\[2pt]
k_4 \qquad H_2O_2
\end{array}
\tag{3}
$$

From Eq. (3), it can be seen that the reaction begins with the adsorption of oxygen molecules ($O_{2(ads)}$) on the surface of the catalyst, constituting the determining stage of the reaction. The preferred reaction pathway is the one that involves a rate constant k_1 for the direct formation of water by the transfer of four electrons ($4e^-$). This pathway is known as the 'direct reduction reaction pathway'. However, under acidic media conditions, a two-electron transfer step ($2e^-$) can also occur to form hydrogen peroxide (H_2O_2) as an intermediate product in a step known as the 'stepwise reduction reaction pathway' which involves rate constants k_2 or k_3 (Paulus et al., 2010). H_2O_2 is not a desirable by-product due to its strong oxidizing power that causes long-term degradation of the fuel cell components, in particular, the proton exchange membrane (Guilminot et al., 2007; Iiyama et al., 2010; Houchins et al., 2012). The use of Pt-based catalysts greatly reduces the formation of peroxide even at low potentials.

As catalysis is a surface-based phenomenon, the surface contact area between the gases and the catalyst in fuel cells must be maximized; therefore, metal nanoparticles with sizes in the range of 2–50 nm are preferentially used. Due to the high cost of highly efficient noble metal catalysts, strategies have been adopted to optimize the distribution and content amount of the noble metal in the nanoparticles, including the formation of alloys with non-noble metals like Cu, Ni, Co, Fe, Mn (Antolini et al., 2007), synthesizing core-shell structures where the non-noble metal forms the core and the noble metal forms a shell (Gilroy et al., 2018; Reyes-Rodríguez et al., 2013; Shao et al., 2016), or to synthesize polyhedral bimetallic nanoparticles with high symmetry and faceting (Long et al., 2011; Chen et al., 2014; Reyes-Rodríguez et al., 2019). The incorporation of one or more platinum-alloyed metals is known to confer new intrinsic properties on the catalytic material. At the structural level, alloying metals decrease the interatomic distances between Pt atoms (geometric effects) by modifying the cell parameters of the crystal structure due to the interaction of metals with different cell sizes. At the electronic level, electronic structures with vacancies in the $5d$ orbitals are promoted. Thus, for a reduction process, such as ORR, the donation of electrons occurs from the surface of the electrode to the antibonding π orbitals of the oxygen molecule. This process facilitates the adsorption processes, the formation of OH species on the surface of the alloy particles, and the resulting dissociation of the O_2 molecule (Lee and Do, 2009; Bing et al., 2010; Özaslan, 2012). The enhanced ORR activity of these bi-, tri- or multi-metallic catalysts is a reason for the significant interest in their development.

The catalytic activity parameters of a catalyst towards ORR are *specific activity* (SA) and *mass activity* (MA) (Kinoshita, 1990). The former indicates the amount of current obtained per unit of active surface area of the material and has units of $\mu A\ cm^{-2}$. The latter indicates the amount of current obtained per mass of the material and has units of $mA\ g^{-1}$.

Determining the catalytic properties of synthesized nanomaterials towards ORR is a meticulous task that must be carried out to ascertain the viability of a candidate material for use in a fuel cell device. Electrochemical evaluations should follow a standardized protocol to ensure comparability and reproducibility of results across laboratories and against widely accepted reference benchmarks, like those reported for known nanocatalysts, such as commercial Pt/C Etek. Although a universal standardized protocol for evaluating catalysts towards ORR does not currently exist, a commonly cited guideline is the testing protocols reported by the American Department of Energy (DOE)[2,3]. Additional catalyst evaluation protocols have been reported in this research field over the last few decades; consult the following references for additional details (Markovic et al., 2001;

---

[2]For more information, refer to the slides of the DOE's Webinar titled *Testing Oxygen Reduction Reaction Activity with the Rotating Disc Electrode Technique* from 12 March, 2013, at: https://www.energy.gov/sites/prod/files/2014/03/f12/webinarslides_ rde_technique_031213.pdf

[3]Refer to the Hydrogen and Fuel Cell Technologies Office Webinars from the U.S. Department of Energy at: https://www.energy.gov/eere/fuelcells/hydrogen-and-fuel-cell-technologies -office-webinars

Schmidt et al., 2001; Paulus et al., 2001, 2002; Mayrhofer et al., 2005; Mayrhofer et al., 2008; Garsany et al., 2010; Van Der Vliet et al., 2010; Garsany et al., 2011; Katsounaros et al., 2013; Nrel et al., 2013; Hodnik et al., 2015; Shinozaki et al., 2015; Kocha et al., 2017.

This chapter aims to offer a step-by-step guide for the electrochemical characterization of catalysts towards ORR, taking the DOEs and other reported guidelines as a foundation, in order to contribute to the efforts to homologate catalytic parameters and standardize best practices. In this way, this chapter seeks to provide new and existing researchers with a set of tools and instructions to get started or refine their catalyst characterization experimentation while achieving more reproducible results.

3.2 THE ELECTROCHEMICAL CHARACTERIZATION PROTOCOL

The electrochemical evaluation of active nanomaterials towards ORR can be divided in two parts: a basic characterization protocol and a complementary protocol. The basic protocol consists of the use of techniques based on cyclic voltammetry, steady-state polarization curves, and CO stripping to determine the electrochemical surface area (ECSA). The complementary protocol includes approaches to measure the amount of hydrogen peroxide formed during ORR on a rotating ring-disk electrode (RRDE) configuration and a durability testing approach based on accelerated electrochemical degradation. Tables 3.1 and 3.2 show a summary of the parameters used in the basic and the complementary electrochemical characterization protocols, respectively. Each electrochemical technique is described in detail in the following pages.

It is worth mentioning that the protocols described below correspond to electrochemical testing using a three-electrode cell configuration in an acid medium. Acid media are typically used in research because most types of portable fuel cells are based on a proton exchange membrane (PEMFC) architecture that normally operates in an acidic environment. Analogous versions of this protocol for other media, such as neutral or alkaline electrolytes, can be implemented with minor adaptations, such as electrochemical cell design, glassware cleaning procedures, variations in the potential sweeping range, etc.

3.3 THE THREE-ELECTRODE CELL

Various designs and configurations are reported in the literature for three-electrode cells, from custom designs to standard commercial options. Figure 3.1 shows a typical configuration of a three-electrode cell.

Table 3.1 Parameters used in the basic electrochemical characterization protocol towards ORR [Adapted from (Reyes-Rodríguez et al., 2019)]

Basic Electrochemical Characterization Protocol

1. Cyclic Voltammetry for Preconditioning

Potential Profile

1.2 V vs. RHE; 0.1 V; 0.05 V; 50 or 100 mV s^{-1}; 30 or 100 Cycle; 0.1 M HClO$_4$ under N$_2$ saturatio.

Measurement parameters

- Electrolyte: 0.1 M HClO$_4$
- Temperature: 25–30°C
- Saturation gas: N$_2$
- Potential window: 0.05–1.2 V vs. RHE
- Scan rate: 50 or 100 mV s^{-1}
- RDE rotation rate: 0 rpm
- Total potential scans: 30–100 cycles

2. Determination of Ohmic drop by EIS

Profile

Applying 10 mV Amplitude at OCP or capacitive region; 10 kHz; 1 Hz; 1 Cycles; 0.1 M HClO$_4$ under N$_2$ or O$_2$ saturation.

Measurement parameters

- Electrolyte: 0.1 M HClO$_4$
- Temperature: 25–30°C
- Saturation gas: N$_2$ or O$_2$
- Frequency window: 10 kHz–1 Hz
- AC Perturbation: 10 mV at OCP or potential in capacitive region
- RDE rotation rate: With (e.g. 1600 rpm) or without rotation
- Total scans: 1 cycle

3. Corrected Cyclic Voltammetry

Potential Profile

1.2 V vs. RHE; 0.1 V; 0.05 V; 20 or 50 mV s^{-1}; 3 Cycles; 0.1 M HClO$_4$ under N$_2$ saturatio.

Measurement parameters

- Electrolyte: 0.1 M HClO$_4$
- Temperature: 25–30°C
- Saturation gas: N$_2$
- Potential window: 0.05–1.2 V vs. RHE
- Scan rate: 20 and 50 mV s^{-1}
- RDE rotation rate: 0 rpm
- iR compensation: Yes
- Total potential scans: 3 cycles

4. Inert Background (b.g.) under steady-state regime

Potential Profile

1.05 V vs. RHE; 1.045 V; 0.05 V; 20 mV s^{-1}; 3 Cycles; 0.1 M HClO$_4$ under N$_2$ saturation.

Measurement parameters

- Electrolyte: 0.1 M HClO$_4$
- Temperature: 25–30°C
- Saturation gas: N$_2$
- Potential window: 1.05–0.05 V vs. RHE
- Scan rate: 20 mV s^{-1}
- RDE rotation rates: 400, 900, 1600, and 2500 rpm
- iR compensation: Yes
- Total potential scans: 3 cycles

5. CO stripping		6. Steady-state polarization curves for ORR	
Potential Profile	**Measurement parameters**	**Potential Profile**	**Measurement parameters**
1.2 V vs. RHE 0.1 V 0.05 V 20 mV s⁻¹ 3 Cycles 0.1 M HClO₄ under CO saturation by 300 s and then evacuation with N₂ saturation by 600 s.	Electrolyte: 0.1 M HClO₄ Temperature: 25–30°C Adsorption potential: 0.1 V vs. RHE Adsorption gas: CO Absorption time: 300 s Saturation gas: N₂ N2 evacuation time: 600 s Potential window: 0.05–1.2 V vs. RHE Scan rate: 20 mV s⁻¹ RDE rotation rate: 0 rpm iR compensation: Yes Total potential scans: 3 cycles	1.05 V vs. RHE 1.045 V 0.05 V 20 mV s⁻¹ 3 Cycles 0.1 M HClO₄ under O₂ saturation Rotation rates: 400, 900, 1600, and 2500 rpm.	Electrolyte: 0.1 M HClO₄ Temperature: 25–30°C Saturation gas: O₂ Potential window: 1.05–0.05 V vs. RHE Scan rate: 20 mV s⁻¹ RDE rotation rates: 400, 900, 1600, and 2500 rpm iR compensation: Yes Total potential scans: 3 cycles

Note: EIS: Electrochemical Impedance Spectroscopy. OCP: Open Circuit Potential. RDE: Rotating Disk Electrode. RHE: Reversible Hydrogen Electrode. RRDE: Rotating Ring-Disk Electrode.

Table 3.2 Parameters used in the complementary ORR electrochemical characterization protocol [Adapted from (Reyes-Rodríguez et al., 2019)]

Complementary Electrochemical Characterization Protocol

7. Determination of collector factor in RRDE

Potential Profile	Measurement parameters
E_{Ring}: 1.55 V vs. RHE 1.05 V vs. RHE 1.045 V 0.05 V 20 mV s^{-1} 3 Cycles	RRDE configuration Electrolyte: 10 mmol of $K_3Fe(CN)_6$ in 0.1 M $HClO_4$ Temperature: 25–30°C Saturation gas: N_2 Potential window: 1.05–0.05 V vs. RHE Fixed potential in Ring: 1.55 V vs. RHE Scan rate: 20 mV s^{-1} RDE rotation rates: 400, 900, 1600, and 2500 rpm iR compensation: Yes Total potential scans: 3 cycles
10 mmol of $K_3Fe(CN)_6$ in 0.1 M $HClO_4$ under N_2 saturation Rotation rates: 400, 900, 1600, and 2500 rpm.	

8. Determination of H_2O_2 formation by RRDE

Potential Profile	Measurement parameters
E_{Ring}: 1.2 V vs. RHE 1.05 V vs. RHE 1.045 V 0.05 V 20 mV s^{-1} 3 Cycles	RRDE configuration Electrolyte: 0.1 M $HClO_4$ Temperature: 25–30°C Saturation gas: O_2 Potential window: 1.05–0.05 V vs. RHE Fixed potential in Ring: 1.2 V vs. RHE Scan rate: 20 mV s^{-1} RDE rotation rates: 400, 900, 1600, and 2500 rpm iR compensation: Yes Total potential scans: 3 cycles
0.1 M $HClO_4$ under O_2 saturation Rotation rates: 400, 900, 1600, and 2500 rpm	

9. Durability test by accelerated electrochemical degradation

Potential Profile	Measurement parameters
1.0 V vs. RHE 0.65 V 0.6 V 100 mV s^{-1} 10,000 Cycles	Electrolyte: 0.1 M $HClO_4$ Temperature: 25–30°C Saturation gas: O_2 Potential window: 0.6–1.0 V vs. RHE Scan rate: 100 mV s^{-1} RDE rotation rate: 0 rpm Total potential scans: 10,000 cycles, performing the Basic Electrochemical Protocol after: 1000, 3000, 5000, and 10,000 cycles
0.1 M $HClO_4$ under O_2 saturation. Basic Electrochemical Protocol after 1000, 3000, 5000, and 10,000 cycles.	

Note: EIS: Electrochemical Impedance Spectroscopy. *OCP:* Open Circuit Potential. *RDE:* Rotating Disk Electrode. *RHE:* Reversible Hydrogen Electrode. *RRDE:* Rotating Ring-Disk Electrode.

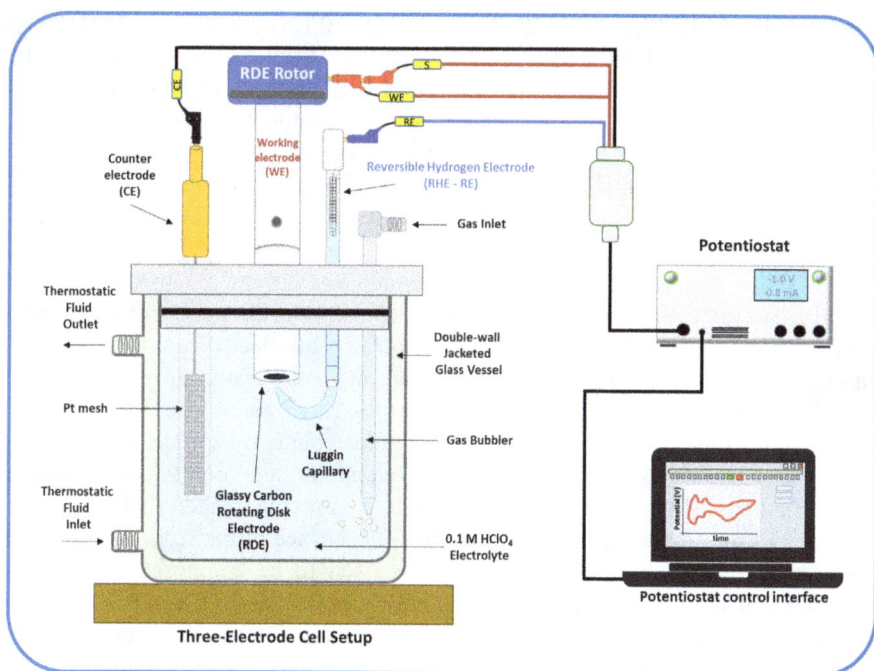

Figure 3.1   Basic configuration of a three-electrode cell for electrochemical measurements.

For best performance, cells usually include a double-walled glass container with a lid. The double wall or jacketing allows the recirculation of a thermostatic fluid to maintain the electrolyte temperature close to the standard room temperature (20°C). The cell as a whole or at least its inner chamber and submerged components must regularly undergo a deep cleaning process. For example, this can be chemical cleansing by placing the submersible components (usually overnight) in one of the following agents: (i) aqua regia (conc. HNO_3/HCl – 1:3 v/v ratio), (ii) sulfuric-nitric acid mixture (conc. HNO_3/H_2SO_4 – 1:1 v/v ratio), (iii) piranha solution (conc. $H_2SO_4/30$ wt.% H_2O_2 – 3:1 v/v ratio), (iv) hot conc. HNO_3 or H_2SO_4, (v) acidified potassium permanganate solution ($KMnO_4$), or (vi) Nochromix detergent solution. The cell and its components should then be rinsed, immersed, and boiled in deionized water (DI) (>18.2 MΩ·cm, TOC < 5 ppb) from a nano-pure water system or similar. This water rinsing process should be repeated between three and six times by changing the DI water each time. Before starting the experiments, the cell should be rinsed two or three times with the working electrolyte (Garsany et al., 2010; Shinozaki et al., 2015). At the conclusion of the experiments or while the cell and its components are not in use, storage should be done by placing the cell and submerged components in DI water, inside a container with a lid, to avoid contamination when exposed to air.

The electrochemical cell usually uses an acid solution as an electrolyte. A few years ago, sulfuric acid (H_2SO_4) was the most commonly used electrolyte, but its bisulfate ions (HSO_4^-) are strongly adsorbed on the active sites of the catalyst, causing a significant decrease in the catalytic activity (Markovic et al., 1997;

Wang et al., 2004; Wang et al., 2021). Currently, a 0.1 M perchloric acid ($HClO_4$) solution is a standard choice for electrolyte as the $HClO_4$ is not adsorbed on the catalyst, and it also simulates the important role of the widely used Nafion® ionomer employed in fuel cells as the conductive proton exchange membrane. It should be remembered that Pt nanoparticles are highly susceptible to '*poisoning*' by impurities (in the order of ppm-ppb), such as anions like sulfates, chlorides, or nitrates. Any trace of these ions being adsorbed on the Pt surface of catalysts can adversely affect the ORR catalytic activity. For this reason, it is advisable to use $HClO_4$ of high purity ($\geq 70\%$), preferably subjected to double distillation, with few traces—kept in the range of 0.1–10 ppm—of chloride ions, sulfate, nitrate, phosphate, etc. (Shinozaki et al., 2015). Naturally, all electrolyte solutions used in electrochemistry must be prepared with ultrapure DI water.

The electrochemical testing setup requires three electrodes. They are the working electrode, the reference electrode, and the counter-electrode. The counter-electrode (CE), also known as an auxiliary electrode, may consist of a graphite bar, a platinized Pt spiral wire, or a Pt mesh. It is recommended that the CE be at least 10 times greater in surface area than the surface of the working electrode (Westbroek, 2005). In general, the CE is chosen to be composed of an inert material that does not produce substances by electrolysis that could potentially reach the surface of the working electrode and promote interfering reactions (Bard and Faulkner, 2001; Zoski, 2007).

The reference electrode (RE) consists of a reversible redox electrochemical pair that facilitates the calculation of a known potential from the Nernst equation. The purpose of the RE is to provide a stable potential for the controlled regulation of the working electrode potential (Bard and Faulkner, 2001; Zoski, 2007). While the ideal RE is a non-polarizable electrode (i.e. its potential does not change regardless of the flow of current circulating through it), in reality, an actual RE does experience small electrical currents that can slightly modify its equilibrium potential. To be functional, the RE must have the ability to easily recover its equilibrium potential after experiencing a perturbation. A commonly used RE is the silver chloride-silver (AgCl/Ag) electrode in 3.5 M KCl as electrolyte support. Its equilibrium potential is +0.205 V *vs*. RHE. Another commonly used RE is the hydrogen reference electrode, which is probably the most widely used for ORR electrochemical characterization because it is not susceptible to generate contamination traces from anion leakage of an internal support electrolyte (for example, KCl in the AgCl/Ag). Hydrogen RE uses the same electrolyte ($HClO_4$) of the cell and its defined potential as 0.000 V *vs*. RHE enables the direct measurement of potentials on the standard electrode (SHE) scale or the reversible hydrogen (RHE) scale without the need for additional conversions. A simple way to build a hydrogen reference electrode in the laboratory (Eggen, 2009) is by sealing a platinized wire or Pt mesh inside a glass capillary filled with the working electrolyte solution and generating a hydrogen bubble by an electrolysis procedure that displaces part of the electrolyte. Note that part of the Pt mesh must be in contact with both the gas and the electrolyte to allow ionic contact. Traditionally, the RHE is attached to the electrochemical cell by means of a glass bridge piece which contains the same type of acidic electrolyte solution. The solutions between the bridge and the cell are separated by means of a porous membrane (Garsany et al., 2010; Zoski, 2007), as illustrated in Fig. 3.1. Additionally, a Luggin capillary must be placed such that its tip is as close as possible to the surface of the working electrode, thus reducing the electrical resistance effect of the solution. It is recommended to place the tip of the Luggin capillary in a position perpendicular to the surface of the working electrode while maintaining a separation between 5 and 10 mm. A distance of at least 2x on the outer diameter of the capillary tip should be sufficient to avoid the shielding effect that appears when the capillary is placed too close to the working electrode. This shielding effect results in a partial blockage of the electrical current path of the solution, causing non-uniform current densities on the surface of the working electrode (Bard and Faulkner, 2001; Van Der Vliet et al., 2010).

The working electrode (WE) is a rotating disk electrode (RDE) normally constructed from a cylinder of glassy carbon embedded in an insulating jacket, such as Teflon (PTFE), polyether-ether-ketone (PEEK), or an epoxy resin. When bi-potentiostat measurements are required, a rotating ring-disk electrode (RRDE) is used. An RRDE consists of a ring of an inert metal placed around the glassy carbon disk that allows the monitoring of secondary electrochemical processes by imposing a working potential different from that of the disk. These electrodes can be bought commercially or be custom made in the laboratory. The results present in this work were carried out using a custom-made RDE with a glassy carbon disk of 6 mm in diameter and a surface area of 0.283 cm^2.

During testing, the electrochemical reactions under study are carried out in the WE, so it is imperative that the surface of the RDE is well defined, smooth, and extremely clean. It is recommended that the surface of WE be inspected under an optical microscope to verify that the surface of the glassy carbon does not present considerable scratching. If significant scratching is observed, the surface of the RDE should be polished using fine-grained sandpaper (SiC-based) (sequential grit 1000, 3000, 5000, 7000, and 10,000) until the surface is as smooth as possible. Next, the RDE must be mechanically polished using 0.05 mm alumina-water slurry in a polishing cloth pad by making '*figure-eight motions*' until a mirror-like finish is achieved (Elgrishi et al., 2018). The surface of the RDE should then be rinsed in a DI water ultrasonic bath for about 30 seconds at least three times to ensure the removal of both sandpaper and alumina particles left over from the polishing process.

The glassy carbon surface is very reactive once it is mechanically activated by polishing, and can easily adsorb impurities; therefore, is recommendable to carry out a pretreatment or electrochemical activation step. To do this, the surface of the RDE should be immersed in the 0.1 M HClO$_4$ electrolyte under inert gas saturation, preferentially in an electrochemical cell specially designed for this pretreatment, and cyclic voltammetry scans should be carried out along a wide potential window (e.g. −0.05–1.3 *vs.* RHE) at different scan rates (e.g. 20, 50, 100, 200 mV s^{-1}) to eliminate any adsorbed species acquired during the polishing procedure. These sweeps should be repeated until the voltammetric profiles overlap consecutively and no redox peaks are observed. The characteristic voltammetric profile for a glassy carbon resembles a constant capacitive signal in the current range of 10^{-7} A without any redox peaks along the entire potential window. Finally, the electrode should be rinsed once more with ultrapure DI water and dried under inert gas flow.

Another important component in the electrochemical cell is the glass bubbler which supplies the flow of inert gas (e.g. nitrogen, carbon monoxide or oxygen, depending on the test to be developed) and remains in contact with the electrolyte. The bubbler maintains the gaseous saturation of the electrolyte and ensures a controlled atmosphere inside the electrochemical cell chamber. Care should be taken to prevent backflow of electrolyte into the bubbler, as this could lead to cross-contamination of chemical residues between experiments.

3.4   THE CATALYTIC INK

Figure 3.2 shows the preparation procedure of a catalytic ink. Preparation begins with the dispersion of a few milligrams of the catalytic powder in a solution usually formulated with DI water (74.6 v/v %), Nafion® 5 wt.% (0.4 v/v %) ionomer solution, and 2-propanol (25 v/v %). Several formulations of the dispersant solution can be found in the literature (Garsany et al., 2010; Paulus et al., 2001), but the most effective 'recipe' will depend on the chemical nature of the catalyst and its degree of hydrophobicity. Therefore, preliminary tests should be carried out to determine an optimal formulation for the deposition of catalytic films on the glassy carbon surface of the RDE. A good strategy is to start by slightly varying the alcohol: water ratio. Formulations with higher alcohol content are ideal for more hydrophobic materials (Garsany et al., 2010).

Once a recipe is obtained, the vial with the catalytic ink should be placed in a cold ultrasonic bath for 20 minutes to achieve homogenization and thorough dispersion of the catalyst. Subsequently, a volume of the catalytic ink is deposited directly on the polished and activated glassy carbon RDE using a micropipette. After the catalytic droplet is deposited, the solvents must be evaporated from the surface of the RDE to leave behind a thin catalytic film. One strategy to cast this film is the 'Rotational Air Drying' method (Kocha et al., 2017) that consists of placing the RDE on the inverted shaft of the RDE's rotor, depositing the catalytic ink on the glassy carbon surface of the RDE while it remains static, and activating rotation at 100 rpm (although static drying is also possible). Care should be taken that the ink is confined to the surface of the glassy carbon and that it doesn't spill onto the insulated rim of the electrode. To accelerate the drying, the rotational speed may be increased to 700 rpm for 15–20 minutes. The result should be a catalyst film with a uniform and well-distributed appearance.

The catalytic loading of Pt (L_{Pt}) on the surface of glassy carbon should preferably be in the range of 7–30 μg Pt cm^{-2} when using catalysts that have a Pt percentage content of 10–50 wt.% in the carbonaceous support material, usually Vulcan Carbon (Cabot Co.). The Pt catalytic load must be adjusted to produce a thin film on the surface of the glassy carbon. Thin films with a thickness in the order of 0.2 μm are ideal for adequate access to the electrochemical surface area of the catalyst because thicker films increase the resistance to mass transport through the catalytic layer (Garsany et al., 2010).

The bottom section of Fig. 3.2 shows optical micrographs for various catalytic ink depositions on the glassy carbon surface of the RDE; note that various 'qualities' of films can be produced. A 'bad film' is often characterized by the appearance of island-shaped accumulations of catalytic material which result in uneven film thickness. These accumulations often lead to exposed portions of glassy carbon surface and have a negative impact on catalytic activity. Additional polishing or increased amount of catalytic material cast on to the glassy carbon may aid in ensuring complete coverage of the glassy carbon substrate and improve film formation.

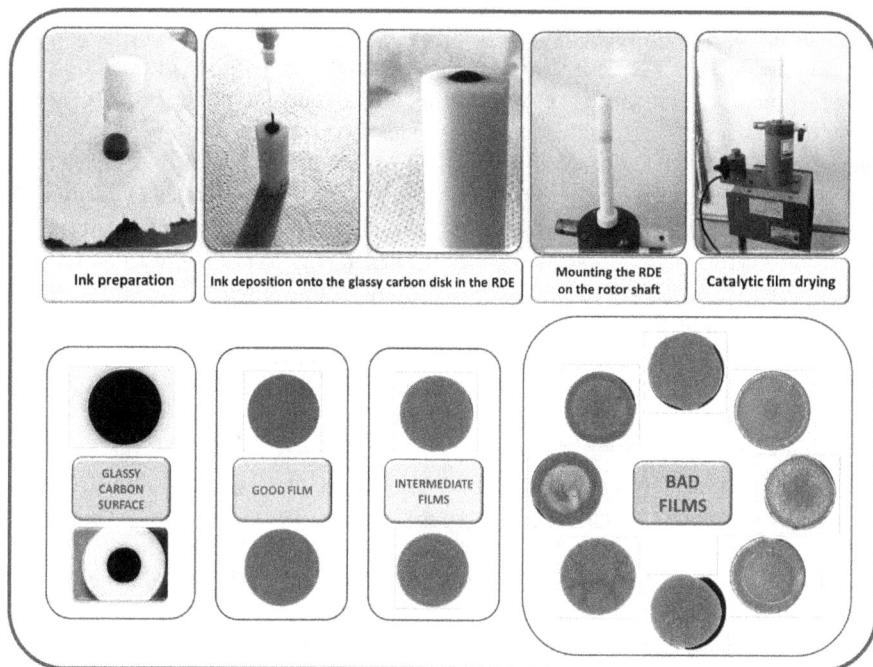

Figure 3.2 (Top) Preparation, deposition, and drying steps of the catalytic film on the glassy carbon surface of the RDE. (Bottom) Optical micrographs for different film qualities formed during the deposition of catalytic ink.

An 'intermediate-quality film' is characterized by having a more uniform film, although faint catalytic material accumulations or small areas of exposed glassy carbon remain. A 'good film' is characterized by a thin and uniform coating of the entire glassy carbon surface by the catalytic material (Garsany et al., 2010); no film spill-over on to the insulating rim should be present. Prior to performing any electrochemical testing, it is important to optimize all precursory steps, from the preparation of the dispersant solution, to ultrasonic dispersion time, and to the selection of an effective drying method.

Once the catalytic film has been successfully deposited on the RDE, it should be carefully moistened with a drop of DI water or the 0.1 M $HClO_4$ electrolyte, paying close attention to avoid scratching or detaching it from the glassy carbon surface. This is a useful step that helps to reduce particle pollution from the air environment (Shinozaki et al., 2015), as well as to prevent introducing interface bubbles between the film and electrolyte solution when submerging the electrode in the solution in the electrochemical cell.

The experimental data shown in this chapter corresponds to the electrochemical evaluation of a commercial catalyst ink composed of 20 wt.% Pt/C Etek with a L_{Pt} of 20.14 mg Pt cm_{geo}^{-2} on the glassy carbon surface of the RDE.

3.5   THE POTENTIOSTAT

Control over the potentials imposed on the WE during electrochemical measurements is carried out using a potentiostat operated either in potentiostatic mode (potential control) or galvanostatic mode (current control). In general, potentiostats have five connection cables clearly identified by labels or colors: one cable for the RE, another for the CE, another for the WE connected in series to the 'S' cable, and the ground cable. In an electrochemical system with three-electrode configuration—the most commonly used in electrochemistry— an electrical current flow is established between CE and WE. While difference in potential between CE and WE is controlled, the actual potential measurement is made between the S and RE electrodes (Bard and Faulkner, 2001)[4]. To ensure adequate measurements, a Luggin capillary is often used to reduce the separation between the RE and the WE; additionally, the WE and S cables must be always connected together. The potentiostat has a high impedance system at the input, so a negligible current can pass through the RE. This does not prevent the potential at the RE from remaining constant and equal to its open circuit value (OCP).

Generally, potentiostats always have a master power switch and a cell on/off switch to control electrical flow to electrodes. Although many manufacturers have systems that allow user manipulation of the electrical probes while the master switch is on, it is recommended that the cell power switch is turned off while setting up the experiment to prevent residual currents from injuring the user or damaging the WE. It is also convenient to connect the RE first, followed by the CE, and finally the WE to avoid any power spikes that could damage the film in WE. Just before starting the experiment, the cell power switch should be turned on to close the circuit.

3.6   RHE REFERENCE ELECTRODE PREPARATION

The RHE should be prepared by an electrolytic procedure using an external power supply or the potentiostat in a two-electrode configuration. In this scenario, the RE and the CE cables are connected together to the Pt mesh to act as an anode, and the WE and S cables are connected to the RHE's Pt wire or mesh to act as

---

[4]Additional information available from the Autolab potentiostat manufacturer notes at https://www.metrohm.com/en/applications/application-notes/autolab-applikationen-anautolab/an-ec-008.html

a cathode. A potential of –3 to –10 V is applied to the Pt wire in the RHE filled with the 0.1 M HClO$_4$ electrolyte solution to generate a hydrogen bubble inside the glass capillary (Eggen, 2009; Garsany et al., 2010). When sufficient hydrogen has been produced, the electrolysis reaction should be stopped to ensure that Pt wire or mesh remains in physical contact with both the hydrogen bubble and the acid solution.

3.6.1   Electrochemical Activation of the Catalyst

Once the RDE is immersed in the electrolyte and the RHE has been prepared, the activation or pre-conditioning of the catalyst by means of cyclic voltammetry scans is carried out as the first part of the experiment. The electrolyte must be saturated with an inert gas (purging for at least 10 minutes before the test) prior to measuring a voltammetric base profile (*I-E* plot). Potential scan cycles are performed over a window of 0.05–1.2 V *vs.* RHE in the anodic-cathodic scanning direction, with a scan rate of 50 or 100 mV s^{-1}, and without rotation of the RDE (Table 3.1). The immersion potential of the WE is set at 0.1 V *vs.* RHE to observe possible oxidation peaks during the first scan which, in the case of Pt-alloy catalysts, are attributed to oxidation of non-noble metals, or to the oxidation of surface organic matter in Pt-catalysts obtained by chemical synthesis using organic precursors.

Figure 3.3a shows a voltammetric profile for commercial 20 wt.% Pt/C Etek nanoparticles during their electrochemical preconditioning. The first cyclic sweep is characterized by having a small current response that gradually increases as more cycles are completed until a stable profile is reached. In this way, activation of the catalyst can be understood in terms of a tuning of the peaks associated to redox processes, and those associated to the absorption and desorption of hydrogen and oxygen. In general, as the number of potential scanning cycles increases, these peaks increase their current response and even shift a few millivolts until settling into a characteristic voltammetric profile.

The criterion for establishing the conclusion of the pre-conditioning process is a response overlap of consecutive scans. Generally, 30 to 50 cycles are necessary for pre-conditioning, but much depends on the nature of the catalyst.

Once the activation is finished, the electrolyte solution should be replaced. The electrochemical cell should be rinsed with DI water, followed by two or three rinses with fresh electrolyte. The surface of the RDE should also be carefully rinsed with fresh electrolyte solution. In an ideal setup, pre-conditioning should be carried out using a dedicated electrochemical cell and main electrochemical characterization should be performed in a separate cell.

3.6.2   OHMIC DROP COMPENSATION

The cyclic voltammetry profile constitutes an electrochemical footprint of a catalyst, and is related to redox processes, structural nature of the nanomaterials, chemical composition, particle size, etc. However, there are other electrochemical characterization factors that play a role in the voltammetric profile and include the working electrolyte, its conductivity, the presence of impurities, gas saturation, physical geometry of the cell, the shape of the electrodes, and the distance between them.

Every electrolytic solution will have an associated intrinsic resistance called *solution resistance* (R_s), defined as the opposition to the flow of current through the electrolyte between two electrodes. R_s depends on the ionic concentration of the electrolyte, type of ions, ionic mobility, and the temperature of the medium.

The solution resistance can be divided into two components, the *compensated resistance* (R_c) that is established between RE and CE, and the *uncompensated resistance* (R_u) that is measured between RE and WE (shown in inset of Fig. 3.3c). In a three-electrode cell configuration, most of the R_s (the R_u portion) can be compensated if the RE is placed as close as possible to the WE. However, in some cases where an organic or low-conductivity electrolyte is used, it may not be possible to completely account for the total R_s due to the high resistivity of the medium, regardless of the proximity between the RE and the WE. In real-life conditions, because REs are not infinitely small and cannot be placed infinitely close to the WE, absolute compensation for R_s is not possible. The portion of the total R_s that remains uncompensated (R_u) cannot be accounted by the potentiostat (Bard and Faulkner, 2001; Zoski, 2007; Van Der Vliet et al., 2010). Therefore, even with an optimal cell design and well-prepared electrodes, there will always be a contribution from R_u during electrochemical measurements that will cause the potential measured by the potentiostat to be different from the actual potential experienced by the analyte in the electrolyte. This phenomenon is known as Ohmic drop (iR_{drop}) and is associated to a potential difference, obtained by Ohm's Law, that is equal to the current (i) that passes through the resistance R_u (Elgrishi et al., 2018). If i or R_u are sufficiently large (e.g. $R_u > 10$ Ω), the resulting iR_{drop} can be very large and affect the accuracy of the measured data. The negative effects produced by a high iR_{drop} are clearly manifested in the visual shape of voltammetry profiles, for instance, in the form of shifts in peak positions or greater peak-to-peak separations in reversible redox processes (Elgrishi et al., 2018).

The iR_{drop} can be reduced by increasing the ionic concentration (conductivity) of the electrolyte, by decreasing the separation distance between the RE and the WE, by reducing the surface area of the WE, and by reducing the scanning rate of the measurements. In most cases, implementing a combination of these actions can sufficiently reduce iR_{drop} to the point where its impact on the electrochemical measurements becomes negligible (Bard and Faulkner, 2001; Elgrishi et al., 2018; Van Der Vliet et al., 2010). Nevertheless, there may be situations where suboptimal experimental and geometric conditions of the cell and electrodes require the determination of the R_u for active measurement compensation while the experiment is in progress. This active compensation can often be achieved by using potentiostat built-in routines that can approximate R_u to a certain level. The three most common routines for determining iR_{drop} using a potentiostat are Current Interrupt, Positive Feedback, and AC Impedance. Availability of these routines should be confirmed with the potentiostat's manufacturer.

R_u determination by AC Impedance consists in an Electrochemical Impedance Spectroscopy (EIS) measurement where a small electrical perturbation of AC sinusoidal amplitude (e.g. 10 mV) is applied to the cell, at a given potential, in a frequency range of 10 kHz–1 Hz (Bard and Faulkner, 2001; Shinozaki et al., 2015; Zoski, 2007). Great care must be taken in selecting the applied EIS potential to prevent electrochemical reactions or electrolytic processes from damaging the catalytic film in the WE. It is a good strategy to choose the open circuit potential (OCP) for the EIS measurement, as this is the equilibrium potential held by the WE in the electrolyte. Before determining iR_{drop} by EIS, the OCP must be measured and taken into account in the correction. Some potentiostats can measure OCP in real time by showing it on a digital display or in the form of pop-up notifications. Alternatively, it is possible to take a value for potential from the activation voltammetry in a region without intrinsic faradaic processes (associated with some redox reactions). For example, a potential taken in the capacitive region (Figs. 3.3a and 3.3d) of the voltammetric profile for Pt/C Etek nanoparticles (between 0.4–0.46 V *vs.* RHE) would be suitable for this purpose. The feasibility of this approach is dependent on the electrochemical response of the material.

Two characteristic plots are obtained from the EIS experiment: one is the Bode diagram (Fig. 3.3b) and the other is the Nyquist diagram (Fig. 3.3c). The Bode diagram is a representation of the impedance vector of a system that consists of a double-axis graph of impedance modulus (Z) and phase angle *vs.* frequency. During the EIS experiment, the small AC sinusoidal voltage applied produces a proportional AC sinusoidal current at the same frequency but with different amplitude and phase. When the impedance of the cell, at the applied voltage and frequency, is purely resistive in nature, a zero-phase angle between the voltage and the current will be obtained. For phase angles close to +90° there will be an inductor behavior, and for phase angles close to –90° there will be a capacitive behavior (Bard and Faulkner, 2001; Napporn et al., 2018).

The R_u of the system can be determined from the Bode diagram (Fig. 3.3b) as the extrapolated value of the Z module at high frequencies where the phase angle is close to zero. For the purposes of this example, the value obtained was 17.7 Ω.

On the other hand, the Nyquist diagram (Fig. 3.3c) is a representation in the complex plane of the real and imaginary components of the impedance module. Mathematically, it is equivalent to the Z vector (magnitude and phase angle) of the Bode diagram. In the Nyquist diagram, R_u corresponds to the real component of the impedance (Z′) at the point where the imaginary component of the impedance (Z″) is close to zero in the high-frequency zone to the left of the semicircle formed. In this example, the value obtained was also 17.7 Ω.

Once the R_u value has been obtained from any of its EIS representations, it must be applied to perform the Ohmic drop correction, also known as Ohmic compensation ($iR_{comp.}$). This compensation enables more precise results. Usually, only a fraction of the R_u should be considered in the correction, as an overcompensation can lead to *I-E* profiles with noisy oscillations. Therefore, it is recommended to consider a weighted iR_{comp} value between 75–95% of the R_u obtained. R_u values in the range 10–30 Ω are typical for the entire electrochemical cell and RDE setup used in electrochemical characterization systems towards the ORR; the lower the value, the better the results that can be obtained. For some systems with a low R_u value ($< 10\ \Omega$) where the *I-E* curves show currents in the range of µA scale, the magnitude of a iR_{comp} is in the order of a few µV, so the iR_{drop} is usually negligible (Bard and Faulkner, 2001).

The rest of the measurements in this chapter were calculated correcting for iR_{comp}. It is a good practice to measure R_u under similar conditions between consecutive experiments, for example, under inert gas saturation conditions and zero RDE rotation before obtaining the final *I-E* profiles, or when testing under oxygen saturation conditions and RDE rotation to facilitate calculation of the polarization curves in the steady-state regime.

3.6.3 Cyclic Voltammetry with iR_{comp} and Background Compensation for Capacitive Correction

Continuing with the electrochemical characterization protocol, cyclic voltammetry measurements should be carried out under the same conditions as those used for pre-conditioning; that is, in a potential window of 0.05–1.2 V *vs.* RHE, in the anodic-cathodic scanning direction, and with a scan rate of 20 or 50 mV s^{-1}. Fresh 0.1 M HClO$_4$ electrolyte saturated with inert gas and no RDE rotation should be used.

In this case, the difference lies in the fact that iR_{comp} is applied by considering 93% of the previously determined R_u value. Capturing 3–5 voltammetry cycles or measuring until overlapping consecutive scans is observed is recommended. Figure 3.3d shows an iR-compensated voltammetry profile for the Pt/C Etek catalyst film. Characteristic peaks and regions for this material are highlighted: (i) the anodic and cathodic peaks between 0.05–0.4 V *vs.* RHE corresponding to adsorption/desorption of Pt-hydrogen species; note that the cathodic peak around 0.11 V *vs.* RHE and the small shoulder between 0.20–0.30 V *vs.* RHE are mainly associated to the adsorption of hydrogen atoms at the respective (110) and (100) sites of polycrystalline Pt nanoparticles (Clavilier et al., 1996; Maillard et al., 2005); (ii) the capacitive region (0.4 V *vs.* RHE < E < 0.6 V *vs.* RHE) where the double-layer charging process is carried out; and (iii) the region of formation/reduction of platinum oxides (E > 0.7 V *vs.* RHE) where the wide anodic peak around 0.75–0.9 V *vs.* RHE is associated to Pt surface oxidation to form Pt-O and Pt-OH species. Note that the large cathodic peak between 0.6–0.9 V *vs.* RHE and minimum current close to 0.76 V *vs.* RHE is associated to the reduction process of oxidized species previously formed on the Pt. Qualitatively, the displacement of this reduction peak towards more positive potentials (> 0.76 V *vs.* RHE) is an indicator that the material under study is a good candidate towards ORR, when compared to a reference material like Pt/C Etek, because it requires a lower overpotential to carry out the reductive process (Schmidt et al., 1998; Paulus et al., 2002; Arenz et al., 2005; Mayrhofer et al., 2008; Garsany et al., 2010).

Since the electrochemical cell is under inert gas saturation, the next step is to measure the background (b.g.) *I-E* profile that will later aid in the correction of the capacitive effects for the ORR polarization curves in the steady-state regime when the system is saturated with oxygen, as addressed later in this chapter. Measurement of b.g. is done under similar conditions to those of ORR polarization curves, over a potential window of 1.05–0.05 V *vs.* RHE in the cathodic-anodic scanning direction, under inert gas saturation, iR_{comp} correction, and RDE rotation (400, 900, 1600, and 2500 rpm). It is recommended to capture at least three complete scans of the *I-E* profile or to capture until consecutive overlapping scans are observed.

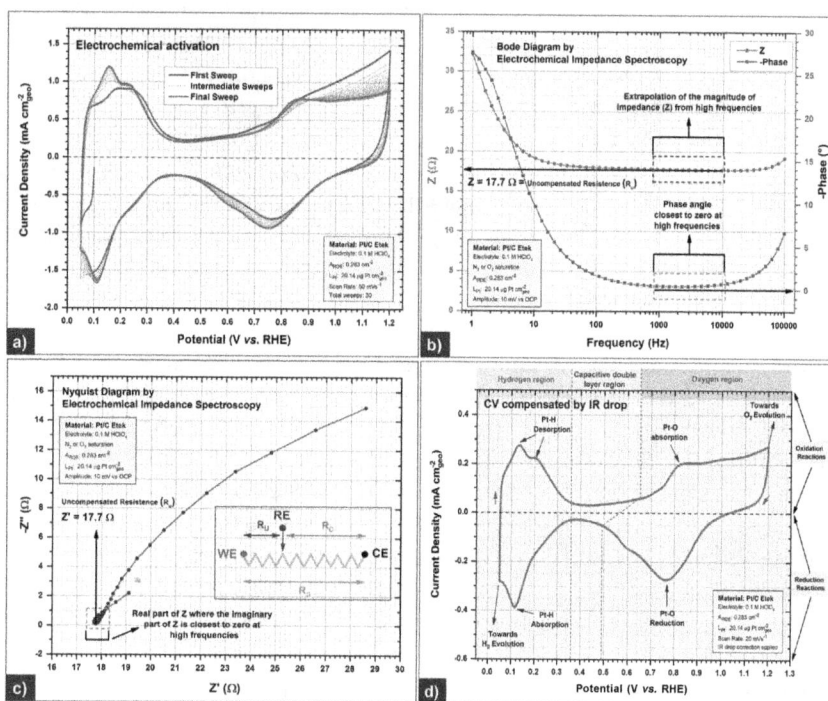

Figure 3.3 (a) Cyclic voltammetry obtained during the 'activation' or pre-conditioning process of a commercial 20 wt.% Pt/C Etek catalyst in a three-electrode cell. (b) Bode diagram obtained by Electrochemical Impedance Spectroscopy (EIS) used for the determination of R_u in the electrochemical cell. (c) Nyquist diagram obtained by EIS for the determination of R_u in the complex space. (d) iR_{comp}-corrected cyclic voltammetry for a commercial 20 wt.% Pt/C Etek catalyst where characteristic peaks and regions of this Pt catalyst are highlighted.

3.6.4 Carbon Monoxide Electro-oxidation (CO Stripping)

The determination of the electrochemical surface area (ECSA) is a very important parameter to calculate to evaluate the electrochemical activity towards ORR. Usually, ECSA is obtained by integrating the area under the *I-E* voltammogram curve (charge) under inert gas saturation in the region between 0.05–0.36 V *vs.* RHE, known as the Hydrogen Under-Potential Deposition (H_{UPD}). This is the adsorption/desorption region where hydrogen atoms are present on the Pt atoms in the catalyst. A specific charge of 210 μC $cm^{-2}{}_{Pt}$ is applied as a conversion factor that corresponds to the charge required for the formation of a monolayer of H atoms on the Pt surface. The determination of the ECSA via H_{UPD} is usually calculated to the degree of a close approximation by considering the limits of the H region bounded by horizontal straight lines that delimit the anodic and cathodic capacitive currents of the *I-E* profile.

One way to corroborate and obtain a more accurate ECSA value is by applying a carbon monoxide (CO) electro-oxidation approach (Maillard et al., 2004; Mayrhofer et al., 2005; Brimaud et al., 2008; Ochal et al., 2011). This approach requires fresh electrolyte in the cell to be saturated under a CO flow for 300–600 seconds while keeping an RDE potential of 0.1 V *vs.* RHE. Under these conditions, the CO molecules are adsorbed on the surface atoms of the active metal (Pt), causing their *'poisoning'*. Next, the electrolyte is purged with an intense flow of inert gas to evacuate the remaining CO that was not absorbed on the catalyst. Usually, the inert gas purge time can set to twice the time used for CO saturation. Afterward, cyclic voltammetry scanning (three cycles) is performed, over a potential window of 0.05–1.2 V *vs.* RHE in the anodic-cathodic direction, with a scan rate of 20 mV s^{-1}, and considering iR_{comp} corrections. Figure 3.4a shows the CO stripping profile of the Pt-Etek/C catalyst. Note that the characteristic anodic peaks for the desorption of Pt-H at low potentials (0.05–0.3 V *vs.* RHE) do not appear in the first scan due to the formation of a baseline close to 0 A. Typically, the characteristic peaks corresponding to CO oxidation to form CO_2 appear in the region 0.64–0.9 V *vs.* RHE. For the rest of the cathodic sweep and subsequent cyclic scans, the profile returns to the characteristic voltammetry for these Pt nanoparticles. When testing, if any oxidation peaks reappear during the second scan, the test must be repeated and care must be taken to ensure the complete purge of the electrolyte with the inert gas to make sure that all remaining CO is fully evacuated.

For ECSA calculations, the CO stripping *I-E* profiles should use units of *ampere* and *volt* for current intensity and potential, respectively. Subtracting the second anodic cycle from the first cycle gives the CO oxidation peaks without capacitive effects. To calculate the CO charge, it is necessary to integrate the area under curve for the CO peaks ($\int i\ dE$) and to divide the resulting value by twice the scanning rate (v) used during CO-stripping (usually 20 mV s^{-1}). Multiplying by conversion factors, the CO charge (Q_{CO}) can be obtained in units of μC as per Eq. (4).

$$Q_{CO} = \frac{\int i\ dE}{v} \tag{4}$$

The baseline obtained from the first CO stripping scan can also be used as a reference to determine the hydrogen charge (Q_H) in the H_{UPD} region more accurately than by using the graphical horizontal straight lines approach, as shown in Fig. 3.4a. A similar mathematical integration treatment is applied as per Eq. (5):

$$Q_H = \frac{\int i\, dE}{v} \tag{5}$$

The difference between both equations lies in the number of electrons transferred during the oxidative process. In the first case, a one-electron transfer is necessary to achieve the desorption of each hydrogen adsorbed on the catalyst; in contrast, in the oxidation of CO to CO_2, a two-electron transfer is required. Therefore, Q_{CO} will be half of Q_H (Mayrhofer et al., 2008).

Next, Q_{CO} or Q_H is divided by a factor of 195 μC cm^{-2}$_{Pt}$ corresponding to the adsorption charge for the formation of a monolayer of H atoms on the metal surface of polycrystalline Pt electrodes (Mayrhofer et al., 2008), as per Eqs. (6) and (7). Thus, the surface area of the catalyst may be calculated in cm^2 from the adsorption of CO (S_{CO}) or H_{UPD} (S_H).

$$S_{CO} = \frac{Q_{CO}}{195\, \dfrac{\mu C}{cm^2}} \tag{6}$$

$$S_H = \frac{Q_H}{195\, \dfrac{\mu C}{cm^2}} \tag{7}$$

Finally, the ECSA of the catalyst is determined by dividing the S_{CO} or S_H by the product of the catalytic Pt load (L_{Pt} in μg Pt cm^{-2}$_{geo}$) in the RDE and multiplying by the geometric area of the RDE (A_{geo} in cm^2), as per Eq. 8.

$$ECSA = \frac{S_{CO} \text{ or } S_H}{L_{Pt} \times A_{geo}} \tag{8}$$

3.6.5   Steady-State Polarization Curves towards ORR

Measurements of steady-state polarization curves to evaluate catalytic activity towards ORR are performed under oxygen saturation conditions of the 0.1 M $HClO_4$ electrolyte. A cyclic voltammetry sweep at 20 mV s^{-1} is used over a potential window of 1.05–0.05 V *vs.* RHE in cathodic-anodic direction, with iR_{comp} correction, and RDE or RRDE rotational rates set to 400, 900, 1600, and 2500 rpm. Figure 3.4b shows the *I-E* curves obtained for the 20 wt.% Pt/C Etek catalyst under discussion.

For the catalytic activity calculations, the previously obtained *I-E* background profile (b.g.) captured under inert gas saturation conditions must be subtracted from the ORR polarization curves to suppress the capacitive effects of the material. The dotted lines in Fig. 3.4b show the original *I-E* measured; note that small capacitive effect and slightly higher current densities in the hydrogen region (0.05–0.3 V *vs.* RHE) are still appreciated. Also shown at the top of Fig. 3.4b is the voltammetry profile of the background curve, obtained under identical sweeping conditions as the ORR polarization window, but measured under inert gas saturation of the electrolyte. The continuous-line *I-E* plots in this figure were drawn after performing the capacitive subtraction and correspond to the various rotational speeds indicated. By convention, only the anodic sweep should be captured and presented for comparison purposes, since this sweep tends to shift towards positive potentials.

ORR is an electrochemical reaction associated to the electrodic kinetics in the electrode-electrolyte interface where charge transfer and mass transfer processes occur. Polarization curves are classified in three regions, as shown in Fig. 3.4b.

Region I describes the charge transfer process that occurs in the immediate vicinity of the electrode surface. The electrochemical reactions in the electrodes occur either by oxidation processes where electrons are lost, or by reduction processes where electrons are gained. The general way to represent an electrochemical reaction is:

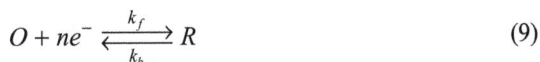

$$O + ne^- \underset{k_b}{\overset{k_f}{\rightleftharpoons}} R \tag{9}$$

where O refers to an oxidized species, ne^- is the number of moles of electrons transferred, R is a reduced species, k_f is the rate constant in the forward direction (cathodic process) of the reaction, and k_b is the backward rate constant (anodic process).

The current intensity magnitude (i) obtained in each direction of reaction is given in terms of the number of moles of electrons transferred times the Faraday constant (F), and is equal to 96,485 C mol^{-1} multiplied by the surface area of the electrode (A) times the reaction rate (r) of the respective cathodic or anodic process, as shown in Eqs. (10) and (11):

$$i_c = nFAr_f \tag{10}$$

$$i_a = nFAr_b \tag{11}$$

where i_c is the cathodic current and i_a is the anodic current. The reaction rates (r) for each direction are given by the respective rate constant and the corresponding concentration of the oxidized species (C_O) or the reduced species (C_R) (Bard and Faulkner, 2001; Zoski, 2007):

$$r_f = k_f C_O \tag{12}$$

$$r_b = k_b C_R \tag{13}$$

The net reaction rate is calculated by:

$$r_{net} = r_f r_b \tag{14}$$

Therefore, the total current in the electrode is:

$$i = i_c - i_a = nFA[k_f C_O - k_b C_R] \tag{15}$$

The Faradic current (associated with redox processes) involves mass transport and charge transport processes. Each process limits the overall reaction rate depending on its relative kinetic speed. For example, when a potential close to the equilibrium potential is applied to the electrode, the charge transfer is slow, and it drives the overall reaction speed. This behavior is known as 'activation control' because the reaction rate is governed by an activation energy (Region I of the ORR curve in Fig. 3.4b). On the other hand, as the potential decreases and becomes more negative (increasing in overpotential), the charge transfer is very fast, and the process is now limited by mass transport (Region III). Therefore, the generated current does not depend on the potential and is called the *limit current* or *limit current density* when it is expressed as a function of the unit area. Region II or *mixed* control region involves both processes.

The equation describing the controlled global rate for the charge transfer in Region I, for a reductive process (e.g. ORR), can be derived from Eqs. (10) and (12), and is given by Eq. (16):

$$i_k = nFAk_f C_O^*$$

(16)

where i_k is the kinetic current independent of the rotational rate, k_f is the reaction rate constant for forward (cathodic) charge transfer, and C_O^* is the concentration of oxygen in the bulk electrolyte. Equation (16) can be rewritten in terms of kinetic current density (J_k) by dividing i_k over the electrode area:

$$J_k = nFk_f C_O^*$$

(17)

Charge transfer can only be carried out over very short distances, in the order of 0.5 nm; therefore, a mass transport process is required to allow the reactive species move close to the surface of the electrode where the reactions occur. This process is achieved by rotating the RDE or RRDE at high revolutions in the electrolyte. The rotation produces a perpendicular convective movement that physically draws the reactant species to the surface of the electrode (steady-state). While rotation results in a very uniform concentration of all reactive species in the solution, there is still a critical distance about the surface of electrode where no physical motion of the electrolyte occurs. There, mass transport occurs by a diffusion mechanism (Bard and Faulkner, 2001; Zoski, 2007).

The equation that describes the steady-state process for Region III is:

$$i_d = 0.2nFAD^{2/3} v^{-1/6} C_O^* \omega^{1/2}$$

(18)

Equation (18) is known as the Levich equation, and it is applicable when the charge transfer is so fast that the overall reaction rate is determined primarily by mass transport. Note that i_d is the diffusion limit current, 0.2 is a constant used when the rotational speed of the electrode is expressed in rpm (if expressing in terms of angular frequency, the constant would be 0.62 rad s^{-1}), $n = 4$ is the number of electrons that are transferred by each diatomic molecule of oxygen during reduction, F is Faraday's constant, A is the geometric area of the electrode in cm^2, D is the diffusion coefficient of oxygen in the electrolyte (1.7 × 10^{-5} cm^2 s^{-1} in 0.1 M HClO$_4$), v is the kinematic viscosity of the electrolyte (1.01 × 10^{-2} cm^2 s^{-1} in 0.1 M HClO$_4$), C_O^* is the concentration of dissolved oxygen in the bulk electrolyte (1.26 × 10^{-6} mol cm^{-3} at a pressure of 1 atm), and ω is the rotational rate of the electrode. (Paulus et al., 2001; Schmidt et al., 2001; Yang et al., 2012). In terms of current density (J), Eq. (18) can be expressed as:

$$J_d = 0.2nFAD^{2/3}\,v^{-1/6}C_O^*\,\omega^{1/2} \tag{19}$$

The current density equation that describes the behavior of the polarization curve over the two dominant contributions (Regions I and III) is:

$$\frac{1}{J} = \frac{1}{J_k} + \frac{1}{J_d} \tag{20}$$

Substituting Eq. (19) in Eq. (20) gives:

$$\frac{1}{J} = \frac{1}{J_k} + \frac{1}{0.2nFD^{2/3}v^{-1/6}C_O^*\omega^{1/2}} = \frac{1}{J_k} + \frac{1}{B\omega^{1/2}} \tag{21}$$

All constant terms for J_d in Eq. (19) can be combined for a new expression in terms of B, known as the *Koutecky–Levich slope*. This way, Eq. (21) may be rewritten so that it only depends on the RDE rotational rate.

Substituting the constant values in Eq. (19) returns a value for B of 13.83×10^{-2} mA cm^{-2} rpm$^{-0.5}$ for a four-electron ORR transfer process (Reyes-Rodríguez et al., 2019). Multiplying by the B constant and substituting the squared root of the RDE rotational speed (ω) at 400, 900, 1600, and 2500 rpm in Eq. (19) gives the theoretical J_d values as shown in Table 3.3. Depending on the diameter and surface area of the glassy carbon used in the RDE or RRDE, the expected theoretical i_d for each rotational rate can be calculated.

Table 3.3 Expected theoretical values for J_d and i_d as a function of RDE rotational rate for the 4e$^-$ ORR transfer process

Rpm	J_d (mA cm^{-2})	i_d (A) for EDR with $\phi_{\text{glassy carbon}}$: 5 mm A_{geo}: 0.196 cm^2	i_d (A) for EDR with $\phi_{\text{glassy carbon}}$: 6 mm A_{geo}: 0.283 cm^2
400	2.77	0.000542	0.000783
900	4.15	0.000813	0.001174
1600	5.53	0.001084	0.001566
2500	6.92	0.001356	0.001957

In Eqs. (18) and (19), the C_O^* term associated to the concentration of dissolved oxygen in the bulk electrolyte depends on the saturation pressure of the gas, and this in turn depends on the atmospheric pressure and the altitude of the location where the experiments are carried out. In other words, the concentration of dissolved oxygen in the electrolyte decreases with altitude. A correction for the concentration of dissolved oxygen in the electrolyte can be made by considering the following expression for atmospheric pressure (Perry and Denuault, 2015):

$$C_{Oaq}^* = \frac{C_{Oaq\,@\,760\,\text{mm Hg}}^*\,(P_{bar} - P_{vap})}{760\,\text{mm Hg} - P_{vap}} \tag{22}$$

where C_{Oaq}^* is the concentration of dissolved oxygen in the bulk electrolyte, $C_{Oaq\,@\,760\,\text{mmHg}}^*$ is the concentration of dissolved oxygen in the saturated electrolyte at a pressure of 760 mmHg, P_{bar} is the barometric pressure, and P_{vap} is the vapor pressure of water at 298 K (whose value is 23.76 mmHg).

For Pt-Etek/C nanocatalyst, the reported value in the literature for limit current density (J_d) at 1600 rpm that takes into account the saturation pressure of dissolved oxygen in the 0.1 M HClO$_4$ electrolyte at 100 kPa is 5.7 mA cm^{-2}. ORR experiments conducted in any lab around the world should take this factor into consideration and make the appropriate corrections. For instance, at an altitude of 2,300 meters above sea level (e.g. a lab in Mexico City), electrochemical oxygen reduction experiments would have a B slope of 11.54 × 10^{-2} mA cm^{-2} rpm$^{-0.5}$ due to the low oxygen saturation pressure at this high elevation over sea level. Therefore, even if measuring the same Pt-Etek/C catalyst under identical conditions, a J_d value of only 4.62 mA cm^{-2} would be obtained (Reyes-Rodríguez et al., 2019). This represents a 19% decrease in J_d and a decrease in overall catalytic activity with respect to the maximum theoretically expected J_d value of 5.7 mA cm^{-2} under oxygen saturation conditions at altitudes close to sea level (Paulus et al., 2001).

To standardize and report comparable results for catalytic activity, it is important to always account for partial oxygen pressure when correcting the limit current. In other words, this means correcting the kinetic current by normalizing the data to an oxygen saturation pressure of 100 kPa[5] (Gasteiger et al., 2005; Shinozaki et al., 2015; Xing et al., 2014).

ORR parameters can be obtained graphically by taking the inverse of the total current density ($1/J$) *vs.* the inverse of the square root of rotational rate of the RDE ($1/\omega^{1/2}$) at a potential of 0.9 V *vs.* RHE, as shown in Fig. 3.4d. The intercept to the Y axis of the graph will correspond to the kinetic current density J_k, as per Eq. (21). It is also possible to calculate J_k by separating this term from Eq. (20); remembering that J_d is the limit current and that J corresponds to the measured current density. This approach is a useful way to apply the mass transport correction (Mayrhofer et al., 2008).

Dividing J_k by the active area of Pt (S_{CO}) obtained from CO stripping (Eq. 6) yields the specific current density of the catalyst (SA or J_k) in units of mA cm^{-2} $_{Pt}$, as shown in Eq. (23).

$$J_k = i_k \times \frac{1}{S_{CO}} = \frac{J \times J_d}{J_d - J} \times \frac{A_{geo}}{S_{CO}} \qquad (23)$$

Finally, to determine the mass activity (MA or J_m) in units of A mg^{-1}, the values for the S_{CO} area, Pt catalytic loading L_{Pt}, and the geometric area A_{geo} of the RDE are plugged into Eq. (24) (Garsany et al., 2010; Mayrhofer et al., 2008; Schmidt et al., 2001)

$$J_m = J_k \times \frac{S_{CO}}{L_{Pt} A_{geo}} \qquad (24)$$

---

[5]More information available from the Hydrogen and Fuel Cell Technologies Office Webinars from U.S. Department of Energy at: https://www.energy.gov/eere/fuelcells/hydrogen-and-fuel-cell-technologies-office-webinars

Plotting mass transport-corrected J_k values (Eq. 23) in a logarithmic scale *vs.* potential results in a Tafel curve, as shown in the inset of Fig. 3.4c. If the J_k values are plotted in the potential range of 0.75–1.0 V *vs.* RHE, the Tafel slope can be determined.

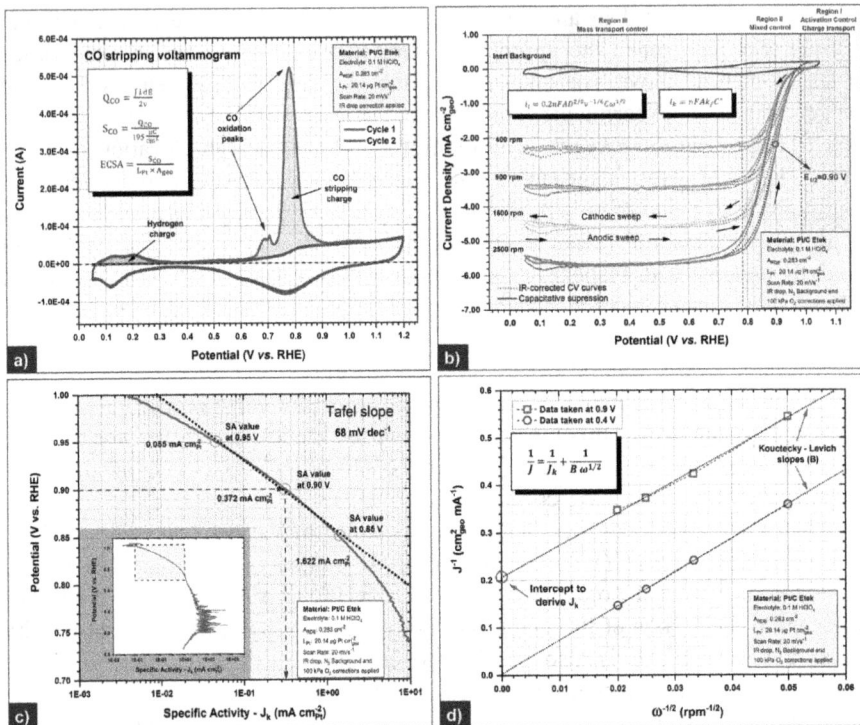

Figure 3.4 (a) CO stripping measurement of a commercial 20 wt.% Pt/C Etek catalyst. (b) Steady-state ORR polarization curves corrected (solid lines) and not corrected (dotted lines) for capacitive effects using the *Background I-E* curve. (c) Tafel slope obtained in the range 0.80–1.0 V *vs.* RHE using mass-transport corrected J_k data. Inset: ORR logarithmic representation of Tafel curve using mass-transport corrected J_k data for the potential interval measured. (d) Koutecky-Levich interpolation obtained from different RDE rotation rates during ORR; data comparison was taken at 0.9 V and 0.4 V *vs.* RHE.

According to the literature, the Tafel slope is related to the interaction of oxygen species adsorbed on the surface of the active metal (Pt) that carry out ORR (Wang et al., 2004). For Pt-based catalysts, it is possible to observe two characteristic Tafel slopes. The first one occurs at a low overpotential (E > 0.85 V *vs.* RHE) and is calculated as RT/F (approximate value of −60 mV dec⁻¹). This Tafel slope is associated to a pseudo 2e⁻ transfer limiting step where an oxygen reduction process that takes place on the Pt surface covered by oxidized species. The second Tafel slope is found at higher overpotentials (E < 0.80 V *vs.* RHE) and is calculated as 2RT/F (approximate value of 120 mV dec⁻¹). In this second case, the Tafel slope is associated to a reduction process that takes place on an oxide-free Pt surface, where the transfer of the first electron is the reaction's determining step (Teran-Salgado et al., 2019).

In Fig. 3.4c, the Tafel slope is calculated as 68 mV dec^{-1} for the 0.85–1.0 V *vs.* RHE interval and is in agreement with the expected theoretical value of 60 mV dec^{-1}. For ORR, this suggests that the catalyst carries out a two-electron transfer process on its oxidized surface. Finally, Fig. 3.4c shows the J_k values for potential values of 0.85, 0.90 and 0.95 V *vs.* RHE, commonly reported when comparing activity to other materials. Generally, the most reported reference value is calculated at 0.9 V *vs.* RHE.

3.6.6 Determination of H_2O_2 by RRDE

The next measurements form part of the *complementary protocols* for ORR electrochemical characterization (Table 3.2). It is known that the oxygen reduction reaction can lead to the formation of hydrogen peroxide (H_2O_2) if the electron transfer is not carried out by a four-electron pathway. The RRDE *(rotating ring-disk electrode)* technique makes it possible to monitor and quantify the amount of H_2O_2 formed during ORR process.

In this technique, a cyclic voltammetry of the catalyst deposited in the glassy carbon of the RRDE must be performed using the Bi-potentiostat measurement mode. This configuration independently controls the potentials applied to the ring and the disk of the electrode. During measurement, the rotation of the electrode generates convection movements of the species in the solution (oxygen molecules), causing them to move toward the surface of the disk. At the surface, they are chemically reduced by the cathodic potential sweep, producing water and some H_2O_2. The products generated in the disk are tangentially driven to the electrode ring and are oxidized due to the independent potential applied. The RRDE configuration maintains both the electrode disk and the collection ring in the same bi-dimensional plane; however, not all species generated on the disk reach the surface of the ring. Therefore, a collection factor is required to determine the collection efficiency of the electrode. The collection factor is identified as N and is obtained from the ratio of ring current (I_R) to disk current (I_D).

$$N = -\frac{I_R}{I_D} \tag{25}$$

The collection factor N is calculated from the electrical currents obtained from the *I-E* polarization curves at the steady-state regime. Different rotational rates and an aqueous solution composed of 10 mmol of K_3Fe (CN)$_6$ in 0.1 M of $HClO_4$ as electrolyte (Fig. 3.5a) are used. During the analysis, a redox pair $Fe^{3+}(CN)_6^{3-}/Fe^{2+}(CN)_6^{4-}$ is established. This is a 1-electron transfer reversible reaction (Paulus et al., 2001). In the disk, Fe^{3+} is reduced to form Fe^{2+} species that are themselves collected by the ring and oxidized back to Fe^{3+} by the 1.5 V *vs.* RHE potential applied. A typical collection factor has a magnitude of 0.2–0.3.

After the polarization curves in the potassium ferricyanide electrolyte have been obtained, the electrolyte must be replaced by fresh 0.1 M $HClO_4$.

Next, polarization curves should be captured at different rotational rates in the bi-potentiostat mode at a ring potential set to 1.2 V *vs.* RHE (Fig. 3.5b) (Paulus et al., 2001; Reyes-Rodríguez et al., 2013). It is worth mentioning that these measurements should be performed using in the same potential window and system conditions as those used to measure ORR in the RDE configuration. The iR_{comp} correction, capacitive subtraction, and partial oxygen correction should also be applied.

Two related plots are shown in Fig. 3.5b. The upper plot shows the H_2O_2 oxidation process carry out on the ring of electrode to form oxygen, and the bottom plot shows the corresponding ORR cathodic sweep carry out in the disk in the RRDE configuration.

Once the electrical currents of the disk and the ring are known, it is possible to plug in their values into Eq. (26) to determine the amount of H_2O_2 generated during the oxygen reduction process. Note in this particular case, the collection factor shown in the inset of Fig. 3.5b has been taken into account.

$$X_{H_2O_2} = \frac{2I_R/N}{I_D + I_R/N} \tag{26}$$

The generation of a small amount of hydrogen peroxide (<5%) is a good indicator that the ORR reaction kinetics mainly occur due to a four-electron transfer toward water formation, and this analysis is in agreement with the Koutecky–Levich slope calculations.

3.7   STABILITY TESTING

Stability tests should be performed to determine how well a candidate material maintains its catalytic activity over time; in other words, stability testing is a window into the material's degradation mechanisms and durability. One type of stability testing is accelerated electrochemical degradation testing (AEDT) (Colón-Mercado et al., 2004; Li et al., 2012; Kim et al., 2016; Wang et al., 2016). AEDT consists of performing 10,000–30,000 cyclic voltammetry scans at a scan rate of 100 mV s^{-1}, in a potential window of 0.6–1.0 V *vs.* RHE, under electrolyte oxygen saturation conditions. This is performed to emulate the operational and degradation conditions of the catalytic material while maintaining an oxidative environment similar to that of the cathode in a fuel cell.

After a certain number of consecutive potential scans, for example: 1000, 3000, 5000, and 10,000 cycles, the electrolyte must be replaced by fresh solution and the entire *Basic Electrochemical Characterization Protocol* (Table 3.1) described in the first section of this chapter should be carried out. As a reminder, this procedure consists of measuring inert gas voltammetry, CO stripping, ORR background, and steady-state ORR polarization curves. In all measurements, the pertinent corrections must be made (iR_{comp}, capacitive subtraction, and normalization by partial oxygen pressure). Figure 3.5c shows the anodic scans of the steady-state ORR polarization curves, taken under oxygen saturation at 1600 rpm after a certain number of cyclic scans as part of an AEDT test. The measured profiles are very similar to each other, although a small shift towards more negative potentials (larger overpotentials) is visible. This shift is attributed to the chemical degradation of the material and the slow erosion of the deposited catalytic film. The Tafel slope plot (inset Fig. 3.5c) presents a magnified view of this degradation after 10,000 cycles for J_k values at 0.90 V *vs.* RHE. The slight degree of change in the Tafel curves is indicative of a very stable catalyst.

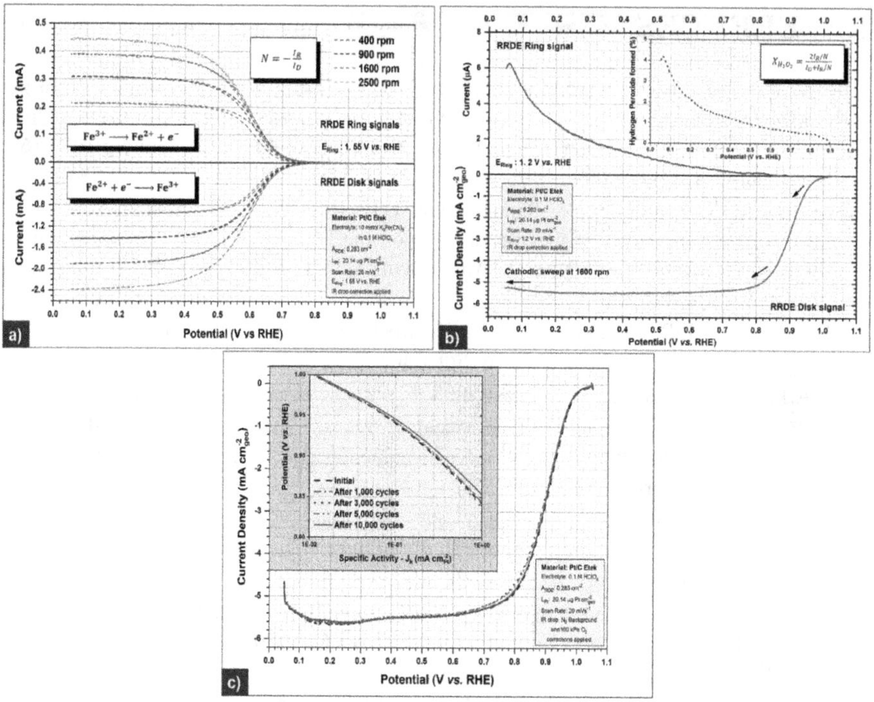

Figure 3.5 (a) RRDE steady-state polarization curves for a potassium ferricyanide redox couple measured to calculate the collection factor, *N*. (b) RRDE steady-state cathodic polarization curve at 1600 rpm taken during chemical reduction on the disk. Oxidation of species is carried out at the ring of the electrode to quantify the amount of hydrogen peroxide formed during the ORR process. Inset shows the percentage of H_2O_2 formed as a function of the applied potential. (c) Anodic ORR steady-state polarization curves at 1600 rpm obtained after several cyclic scans to illustrate the benefit of using accelerated electrochemical degradation testing to monitor the stability of the catalytic material over time.

After stability testing is completed, the general electrochemical characterization of the catalytic material is concluded. The protocol presented in this chapter is a powerful approach that yields conclusive information to determine if a catalyst candidate is viable for use as a cathode in proton exchange membrane fuel cells. The *SA*, *MA*, and *ECSA* parameters, the Tafel slope, or the percent of generated H_2O_2 may also be used to compare the performance of the material under evaluation to other top-performing reference catalysis. This protocol is a snapshot of existing state of the art electrochemical characterization approaches and is meant to serve as a guide to help researchers achieve greater comparability between data taken in the same lab or in different labs across the world. This characterization protocol is meant to be a living guide that researchers can complement with their own novel techniques to push the boundaries of ORR materials discovery.

REFERENCES

Antolini, E., J.R.C. Salgado, R.M. da Silva and E.R. Gonzalez. 2007. Preparation of carbon supported binary Pt–M alloy catalysts (M=first row transition metals) by low/medium temperature methods. Mater. Chem. Phys. 101(2–3): 395–403. https://doi.org/10.1016/j.matchemphys.2006.07.004

Arenz, M., K.J.J. Mayrhofer, V. Stamenkovic, B.B. Blizanac, T. Tomoyuki, P.N. Ross, et al., 2005. The effect of the particle size on the kinetics of CO electrooxidation on high surface area Pt catalysts. J. Am. Chem. Soc. 127(18): 6819–6829. https://doi.org/10.1021/ja043602h

Bard, A.J. and L.R. Faulkner. 2001. Electrochemical Methods—Fundamentals and Applications, 2nd Ed. John Wiley & Sons, Inc.

Bard, A.J., G. Inzelt and F. Scholz (eds) 2008. Electrochemical Dictionary. Springer.

Bing, Y., H. Liu, L. Zhang, D. Ghosh and J. Zhang. 2010. Nanostructured Pt-alloy electrocatalysts for PEM fuel cell oxygen reduction reaction. Chem. Soc. Rev. 39(6): 2184. https://doi.org/10.1039/b912552c

Brimaud, S., S. Pronier, C. Coutanceau and J.M. Léger. 2008. New findings on CO electrooxidation at platinum nanoparticle surfaces. Electrochem. Commun. 10(11): 1703–1707. https://doi.org/10.1016/j.elecom.2008.08.045

Carrette, L., K.A. Friedrich and U. Stimming. 2001. Fuel cells—fundamentals and applications. Fuel Cells. 1(1): 5–39. https://doi.org/10.1002/1615-6854(200105)1:1<5::AID-FUCE5> 3.0.CO;2-G

Chen, C., Y. Kang, Z. Huo, Z. Zhu, W. Huang, H.L. Xin, et al., 2014. Highly crystalline multimetallic nanoframes with three-dimensional electrocatalytic surfaces. Science. New York, N.Y. 343(6177): 1339–1343. https://doi org/10.1126/science.1249061

Clavilier, J., J.M. Orts, R. Gómez, J.M. Feliu and A. Aldaz. 1996. Comparison of electrosorption at activated polycrystalline and Pt(531) kinked platinum electrodes: surface voltammetry and charge displacement on potentiostatic CO adsorption. J. Electroanal. Chem. 404(2): 281–289. https://doi.org/10.1016/0022-0728(95)04365-9

Colón-Mercado, H.R., H. Kim and B.N. Popov. 2004. Durability study of Pt_3Ni_1 catalysts as cathode in PEM fuel cells. Electrochem. Commun. 6(8): 795–799. https://doi.org/10.1016/j.elecom.2004.05.028

Eggen, P.-O. 2009. A simple hydrogen electrode. J. Chem. Educ. 86(3): 352–354. https://doi.org/10.1021/ed086p352.

Elgrishi, N., K.J. Rountree, B.D. McCarthy, E.S. Rountree, T.T. Eisenhart and J.L. Dempsey. 2018. A practical beginner's guide to cyclic voltammetry. J. Chem. Educ. 95(2): 197–206. https://doi.org/10.1021/acs.jchemed.7b00361

Garsany, Y., O.A. Baturina, K.E. Swider-Lyons and S.S. Kocha. 2010. Experimental methods for quantifying the activity of platinum electrocatalysts for the oxygen reduction reaction. Anal. Chem. 82(15): 6321–6328. https://doi.org/10.1021/ac100306c

Garsany, Y., I.L. Singer and K.E. Swider-lyons. 2011. Impact of film drying procedures on RDE characterization of Pt/VC electrocatalysts. J. Electroanal. Chem. 662(2): 396–406. https://doi.org/10.1016/j.jelechem.2011.09.016

Gasteiger, H.A., S.S. Kocha, B. Sompalli and F.T. Wagner. 2005. Activity benchmarks and requirements for Pt, Pt-alloy, and non-Pt oxygen reduction catalysts for PEMFCs. Appl. Catal., B. 56(1–2): 9–35. https://doi.org/10.1016/j.apcatb.2004.06.021

Gilroy, K.D., X. Yang, S. Xie, M. Zhao, D. Qin and Y. Xia. 2018. Shape-controlled synthesis of colloidal metal nanocrystals by replicating the surface atomic structure on the seed. Adv. Mater. 30(25): 1–25. https://doi.org/10.1002/adma.201706312

Guilminot, E., A. Corcella, M. Chatenet, F. Maillard, F. Charlot, G. Berthomé et al., 2007. Membrane and active layer degradation upon PEMFC steady-state operation. J. Electrochem. Soc. 154(11): B1106. https://doi.org/10.1149/1.2775218

Hodnik, N., C. Baldizzone, S. Cherevko, A. Zeradjanin and K.J.J. Mayrhofer. 2015. The effect of the voltage scan rate on the determination of the oxygen reduction activity of Pt/C fuel cell catalyst. Electrocatalysis. 6(3): 237–241. https://doi.org/10.1007/s12678-015-0255-0

Houchins, C., G. Kleen, J. Spendelow, J. Kopasz, D. Peterson, N. Garland, et al., 2012. U.S. DOE Progress towards developing low-cost, high performance, durable polymer electrolyte membranes for fuel cell applications. Membranes. 2(4): 855–878. https://doi.org/10.3390/membranes2040855

Iiyama, A., K. Shinohara, S. Iguchi and A. Daimaru. 2010. Membranes and catalyst performance targets for automotive fuel cells. *In*: W. Vielstich, A. Lamm, H.A. Gasteiger and H. Yokokawa (eds). Handbook of Fuel Cells. John Wiley & Sons, Ltd. https://doi.org/10.1002/9780470974001.f500059

Katsounaros, I., J.C. Meier and K.J.J. Mayrhofer. 2013. The impact of chloride ions and the catalyst loading on the reduction of H_2O_2 on high-surface-area platinum catalysts. Electrochim. Acta. 110: 790–795. https://doi.org/10.1016/j.electacta.2013.03.156

Kim, J.H., J.Y. Cheon, T.J. Shin, J.Y. Park and S.H. Joo. 2016. Effect of surface oxygen functionalization of carbon support on the activity and durability of Pt/C catalysts for the oxygen reduction reaction. Carbon. 101: 449–457. https://doi.org/10.1016/j.carbon.2016.02.014

Kinoshita, K. 1990. Particle size effects for oxygen reduction on highly dispersed platinum in acid electrolytes. J. Electrochem. Soc. 137(3): 845. https://doi.org/10.1149/1.2086566

Kocha, S.S., K. Shinozaki, J.W. Zack, D.J. Myers, N.N. Kariuki, T. Nowicki, et al., 2017. Best practices and testing protocols for benchmarking ORR activities of fuel cell electrocatalysts using rotating disk electrode. Electrocatalysis.. 8: 366–374. https://doi.org/10.1007/s12678-017-0378-6

Lee, M.H. and J.S. Do. 2009. Kinetics of oxygen reduction reaction on Corich core-Ptrich shell/C electrocatalysts. J. Power Sources. 188(2): 353–358. https://doi.org/10.1016/j.jpowsour.2008.12.051

Lee, J., B. Jeong and J.D. Ocon. 2013. Oxygen electrocatalysis in chemical energy conversion and storage technologies. Curr. Appl Phys. 13(2): 309–321. https://doi.org/10.1016/j.cap.2012.08.008

Li, Y., Y. Li, E. Zhu, T. McLouth, C.-Y. Chiu, X. Huang, et al., 2012. Stabilization of high-performance oxygen reduction reaction Pt electrocatalyst supported on reduced graphene oxide/carbon black composite. J. Am. Chem. Soc. 134(30): 12326–12329. https://doi.org/10.1021/ja3031449

Long, N.V., M. Ohtaki, M. Uchida, R. Jalem, H. Hirata, N.D. Chien, et al., 2011. Synthesis and characterization of polyhedral Pt nanoparticles: Their catalytic property, surface attachment, self-aggregation and assembly. J. Colloid Interface Sci. 359(2): 339–350. https://doi.org/10.1016/j.jcis.2011.03.029

Maillard, F., M. Eikerling, O.V. Cherstiouk, S. Schreier, E. Savinova and U. Stimming. 2004. Size effects on reactivity of Pt nanoparticles in CO monolayer oxidation: The role of surface mobility. Faraday Discuss. 125: 357. https://doi.org/10.1039/b303911k

Maillard, F., S. Schreier, M. Hanzlik, E.R. Savinova, S. Weinkauf and U. Stimming. 2005. Influence of particle agglomeration on the catalytic activity of carbon-supported Pt nanoparticles in CO monolayer oxidation. Phys. Chem. Chem. Phys. 7(2): 385–393. https://doi.org/10.1039/b411377b

Markovic, N., H. Gasteiger, P.N. Ross, L. Berkeley and M.S. Division. 1997. Kinetics of oxygen reduction on Pt (hkl) electrodes: implications for the crystallite size effect with supported Pt electrocatalysts. J. Electrochern. Soc. 144(5): 1591–1597. https://doi.org/10.1149/1.1837646

Markovic, N.M., T.J. Schmidt, V. Stamenkovic and P.N. Ross. 2001. Oxygen reduction reaction on Pt and Pt bimetallic surfaces: a selective review. Fuel Cells. 1(2): 105–116. https://doi.org/10.1002/1615-6854(200107)1:2<105::aid-fuce105>3.3.co;2-0

Mayrhofer, K.J.J., M. Arenz, B.B. Blizanac, V. Stamenkovic, P.N. Ross and N.M. Markovic. 2005. CO surface electrochemistry on Pt-nanoparticles: a selective review. Electrochim. Acta. 50(25–26): 5144–5154. https://doi.org/10.1016/j.electacta.2005.02.070

Mayrhofer, K.J.J., D. Strmcnik, B.B. Blizanac, V. Stamenkovic, M. Arenz and N.M. Markovic. 2008. Measurement of oxygen reduction activities via the rotating disc electrode method: From Pt model surfaces to carbon-supported high surface area catalysts. Electrochim. Acta. 53(7): 3181–3188. https://doi.org/10.1016/j.electacta.2007.11.057

Mazloomi, K. and C. Gomes. 2012. Hydrogen as an energy carrier: Prospects and challenges. Renewable Sustainable Energy Rev. 16(5): 3024–3033. https://doi.org/10.1016/j.rser.2012.02.028

Napporn, T.W., Y. Holade, B. Kokoh, S. Mitsushima, K. Mayer, B. Eichberger, et al., 2018. Electrochemical measurement methods and characterization on the cell level. pp. 175–214. *In*: V. Hacker and S. Mitsushima (eds). Fuel Cells and Hydrogen: From Fundamentals to Applied Research. Elsevier. https://doi.org/10.1016/B978-0-12-811459-9.00009-8

Nrel, S.S.K., Y. Garsany, E. Nrl, D.M. Anl, S.S. Kocha, Y. Garsany, et al., 2013. Testing oxygen reduction reaction activity with the rotating disc electrode technique. DOE Webinar. http://energy.gov/eere/fuelcells/downloads/testing-oxygen-reduction-reaction-activity-rotating-disc-electrode

Ochal, P., J.L. Gomez de la Fuente, M. Tsypkin, F. Seland, S. Sunde, N. Muthuswamy, et al., 2011. CO stripping as an electrochemical tool for characterization of Ru@Pt core-shell catalysts. J. Electroanal. Chem. 655(2): 140–146. https://doi.org/10.1016/j.jelechem.2011.02.027

Özaslan, M. 2012. Oxygen electroreduction on core-shell nanoparticle catalysts prepared by selective electrochemical metal dissolution of Pt-Cu and Pt-Co alloys. Doctoral Thesis, Technische Universität Berlin. https://doi.org/10.14279/depositonce-111

Paulus, U.A., T.J. Schmidt, H.A. Gasteiger and R.J. Behm. 2001. Oxygen reduction on a high-surface area Pt/Vulcan carbon catalyst: a thin-film rotating ring-disk electrode study. J. Electroanal. Chem. 495(2): 134–145. https://doi.org/10.1016/S0022-0728(00)00407-1

Paulus, U.A., A. Wokaun, G.G. Scherer, T.J. Schmidt, V. Stamenkovic, N.M. Markovic. and P.N. Ross. 2002. Oxygen reduction on high surface area Pt-based alloy catalysts in comparison to well defined smooth bulk alloy electrodes, Electrochim. Acta. 47(22–23): 3787–3798. https://doi.org/10.1016/S0013-4686(02)00349-3

Paulus, U.A., T.J. Schmidt and H.A. Gasteiger. 2010. Poisons for the O_2 reduction reaction. *In:* W. Vielstich, H.A. Gasteiger, A. Lamm and H. Yokokawa (eds). Handbook of Fuel Cells—Fundamentals, Technology and Applications. John Wiley & Sons. On line.

Perry, S.C. and G. Denuault. 2015. Transient study of the oxygen reduction reaction on reduced Pt and Pt alloys microelectrodes : evidence for the reduction of pre-adsorbed. Phys. Chem. Chem. Phys. 17: 30005–30012. https://doi.org/10.1039/C5CP04667J

Reyes-Rodríguez, J.L., F. Godínez-Salomón, M.A. Leyva and O. Solorza-Feria. 2013. RRDE study on Co@Pt/C cores-shell nanocatalysts for the oxygen reduction reaction. Int. J. Hydrogen Energy. 38(28): 12634–12639. https://doi.org/10.1016/j.ijhydene.2012.12.031

Reyes-Rodríguez, J.L., A. Velázquez-Osorio, D. Bahena-Uribe, A.B. Soto-Guzmán, M.A. Leyva, A. Rodríguez-Castellanos, et al., 2019. Tailoring the morphology of Ni-Pt nanocatalysts through the variation of oleylamine and oleic acid: a study on oxygen reduction from synthesis to fuel cell application. Catal. Sci. Technol. 9(10): 2630–2650. https://doi.org/10.1039/c9cy00419j

Sandstede, G., E.J. Cairns, V.S. Bagotsky and K. Wiesener. 2010. History of low temperature fuel cells. pp. 145–218. *In:* W. Vielstich, H.A. Gasteiger, A. Lamm and H. Yokokawa (eds). Handbook of Fuel Cells—Fundamentals, Technology and Applications. John Wiley & Sons. https://doi.org/10.1002/9780470974001.f104011

Schmidt, T.J., H.A. Gasteiger, G.D. Stab, P.M. Urban, D.M. Kolb and R.J. Behm. 1998. Characterization of high-surface area electrocatalysts using a rotating disk electrode configuration. J. Electrochem. Soc. 145(7): 2354–2358. https://doi.org/Doi 10.1149/1.1838642

Schmidt, T.J., U.A. Paulus, H.A. Gasteiger and R.J. Behm. 2001. The oxygen reduction reaction on a Pt/carbon fuel cell catalyst in the presence of chloride anions, J. Electroanal. Chem. 508(1–2): 41–47. https://doi.org/10.1016/S0022-0728(01)00499-5

Shao, M., Q. Chang, J.-P. Dodelet and R. Chenitz. 2016. Recent advances in electrocatalysts for oxygen reduction reaction. Chem. Rev. 116(6): 3594–3657. https://doi.org/10.1021/acs.chemrev.5b00462

Shinozaki, K., J.W. Zack, R.M. Richards, B.S. Pivovar and S.S. Kocha. 2015. Oxygen reduction reaction measurements on platinum electrocatalysts utilizing rotating disk electrode technique I. Impact of impurities, measurement protocols and applied corrections. J. Electrochem. Soc. 162(10): 1144–1158. https://doi.org/10.1149/2.1071509jes

Teran-Salgado, E., D. Bahena-Uribe, A.M. Pedro, J.L. Reyes-Rodriguez, R. Cruz-Silva, and O. Solorza-Feria. 2019. Platinum nanoparticles supported on electrochemically oxidized and exfoliated graphite for the oxygen reduction reaction. Electrochim. Acta. 298: 172–185. https://doi.org/10.1016/j.electacta.2018.12.057

Van Der Vliet, D., D.S. Strmcnik, C. Wang, V.R. Stamenkovic, N.M. Markovic and M.T.M. Koper. 2010. On the importance of correcting for the uncompensated Ohmic resistance in model experiments of the oxygen reduction reaction. J. Electroanal. Chem. 647(1): 29–34. https://doi.org/10.1016/j.jelechem.2010.05.016

Wang, J.X., N.M. Markovic and R.R. Adzic. 2004. Kinetic analysis of oxygen reduction on Pt(111) in acid solutions: intrinsic kinetic parameters and anion adsorption effects. J. Phys. Chem. B. 108(13): 4127–4133. https://doi.org/10.1021/jp037593v

Wang, X., L. Figueroa-Cosme, X. Yang, M. Luo, J. Liu, Z. Xie, et al., 2016. Pt-based icosahedral nanocages: using a combination of {111} facets, twin defects, and ultrathin walls to greatly enhance their activity toward oxygen reduction. Nano Lett. 16(2): 1467–1471. https://doi.org/10.1021/acs.nanolett.5b05140

Wang, S., E. Zhu, Y. Huang and H. Heinz. 2021. Direct correlation of oxygen adsorption on platinum-electrolyte interfaces with the activity in the oxygen reduction reaction. Sci. Adv. 7(24): 1–16. https://doi.org/10.1126/sciadv.abb1435

Westbroek, P. 2005. Fundamentals of electrochemistry. pp. 3–36. *In*: P. Westbroek, G. Priniotakis and P. Kiekens (eds). Analytical Electrochemistry in Textiles. Woodhead Publishing. https://doi.org/10.1533/9781845690878.1.1

Wroblowa, H.S. and G. Razumney. 1976. Electroreduction of oxygen. J. Electroanal. Chem. Interfacial Electrochem. 69(2): 195–201. https://doi.org/10.1016/S0022-0728(76)80250-1

Xing, W., M. Yin, Q. Lv, Y. Hu, C. Liu and J. Zhang. 2014. Oxygen solubility, diffusion coefficient, and solution viscosity. pp. 1–31. *In*: W. Xing, G. Yin and J. Zhang (eds). Rotating Electrode Methods and Oxygen Reduction Electrocatalysts. Elsevier. https://doi.org/10.1016/B978-0-444-63278-4.00001-X

Yang, D.-S., M.-S. Kim, M.Y. Song and J.-S. Yu. 2012. Highly efficient supported PtFe cathode electrocatalysts prepared by homogeneous deposition for proton exchange membrane fuel cell. Int. J. Hydrogen Energy. 37(18): 13681–13688. https://doi.org/10.1016/j.ijhydene.2012.02.108

Zoski, C.G. (ed.) 2007. Handbook of Electrochemistry. Elsevier.

Chapter 4

Graphene Synthesis via Liquid Phase Exfoliation: Synthetic Approaches toward an Enhanced Nanoparticle Support

Carlos Galindo-Uribe[1]*, José Luis Reyes-Rodríguez[2],
Patrizia Calaminici[1] and Omar Solorza-Feria[1]

[1]Departamento de Química, Centro de Investigación y de Estudios Avanzados
(Cinvestav IPN), Av. IPN 2508, Col. San Pedro Zacatenco,
Alcaldía Gustavo A. Madero, C.P. 07360, Ciudad de México, México.
Email:carlosd.galindo@cinvestav.mx; Tel. +52 1 55 5747 3715.

[2]Escuela Superior de Ingeniería Química e Industrias Extractivas (ESIQIE), Instituto
Politécnico Nacional (IPN). Av. Luis Enrique Erro S/N,
Unidad Profesional Adolfo López Mateos, Zacatenco, Alcaldía Gustavo A. Madero.
Email: jlreyes@ipn.mx; Tel. +52 1 55 3084 9664.
C.P. 07738, Ciudad de México, México.

4.1 INTRODUCTION

Graphene is a bi-dimensional material formed only by carbon atoms in a hexagonal 'honeycomb' structure that is one carbon atom thick (Geim and Novoselov, 2007; Novoselov et al., 2005). Due to this structure, special properties, such as lightweight, excellent heat and electrical conductivity, and mechanical resistance

---

*For Correspondence: Email: carlosd.galindo@cinvestav.mx

(Chen et al., 2018; Raccichini et al., 2015) make them especially attractive to several applications in material science and electronics (Li et al., 2014; Patel et al., 2019; Sumdani et al., 2021). The challenge today is to develop an easy, cheap, scalable and environmentally friendly method to produce massive amounts of high-quality graphene. Liquid phase exfoliation (LPE) has the potential to fulfill these requirements of materials (Xu, 2018; Amiri et al., 2018).

LPE is a general concept that involves different mechanisms and approaches; that also is applicable to different 2D materials (Huo et al., 2015). The LPE is a method that encompasses a different sort of techniques and equipment, but all of these have in common: (1) a liquid media, which can be a pure solvent or a wide array of mixtures, additives and stabilizers, and (2) a starting material, normally pure graphite, however, the use of different starting materials, such as high ordered pyrolyzed graphite (HOPG), expandable graphite (Hamilton et al., 2009) or graphite intercalation compounds (GIC) (Chung, 2016) greatly modifies and improves the yield and quality of the LPE. Thereof, a careful selection of the techniques, liquid media and starting materials is important to optimize the graphene obtained.

Galindo-Uribe et al. proposed a classification of LPE based on their mechanism involved. The first type of mechanism comes from the use of high shear forces over a suitable liquid medium to break the interaction force of the graphene layers inside the graphite structure. The second mechanism is the exfoliation achieved by using molecules entering the inner structure of graphite and forming an intermediary of a GIC. A more detailed insight of both mechanisms is found in the reference (Galindo-Uribe et al., 2022). However, those mechanisms share some factors that are key to understanding the phenomenon of exfoliation. Moreover, these factors could be used advantageously to create a better support suitable for NPs. The present chapter is divided into factors that improve the exfoliation to understand the basic phenomenon of the method. Then a review of the functionalization of graphene is presented and divided into covalent and non-covalent functionalization. Finally, some examples of the use of LPE to form NPs deposited *in-situ* over the graphene, and the use of these methods to form suitable catalysts towards ORR are presented.

4.2   FACTORS THAT AFFECT THE EXFOLIATION

Figure 4.1 (Shen et al., 2015) summarizes the three main processes carried out in the LPE and these are explained in three different cases when use is made of the good, the ordinary, and a bad solvent.

The first process is to submerge the starting material into a suitable liquid medium. The second process is to use a proper technique to exfoliate the material. As shown in Fig. 4.1, the good and the ordinary solvent almost instantly carry out the partial exfoliation of the material on contact, whereas the bad solvent cannot modify the starting material at all. Theoretically, a good and an ordinary solvent can break the interactions of the starting material which becomes completely exfoliate. However, the critical step in LPE relies on the good solvent to replace the interactions inside the material with appropriate interactions to stabilize the exfoliated material, whereas the ordinary solvent cannot and the starting material is formed again (Shen et al., 2015). From this perspective, the selection of a suitable solvent or mixture of solvents is the most important part in the LPE techniques and thereof it is important to review the theories that explain how some solvents and mixtures are more suitable for the preparation of graphene via LPE.

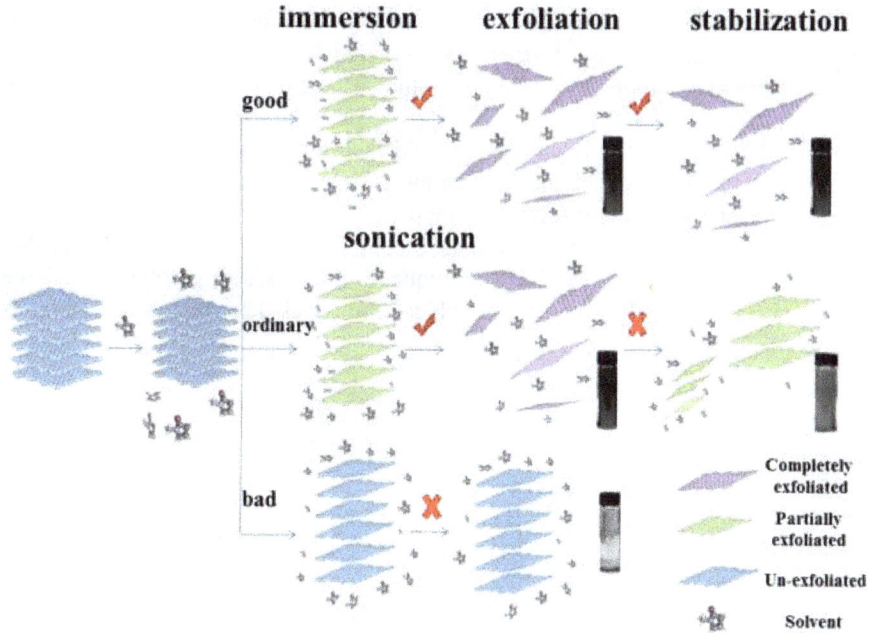

Figure 4.1 Steps of LPE for a good, ordinary, and bad solvent (Reprinted from [Shen et al., 2015]).

4.3 SELECTION OF A SUITABLE SOLVENT AND TECHNIQUE FOR LPE OF GRAPHENE

The first historical approximation for selecting an appropriate solvent for preparation of graphene via LPE is to test the solvents known for carbon nanotubes. It is known that carbon nanotubes and graphene are stabilized in liquid medium via van der Waals interactions. A rigorous mathematical and thermodynamic study (Coleman, 2009) shows that the enthalpy of mixing is minimized when Hildebrand solubility parameter of solvent and solute match. The Hildebrand is calculated as the root square of the total cohesive energy density $E_{C,T}$:

$$\delta_T = \sqrt{\frac{E_{C,T}}{V}} \qquad (1)$$

The Hildebrand parameter only takes into account dispersive interactions, however most molecules also display polar and hydrogen bonding interactions. These interactions are accounts in Hansen parameters, which also affect the performance of a solvent to exfoliate graphene and carbon nanotubes. The relation between Hansen and Hildebrand parameters is:

$$\delta_T^2 = \delta_D^2 + \delta_H^2 + \delta_P^2 \tag{2}$$

where δ_D, δ_H and δ_P are the dispersive, hydrogen-bonding and polar Hansen parameters respectively (Coleman, 2009).

Hernandez et al. tested the behavior of 40 different solvents towards LPE of grapheme. The quantity of graphene exfoliated is contrasted with the Hansen and Hildebrand solubility parameters, and the surface tension of the solvents. The approximation reveals that the best performance is obtained when the Hansen and Hildebrand solubility parameters match; therefore, graphene should have the same values for his Hansen and Hildebrand parameters (Hernandez et al., 2010).

The selection of a good solvent is important in the optimal yield of exfoliated graphene, but also affects the performance of the method. Paton et al. found that the minimum shear rate of LPE technique to successfully exfoliate graphene depends on the surface energy of the solvent employed. The equation that models this behavior is:

$$\dot{\gamma}_{min} = \frac{[\sqrt{E_{S,G}} - \sqrt{E_{S,L}}]^2}{\eta L} \tag{3}$$

where $E_{S,G}$ and $E_{S,L}$ are the surface energies of the graphene and the solvent respectively, η is the dynamic viscosity of the solvent, and L is the length of the obtained graphene (Paton et al., 2014).

As the material exfoliates, energy is required to overcome the interactions between the layers and effectively break the material. This process requires energy to form new surfaces. If the surface interactions are not stabilized via a matching solvent, a higher shear rate is needed to prevent the reaggregation of the material. From this model, the surface tension and energy values of the solvent are used as a good parameter to know if a mixture of solvents is adequate to exfoliate graphene. The best efficiency in exfoliation is obtained in media with energy surface near or equal to the graphene. Using Eq. (1) with the knowledge of the surface energy, the viscosity of the solvents, the length of the obtained graphene, and realizing different experiments varying the shear rate, the surface energy of the graphene has been calculated.

The minimum shear rate found (10^4 s^{-1}) is below the turbulent regime, i.e. any device that causes a turbulent flow has enough energy to exfoliate graphene, e.g. a kitchen blender (Yi and Shen, 2014). These techniques were called 'high shear mixing' (Xu, 2018). In contrast to the exfoliation via sonication, these techniques yield graphene of higher quality with higher energy efficiency. However, the yield to obtain a single layer graphene is low (Amiri et al., 2018). The explanation for the poor performance lies in the mechanism of LPE to obtain graphene.

Figure 4.2 shows the mechanism proposed by Li et al. in which the fissures and kinks of graphite are responsible for the exfoliation process. Such defects are more reactive, allowing the shear forces to act and peel off the graphene layers in the first step of exfoliation. However, when the graphene sheets are about 30 layers thick, bending is more likely instead of forming kinks and defects. Then the predominant mechanisms for graphene exfoliation are the edge tear and the intercalation of the solvent, which typically requires a higher shear rate (Li et al., 2020).

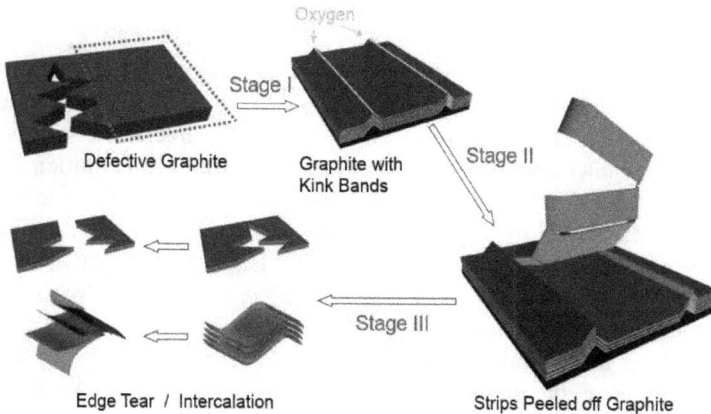

Figure 4.2   The stages for the obtention of graphene via LPE technique (Reprinted from [Li et al., 2020]).

Botto et al. elucidated a mechanism called π-peel which is demonstrated mathematically. The result of this theory predicts a minimum shear rate of $10^8 \ s^{-1}$ to obtain single layer graphene (Botto, 2019). Therefore, both mechanisms are in accordance with the experimental results.

It is important to note that the principal function of the solvent is to stabilize the formed graphene. Other kinds of interactions, such as electrostatic and steric effects, have great impact on the capability of the liquid media to stabilize graphene sheets. In this sense, adding additional species such as ions (Cheng, 2020), polymers (Wajid et al., 2012) and surfactants (Guardia et al., 2011) improve the maximum concentration of the graphene dispersed in the medium, although the difficulty to remove the additives from the surface of graphene increases. These interactions are so strong that some authors refer to them as 'non-covalent functionalization'.

4.4   NON-COVALENT FUNCTIONALIZATION OF GRAPHENE

The non-covalent functionalization has central importance in the LPE methods because it can improve significantly the capacity to exfoliate graphene in a specific solvent. As shown in Figure 4.1, the only difference between a regular and a good solvent is its capability to disperse and stabilize the newly formed graphene flakes. The functionalization process breaks the interactions between graphite sheets and hinders the formation of the graphitic structure again. Furthermore, at the same time, a functionalized graphene performs better as a catalyst support than pristine graphene due to the anchoring interactions with NPs, achieving an effective attachment (Georgakilas et al., 2012).

There is a plethora of different interactions that can occur over the surface of graphene (Kumar et al., 2022); however, there are three of special importance for the LPE:

4.4.1   Cation-π Interaction

The cation-π interaction is an electrostatic interaction between a positive charge and a π electron donor that is neutral in charge. This kind of interaction is strong but depends on the nature of the cation: the interaction is stronger when the cation is smaller and with a greater charge. Using the definition of cation hardness (Pearson, 1968), it is concluded that harder cations can interact better with π electrons present in graphene. Furthermore, for transition metals that have available d-orbitals such as Au^+, Pt^{2+}, Pd^{2+} and Hg^{2+}, the interaction with the π orbital is off-centered due to the back-donation of π electrons. For the cations that do not have d-orbitals available, such as Na^+, Cu^+, and $Ag+$, the interaction is centered over the orbital (Georgakilas et al., 2012).

An example of the stabilization of graphene is use of NaOH to improve twice the exfoliation yield in the case of N-methyl-2-pirrolidone (an already good solvent) to 20-fold increase in the case of cyclohexanone (Liu and Wang, 2011). It is noted that if the filtered graphene formed with this approach is redispersed in water, the pH of the solution becomes alkaline, which is an indication of the functionalization of graphene with cations.

Nonetheless, in the LPE method, the sizes of the cations are most important than the hardness, as performance is better as the size of the ions increases—a result counterintuitive taking into consideration the strength of the interactions. Wang et al. (2017b) exfoliate graphene using different salts. Two salts with the same anion, such as LiCl and KCl, were compared to find the performance of the cation used. In this comparison, the bigger size of K^+ improves the exfoliation of graphene. Furthermore, the performance of the anion is revised using LiCl and Li_2SO_4. Surprisingly, the bigger SO_4^{2-} anion improves the exfoliation. The mechanism proposed first involves the intercalation of cations inside the graphite structure. The size of the cation plays a role in opening the distance between graphene sheets, thus weakening the interactions between layers. Then the positive-charged cation-graphite 'sandwich' attracts the anion. Though the anion can interact with the π electrons (Kumar et al., 2022), this interaction is weak, and the intercalation of the anion is mostly electrostatically driven. The entrance of the anions weakens even more the interactions between graphene sheets, breaking them away (Wang et al., 2017a).

The oxygen groups hinder the entrance of the cations, Therefore, a reduction in the surface of the graphite promotes the exfoliation (Yang et al., 2018).

4.4.2 π-π Interaction

The π-π interaction is one of the most useful interactions for graphene because it extends the research to the functionalization of multiple molecules and polymers with potential biological importance. Although this type of interaction is weaker than the cation-π interaction, it becomes much stronger in macromolecules with multiple aromatic rings.

The π-π interactions have two requirements: the existence of π orbitals on both molecules, and the planarity of both molecules, which is frequently seen as stacking molecules. Also it is important to note that the pure π-π interaction is repulsive in nature but is frequently seen along with the π-σ interaction, which is attractive (Hunter and Sanders, 1990). When the carbon atoms of the ring are substituted by electroattractive atoms like O or N, the electrons withdraw from the π orbital, and make it more accessible for another π orbital to interact with it. Conversely, if electron-donor groups are substituted into the aromatic ring, the repulsion becomes greater (Georgakilas et al., 2012; Kumar et al., 2022).

It is noted that there is always a coupling between different interactions that can help to make the overall interaction stronger. In the case of LPE, this coupling is seen between melamine molecules in the ball-milling process to obtain graphene. In this experiment, four different molecules along melamine are used and compared for the LPE method. It is observed that the interaction between hydrogen bonds, the N-atoms in the ring, and the N-substituents over the C atoms promotes the best coupling to correctly stabilize graphene flakes. It is important to note that a simple rinse with water removes completely the melamine (Leon et al., 2014).

4.4.3 Steric Forces

The bigger polymers and macromolecules can interact and functionalize the graphene flakes and at the same time, with their hydrocarbon chains interact with each other through repulsive interactions that can dominate the electrostatic and van der Waals forces at short distances (Wang et al., 2016).

An interesting case study is how the surfactants adhere to graphene surface using their polar part of their molecule. For ionic tensioactives, the last also means that an induced charge is imposed on the graphene flakes forming a double layer. Therefore, the Derjaguin–Landau–Verwey–Overbeek (DLVO) theory is applied to understand the electrostatic interactions. A complete mathematical formalism is found in the references (Smith et al., 2010; Wang et al., 2016).

The graphene dispersed in the liquid medium increases as the concentration of a single addition of the ionic surfactant increases, until a maximum is reached. The increase in the concentration of the ionic surfactant diminishes the concentration of graphene. This behavior is explained with a mechanism in which a surfactant molecule adsorbs over the graphene interface, effectively imposing a charge to the graphene flake. The electrostatic repulsion maintains the graphene exfoliated; however, these interactions become less effective as van der Waals attractions increase with concentration, until the graphene no longer can stay exfoliated and reaggregates (Wang et al., 2016).

This mechanism contrasts with the non-ionic surfactants, which, even if some charged impurities are still present (Smith et al., 2010), mainly operate with steric repulsion between them. The behavior of the surfactants allows the graphene concentration to increase as surfactant concentration increases, until a maximum concentration of graphene is maintained. The increase in the concentration of the tensioactive does not affect its concentration at all because the surface of the graphene is fully occupied by adsorbed surfactant molecules (Wang et al., 2016). Furthermore, most polymers, such as polyvinylpyrrolidone (PVP), use the same mechanism to stabilize graphene (Wajid et al., 2012).

From the work of Smith et al., is necessary to aggregate an adequate quantity of tensioactive to help the LPE process. A good approximation was given by Notley, aggregating first enough surfactant to lower the surface tension of the water to 40 mJ/m (which is near the graphene surface tension (Hernandez et al., 2010) and then continuously aggregate surfactant, as needed. As the exfoliation process continues, the newly formed graphene adsorbs the tensioactive in its new-formed surface, depleting the surfactant available in the solution; thereby increasing the surface tension and hindering the exfoliation of new graphene. However, adding enough surfactant to maintain the correct surface tension of the solution, the exfoliation process can continue indefinitely (Notley, 2012).

4.5 COVALENT FUNCTIONALIZATION OF GRAPHENE

When a molecule is covalently attached to a carbon atom, changes occur in hybridization from sp^2 to sp^3, breaking the overlapping of the π orbitals of graphene and warping the structure. Therefore, it is important to note that any covalent functionalization leads to a change in characteristics of graphene, such as band gap, to increase the dispersability in common organic solvents and water and improve the properties of graphene as a catalyst, support, and sensors (Quintana et al., 2014).

The first and most important covalent functionalization of graphene is the formation of GO (graphene oxide). Several methods have been deployed to oxidize graphite and form GO, such as the modified Hummers method (Hummers Jr and Offeman, 1958). The original article of the Hummers method has about 30,000 cites at the time of writing this chapter. Although the oxidative process is a synthetic method to obtain GO from graphite, it is a good example of oxygen functionalization. Most of the properties of graphene are lost in its oxidation process (Su et al., 2009). Furthermore, the exact structure of GO is still not totally known because it is very complex and depends heavily on the oxidation process (Khan et al., 2015).

Despite these problems, GO possesses some advantages over pristine graphene. The oxygen functional groups and sp^3 hybridized carbons can undergo different functionalization reactions (Georgakilas et al., 2012). Moreover, the oxygen and hydroxide functional groups impose hydrophilic behavior due to hydrogen bonds. Other advantages of having functional groups and defects over the surface are that they can anchor the NPs more efficiently than graphene, and for this reason, most of the graphene-NPs studies focus over rGO and GO (Khan et al., 2015; Amiri et al., 2018).

Moreover, it is important to note that LPE methods also slightly oxidize the resultant graphene sheet. The oxides are mostly on the borders and are very difficult to be detected.

Most of the covalent functionalization methods of graphene use the well-known sp^2 chemistry of graphene to insert different molecules on to its surface. There are two main reactions that have been used to functionalize LPE graphene:

4.5.1 Cicloaddition

Due to the double-bonded nature of graphene, pericyclic reactions are of special interest for graphene. In general, the reactions involve the concerted movement of π electrons to form new σ bonds.

One of the most important reactions is the 1,3 dipolar addition. An 1,3 dipolar molecule interacts with graphene, which acts as a dipolarophile and forms a five-atom ring perpendicular to the surface of graphene. Quintana et al. designed an experiment in which it can selectively functionalize all the bulk of the graphene sheet or only the oxidized borders. The bulk graphene can be functionalized using a 1,3 dipolar reaction which involves paraformaldehyde and a glycine molecule which is an N-alkylated molecule with a protected carboxylic acid. Once unprotected, the carboxylic acid can react with a special molecule that has three amine groups, with two of them protected to prevent the polymerization of graphene, labeled as dendron 2. The free amine group reacts with carboxylic acids to form an amide. The interesting part is that dendron 2 can also react with the carboxylic acids found over the edges of exfoliated graphene. Once unprotected, the other amine groups can interact, for example, with Au NPs. When dendron 2 reacts with a functionalized graphene, the Au NPs are found in the bulk and edges, whereas when dendron 2 reacts, pristine graphene oxidation of the edges is performed (Quintana et al., 2011).

Tsoufis et al. used a similar approach to synthesize and support Fe_3O_4 NPs over graphene. N-(hydroxyphenyl)-glycine and paraformaldehyde were used to form functionalized graphene. Dendron 2 was also used in this synthesis to form the final functionalized graphene. A wet method is used to form NPs, in which a solution of $Fe(NO_3)_3$ is used as the precursor of the NPs. The mechanism for the formation of NPs is associated with the functional groups attached to graphene and which interact with the Fe^{3+} ions before getting reduced and deposit the Fe_3O_4 NPs over graphene. These NPs have magnetic response (Tsoufis et al., 2015).

4.5.2 Diazonium Functionalization

In benzene chemistry, diazonium salts are an essential and useful synthetic method capable of adding different functional groups via an electrophilic reaction, even creating new σ carbon bonds and are of industrial importance (Chen et al., 2022). The same reactions are used in graphene functionalization with molecules.

Hadad et al. functionalized graphene with 4-amino-N,N,N-trimethylbenzene ammonium iodide. A diazonium salt is formed *in-situ* with the graphene suspension using 3-methylbutil nitrite. The resulting functionalized graphene has positive-charged ammonium groups which are used to anchor Ru polyoxometalates (Ru-POM) to catalyze the peroxide decomposition (Hadad et al., 2014).

The examples discussed give an insight into some functionalization mechanisms. Some authors treat the direct interaction with NPs as a special class of functionalization. In the next section, the direct interaction of NPs with graphene and the formation of NPs over graphene will be discussed.

4.6   GRAPHENE AS NANOPARTICLE SUPPORT

Supporting of NPs could be treated as a type of functionalization in the sense that graphene interacts synergetically with the NP. Depending on the nature of the NP, the interfacial charge transference can be controlled towards or from graphene, enhancing their catalytic properties (Li et al., 2014). The preparation of NP can be explained by using two different mechanisms, '*ex-situ* hybridization', which involves two different syntheses for the NPs and graphene separately, and '*in-situ* crystallization' which involves the nucleation and formation of the NP over the graphene flake (Khan et al., 2015). Although most of the graphene-NP composites had been done over GO or rGO (Liu et al., 2014a), because of their defects they serve as an anchor for NPs. The exfoliated graphene still possesses some defects, such as ruptures and few oxidation sites (Wang et al., 2017a) serving as anchors for NPs or their precursors. An insight into both methods is discussed along with their applications.

4.6.1   *Ex-situ* Hybridization

The nature of the graphene obtained during the LPE has much fewer defects than GO, reducing significantly the quantity of possible anchors where small NPs can deposit. Nevertheless, Liu et al., 2014b used graphene exfoliated with Cyanoethyl-2-ethyl-4-methylimidazole (2E4MZ-CN) in acetonitrile, with Ag nanowires (AgNW) to form an AgNW/graphene composite. It is interesting that XRD analysis reveals a peak at 26° 2Θ that indicates some restacking of the graphene as prepared. The same peak diminishes when the graphene is sonicated and combined with AgNW and almost disappears when the proportion of AgNW:graphene is 4:1. From this result, it is concluded that the graphene can be stabilized using NPs and nanowires. The resultant composites were used as a filler for an epoxy-conductive polymer (Liu et al., 2014b).

The low interaction between formed NPs and graphene can be used advantageously, by using NPs to exfoliate and stabilize the formed flakes. Hadi et al. used Fe_3O_4 NPs to help the exfoliation of NPs. The proposed mechanism involves launching of NPs at great speeds due to bursting bubbles of the sonication process. When these NPs hit the graphite edges, an 'effect wedge' appears that exfoliates the graphite. After exfoliation, the NPs are separated by using magnetic forces. The resulting analysis shows that no NP were deposited over the surface of graphene (Hadi at al., 2018).

4.6.2   *In-situ* Crystallization

The few defects of LPE graphene can be used in wet synthesis approach as a seed for the crystallization and growth of NP. Bourlinos et al. exfoliated graphene in perfluorinated aromatic molecules. The resultant graphene flakes were redispersed in pyridine and mixed with solutions of $HAuCl_4$ in DMF, and $NaBH_4$ as a reductant (Bourlinos et al., 2009).

Yang et al. reported the *in-situ* crystallization of $NiCO_2$ NPs over grapheme by adding the LPE graphene in a solution of $Ni(NO_3)_2$ and $Co(NO_3)_2$ added stoichiometrically in the presence of urea. The resulting $NiCO_2$/graphene have microwave-absorbing properties (Yang et al., 2017). Truong-Huu et al. reported the formation of Pd NPs that can be used as a catalyst for liquid-phase hydrogenation reaction in a simple wet impregnation method. They used a solution of $Pd(NO_3)_2$ to impregnate the graphene and carbon nanotubes before calcination. The Pd/graphene catalyst showed superior performance to the hydrogenation reaction than the equivalent Pd/carbon nanotubes (Truong-Huu et al., 2012).

One of the advantages of the LPE method is that graphene is already dispersed in a solvent that can be suitable for the formation of NPs. Moreover, the molecules that functionalize and help the stabilization of graphene can be used to help the growth and anchoring of NP over the graphene flake. Liu et al. described that the graphene suspension stabilized by 2E4MZ-CN can be used to grow Ag NPs. They described a mechanism in which the precursor of Ag, silver acetate, first forms a complex with 2E4MZ-CN. The complex $Ag(2E4MZ-CN)_2$ then is reduced through a heating treatment at 150°C, to form Ag NPs/graphene composites. The 2E4MZ-CN molecules serve as a bridge between the graphene platelets and Ag NPs (Liu et al., 2012).

The production of NPs via a wet method and the LPE of graphene share the need for a stabilizer or a surfactant and a suitable liquid media. Moreover, the NPs can help to stabilize the graphene sheets. Therefore, a combination of NPs formed and at the same time exfoliating graphene would be feasible. The theoretical considerations and an example of a simultaneous LPE of graphene and formation of NPs is given below.

4.7 SIMULTANEOUS FORMATION OF NANOPARTICLES IN LPE OF GRAPHENE

As noted above, LPE of graphene and NPs formation can share all the requirements for their mechanism. Upadhyay et al. demonstrated that a suitable medium can be used for both methods. They used a mixture of different proportions of Tween 80 (a tensioactive) and hexane to produce a gel. Molecularly talking, the gel is formed because Tween 80 and hexane ensemble, forming giant rod-shaped micelles. Inside these micelles, the reaction is performed. These micelles can be tuned in varying different proportions of hexane and Tween 80, to promote different reactions: (1) With $AuCl_3$ are formed Au NPs of different shapes and sizes. The temperature and the hexane that is added help to tune the morphology of NPs. Tween 80 also serves as a reductant in this synthesis. (2) With $AgNO_3$ and ascorbic acid as reductants, spherical Ag NPs were synthesized. (3) With graphite, a spontaneous exfoliation is performed. The mechanism proposes that a giant, transient network of rod micelles diminishes the interaction among graphite layers. The hexane facilitates the uniform dispersion of graphite. The driving force, as discussed above, is the repulsion of the surfactant attached to graphene layers. It is noted that all NPs and graphene products are recuperated by just dissolving the gel in hexane-water mixtures (Upadhyay et al., 2016).

Taking again the explanation for a good, normal, and bad solvent for LPE (Shen et al., 2015), a proposed simultaneous mechanism of exfoliation and NP seeding can be elucidated, i.e. as soon as a graphene flake is formed, it starts to nucleate a new NP. Such a mechanism would have the following advantages: (1) the graphene/NP composite would not aggregate again because of the NP stabilization, (2) the LPE graphene would conserve all its conductivity and structural properties, (3) a one-pot one-step synthesis is required from convert graphite and NP precursors into a graphene/NP composite. To our best knowledge, no attempt to use the proposed mechanism has been made. Herewith, a simultaneous LPE and formation of PtCu NPs technique is described.

4.7.1   Graphene/PtCu Composite Simultaneous Synthesis

As we have discussed, a careful selection of the LPE technique, solvent and NP precursors was selected. For LPE technique, high shear mixing is one of the most efficient routes to exfoliate graphene. However, the disadvantage of the technique is that although the production of graphene starts at shear forces under turbulent regime, the shear forces required to exfoliate graphene to a single layer are much bigger (Botto, 2019). Therefore, the yield of single-layer graphene is minimal. Nevertheless, it is possible that the simultaneous NP synthesis can help the exfoliation. As noted above, any mixer that promotes turbulence can exfoliate graphite to a few-layered graphene. To maintain the simplicity of this synthesis, an Oster reversible kitchen blender with an aluminum jar and without any modification is used as a high shear mixer.

As a solvent, monoethyleneglycol (EG) is selected because it serves as a shape agent and a reductant to the famous polyol method to synthesize NPs (Fiévet et al., 2018). The boiling point of EG is 197°C, which is perfect for this method. The used precursors are tetrabutylammonium hexachloroplatinate (TBAH) and tetrabutylammonium dibromocuprate (TBAC). These precursors have a large cation and anion which can help with the exfoliation process as discussed above. Furthermore, the tetrabutylammonium cation have carbon chains that can interact with graphene in a way similar to a tensioactive, although these interactions are not confirmed yet. Fine powder graphite was selected as graphene precursor to accelerate the dispersion and exfoliation process.

4.7.2   Experimental Section

Graphite powder (<20 μm, synthetic), K_2PtCl_6 (≥99.9%), CuBr (99.9%), ascorbic acid (99%), and 99% tetrabutylammonium bromide (TBAB) were purchased from Sigma Aldrich; while EG, ethanol, acetone and $CHCl_3$ were purchased from Meyer.

TBAH were synthesized by using the method found elsewhere in literature (Iovel et al., 1987). 0.2423 g of K_2PtCl_6 (0.5 mmol) was dissolved in 20 ml of mili-Q water in an amber flask. 0.3223 g of TBAB (1 mmol) was dissolved in 50 ml of $CHCl_3$ and poured over the previous K_2PtCl_6 solution. The mixture was left for five hours under vigorous magnetic stirring to promote the contact over the phase, or until the water phase becomes colorless. Then the phases are separated and the $CHCl_3$ is distilled. A recrystallization using acetone yields the final product with 90% of yield. XRD monocrystal analysis was used to check that the correct product was obtained.

TBAC was synthesized by adding 0.1435 g (0.1 mmol) CuBr powder to a 50 ml acetone solution containing 0.3223 g of TBAB (1 mmol). The reaction proceeds almost instantly. The solution is distilled and recrystallized by using ethanol. The yield of the reaction is 90%. XRD monocrystal analysis was used to check that the correct product was obtained.

In a typical reaction, 0.0893 g of TBAH (0.1 mmol), 0.0155 g of TBAC (0.033 mmol), 10 mg of graphite and 100 ml of EG were added to a flask and heated for 30 min. at 110°C under magnetic stirring to remove water from the solution. The resultant suspension was added to the jar of the blender. The mixture was blended for 30 min. to exfoliate and mix the graphene solution. After this period, 0.0200 g of ascorbic acid was added directly to the jar, without turning off the blender, to carry out the reduction and another 30 min. of blending was given to the mix. After the blending, the solution was centrifuged at 12,000 rpm for 10 min. to precipitate the composite. After all composite was precipitated, it was washed twice with ethanol to remove all the EG of the surface. The precipitate was dried at 80°C in a conventional oven.

4.7.3   Characterization

XRD

XRD is shown in Fig. 4.3. The peak around 26° shows that the graphite still re-aggregates after the exfoliation process. There is a small peak at 38.5° that shows some Al contamination, possibly from the blender jar. The peaks around 40° and 47° show multiple phases of metal NPs of Pt and Cu of different compositions not in a single phase. Although our goal was to obtain a Pt_3Cu NPs, it is safer to refer to the NPs formed as PtCu NPs.

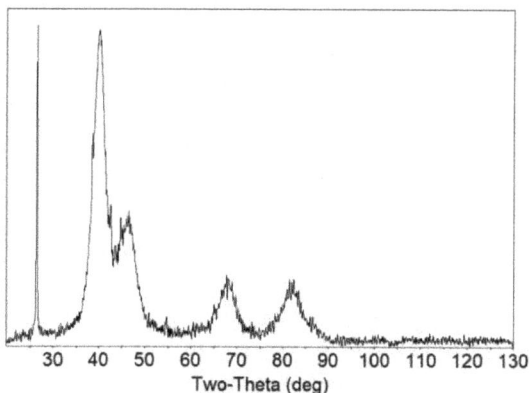

Figure 4.3   XRD of composite.

SEM-EDS

Figure 4.4(a) shows the surface of graphene composite with multiple round structures of PtCu NPs. These structures have a size of about 150 to 200 nm and have agglomerated each other in some zones of the graphene flakes. Figure 4.4(b) shows in detail the edges of the agglomerated graphene sheets. Plenty of edge defects of the graphene are observed and it is found that the NPs prefer to bond to edge defects than in the bulk area of grapheme sheets.

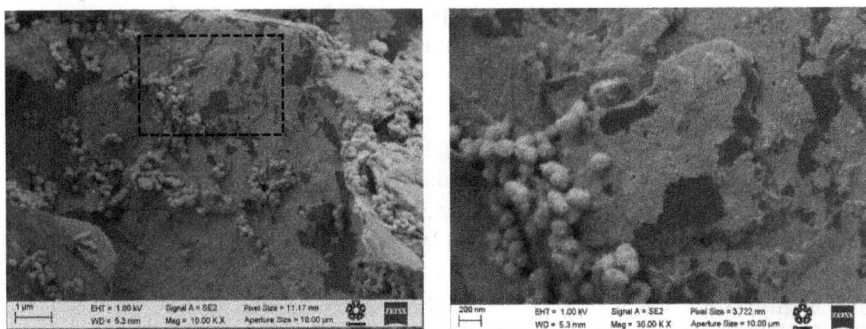

Figure 4.4 (a) SEM micrograph of the obtained graphene composite, (b) magnification of the indicated zone of micrograph showing the NPs and graphene defects. Unpublished results.

Table 4.1 presents the EDS results. Considering that EDS is a semi-quantitative method, some other elements with less than 0.6% of weight were omitted. From EDS analysis, the next composition is obtained:

Table 4.1 EDS elemental composition for the obtained graphene composite

Element	Normalized (wt. %)	Normalized (at. %)	σ (wt. %)
Platinum	53.47	8.28	1.47
Copper	6.44	3.06	0.21
Carbon	29.45	74.08	3.52
Oxygen	3.8	7.18	0.64
Aluminium	5.67	6.35	0.28

From Table 4.1, the aluminum contamination is confirmed. There is still some oxygen left in the system. This is an indication of an overly aggressive LPE method (similar to the extensive periods of sonication (Bracamonte et al., 2014), or not enough reductant. The atomic relation between Pt and Cu is about 2.7, which is near the initial objective of Pt$_3$Cu. The weight relation of NPs/graphene is about 2, which is too much for the typical weight percentage of catalyst.

TEM

Figure 4.5 shows a few-layer functionalized graphene of about 600 nm of length with PtCu NPs in HAADF mode. It is shown that tiny NPs were deposited along with the bigger NPs. These NPs were not evenly distributed over the entire flake, showing preference for some regions. The lower window of Fig. 4.5 offers a detailed image over these bigger NPs. It shows that the once thought single NP is actually hundreds of small NPs of about 2 nm.

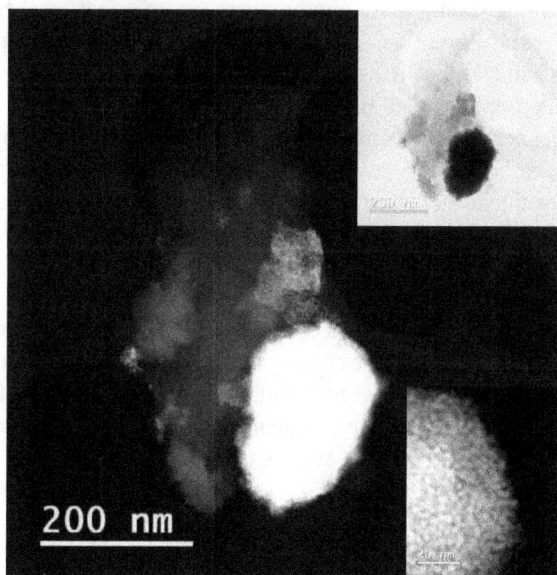

Figure 4.5 HAADF-TEM micrograph of the graphene composite. Upper window: TEM micrograph showing the same zone. Lower window: A zoom over the NP shows that it is actually thousands of smaller NPs. Unpublished results.

4.7.4 Discussion

From XRD, TEM and SEM analysis, it is found that graphene exfoliation was incomplete. However, from TEM images, it is found that the NPs agglomerate each other by thousands, probably a sintering process occurring at the temperatures of the reaction. Although a precise value of the temperature reached in the reaction was not obtained, a sweet odor was detected, suggesting that the temperature was near the ebullition temperature of EG. NPs from non-functionalized graphene tend to roll and aggregate each other (Liu et al., 2014a). From Fig. 4.5. some zones of graphene have evenly distributed small NPs. These zones could be explained that they are graphene nanosheets that reaggregate each other. Since most of the NPs sintered into bigger NPs with sizes similar to graphene flakes, the smaller NPs act as a bridge between the graphene sheets. In SEM analysis, these small NPs were not found because they are 'sandwiched' between graphene layers, whereas the external NPs suffered from agglomeration. The surface NPs probably were 'swept' for the liquid medium. From EDS analysis, aluminum contamination is found, probably from the blender jar, Also the quantity of.% of Pt and Cu is similar to the original stoichiometry of $3:1$. However, XRD shows not the multiple phases of Pt and Cu. An explanation is that the agglomeration is a fast reduction step, not giving enough time to the NPs to form a consolidate phase. More experiments are required to confirm these hypotheses.

4.8 CONCLUSION

The LPE is a robust method to obtain graphene of great quality; however, it is important to carefully select the appropriate medium, starting material and technique that is suitable for the use that the graphene would serve. The appropriate non-covalent functionalization in LPE helps to obtain a higher concentration and more stable graphene suspension which can greatly improve the ability of the graphene to anchor NPs and molecules. The covalent functionalization of graphene introduces new functional groups that can be used as an anchor to NPs. The LPE graphene as a NPs support has some advantages over the traditional GO-rGO support, such as being already dispersed, and in some cases the dispersion is used directly to support or to *in-situ* crystallize the NPs. The conditions of the LPE of graphene (a suitable solvent, a tensioactive, and the NPs precursors can promote graphene exfoliation) are the same as needed to synthesize NPs—a situation that can be used to simultaneously exfoliate graphene and form NPs on their surface. With a simple system using a kitchen blender, and for the first time, it is demonstrated that such an approach is possible. These results demonstrate that LPE exfoliation of graphene is an excellent method to create graphene/NP composites.

ACKNOWLEDGEMENTS

C.D. Galindo-Uribe acknowledge CONAHCyT. The doctoral fellowship number is 864427. The authors also acknowledge the Laboratorio Avanzado de Nanoscopía Electrónica (LANE) for the SEM and TEM images.

REFERENCES

Amiri, A.N., Md. Naraghi, G. Ahmadi, M. Soleymaniha and M. Shanbedi. 2018. A review on liquid-phase exfoliation for scalable production of pure graphene, wrinkled, crumpled and functionalized graphene and challenges. FlatChem 8: 40–71.

Botto, L. 2019. Toward nanomechanical models of liquid-phase exfoliation of layered 2D nanomaterials: analysis of a π−peel model. Front. Mater. 6: 302.

Bourlinos, A.B., V. Georgakilas, R. Zboril, T.A. Steriotis and A.K. Stubos. 2009. Liquid-phase exfoliation of graphite towards solubilized graphenes. Small. 5(16): 1841–1845.

Bracamonte, M.V., G.I. Lacconi, S.E. Urreta and L.E.F. Foa Torres. 2014. On the nature of defects in liquid-phase exfoliated graphene. J. Phys. Chem. C. 118(28): 15455–15459.

Chen, K., Q. Wang, Z. Niu and J. Chen. 2018. Graphene-based materials for flexible energy storage devices. J. Energy Chem. 27(1): 12–24.

Chen, J., X. Xie, J. Liu, Z. Yu and W. Su. 2022. Revisiting aromatic diazotization and aryl diazonium salts in continuous flow: Highlighted research during 2001–2021. React. Chem. Eng. 7(6): 1247–1275.

Cheng, Z.L. Y.-C. Kong, L. Fan and Z. Liu. 2020. Ultrasound-assisted Li$^+$/Na$^+$ co-intercalated exfoliation of graphite into few-layer graphene. Ultrason. Sonochem. 66: 105108.

Chung, D.D.L. 2016. Graphite intercalation compounds. pp. 3641–3645. *In*: K.H.J. Buschow (ed.). Encyclopedia of Materials: Science and Technology. Elsevier, Oxford, UK.

Coleman, J.N. 2009. Liquid-phase exfoliation of nanotubes and graphene. Adv. Funct. Mater. 19(23): 3680–3695.

Fiévet, F., S. Ammar-Merah, R. Brayner, F. Chau, M. Giraud, F. Mammeri, et al., 2018. The polyol process: a unique method for easy access to metal nanoparticles with tailored sizes, shapes and compositions. Chem. Soc. Rev. 47(14): 5187–5233.

Galindo-Uribe, C.D., O. Solorza-Feria and P. Calaminici. 2022. Revisión sobre la síntesis de grafeno por exfoliación en fase líquida: Mecanismos, factores y técnicas. Uniciencia. 36(1): 1–14.

Geim, A.K. and K.S. Novoselov. 2007. The rise of graphene. Nat. Mater. 6: 183–191.

Georgakilas, V., M. Otyepka, A.B. Bourlinos, V. Chandra, N. Kim, K.C. Kemp, et al., 2012. Functionalization of graphene: covalent and non-covalent approaches, derivatives and applications. Chem. Rev. 112(11): 6156–6214.

Guardia, L., M.J. Fernández-Merino, J.I. Paredes, P. Solís-Fernández, S. Villar-Rodil, A. Martínez-Alonso, et al., 2011. High-throughput production of pristine graphene in an aqueous dispersion assisted by non-ionic surfactants. Carbon. 49(5): 1653–1662.

Hadad, C., X. Ke, M. Carraro, A. Sartorel, C. Bittencourt, G. Van Tendeloo, et al., 2014. Positive graphene by chemical design: tuning supramolecular strategies for functional surfaces. Chem. Commun. 50(7): 885–887.

Hadi, A., J. Zahirifar, J. Karimi-Sabet and A. Dastbaz. 2018. Graphene nanosheets preparation using magnetic nanoparticle assisted liquid phase exfoliation of graphite: the coupled effect of ultrasound and wedging nanoparticles. Ultrason. Sonochem. 44: 204–214.

Hamilton, C.E., J.R. Lomeda, Z. Sun, J.M. Tour and A.R. Barron. 2009. High-yield organic dispersions of unfunctionalized graphene. Nano Lett. 9(10): 3460–3462.

Hernandez, Y., M. Lotya, D. Rickard, S.D. Bergin and J.N. Coleman. 2010. Measurement of multicomponent solubility parameters for graphene facilitates solvent discovery. Langmuir. 26(5): 3208–3213.

Hummers Jr., W.S. and R.E. Offeman. 1958. Preparation of graphitic oxide. J. Am. Chem. Soc. 80(6): 1339–1339.

Hunter, C.A. and J.K.M. Sanders. 1990. The nature of. pi.-. pi. interactions. J. Am. Chem. Soc. 112(14): 5525–5534.

Huo, C., Z. Yan, X. Song and H. Zeng. 2015. 2D materials via liquid exfoliation: a review on fabrication and applications. Sci. Bull. 60(23): 1994–2008.

Iovel, I.G., Y.S. Goldberg, M.V. Shymanska and E. Lukevics. 1987. Quaternary onium hexachloroplatinates: novel hydrosilylation catalysts. Organometallics. 6(7): 1410–1413.

Khan, M., M.N. Tahir, S.F. Adil, H.U. Khan, M.R.H. Siddiqui, A.A. Al-Warthan, et al., 2015. Graphene based metal and metal oxide nanocomposites: synthesis, properties and their applications. J. Mater. Chem. A. 3(37): 18753–18808.

Kumar, Y.B., R.K. Rawal, A. Thakur and G.N. Sastry. 2022. Reversible and irreversible functionalization of graphene. *In:* F.H. Tandabany Dinadayalane (ed.). Properties and Functionalization of Graphene: A Computational Chemistry Approach (Vol. 21), Elsevier, Cambridge, USA.

Leon, V., A.M. Rodriguez, P. Prieto, M. Prato and E. Vazquez. 2014. Exfoliation of graphite with triazine derivatives under ball-milling conditions: Preparation of few-layer graphene via selective noncovalent interactions. ACS Nano. 8(1): 563–571.

Li, Q., N. Mahmood, J. Zhu, Y. Hou and S. Sun. 2014. Graphene and its composites with nanoparticles for electrochemical energy applications. Nano Today. 9(5): 668–683.

Li, Z., R.J. Young, C. Backes, W. Zhao, X. Zhang, A.A. Zhukov, et al., 2020. Mechanisms of liquid-phase exfoliation for the production of graphene. ACS Nano. 14(9): 10976–10985.

Liu, W.W. and J.N. Wang. 2011. Direct exfoliation of graphene in organic solvents with addition of NaOH. Chem. Commun. 47(24): 6888–6890.

Liu, K., L. Liu, Y. Luo and D. Jia. 2012. One-step synthesis of metal nanoparticle decorated graphene by liquid phase exfoliation. J. Mater. Chem. 22(38): 20342–20352.

Liu, M., R. Zhang and W. Chen. 2014a. Graphene-supported nanoelectrocatalysts for fuel cells: Synthesis, properties, and applications. Chem. Rev. 114(10): 5117–5160.

Liu, K., S. Chen, Y. Luo and L. Liu. 2014b. Hybrid of silver nanowire and pristine-graphene by liquid-phase exfoliation for synergetic effects on electrical conductive composites. RSC Adv. 4(79): 41876–41885.

Notley, S.M. 2012. Highly concentrated aqueous suspensions of graphene through ultrasonic exfoliation with continuous surfactant addition. Langmuir. 28: 14110–14113.

Novoselov, K.S., D. Jiang, F. Schedin, T.J. Booth, V.V. Khotkevich, S.V., Morozov, et al., 2005. Two-dimensional atomic crystals. Proc. Natl. Acad. Sci. 102(30): 10451–10453.

Patel, G., V. Pillai and M. Vora. 2019. Liquid phase exfoliation of two-dimensional materials for sensors and photocatalysis—a review. J. Nanosci. Nanotechnol. 19: 5054–5073.

Paton, K.R., E. Varrla, C. Backes, R.J. Smith, U. Khan, A. O'Neill, et al., 2014. Scalable production of large quantities of defect-free few-layer graphene by shear exfoliation in liquids. Nat. Mater. 13(1): 624–630.

Pearson, R.G. 1968. Hard and soft acids and bases, HSAB, Part 1: Fundamental principles. J. Chem. Educ. 45(9): 581.

Quintana, M., A. Montellano, A.E.R. Castillo, G.A. Tendeloo, C. Bittencourt and M. Prato. 2011. Selective organic functionalization of graphene bulk or graphene edges. Chem. Commun. 47(33): 9330–9332.

Quintana, M., J.I. Tapia and M. Prato. 2014. Liquid-phase exfoliated graphene: functionalization, characterization, and applications. Beilstein J. Nanotechnol. 5(1): 2328–2338.

Raccichini, R., A. Varzi, S. Passerini and B. Scrosati. 2015. The role of graphene for electrochemical energy storage. Nat. Mater. 14(3): 271–279.

Shen, J., Y. He, J. Wu, C. Gao, K. Keyshar, X. Zhang, et al., 2015. Liquid phase exfoliation of two-dimensional materials by directly probing and matching surface tension components. Nano Letters. 15(8): 5449–5454.

Smith, R.J., M. Lotya and J.N. Coleman. 2010. The importance of repulsive potential barriers for the dispersion of graphene using surfactants. New J. Phys. 12(12): 125008.

Su, C.Y., Y. Xu, W. Zhang, J. Zhao, X. Tang, C.H. Tsai, et al., 2009. Electrical and spectroscopic characterizations of ultra-large reduced graphene oxide monolayers. Chem. Mater. 21(23): 5674–5680.

Sumdani, M.G., M.R. Islam, A.N.A. Yahaya and S.I. Safie. 2021. Recent advances of the graphite exfoliation processes and structural modification of graphene: a review. J. Nanopart. Res. 23(11): 1–35.

Truong-Huu, T., K. Chizari, I. Janowska, M.S. Moldovan, O. Ersen, L.D. Nguyen, et al., 2012. Few-layer graphene supporting palladium nanoparticles with a fully accessible effective surface for liquid-phase hydrogenation reaction. Catal. Today. 189(1): 77–82.

Tsoufis, T., Z. Syrgiannis, N. Akhtar, M. Prato, F. Katsaros, Z. Sideratouk, et al., 2015. *In-situ* growth of capping-free magnetic iron oxide nanoparticles on liquid-phase exfoliated graphene. Nanoscale. 7(19): 8995–9003.

Upadhyay, R.K., P.R. Waghmare and S.S. Roy. 2016. Application of oil-swollen surfactant gels as a growth medium for metal nanoparticle synthesis, and as an exfoliation medium for preparation of graphene. J. Colloid Interface Sci. 474: 41–50.

Wajid, A.S., S. Das, F. Irin, H.T. Ahmed, J.L. Shelburne and D. Parviz. 2012. Polymer-stabilized graphene dispersions at high concentrations in organic solvents for composite production. Carbon. 50(2): 526–534.

Wang, S., M. Yi and Z. Shen. 2016. The effect of surfactants and their concentration on the liquid exfoliation of graphene. RSC Adv. 6(61): 56705–56710.

Wang, Y.Z., T. Chen, X.F. Gao, H.H. Liu and X.X Zhang. 2017a. Liquid phase exfoliation of graphite into few-layer graphene by sonication and microfluidization. Mater. Express. 7(6): 491–499.

Wang, S., C. Wang and X. Ji. 2017b. Towards understanding the salt-intercalation exfoliation of graphite into graphene. RSC Adv. 7(82): 52252–52260.

Xu, Y., H. Cao, Y. Xue, B. Li, and W. Cai. 2018. Liquid-phase exfoliation of graphene: an overview on exfoliation media, techniques, and challenges. Nanomaterials. 8(11): 942.

Yang, R., B. Wang, J. Xiang, C. Mu, C. Zhang, F. Wen, et al., 2017. Fabrication of NiCo$_2$-anchored graphene nanosheets by liquid-phase exfoliation for excellent microwave absorbers. ACS Appl. Mater. Interfaces. 9(14): 12673–12679.

Yang, L., F. Zhao, Y. Zhao, Y. Sun, G. Yang, L. Tong, et al., 2018. Enhanced exfoliation efficiency of graphite into few-layer graphene via reduction of graphite edge. Carbon. 138: 390–396.

Yi, M. and Z. Shen. 2014. Kitchen blender for producing high-quality few-layer graphene. Carbon. 78: 622–626.

Chapter 5

Design and Development of Membrane Electrode Assemblies for an Improved ORR towards High-performance PEM Fuel Cells

Veeman Sannasi[1,2], Karmegam Dhanabalan[1], Kanalli V. Ajeya[1],
Pham Tan Thong[1] and Ho-Young Jung[1,2,3]*

[1]Department of Environment and Energy Engineering,
Chonnam National University, 77 Yongbong-ro, Buk-gu,
Gwangju, 61186, Republic of Korea.

[2]Center for Energy Storage System, Chonnam National University,
77 Yongbong-ro, Buk-gu, Gwangju, 61186, Republic of Korea.

[3]Specialized Research Center for Hybrid Power Pack,
Chonnam National University, 77 Yongbong-ro, Buk-gu,
Gwangju, 61186, Republic of Korea.

5.1 INTRODUCTION

The extensive use of fossil fuels for energy demand increased with the growing population, modernization of lifestyle, and miniaturization of portable electronics.

*For Correspondence: Email: jungho@chonnam.ac..kr

It is well known that in the fast-growing energy-consuming market, no single energy source can fulfill all energy demands, albeit the recent focus on renewable energies, such as geothermal, tidal, wind, and solar energy by various countries for energy application. The drawback of such alternative energy is their intermittent energy supply. Though great interest has been paid to alternative energy in recent times, their low share compared to fossil fuels leads us to depend on fossil fuels to fulfill the requirements, thus ultimately increasing the atmospheric CO_2, which causes greenhouse gas emissions and global warming (Victor et al., 2014). On the other hand, intermittent alternative energy needs to be stored by proper storage devices, such as batteries, supercapacitors, and fuel cells combined with electrolyzers.

In the energy consumption sector, electricity and transportation consume more than others. In 2017, approximately 19.5% of total energy consumption devoted on transportation in the Republic of Korea (Boo, 2021). At the same time, fossil fuel usage for transportation could be minimized by focusing on alternative energy vehicles, including electric ones. Electric vehicles' low-carbon transport routes are classified as battery electric vehicles (BEV) and fuel cell electric vehicles (FCEV). Among these fuel cells and batteries, fuel cells are used as energy conversion devices rather than energy storage, like lithium-ion batteries. Presently various countries have invested in electric vehicles to minimize fossil fuel usage, albeit this has shortcomings such as the large scale of power source, long charging time, short cruising range, limited promotion, cost, and potential safety that needs improvement. Since 2015, various FCEVs have been introduced globally by prominent corporations, like Hyundai Nexo, Hyundai ix35, Toyota Mirai, and Honda clarity. These FCEVs further required modifications, and improvements in their components, making design to attain long, short drive heavy- as well as light-duty (Cano et al., 2018; Bethoux, 2020; Cullen et al., 2021).

Hydrogen production effectively fulfills the required energy demands without much environmental concern and stores intermittent energy from alternative energy sources. Hydrogen gas is well known as a clean energy carrier due to its unique properties and a high abundance of 75% normal matter by mass and 90% normal matter by a number of atoms. Until the last decade, though the majority of hydrogen has been used as feedstock for chemical products, such as methanol, ammonia, and petroleum refining in industries, recent developments indicate that by a suitable mechanism, like a fuel cell, hydrogen can be better used for other energy fields, such as power generation, transportation, and militarized equipment. Fuel cells convert chemical energy, utilizing fuel as an oxidant, into electrical energy if they can be fed chemical energy.

The first fuel cell was invented separately by Welsh physicist Sir William Grove in 1838 and by German physicist Christian Friedrich Schonbein in 1839. Following the invention of hydrogen-oxygen fuel cells by Francis Thomas Bacon in 1932 and alkaline fuel cells by him also in 1960, fuel cells are widely used for primary and backup power applications. Fuel cells can be classified according to their electrolyte, fuel feed, and catalyst, such as polymer electrolyte fuel cells (PEM), alkaline fuel cells (AFC), solid oxide fuel cells (SOFC), direct methanol fuel cells (DMFC), direct formic acid fuel cells (DFFC), molten carbonate fuel cells (MCFC), phosphoric acid fuel cells (PAFC) and recently unitized regenerative fuel cells (URFC). Further fuel cells are classified into high-, mid-, and low-temperature fuel cells by operating temperature conditions (Fan et al., 2021).

Among the different fuel cell technologies, the proton exchange membrane (PEM) fuel cell is versatile in different applications, such as in portable systems, small stationery, transportation, and power generation due to its less carbon emission and carbon footprint and neutrality. Furthermore, PEMFC's flexibility of operating temperature, compact size, low weight and volume, high efficiency, and ability to work in start-stop conditions make it a versatile technology. Albeit the over-welcome of the PEMFCs, there is room to improve their efficiency and applicability to emulate fossil fuel as primary energy source, such as its capital and operational cost of FCEV are higher than gasoline vehicles.

A large share was spent for electrocatalysts in the fuel cells' capital cost, thus mainly occupied by precious platinum group elements, such as platinum, palladium and materials containing these two along with other elements like carbon-supported PGM, alloys and PGM catalysts. However, commercial PEM fuel cells are made up of costly platinum group metal (PGM) catalysts due to their comparatively high oxygen reduction reaction (ORR) rate than other non-PGM catalysts. The slow ORR reaction rate is the main limiting factor in the fuel cell technology; thus somehow managed by using platinum electrocatalyst, which has good oxygen absorption. Although in recent times a lot of improvements made to minimize the amount of PGM electrocatalysts by moving towards non-PGM catalysts, they are not up to PGM materials, and a lot of investments are to be made on a research scale (Tellez-Cruz et al., 2021).

One of the most important components in PEMFC is the polymer electrolyte membrane (PEM) that transports the proton from the anode to the cathode for completion of the fuel cell reaction. Nafion, a perfluorinated sulfonic acid polymer, is well known as a typical PEM for PEMFC application because it can give high proton conductivity, good chemical/mechanical stability, and good thermal stability. However, it shows some drawbacks, such as high permeability and high-cost issues. Even though some alternative polymer electrolyte membranes, such as hydrocarbon-based membranes, organic/inorganic composite membranes, and reinforced membranes, have been suggested for application to PEMFC, there is no promising candidate due to chemical degradation, nanoparticle dispersion issue, and interfacial adhesion between ionomer and substrate. To find some alternative PEM to Nafion, huge amounts of energy, such as money, time, and human power have been focused on this area.

Among the fuel cell components, membrane electrode assembly comprising of a mesoporous transport layer, gas diffusion layer, electrocatalyst, and PEM membranes have a high percentage of share in efficiency, durability as well as capital cost than the other components like an end plate and bipolar plates (Sadhasivam et al., 2020; Sadhasivam et al., 2021). Though each component in the membrane electrode assembly triggers the performance of PEM fuel cells and a vast amount of research is focused on individual components, stacking the membrane electrode assembly also has a unique role in the performance. In recent times, considerable attention has been given to MEA to improve its performance towards high-performance PEM fuel cells. From the viewpoint of this component issue in MEA, it is necessary to discuss MEA technology in recent years.

5.2 FUEL CELLS

Generally, the term 'fuel cell' refers to an electrochemical cell that can be used to produce electricity through the use of fuels; in other words, fuel cells are electrochemical energy conversion devices that use oxidants as an aid to convert the chemical energy of fuel into electricity when it is either gaseous or liquid. Fuel cells fall among the three electrochemical technologies that include energy conversion, synthesis, and energy storage. The main difference between electrochemical synthesis and the other two electrochemical technologies is in electrochemical synthesis, where electrical energy is applied to derive chemical synthesis in which the conversion of electrical energy produces chemical energy. In contrast, electrical energy is produced from chemical energy in electrochemical conversion and storage systems. In energy storage systems, we can draw electricity until the stored energy vanishes and recharged in the case of secondary batteries. In electrochemical fuel cells, we can get electricity if the fuel is fed. Any substance that tends to chemical oxidation and is galvanically burned can be used as fuel at the anodic part of the fuel cell, whereas a substance that can be reduced at the cathodic part of a fuel cell is used as an oxidant. Commonly hydrogen is used as fuel. In most fuel cell types, hydrogen is used as fuel due to its availability in terrestrial applications of hydrocarbon, and high energy density in special applications. Oxygen is the most used oxidant due to its availability and usage in a closed environment. The hydrogen combustion of a simple fuel cell is split into two electrochemical half-cell reactions indicated in the equation:

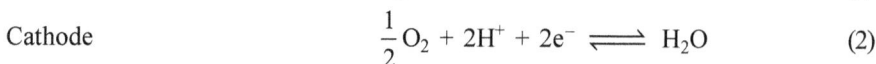

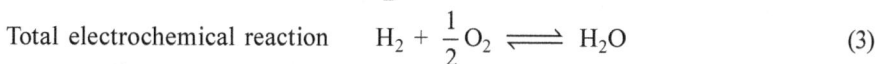

Anode $\qquad\qquad H_2 \rightleftharpoons 2H^+ + 2e^-$ $\qquad\qquad$ (1)

Cathode $\qquad\qquad \dfrac{1}{2}O_2 + 2H^+ + 2e^- \rightleftharpoons H_2O$ $\qquad$ (2)

Total electrochemical reaction $\qquad H_2 + \dfrac{1}{2}O_2 \rightleftharpoons H_2O$ $\qquad$ (3)

When the above two half-cell reactions are separated, the transfer of electrons from the feed fuels is forced to flow through an external circuit. The reaction will be completed once the flowing electrons are converted into useful work. In a typical fuel cell, the separation of half-cells was done by normal electrolytes that selectively allow ions and not electrons. In the model fuel cell demonstrated by Grove, he used sulfuric acid as an electrolyte and platinum as an electrode, thus use was made with a modification of carbon-supported platinum as an electrode and perfluoro-substituted sulfonic acid (PFSA) as electrolyte membranes nowadays. Five major types of fuel cells classified based on electrolytes are (i) PEMFC, (ii) AFC (iii) PAFC (iv) SOFC (v) MCFC.

Further, depending on the temperature, they are classified into three types:

1. Low-temperature fuel cell which operates at the range of 10–80°C (PEMFC, DMFC, DFFC, AFC).

2. Intermediate fuel cell operating at a temperature range of 120–200°C (PAFC, high-temperature PEMFC).
3. High-temperature fuel cell which operates at 650–1000°C range (direct and indirect MCFC, direct SOFC and indirect SOFC, and hybrid SOFC/MCFC/ gas turbines).

Further, due to the researcher's great interest and recent developments, a variety of fuels have been studied for fuel cells, such as;

(i) Other carbon compound fuel cells (acetylene and higher hydrocarbon-oxygen fuel cell, ethanol fuel cell).

(ii) Metal hydrogen fuel cell (Zn/O_2, Al/O_2).

(iii) Biochemical fuel cell (heart-energy metabolism, glucose/oxygen, living metabolism).

(iv) Nitrogenous fuel cells (ammonia/oxygen, hydrazine/oxygen fuel cell).

(v) Regenerative fuel cells (electrical and radiochemical fuel cells, solar-photovoltaics electrical such as hydrogen/bromine and hydrogen/chlorine).

Though there are so many classifications of fuel cells, the most common classifications are based on operating temperature, fuel, and electrolytes (Winter and Brodd, 2004; Sharaf and Orhan, 2014). Different types of fuel cells with their operating temperature, transfer species, electrode reactions and power density are given in Table 5.1

5.2.1 Polymer Electrolyte Membrane Fuel Cell

Among all the fuel cells installed and applied in technology, more than 80% of units were proton exchange membrane fuel cells (PEMFC). The name demonstrates that these fuel cells contain polymer electrolyte membranes that selectively pass protons. PEMFCs show their broad applicability in micro-portable to large-scale stationary stations. Recently, many companies have directed their interest in developing PEMFCs, making PEMFC the only mature technology among the fuel cell technologies and requiring the least infrastructure. Moreover, PEMFCs have the advantages of a fast startup, low cost, the ability to work for larger thermal cycles, and high hydrogen conversion efficiency (Litster and McLean, 2004). Though the working temperature of PEMFC typically lies in the range of 70–80°C, it requires to spread its applicability between 120–150°C for power generation and automotive applications, in which this temperature range is required for high carbon monoxide (CO) tolerance and efficient heat exchange (Shao et al., 2007).

Currently, the challenges for PEMFCs have improved performance, searching for an effective and low-cost electrocatalyst for oxygen reduction reaction, water management, cost of full stack, freeze start, and thermal management. In particular, improvements are needed in MEA and water management. During the PEM fuel cell operation, water is produced in the cathodic part, which wets the membrane and conducts the proton. At the same time, the produced excess water can block the GDL's pores and restrict oxygen transport towards an electrocatalyst (Habib et al., 2021; Qu et al., 2022). The PEMFC's basic design is shown in Fig. 5.1, in which hydrogen gas was passed at negative side and air was passed at positive electrode side to react with corresponding electrocatalyst to give electricity.

Table 5.1 Fuel cell type and characteristics

Fuel cell type	Operating temperature (°C)	Transfer species/ electrolyte	Electrodes/reactions	Power densities (mW/cm²)
Alkaline (AFC)	60–220	Liquid KOH/OH⁻	Anode $H_2 + 2OH^- \rightarrow 2H_2O + 2e^-$ Cathode $1/2O_2 + 2H^+ + 2e^- \rightarrow H_2O$	150–400
Polymer (PEM)	80	Polymer membrane/H⁺	Anode $H_2 \rightarrow 2H^+ + 2e^-$ Cathode $1/2O_2 + 2H^+ + 2e^- \rightarrow H_2O$	500–2500
Direct methanol (DMC)	60–130	Polymer membrane/H⁺	Anode $CH_3OH + H_2O \rightarrow 6H^+ + 6e^- + CO_2$ Cathode $3/2O_2 + 6H^+ + 6e^- \rightarrow 3H_2O$	50–100
Phosphoric acid (PAFC)	200	Liquid H_3PO_4/H⁺	Anode $H_2 \rightarrow 2H^+ + 2e^-$ Cathode $1/2O_2 + 2H^+ + 2e^- \rightarrow H_2O$	150–300
Molten carbonate (MCFC)	650	Molten carbonate/CO_3^{2-}	Anode $H_2 + CO_3^{2-} \rightarrow H_2O + CO_2 + 2e^-$ $CO + CO_3^{2-} \rightarrow 2CO_2 + 2e^-$ Cathode $1/2O_2 + CO_2 + 2e^- \rightarrow CO_3^{2-}$	100–300
Solid oxide (SOFC)	600–1000	Ceramic/O^{2-}	Anode $H_2 + O^{2-} \rightarrow H_2O + 2e^-$ $CO + O^{2-} \rightarrow CO_2 + 2e^-$ $CH_4 + 4O^{2-} \rightarrow 2H_2O + CO_2 + 8e^-$ Cathode $1/2O_2 + 2e^- \rightarrow O^{2-}$	250–500

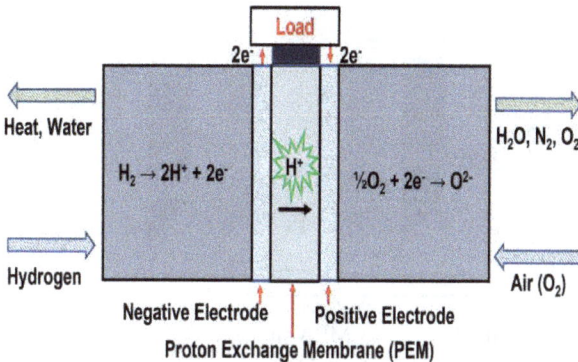

Figure 5.1 Schematic presentation of the PEMFC and its working principle.

5.2.1.1 Applications of PEM Fuel Cell

The requirement of PEMFC is highly focused on its usage in the backup power applications, transportation applications, and stationary applications. Various manufacturers, such as ReliOn, Ballard, Plug, and Nuvera have developed commercial backup power fuel cells using PEMFC with compressed hydrogen fuel cells to achieve a power level between 1–500 kW. In transportation applications, PEMFC is used in all types of transportation, including automobiles, buses, scooters, bicycles, and utility vehicles. The world's leading automobile manufacturers including Hyundai, Toyota, Honda, Volkswagen, Benz, BMW, Ford, and GM demonstrated PEMFCs for vehicles and developed new technologies. Recently most countries have emphasized the development and commercialization of FCEVs for heavy-duty vehicles (HDV) (Cullen et al., 2021)

5.2.1.1a Membrane Electrode Assembly (MEA)

The fuel cell's efficiency and stability mainly depend on the membrane electrode assembly (Sassin et al., 2017). The preparation technologies of MEA influence the durability, performance, and cost of the PEMFC through its degradation, electrochemical reactions, and mass transport through microchannels, and production method. MEA comprises initially of three components such as membrane and anode electrocatalyst and cathode electrocatalyst and further extended to four main components: a polymer electrolyte membrane (PEM), a catalyst layer (CL), a mesoporous layer (MPL), and gas diffusion layers (GDL) (Pettersson et al., 2006). The typical MEA's schematic diagram is shown in Fig. 5.2. It consists of three categories: functional, structure, and transition. The functional region that consists of PEM, CL, MPL, and GDL can generate electric power directly related to the performance of the PEMFC. The structure region, which includes adhesives, seals, and resin frames, does not influence the performance of PEMFCs but can

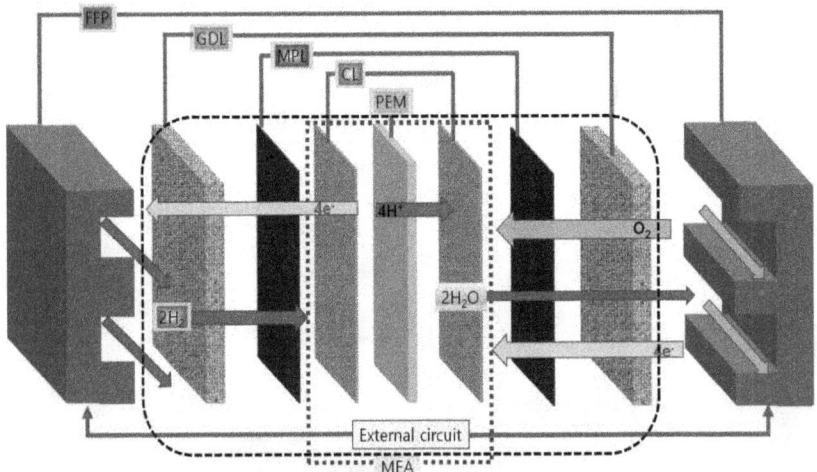

Figure 5.2 Schematic diagram of MEA in a PEMFC (reproduced with permission from Elsevier [Xing et al., 2019]).

affect the mechanical stability and durability of PEMFCs. The transition region is a gap between the functional and structure regions and should be overlapped for the high lifetime and performance of the PEMFCs.

During hydrogen dissociation, the solid acidic polymer membrane (polymer electrolyte membrane) allows protons to pass through it and react with the catalyst on each electrode. The proton conduction of the polymer membranes at higher water-uptake levels can be followed by Grotthus (hopping) mechanism in which a hydronium ion is formed by combining the proton formed at the anode with a water molecule. On the other vehicular (diffusion) mechanism, the hydronium ions pass through the water at low water-uptake levels of PEM (Bhosale et al., 2020). To achieve high performance in the PEMFC, the membrane should have higher chemical, electrochemical, and mechanical stability including higher proton conductivity in low humid conditions.

The thickness of the membrane is an essential factor, thus affecting the power density and stability of the PEMFCs; use of thin membranes causes fuel cross-over, mechanical damage, and electrochemical instability, whereas use of thick membranes causes self-humidification and limits proton, and water transport. Various modifications have been done to improve the membranes' properties by reinforcing the perfluorosulfonic acid membranes (PFSA) with another stable polymer substrate, polytetrafluoroethylene (PTFE).

Further, the stability of the thin membranes can be improved by compositing them with inorganic materials like Cerium salts (Jiao et al., 2021). Apart from perfluorinated PFSA polymers, which dominate in the PEMFC as membranes nowadays, various other polymers, including sulfonated poly(ether sulfone) (SPES), sulfonated polyimides (SPI), sulfonated poly(ether ether ketone)s (SPEEK) and sulfonated polyphenylene-based PEMs were used due to their promising high proton and stability in higher humid and temperature conditions. However, the hydrocarbon-based membranes suffered from low proton conductivity at low humidity conditions; some of the membranes, such as sulfonated polysulfones, interpenetrating network membranes, and phosphoric acid, show very good stability or performance. Furthermore, membranes with orientation, particularly through the plane, showed high water retention behavior, and the incorporation of ferrocyanides favors proton conduction with less free-radical degradation.

Due to the direct interaction of the catalyst layer with the gas diffusion layer and membrane, developing a catalyst layer for efficient fuel cells has been of great interest over the past few years. The performance of the catalyst layer also affects the power density of the PEMFCs. To achieve a high-power density, the PEMFC must have over 0.8 V cell voltage and 4.4 A cm^{-2} current density which requires modifications in catalyst activity and the catalyst layer. One can improve the mass or specific activity by careful design of the catalyst as nanocage, nanowires, nano frames, and core-shell like novel architectures. The highest mass activity of 13.6 A and specific activity of 11.5 mA cm^{-2} were reported per milligram Pt (mg$_{pt}$) for ultrafine jagged platinum nanowires in the rotating disk electrode setups (Jiao et al., 2021).

The gas diffusion layer including the mesoporous layer is composed of a porous material that effectively removes reactants and products from electrodes.

As gas diffusion layers in PEMFCs, materials that are electrically conductive and porous have been employed, such as carbon cloth or carbon paper. The essential criteria for a gas diffusion layer are high electronic conductivity, low cost, ease of preparation, ease of modification of porosity, and chemically/mechanically stable; at the same time, it can well interact with the adjacent PEMFC components. Typically available gas diffusion layers have the different thicknesses (0.0017–0.04 cm), porosity (70–80%) and density (0.21–0.73 g cm^2). The GDL is also part of the water management of the PEMFC due to its role as liquid water exit at the cathode. The thickness of the GDL directly affects the electrical resistance, gas permeability, and water drainage of the PEMFC. Four different GDL configurations are used in PEMFCs, which are shown in Fig. 5.3, such as two layers consisting carbon cloth (CC) and a mixture of carbon cloth and carbon-polymer (CCCP), only one layer of a mixture of carbon cloth and carbon-polymer (CCCP), two layers of CPCC and carbon-polymer (CP), and a structure containing CP–CPCC–CC–CPCC, all the four layers in the order (Fan et al., 2021).

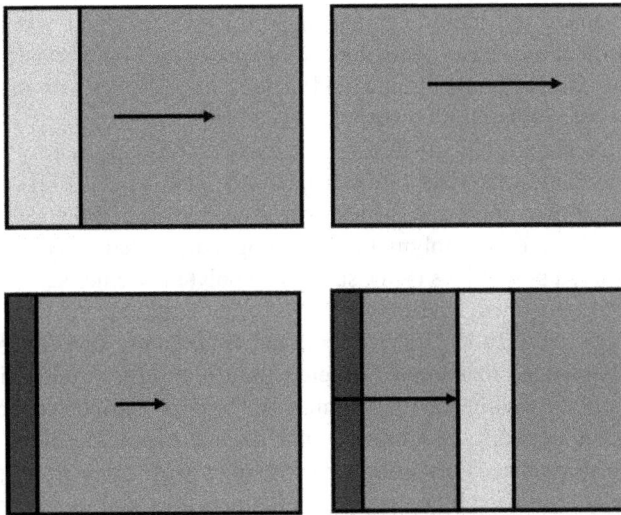

Figure 5.3   Structural configuration of GDL (reproduced with permission from Elsevier [Sitanggang et al., 2009]).

5.3 OXYGEN REDUCTION REACTION (ORR)

Apart from the functions of a PEM, PEMFCs performance is highly affected by the properties of catalyst layers where the prenominal oxygen reduction reactions (ORR) and hydrogen oxidation (HOR) occur. Among the HOR and ORR reactions, ORR controls the reaction rate of the overall PEMFCs and one of the challenges for achieving high-performance PEMFCs relies on overcoming the sluggish ORR rate of at least four to six orders slower than the counterpart HOR rate. The mechanistic equation for the ORR taking place at the cathode side of a PEMFC is given as follows:

$$O_2 + 4H^+ + 4e^- \rightarrow 2H_2O \quad E^0 = +1.229 \text{ V } vs. \text{ RHE} \quad (4)$$

$$O_2 + 2H^+ + 2e^- \rightarrow H_2O_2 \quad E^0 = +0.682 \text{ V } vs. \text{ RHE} \quad (5)$$

$$H_2O_2 + 2H^+ + 2e^- \rightarrow 2H_2O \quad (6)$$

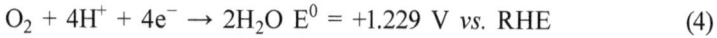

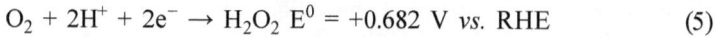

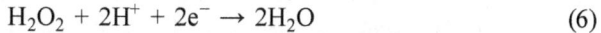

The exact mechanistic pathway for the ORR reaction is not yet revealed completely, but the most accepted pathways were the 2-electron and 4-electron pathways described in the above equation. In the $2e^-$ ORR pathway, the Pt atoms' surface adsorps the O_2 molecules and then forms HOOH when combined with two hydrogen protons. The unstable intermediate then may follow either reduction to form H_2O, oxygen formation by re-oxidation, or detached as H_2O_2. Two possible pathways were postulated for the $4e^-$ mechanism, (i) oxygen binding pathway and (ii) oxygen dissociation pathway. In the oxygen-dissociation pathway, the O–O bond of the adsorbed oxygen molecule at the Pt atom's surface breaks into two atomic oxygen in the adsorbed state, thus forming OH when combined with one hydrogen and then forming H_2O when combined with another proton. On the other hand, an intermediate OOH was formed by the adsorption of an oxygen molecule at the Pt surface and then broken into OH by breaking the O–O bond, and finally, the formed OH is reduced to form H_2O. Detailed reaction equations for the $2e^-$ and $4e^-$ mechanistic pathways are given below (Xie et al., 2021):

$2e^-$ pathway

$$O_2 + {}^* \rightarrow O_2^*$$
$$O^* + H^+ \rightarrow OOH^*$$
$$OOH^* + H^+ \rightarrow H_2O_2^*$$
$$H_2O_2^* + {}^* \rightarrow 2OH^*$$
$$2OH^* + 2H^+ \rightarrow 2H_2O$$

$4e^-$ Oxygen-binding pathway

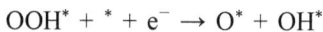

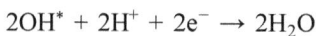

$$O_2 + {}^* \rightarrow O_2{}^*$$
$$O_2^* + H^+ \rightarrow OOH^*$$
$$OOH^* + {}^* + e^- \rightarrow O^* + OH^*$$
$$O_2^* + H^+ + e^- \rightarrow OH^*$$
$$2OH^* + 2H^+ + 2e^- \rightarrow 2H_2O$$

$4e^-$ oxygen-dissociation pathway

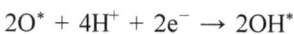

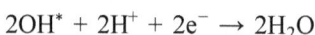

$$O_2 + {}^* \rightarrow O_2^* \rightarrow 2O^*$$
$$2O^* + 4H^+ + 2e^- \rightarrow 2OH^*$$
$$2OH^* + 2H^+ + 2e^- \rightarrow 2H_2O$$

In PEMFCs, the $2e^-$ pathway is considered as an unavoidable side reaction formed due to defects, interparticle distance, or active site densities. Though the $2e^-$ pathway affects the performance of PEMFC by affecting the membranes like $4e^-$ pathways, its production yield is very low ($\leq 2\%$) (Tian et al., 2020). To overcome ORR's sluggish behavior, various high-active electrocatalysts, particularly platinum-based catalysts, are used in PEMFCs. Apart from the PGM electrocatalysts, researchers also showed great interest in non-PGM electrocatalysts and metal-free electrocatalyst (Li et al., 2019).

5.3.1   Fabrication of Membrane Electrode Assembly (MEA)

The fabrication of membrane electrode assembly consists of different interfacial junctions like membrane ‖ CL (anode), membrane ‖ CL (cathode), CL (anode) ‖ GDL, CL (cathode) ‖ GDL, and all the interfacial junctions will play an important role in PEMFCs overall performance.

5.3.2   Preparation of Polymer Membranes

Albeit the fuel cell's performance depends on the chemical structure and properties of the polymer, the thickness and the method of polymer coating also imparts its properties. In particular, the polymer membranes' thickness and mechanical properties depend on how they are prepared. Two types of methods have been used extensively for the preparation of membranes: solution casting and melt extrusion methods (Mokrini et al., 2010). In the melt extrusion method, the polymer or composite was fed into an extruder equipped with a film line at high temperatures. In this method, we can adjust the thickness of the membranes by modifying the gap between the film line and die; screw RPM calendar rolls' speed, and torque. This melt-extrusion method has been well used for preparing membranes from the 1960s to 2000s and is well suited for composite membranes. Whereas in the solution casting method, in which a polymer is dissolved, an adequate amount of high boiling solvent to get preferable viscosity is used for casting to get the membranes. Further, in solution casting, the thickness of the membrane is tuned by using a suitable doctor blade. Once the membranes are prepared, they should be pre-treated to remove any impurities and unreacted functional groups. Oxidizing agents, such as HNO_3 (~35 wt.%), H_2SO_4 (~1 M), and H_2O_2 (~5 wt. %) were used for pretreatment in which H_2SO_4 can improve the membrane's proton content as well (Roh et al., 2016a).

Though various polymer materials are used for the membrane purpose, three major group polymers used are poly(perfluorosulfonic acid) (PFSA)-based polymers containing short and long chains. Among the PFSA-based polymers, Nafion is extensively used in all membrane technologies due to its standout performance, such as proton conductivity (0.1 to 0.25 S/cm) at moderate temperature range up to 90°C and stability. The second type of PEM polymers contains poly(arylene ether ketone) (SPAEK) and sulfonated (ether ether ketone) (SPEEK) like sulfonated ether group polymers. The high thermal stability with high ion exchange capacity (1.75–2.05 meq/g) makes it viable for intermediate-temperature PEMFC. The third significant polymer used in PEMFCs is based on polybenzimidazole (PBI) derivatives. These polymers having high glass transition temperature and acidic-basic resistivities make them prominent polymer membranes for high-temperature PEMFCs. Apart from these prominent polymer groups, other polymers, including polyphenylene oxides (PPO), poly(aromatic), and poly(heteroaromatic) polymers also have been focused on recently.

5.3.3 Preparation of Catalyst Layer

Apart from the chemical composition and morphological structure of the anode and cathode electrocatalyst materials which showed good performance in three-electrode RDE systems, the catalyst layer formation also affects the performance of its real PEMFC application. The catalyst layers for the PEMFCs can be fabricated by two methods: catalyst-coated substrate/gas diffusion electrode (CCS/GDE) and catalyst-coated membrane (CCM). A schematic diagram of CL preparation process is shown in Fig. 5.4. In CCM, the most used and developed method for preparing the catalyst layer consists of directly coating the prepared catalyst slurry on the PEM and then drying. Compared with the CCS/GDE, CCM showed less membrane ∥ CL contact resistance. In some cases, a membrane expansion was observed during the coating of wet catalyst on PEM membrane, which is a transfer printing method in which an inert substrate like polyimide (PI) or polyethylene terephthalate (PET) film is used as a substrate for catalyst coating; then it is hot pressed with PEM and finally peeled off from the catalyst layer. In the CCM method, a very low amount of catalyst (0.8 mg cm^{-2} Pt loading at the cathode and 0.4 mg cm^{-2} Pt loading at the anode) is required compared with CCS (Pt loading 4 mg cm^{-2}). Apart from the conventional drop cast method, inject printing and spray coating methods are also used to prepare CCM electrodes.

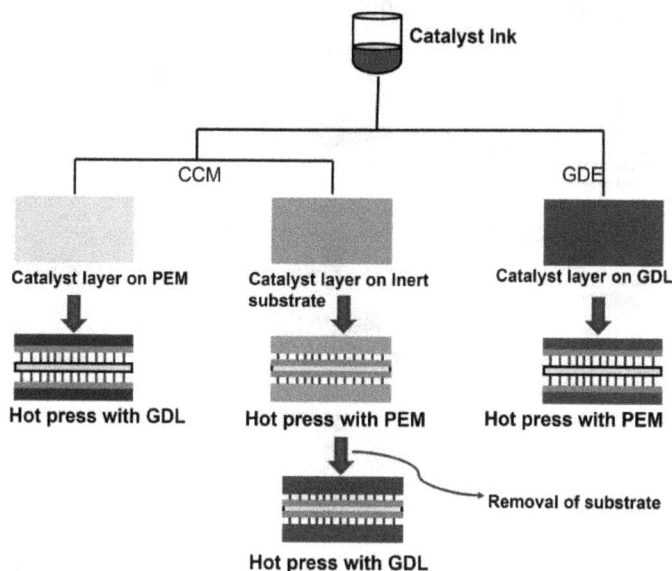

Figure 5.4 A schematic diagram of CL preparation process (modified with permission from Elsevier [Shahgaldi et al., 2018]).

Whereas in CCS/GDE process, the catalyst slurry was coated on the substrate that might be either a solid substrate or a gas diffusion layer (Shahgaldi et al., 2018). The catalyst slurry was mixed with the corresponding ionomer to improve the proton path and enhance the adhesiveness between the membranes and GDL. The catalyst materials sometimes use support material that has high electrical conductivity, porosity, corrosion resistivity, and water-handling ability. In most of the cases, carbon in particular, Vulcan XC-72 was used as support for the catalyst in PEMFCs even though a lot of other supports are also available and reported in the literature, such as teflonized carbon paper, CNTs, TiCN, PtRu, antimony doped tin oxide (ATO), Ta-TiO$_2$, TiO$_2$, SiC, and polypyrrole (Sitanggang et al., 2009).

5.3.4 Preparation of Gas Diffusion Layer and Mesoporous Layer

Apart from acting as an electron conductor between the catalyst layer and conductor, GDL in PEMFCs also allows fuel diffusion in the PEMFC. Furthermore, besides these primary functions, GDL also passes the formed water to the oxygen side, which requires high hydrophobicity. Typically GDL layers are coated with a low amount of Teflon to improve the hydrophobicity, and the amount of Teflon concentration is optimized as 15 wt.%. Normally the gas diffusion material is pretreated with PTFE, ammonia, and water for the teflonization, wherein the modification of GDL with the addition of carbon improves its contact with the catalyst and favors good water transport and is termed as a mesoporous layer. Figure 5.5 shows the typical representation of CCE using the brush method. Though various materials were reported, mostly carbon materials with PTFE are used for MPL (Sadhasivam et al., 2016; Bhosale et al., 2020).

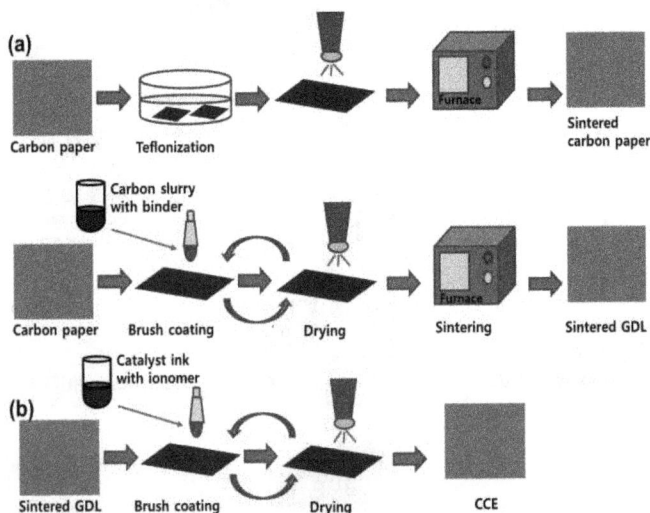

Figure 5.5 Schematic representation of (a) fabrication of GDL layer with MPL and (b) preparation of catalyst-coated electrode (CCE), using the brush coating method (modified with permission from Elsevier [Bhosale et al., 2020]).

5.3.5 Assembly of MEA

Once all the components are prepared, the final step of the MEA fabrication is sandwiching the individual components. Various methods have been used to sandwich the components, such as ionomer dispersion, conventional decal, and low temperature decal method, platen-based press and roll press methods. When using hot pressing methods, care should be taken to avoid the coverup of the catalyst layer on the other layers.

Figure 5.6 shows the conventional and low temperature decal method of MEA fabrication. In the conventional decal method, the catalyst-coated substrate and the salt form of the membrane are combined by a hot press (at more than 130°C). After peeling off the substrate and acid treatment, MEA with the H$^+$ form of the membrane was obtained, while in the low temperature decal method, in the first step, the carbon layer was coated on the substrate and then the catalyst layer was coated on the carbon layer. A solution of ionomer is coated on the prepared catalyst layer and combined with the H$^+$ form of the membrane using a hot press. Finally, the substrate is peeled off, giving the required MEA. Apart from the conventional flat decal method, the roll press decal method was also used to obtain well-uniformed MEA in which pressure-controlled rotating rollers press all the electrode components and membrane. A typical roll press machine and comparative MEA obtained by using the decal method and roll press methods are shown in Fig. 5.7.

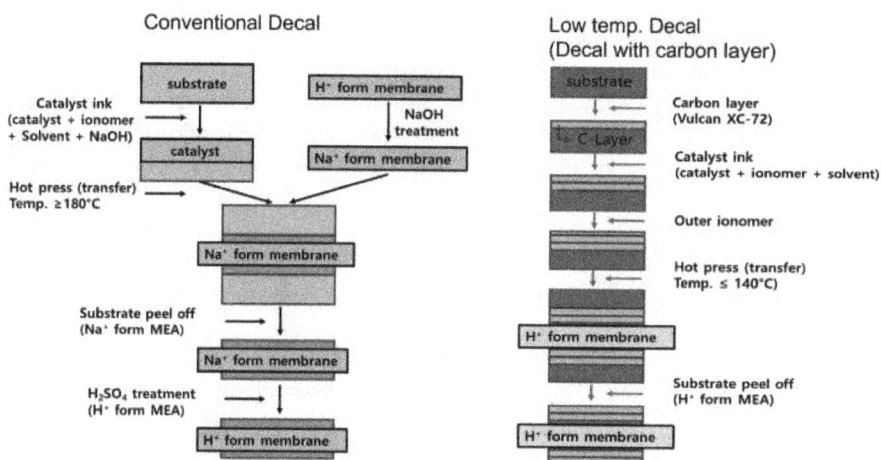

Figure 5.6 Conventional and low temperature decal method of MEA fabrication (modified with permission from Elsevier [Saha et al., 2010]).

MEAs can also be prepared by spray coating in which the cathode is coated on the substrate by using a gasket followed by spray deposition of ionomer solution, spray coating of anode catalyst, and finally, the GDL layer is attached on the anode side (Bhosale et al., 2020). Schematics of the process for spray-coated MEA is given in Figure 5.8.

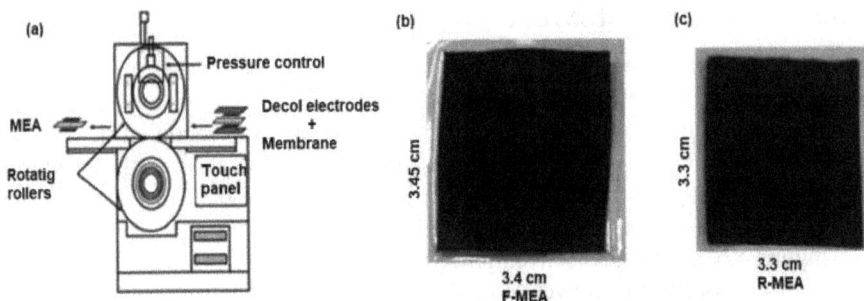

Figure 5.7 Schematic diagram of a roll-press machine; photographs of MEAs; presenting changes in geometrical s shape with © flat decal method and (d) roll press (reproduced with permission from Elsevier [Mehmood and Ha, 2013]).

5.4 PROBLEMS IN CURRENT TECHNOLOGY

5.4.1 Cost and Degradation of the Electrocatalyst

Various medications have been developed to improve the cathodic ORR in PEMFCs by modifying the structure and morphology of platinum-based electrocatalysts, such as alloying, core-shell, carbon supports, shape control in nanostructures, Pt nano, as well as Pt-coated whiskers. US-based Department of Energy reveals that in large-scale PEMFCs, electrocatalyst occupies about 50% of the total cost. Although this modification yielded benefits by employing catalytic activity and cost saving by low platinum loading, it is still not viable for the technology level due to its inability to meet the end users' targets, such as power density, cost, and durability for automobile industries. Though the state of the art of PGM-based catalyst showed high mass activity in a half cell as well as in an H_2/O_2 full cell (over 13 A $mgPt^{-1}$ at 0.9 V [vs RHE] and up to 1.77 A $mgPt^{-1}$ at 0.9 V voltage) which is much higher than the DOEs target (0.44 A $mgPt^{-1}$). Whereas PGM-free M–N–C catalysts showed good current density (Fe–N–C cathode, >30 mA cm^{-2} at 0.9 V in fuel cell MEAs than the DOE target (44 mA cm^{-2}) and even closer to commercial Pt/C, the stability of the PEMFCs is in consideration (Shao, 2019; Du et al., 2021). Though various methods have been developed to minimize the cost of the electrocatalyst and MEA components, it further affects its performance in other ways, such as when using PGM alloys and nanomaterials, metal dissolution and a decrease in electrochemical surface area (ECSA) occurs (Mehmood and Ha, 2013). Wherein using low loading PGM catalyst can increase the mass transfer resistance; for example, manifold increase in total oxygen transport resistance (43.1%) and local oxygen transport resistance (154.9%) was reported for low-PGM MEA (0.05 mgPGM cm^{-2}) than MEA loaded with high PGM (0.15 mgPGM cm^{-2}) that showed 12.3% and 95.1% of total and local oxygen resistance (Borup, 2019).

Carbon corrosion is the other factor that affects the performance of the MEA and the fuel cells' stability. It directly decreases the electrocatalyst's performance by degrading the catalyst-support interaction. Further, carbon corrosion influences the ionomer orientation and hydrophobicity of the membrane. In the normal temperature PEMFC, managing liquid water is a concern because it affects the performance.

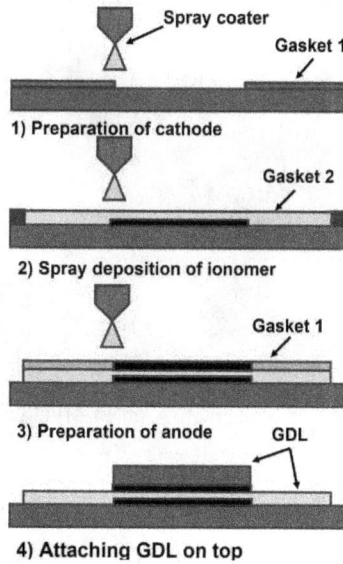

Figure 5.8 Schematics of the process of spray-coated MEA (modified with permission from Elsevier [Bayer et al., 2016]).

5.4.2 Degradation of PEM

During the operation of PEMFCs, the PEM in the MEA suffers from chemical and mechanical degradation. The degradation of the PEM is caused by the deterioration of structural integrity and microstructure of the materials that lead to the formation of local thinning, pinholes, and cracks. While the mechanical degradation of the membranes leads to delamination, wrinkles, cracks, or pinholes, thus formed by the changes in humidity during start/stop cycles, change in temperature, and uneven assembly (Xing et al., 2021). The chemical degradation of the PEM was caused by its thickness and equivalent weight. For example, the proton conductivity of the PEM was improved by the low equivalent weight/high sulfonic acid groups, which lead to poor chemical stability because higher sulfonic acids can easily be affected by free radicals, such as hydroxy ($OH^\cdot$), hydrogen (H^-) and hydroperoxyl ($HOO^\cdot$).

5.4.3 Fabrication Method

Though the CCS method has been used extensively due to its ease of fabrication and ability to produce on a large scale, recent studies indicate that the low performance of CCS methods consists of heterogeneous catalyst dispersion and difficulties in mass transport. Further, it suffered from the high cost of Pt loading. Initial studies of the CCM fabrication indicate good performance over the CCS method due to improved conductivities in MEA, while CCM methods face the difficulties of short stability and swelling issues that consequently affect the catalyst layer and performance (Lim et al., 2021). Figure 5.9 shows typical MEA, unit cell and PEMFC stack.

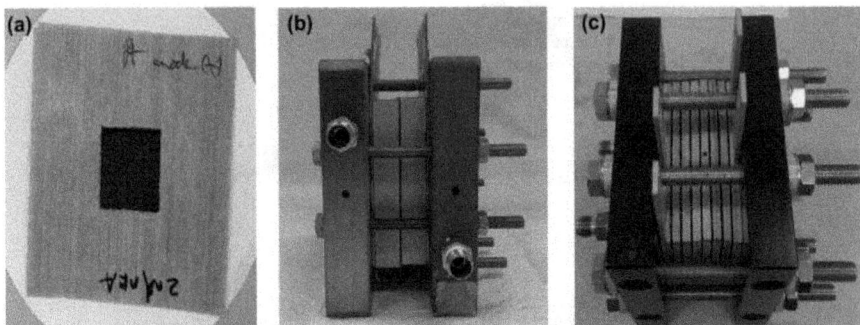

Figure 5.9   Photo-images of typical MEA (a) PEMFC unit cell (b) and stack (c).

5.5   RECENT ADVANCEMENTS IN MEA

Due to the influence of GDL and MPL in the oxygen transport and performance of the PEMFCs, various trials have been devoted to developing GDL and MPL, including laser perforated GDL, catalyst layer with a large surface area can decrease the amount of Pt loading and consequently reduce the cost of the MEA (Manahan and Mench, 2012). Apart from the conventional methods, electrospinning has recently been used as a versatile method to prepare nanofibers for catalyst support and free-standing composite nanofiber mat with controllable composition and diameter. For the preparation of electrocatalyst support, various polymers, like polyvinyl alcohol (PVA) and polyacrylonitrile (PAN), were used as a polymer matrix, and metal salts were used as metal sources. Schematics of nanoparticle fiber formation by electrospinning method are shown in Fig. 5.10. Pristine polymer or combination with metal precursor was electrospunned, and the prepared electrospun mat with or without metal oxide was then pyrolyzed to give carbon or metal oxide nanofiber that can be used as support for electrocatalyst. For the free-standing composite electrospun fiber preparation, two methods were used—either the ionomer was mixed with polymer matrix and catalyst ink and then electrospunned, or the catalyst precursor was deposited using the electrospray during electrospun the ionomer with polymer matrix. Various electrospun supports and free-standing mats have been reported recently, showing improved efficiency and power density in the PEMFCs, as shown in Table 5.2. These electrospun polymer/solid electrode materials are high in focus (Waldrop et al., 2020).

Klingele et al. reported a high-power density of 1.29 W/cm^2 for MEA that was prepared by a thin layer of polymer solution sprayed on top of the GDE. It showed 39% higher performance than the Nafion® NR211 with the same thickness. The high performance was caused by the enhancement of oxygen diffusivity and reduction of mass transport resistance (Liu et al., 2022), while direct membrane deposition and ionomer overlayer on the GDE also improved the contact area between GDE and PEM, reducing the charge-transfer, ionic, and mass transport resistances (Klingele et al., 2016; Vierrath et al., 2016).

Table 5.2 Summary of the power density of different cathode morphologies (Waldrop et al., 2020)

Cathode type	Pt on fiber eCNF support	Pt on fiber Nb-SnO$_2$ support	PtCo/C fiber cathodes	Pd/C@Pt skin fiber cathodes	Simultaneous Nafion spin/ Pt/C spray
Pt loading (mg$_{Pt}$/cm^2)	0.157	0.5	0.1	0.019	0.052
Power density at 0.65 v (mW/cm^2)	670	260	751	325	390
Feed gases	H$_2$/air	H$_2$/O$_2$	H$_2$/air	H$_2$/air	H$_2$/air
Operating temperature (°C)	68	80	80	65	80
Pressure (kPa$_{abs}$)	150	200	N/A	N/A	272

5.5.1 Mitigation of Mechanical Failure

The mechanical damage caused by heterogeneous stress in membrane and frame junctions could be minimized by the edge protection method, wherein mechanical damage caused by the thin membrane that leads to low strength and durability could be minimized by using stable materials as a composite with PEM and reinforced membranes containing stable, inert substrate, nanofillers such as carbon nanofibers, carbon nanotubes, clay, and inorganic particles. There were contradictory reports on the role of hot pressing in MEA preparation. Some of the results indicate that hot pressing can enhance the membrane-catalytic layer contact, the porosity, and the thickness of the membranes. While other reports showed that hot pressing makes more accumulation of water in the catalyst layer as well GDL due to the porosity decrease.

Further, the chemical degradation of membranes by peroxide free radicals was minimized by adding either hydrogen peroxide decomposition or radical scavenging materials. The chemical degradation of the PEM was decreased by doping it with heteropoly acids and doping various oxide materials, including CeO$_2$, MnO$_2$, TiO$_2$ and ZrO$_2$ with Pt catalyst. In particular, CeO$_2$ acted as an excellent free radical scavenger when doped with Nafion and commercialized as Nafion XL. Apart from CeO$_2$, CeO$_2$-SiO$_2$, and halloysite nanotubes have also been used to eliminate chemical degradation. Another way to eliminate the chemical degradation was by cross-linking the PEM, which effectively restricts the polymer chain's movement and free radical diffusion (Hiroki and LaVerne, 2005; Baker et al., 2015; Xing et al., 2021).

The drawbacks raised by the nature of electrocatalysts and their cost is reduced by lowering the amount of PGM in electrocatalysts and preparing stable PGM materials through dealloying and Pt-skin structure. Stability issues of the PGM electrocatalyst were effectively reduced by the preparation of highly ordered core@shell structures, such as PtCo@Pt, tungsten-doped PtCo@Pt, PtNi@Pt,

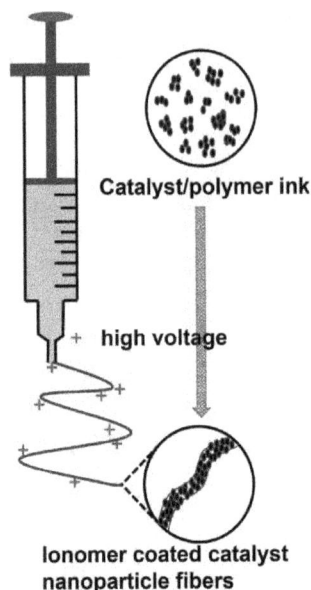

Figure 5.10　Schematics of nanoparticle fiber formation by electrospinning method (modified with permission from Elsevier [Waldrop et al., 2020]).

etc. while carbon corrosion in the MEA was suppressed by the developments of rationally designed 3D carbon structures. Its small pore size gives good stability and non-carbon materials, such as doped RuO_2 with TiO_2 and TiO_2 with Nb as catalyst supports. The ink formulation, nanoscale catalyst layer, and deposition method are the crucial factors that affect the performance of the cathode reaction in fuel cells. Different models, such as nanoscale thin film (NSTF), are an effective way to enhance the local mass transfer. To minimize the cost of PGM electrocatalysts, various PGM-free electrocatalysts, including M–N–C (M = Fe, Co, Mn), were developed recently and found to be effective in ORR as well as in minimizing the free radical degradation of ionomers/membranes. Among the M-N-Cs, apart from Fe-N-C, all others showed better stability. Due to the better catalytic property of PGM materials, mixing PGM material with non-PGM particles and adding a free radical scavenger (CeO_2) is an effective method to increase the ORR and decrease the electrocatalyst's cost (Roh et al., 2016b; Sadhasivam et al., 2018; Du et al., 2021).

The lower proton conductivities of the poly(aromatic) polymers can be overcome by the addition of MOF, such as sulfonated MiL-101 as a filler (proton conductivity 0.3 S/cm). Other MOFs based on Zr, chromium, and various ionic liquids, are also used to improve the proton conductivities (Tellez-Cruz., et al., 2021). Recently, many studies, particularly in CCM, revealed that incorporating of the Nafion ionomer solution in catalyst fabrication can increase performance as it acts as a bridge between the catalyst and membrane and aids the dispersity of Pt which intensifies the Pt deployment. Despite the importance of optimizing Nafion ionomer content within the catalyst layer, the low amount of Nafion ionomer leads to weak interactions. At the same time, a higher quantity prevents gas permeability and increases mass transfer resistance. Researchers recently found that adding Nafion with NaOH in the electrocatalyst showed the best performance. The fuel cells' performance is also affected by the preparation method of catalyst ink and solvents used; typically, water and IPA have been used for CCM preparation, and it was found that the content of IPA triggers the morphology of the ionomer and the surface area of the catalyst (Lim et al., 2021).

5.6 CONCLUSION

Leaving the world green with enormous resources, new technologies that fulfil their energy, and foreseeing the evolution requirements of our next generation, is an immediate duty for us as a research community and as ancestors. Hence, with the excessive usage of fossil fuels and modernization, we have already polluted the world as much as we can, leading to diminishing of the available fossil sources, climate change, and global warming. Finding renewable energy sources and effective energy usage is an immediate solution to minimize the above problems. Thanks to the recent awareness of environmental concerns and the various governments' futuristic plans, many new technologies have arisen to fulfill the energy requirements without much environmental effect. Hydrogen technology and fuel cells, particularly PEMFCs, can be extensively used as energy storage/conversion systems for high energy-demand applications, such as transport and electricity sectors. Albeit they showcase the excellent performance of PEMFC compared with their counterparts, there was ample room to improve its efficiency and utility. In particular, the oxygen reduction reaction is a slow step in the fuel cell reactions and sluggish to limit its performance. To overcome the slow ORR reaction, a large amount of platinum group catalysts (PGM) have been used in PEM fuel cells, leading to high fuel cell costs. Apart from the development of electrocatalysts, the other components of fuel cells, such as the current collector, flow field, GDL, PTL, and membrane, also affect the performance. Membrane electrode assembly, thus consisting of CL, GDL, MPL, and membrane, affects the cost and performance. Though the electrocatalyst layer directly affects the ORR but cannot omit the role of MEA in ORR as well as overall fuel cell performances. The type and method of polymer membrane preparation, type of GDL, and different types of catalyst-coated methods and methods of MEA preparations play a critical role in the performance of the fuel cell. The current technology of PEM fuel cells suffered from the cost of membranes, cost and loading of PGM catalyst, low ORR performance of other catalyst layers, and degradation of membranes during the operations.

The cost problem of PGM-based electrocatalysts was minimized by emerging metal-free catalysts, non-precious metal catalysts, single-atom doped metal catalysts, and carbon-doped metal catalysts. Further, perforated/laser-perforated GDL and sputtered MPL on GDL were investigated to improve fuel cell performance. Recent studies indicate that the resistivity among the electrocatalyst and substrate/membrane in the CCM or CCS method were minimized by using electrospinning electrocatalysts. The edge protection method could minimize the degradation of membranes and MEA. The MEA was improved by using free radical scavengers and cross-linking. Overall, the performance of ORR and fuel cell's performance is predominantly covered by MEA components. There were significant improvements in ORR and cell performances achieved in recent times by modification of MEA components and new fabrication techniques. Hence, apart from the individual components of MEA, the design, fabrication, and processing of MEA are also essential to achieve PEMFC to fulfill the following generation's energy requirements.

ACKNOWLEDGMENTS

This study was financially supported by Agency for Defense Development by the Korean Government (UD2200061D) and the Korea Institute of Energy Technology Evaluation and Planning (KETEP) and the Ministry of Trade, Industry & Energy (MOTIE) of the Republic of Korea (No. 20213030040590).

REFERENCES

Baker, A.M., D. Torraco, E.J. Judge, D. Spernjak, R. Mukundan, R.L. Borup, et al., 2015. Cerium migration during PEM fuel cell assembly and operation. ECS Trans. 17: 1009–1015.

Bayer, T., H.C. Pham, K. Sasaki and S.M. Lyth. 2016. Spray deposition of Nafion membranes: Electrode-supported fuel cells. J. Power Sources. 327: 319–326.

Bethoux, O. 2020. Hydrogen fuel cell road vehicles: state of the art and perspectives. Energies. 13: 5843.

Bhosale, A.C., P.C. Ghosh and L. Assaud. 2020. Preparation methods of membrane electrode assemblies for proton exchange membrane fuel cells and unitized regenerative fuel cells: a review. Renewable Sustainable Energy Rev. 133: 110286.

Boo, K-J. 2021. Republic of korea country report. pp. 134–150. *In:* P. Han and S. Kimura (eds). Energy Outlook and Energy Saving Potential in East Asia 2020. Jakarta. ERIA.

Borup, R. 2019. Fuel cell performance and durability consortium, US DOE 2019. Annual Merit Review Proceedings.

Cano, Z.P., D. Banham, S. Ye, A. Hintennach, J. Lu, M. Fowler, et al., 2018. Batteries and fuel cells for emerging electric vehicle markets. Nat. Energy. 3: 279–289.

Cullen, D.A., K.C. Neyerlin, R.K. Ahluwalia, R. Mukundan, K.L. More, R.L. Borup, et al., 2021. New roads and challenges for fuel cells in heavy-duty transportation. Nat. Energy. 6: 462–474.

Du, L., V. Prabhakaran, X. Xie, S. Park, Y. Wang and Y. Shao. 2021. Low-PGM and PGM-free catalysts for proton exchange membrane fuel cells: stability challenges and material solutions. Adv. Mater. 33: 1908232.

Fan, L., Z. Tu and S.H. Chan. 2021. Recent development of hydrogen and fuel cell technologies: a review. Energy Rep. 7: 8421–8446.

Habib, M.S., P. Arefin, M.A. Salam, K. Ahamed, M.S. Uddin, T. Hossain, et al., 2021. Proton exchange membrane fuel cell (PEMFC) durability factors, challenges, and future perspectives: a detailed review. Mat. Sci. Res. India. 18: 217–234.

Hiroki, A. and A. LaVerne. 2005. Decomposition of hydrogen peroxide at water-ceramic oxide interfaces. J. Phys. Chem. B. 109: 3364–3370.

Jiao, K., J. Xuan, Q. Du, Z. Bao, B. Xie, B. Wang, et al., 2021. Designing the next generation of proton-exchange membrane fuel cells. Nature. 595: 361.

Klingele, M., B. Britton, M. Breitwieser, S. Vierrath, R. Zengerle, S. Holdcroft, et al., 2016. Membrane electrode assemblies for PEM fuel cells: a review of functional graded design and optimization. Electrochem. Commun. 70: 65–68.

Li, Y., Q. Li, H. Wang, L. Zhang, D.P. Wilkinson and J. Zhang. 2019. Recent progresses in oxygen reduction reaction electrocatalysts for electrochemical energy applications. Electrochem. Energy Rev. 2: 518–538.

Lim, B.H., E.H. Majlan, A. Tajuddin, T. Husaini, W.R.W. Daud, N.A.M. Radzuan, et al., 2021. Comparison of catalyst-coated membranes and catalyst-coated substrate for PEMFC membrane electrode assembly: a review. Chin. J. Chem. Eng. 33: 1–16.

Litster, S. and G. McLean. 2004. PEM fuel cell electrodes. J. Power Sources. 130: 61–76.

Liu, Q., F. Lan, J. Chen, C. Zeng and J. Wang. 2022. A review of proton exchange membrane fuel cell water management: membrane electrode assembly. J. Power Sources. 517: 230723.

Manahan, M.P. and M.M. Mench. 2012. Laser perforated fuel cell diffusion media: engineered interfaces for improved ionic and oxygen transport. J. Electrochem. Soc. 159: F322.

Mehmood, A. and H.Y. Ha. 2013. Parametric investigation of a high-yield decal technique to fabricate membrane electrode assemblies for direct methanol fuel cells. Int. J. Hydrogen Energy. 38: 12427–12437.

Mokrini, A., N. Raymonda, K. Thebergea, L. Robitaillea, C. Del Riob, M.C. Ojedab, et al., 2010. Properties of melt-extruded *vs.* solution-cast proton exchange membranes based on PFSA nanocomposites. ECS Trans. 33: 855–865.

Pettersson, J., B. Ramsey and D. Harrison. 2006. A review of the latest developments in electrodes for unitized regenerative polymer electrolyte fuel cells. J. Power Sources. 157: 28–34.

Qu, E., X. Hao, M. Xiao, D. Han, S. Huang, Z. Huang, et al., 2022. Proton exchange membranes for high temperature proton exchange membrane fuel cells: challenges and perspectives. J. Power Sources. 533: 231386.

Roh, S.H., T. Sadhasivam, T.-H. Kim, J.-H. Park and H.-Y. Jung. 2016a. Carbon-free $SiO_2@SO_3H$ supported Pt bifunctional electrocatalyst for unitized regenerative fuel cells. Int. J. Hydrogen Energy. 41: 20650.

Roh, S.H., S.-G. Rho, S.-C. Kim, J.-Y. Kim and H.-Y. Jung. 2016b. The effect of nano-morphology modification using an amphiphilic polymer on the proton conductivity of composite membrane for a polymer membrane-based fuel cell. J. Nanosci. Nanotechnol. 6: 2092.

Sadhasivam, T., S.-H. Roh, T.-H. Kim, K.-W. Park and H.-Y. Jung. 2016. Graphitized carbon as an efficient mesoporous layer for unitized regenerative fuel cells. Int. J. Hydrogen Energy. 41: 18226.

Sadhasivam, T., P. Gowthami, S.-H. Roh, M.D. Kurkuri, S.C. Kim and H.-Y. Jung. 2018. Electro-analytical performance of bifunctional electrocatalyst materials in unitized regenerative fuel cell system. Int. J. Hydrogen Energy. 43: 18169.

Sadhasivam, T., K.V. Ajeya, Y.A. Kim and H.-Y. Jung. 2020. An experimental investigation of the feasibility of Pb based bipolar plate material for unitized regenerative fuel cells system. Int. J. Hydrogen Energy. 45: 13101.

Sadhasivam, T., H.-Y. Jung, J.-H. Wee, Y.A. Kim and S.-H. Roh. 2021. A new strategy of carbon – Pb composite as a bipolar plate material for unitized regenerative fuel cell system. Electrochim. Acta. 391: 138921.

Saha, M.S., D.K. Paul, B.A. Peppley and K. Karan. 2010. Fabrication of catalyst-coated membrane by modified decal transfer technique. Electrochem. Commun. 12: 410-413.

Sassin, M.B., Y. Garsany, B.D. Gould and K.E. Swider-Lyons. 2017. Fabrication method for laboratory-scale high-performance membrane electrode assemblies for fuel cells. Anal. Chem. 89: 511−518.

Shahgaldi, S., I. Alaefour and X.G. Li. 2018. Impact of manufacturing processes on proton exchange membrane fuel cell performance. Appl. Energy. 225: 1022–1032.

Shao, Y., G. Yin, Z. Wang and Y. Gao. 2007. Proton exchange membrane fuel cell from low temperature to high temperature: material challenges. J. Power Sources. 167: 235–242.

Shao, Y. 2019. Highly active and durable PGM-free ORR electrocatalysts through the synergy of active sites. US DOE 2019 Annual Merit Review Proceedings.

Sitanggang, R., A.B. Mohamad, W.R.W. Daud, A.A.H. Kadhum and S.E. Iyuke. 2009. Fabrication of gas diffusion layer based on x–y robotic spraying technique for proton exchange membrane fuel cell application. Energy Convers. Manage. 50: 1419–1425.

Sharaf, O.Z. and M.F. Orhan. 2014. An overview of fuel cell technology: Fundamentals and applications. Renewable Sustainable Energy Rev. 32: 810–853.

Tian X., X.F. Lu, B.Y. Xia and X.W. Lou. 2020. Advanced electrocatalysts for the oxygen reduction reaction in energy conversion technologies. Joule. 4: 45–68. https://doi.org/10.1016/j.joule.2019.12.014

Tellez-Cruz, M.M., J. Escorihuela, O. Solorza-Feria and V. Compañ. 2021. Proton exchange membrane fuel cells (PEMFCs): advances and challenges. Polymers. 13: 3064.

Victor, D.G., D. Zhou, E.H.M. Ahmed, P.K. Dadhich, J.G.J. Olivier, H-H. Rogner, et al. 2014. pp. 111–150. *In:* O. Edenhofer, R. Pichs-Madruga, Y. Sokona, E. Farahani, S. Kadner, K. Seyboth, et al., (eds). Climate Change 2014: Mitigation of Climate Change. Contribution of Working Group III to the Fifth Assessment Report of the Inter-Governmental Panel on Climate Change. Cambridge University Press, New York.

Vierrath, S., M. Breitwieser, M. Klingele, B. Britton, S. Holdcroft, R. Zengerle, et al., 2016. The reasons for the high power density of fuel cells fabricated with directly deposited membranes. J. Power Sources. 326: 170–175.

Waldrop, K., R. Wycisk and P.N. Pintauro. 2020. Application of electrospinning for the fabrication of proton-exchange membrane fuel cell electrodes. Curr. Opin. Electrochem. 21: 257–264.

Winter, M. and R.J. Brodd. 2004. What are batteries, fuel cells, and supercapacitors? Chem. Rev. 104: 4245–4270.

Xie, V., T. Chu, T. Wang, K. Wan, D. Yang, B. Li, P. Ming, et al., 2021. Preparation, performance and challenges of catalyst layer for proton exchange membrane fuel cell. Membranes. 11: 879.

Xing, L., W. Shi, H. Su, Q. Xu, P.K. Das, B. Mao, et al., 2019. Membrane electrode assemblies for PEM fuel cells: a review of functional graded design and optimization. Energy. 177: 445.

Xing, Y., H. Li and G. Avgouropoulos. 2021. Research progress of proton exchange membrane failure and mitigation strategies. Materials. 14: 2591.

Chapter 6

An Overview of Energy-efficient and Sustainable Oxygen Reduction Reaction Cathode in Microbial Fuel Cells

Manju Venkatesan[1,3]*, Chiranjeevi Srinivasa Rao Vusa[2]*,
Annamalai Senthil Kumar[1], Jong Pil Park[3],
Sathish Kumar Kamaraj[4]* and Alberto Alvarez-Gallegos[5]

Email: Chiranjeevi Srinivasa Rao Vusa (Chiru.vusa@gmail.com);
Manju Venkatesan (Manju.chem001@gmail.com);
Sathish kumar Kamaraj (Sathish.bot@gmail.com; Sathish.k@Ilano.tecnm.mx)

[1]Department of Chemistry, School of Advanced Sciences,
Vellore Institute of Technology University,
Vellore, Tamil Nadu, India.

[2]Department of Chemical Engineering,
Indian Institute of Technology,
Kanpur, Uttar Pradesh, India.

[3]Department of Food Science and Technology, Chung-Ang University,
The Republic of Korea.

[4]TecNM-Instituto Technologico El Liano Aguascalientes (ITEL),
Laboratorio de Medio Ambiente Sostenible,
20330 Aguascalientes, Ags. Mexico.

[5]Centro de Investigacion en Ingenieria y Ciencias Aplicadas,
Universidad Autonoma del Estado de Morelos, Mexico.

*For Correspondence: Email: chiru.vusa@gmail.com

6.1 INTRODUCTION

Microbial energy generation technologies produce fuel and electricity from organic waste via a biochemical pathway followed by electrochemical reactions utilizing microbes modified electrodes (Choudhury et al., 2017; Mashkour et al., 2021, Boas et al., 2022; Dange et al., 2022). The microorganisms make fuels from raw organic materials via microbial catalyzed oxidization/reduction, which transfers the electrons to the electrode to produce electricity. The first microbial energy generation system, i.e. microbial fuel cell (MFC) was reported by Potter MC in 1991 (Potter and Waller, 1911). He used *Saccharomyces cerevisiae* bacteria to disintegrate organic matter and found the liberation of electrical energy during the disintegration of organic compounds by microbes. Then Barnett Cohen developed half-cell MFCs in 1931 (Cohen, 1931). The first successful MFC was developed by Suzuki et al. in 1976 (Karube et al., 1976). He connected the half-cell MFCs in series and produced 35 volts with only a current of 2 mA. They used the whole cell of *Clostridium butyricum* IFO 3847, and the immobilized cells continuously evolved hydrogen from glucose under aerobic conditions. After that, various MFC designs were developed for different applications and extensively studied to understand the principle and to increase performance (Africa, 2016; Kumar et al., 2018; Mashkour et al., 2021; Boas et al., 2022). MFC-based technologies are significant because they produce electricity in treating wastewater discharged from industrial, agricultural, and municipal sectors. The significance of the MFC device is microorganisms that perform the dual duty of making fuel out of organic compounds present in effluents and generates power. Henceforth, MFCs appeared as environmentally friendly electrical devices and rapidly evolved as a sustainable system due to their potential to recover energy in wastewater treatment (Apollon et al., 2022a).

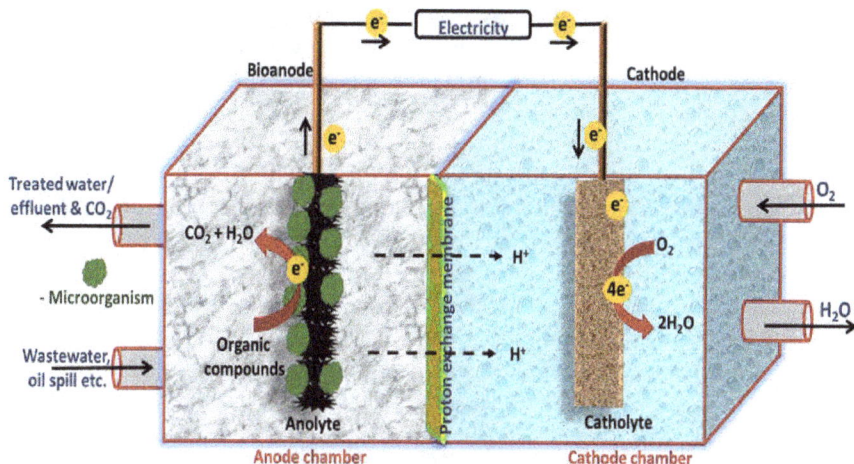

Figure 6.1 The basic principle of double chamber MFC (DCMFC).

The basic working principle of MFC is based on the principles of microbial physiology coupled with electrochemistry (Allen and Bennetto, 1993; Mashkour et al., 2021; Apollon et al., 2022a). The schematic representation of DCMFC is shown in Fig. 6.1, where the MFC consists of two electrodes, an anode, and a cathode, separated by a proton exchange membrane (PEM). The electrons (e^-) metabolically generated by the electroactive microorganisms on the anode surface will be transferred to the cathode via electron machinery. The transferred e^- are transported to the cathode surface via an external electrical connection to generate electricity. The protons are transported through the electrolyte via a membrane to reach the cathode. The electrons that reach the cathode surface react with protons and an electron acceptor (O_2) to produce water/fuel. The advantages of MFC are (i) degradation of organic matter to produce fuel and electricity using microorganisms as catalysts, (ii) it has high conversion efficiency compared to enzymatic fuel cells and harvests 90% of the electrons from the bacterial electron transport system. Several MFC-based systems have been developed for electricity production, wastewater treatment, fuel synthesis, desalinization, bioremediation, metal recovery, self-powered biosensor, etc., owing to their versatility (Choudhury et al., 2017; Kamaraj et al., 2019 ; Apollon et al., 2020, 2021; Mashkour et al., 2021; Boas et al., 2022; Dange et al., 2022; Apollon, et al., 2022b). Various designs and configurations are also developed to increase the power output. They can be gathered into different main categories: (i) single-chamber MFC (SCMFC), (ii) dual-chamber MFC(DCMFC), (iii) stacked MFC, (iv) Up-flow MFC, and (v) microfluidic MFC. Several factors affect the performance of MFCs, such as microorganisms, biofilm on the anode and its nature (mixed or single), growth, thickness, cathode, membrane, electrolyte, temperature, pH, and media in anode/cathode, and number of MFCs connected in either series and/or parallel.

The development of a cathode is considered a crucial factor because it should have the ability to receive the electrons produced at the anode via an external circuit and high electrocatalytic activity toward an electron acceptor (Dange et al., 2022). Also, it should have high conductivity, high surface area, be non-corrosive, and not be affected by microbes. Various electron acceptors, including hexacyanoferrate(IV), permanganate, persulfate, manganese oxide, chromate, copper, oxygen, etc., have been used in MFC (Ucar et al., 2017; Satish Kumar et al., 2012a). Using O_2 as an electron acceptor for MFC is most convenient because of its availability, sustainability, and high redox potential. The oxygen can be provided in the cathode compartment by either an air cathode or by purging the gas into water. Also, the O_2 is reduced to a water molecule which will not affect the biofilm and electrodes. But the kinetics of oxygen reduction reaction (ORR) are slow and sluggish, and the ORR at the cathode is considered a rate-limiting step in MFC. Hence, an efficient catalyst is required to reduce the cathodic overpotential and should be stable for the long term in complex media. Various catalysts have been developed to use as a cathode in MFC to improve the power output (Erable et al., 2012; Yuan et al., 2016; Boukoureshtlieva et al., 2021; Dange et al., 2022; Santoro et al., 2022). This chapter includes the fundamentals of ORR and an overview of a diverse group of catalysts developed for ORR cathode in MFC. The catalysts are classified as chemical, enzymatic and microbial. The chemical catalysts are noble and non-noble metals, metal oxides, metal hydroxides, metal carbides, alloys, metal-macrocycles, metal-nitrogen-carbon (M-N-C), carbon, and composites. The enzymatic catalysts are oxidoreductases (laccase and bilirubin oxidase), while the microbial catalysts are microorganisms that are sensitive to ORR. This chapter also includes the constraints and future perspectives of ORR cathode in MFC. This chapter is expected to deliver the fundamentals of ORR in MFCs, an overview of various cathode catalysts used in MFCs to improve the power output and build economically sustainable MFC devices for real-time applications.

6.1.1 Fundamentals of Oxygen Reduction Reaction in Microbial Fuel Cells

The oxygen reduction reaction (ORR) is one of the critical steps in efficient energy conversion techniques, especially in fuel cells (Erable et al., 2012; Dange et al., 2022; Santoro et al., 2022). Also, we all know that ORR is one of the factors that could limit the power output of MFCs. Oxygen reduction in aqueous solution occurs mainly through different pathways: (i) a four-electron reduction pathway from O_2 to H_2O and (ii) a two-electron pathway from O_2 to H_2O_2. The most accepted mechanism of ORR was first proposed by Damjanovic et al. and later modified by Wroblowa et al. (Wroblowa, Yen-Chi-Pan, and Razumney, 1976; Damjanovic et al., 979). (Damjanovic et al., 1967). The reaction mechanisms and steps were simplified by Wroblowa et al. as shown below in Fig. 6.2(A), making it easier to understand the complicated reaction pathway of oxygen reduction.

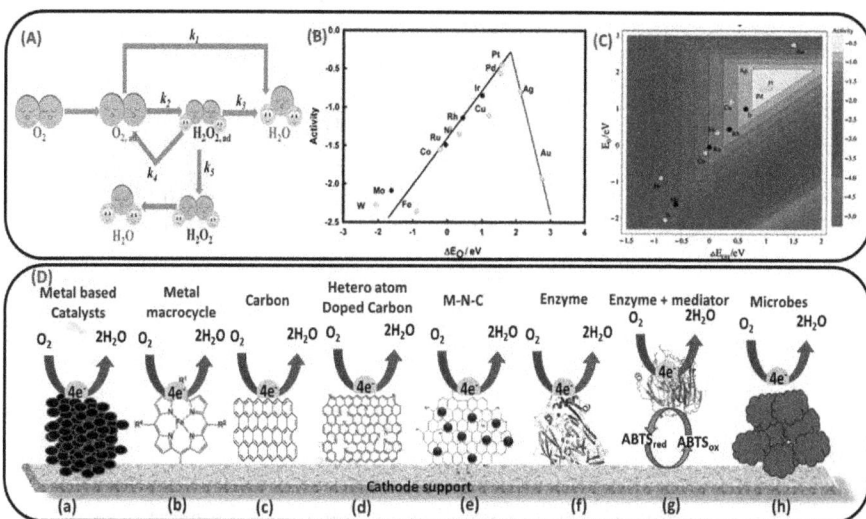

Figure 6.2 (A) The mechanism of ORR proposed by Wroblowa et al. (B) The plot of oxygen activity vs. the oxygen binding energy, and (C) ORR activity plotted as a function of both the O and the OH binding energy (reprinted with permission from [Nørskov et al., 2004]; copyright 2022 American Chemical Society). (D) Different catalysts developed for ORR.

As shown in the above Fig. 6.2, there may be different steps (i) direct $4e^-$ reduction of O_2 to H_2O (k_1), (ii) series $2e^-$ reduction of O_2 to $H_2O_{2,ad}$ (k_2), (iii) series $4e^-$ reduction of O_2 to H_2O through $H_2O_{2,ad}$ intermediate ($k_2 + k_3$),

(iv) series $2e^-$ reduction of O_2 to $H_2O_{2,ad}$ followed by catalytic/chemical decomposition of $H_2O_{2,ad}$ to O_2 ($k_2 + k_4$), and (v) series $2e^-$ reduction of O_2 to $H_2O_{2,ad}$ followed by desorption of $H_2O_{2,ad}$ to the bulk electrolyte ($k_2 + k_5$). The ORR follows different pathways, as shown in Table 6.1, depending on the electrocatalysts and electrolytes used. In addition to that, there are other pathways, such as:

(a) A 'parallel' pathway that is a combined direct and series pathway.
(b) An 'interactive' pathway includes the diffusion of species from a 'series' path into a 'direct' path.
(c) An associative mechanism involves the adsorption of molecular O_2 and direct proton/electron transfer to it and OOH, which breaks into O and OH.
(d) A dissociative mechanism involves the splitting of the O–O bond in O_2 followed by hydrogenation of atomic O to OH and H_2O.

Table 6.1 ORR overall pathways and mechanism (Erable et al., 2012; Yuan et al., 2016)

Conditions	Equations	Potential (E^0)
Acidic	(i) Direct four-electron reduction $$O_2 + 4e^- + 4H^+ \leftrightarrows 2H_2O$$ (ii) Indirect reduction - Peroxide pathway $$O_2 + 2e^- + 2H^+ \leftrightarrows H_2O_2$$ Followed by peroxide reduction or chemical decomposition $$H_2O_2 + 2e^- + 2H^+ \leftrightarrows 2H_2O$$ (or) $$2H_2O_2 \leftrightarrows 2H_2O + O_2$$	+1.229 V *vs.* SHE +0.680 V *vs.* SHE +1.776 V *vs.* SHE
Alkali	(i) Four electron reduction pathway $$O_2 + 4e^- + 2H_2O \leftrightarrows 4OH^-$$ (ii) Hydrogen peroxide pathway $$O_2 + 2e^- + H_2O \rightarrow HO_2^- + OH^-$$ Followed by peroxide reduction or chemical decomposition $$HO_2^- + 2e^- + H_2O \rightarrow 3OH^-$$ (or) $$2HO_2^- \rightarrow 2OH^- + O_2$$	+0.401 V *vs.* SHE +0.080 V *vs.* SHE +0.867 V *vs.* SHE

The availability, sustainability, and oxygen redox potential make it a ubiquitous final electron acceptor for many redox processes. The kinetics of ORR are fast on materials readily available over the earth's surface. Notably, the Pt-based catalysts exhibit good activity towards ORR, but the adsorbed oxygen and hydroxyl are found to be stable intermediates. The density functional theory was utilized to understand the ORR on Pt. The DFT model predicts a volcano-shaped relationship between the rate of the cathode reaction and the oxygen adsorption energy, as shown in Fig. 6.2(B). Metals with lower oxygen-binding energy than Pt should have a higher rate of oxygen reduction. The activity of a metal's surface depends on oxygen-binding energy and –OH-binding energy, as shown in Fig. 6.2(C). The materials which exhibit fast kinetics cannot be used for MFC applications. The rapid kinetics may produce a higher amount of reactive oxygen species (ROS) in the MFC chamber, resulting in increased oxidative stress in living organisms used in an anode. Researchers started developing new catalysts for ORR. It is tough to break O=O electrochemically due to its high binding energy of 498 kJ/mol. Hence, various electrocatalysts, including noble metals, graphite, carbon, metal oxides, enzymes, microbes, and composite materials have been developed (Fig. 6.2D) to increase the electrochemical activity of ORR by lowering the energy barrier via bond activation and cleavage. The catalyst, which follows the direct $4e^-$ pathway to produce water, is desirable for better performance because most of the surface is affected by the intermediates formed during ORR. Most of these developed catalysts exhibit ORR only in acidic or alkali medium except in pure carbon-based materials.

The working conditions of MFCs are different and it contains complex media, i.e. microbe culture medium, molecules, and charged species. Also, the pH of the solutions in MFC is usually neutral. Hence, the cathode should be active towards ORR in a neutral medium for MFC. Recently, enzymes and microbe-based catalysts have been given importance to use at the cathode for ORR as they can reduce oxygen in a neutral medium. The main criteria for an ideal ORR cathode are high catalytic activity, durability in complex media (wastewater and culture medium), anti-fouling ability, and restoration of catalytic activity after usage and cleaning. The section below discusses the different types of catalysts used as cathodes in MFC.

6.1.1.1 Various Cathode Catalysts in MFC

The microbial fuel cell is a multidisciplinary area of research. One could understand the concepts of different fields, including electrochemistry, materials science, and biotechnology to improve the performance of MFC. Many studies have been conducted to increase MFC's power output and efficiency. The performance of MFC depends on the efficiency, activity, and stability of components, such as anode, cathode, membrane, microbes, biofilm, electrolyte, media, temperature, pH and gas diffusion electrode, etc. The development of an anode and cathode are crucial factors, and the electrodes play a significant role in MFC applications. Few studies have been conducted to develop a cathode compared to an anode development. Usually, the ideal electrodes should be highly active, conductive, and stable in the electrolyte. Also, it should have a high surface area and high conductivity. In addition to the above properties, the electrode should be biocompatible and have bacteria adhesion properties to use as an anode. At the same time, the cathode should be non-corrosive and fouling resistant because we know that the electrolytes of MFC contain charged species, microorganisms, and molecules which can bind/adsorb on the cathode and can deactivate the cathode. The role of the cathode is to reduce an electron acceptor by receiving metabolically-generated electrons from the anode through an external circuit. The oxygen is an ideal electron acceptor for MFC to produce electricity and treat wastewater simultaneously. As discussed above, the kinetics of ORR at the cathode are poor and sluggish in neutral media due to lack of H^+ and OH^-, which lead to high overpotential.

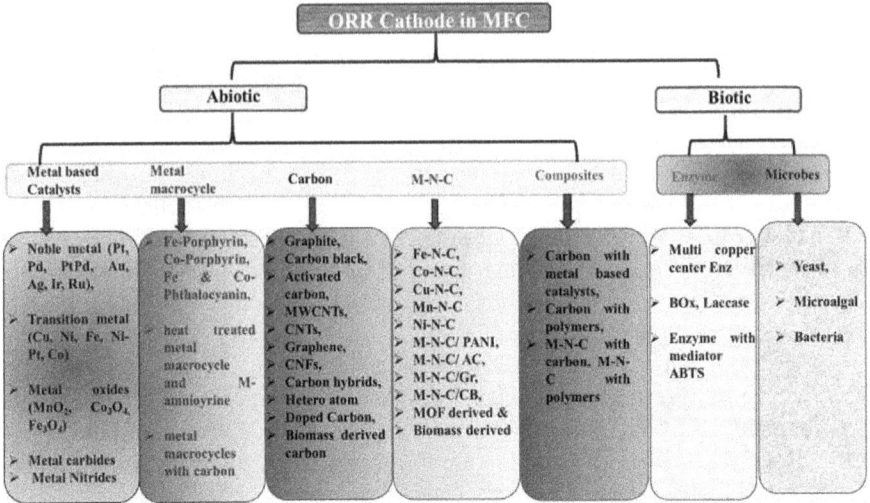

Figure 6.3 (A) Different categories of electrocatalysts. An example of MFC was used (B) chemical catalyst, (C) enzymatic, and (D) microbial catalyst at a cathode.

Different catalysts have been developed to lower the overpotential and increase the kinetics of ORR at the cathode (Yuan et al., 2016; Gouse et al., 2020; Boukoureshtlieva et al., 2021; Mashkour et al., 2021; Santoro et al., 2022). They are categorized as biotic and abiotic (Gao et al., 2020). The biotic catalysts are enzymes and microbes-based catalysts, whereas the abiotic are chemical catalysts. Figure 6.3 shows different categories of catalysts used at the cathode in MFCs. The abiotic catalysts are based on metals, carbon, polymers, and their composites while the biotic catalysts are based on enzymes, cells, and microbes. The enzymatic catalysts are oxidoreductase enzymes laccase and bilirubin oxidase because they have the ability to reduce O_2 to H_2O via $4e^-$ pathway. The microbial catalysts are reported recently which are microbes whose enzymes or surface proteins are electrochemically active towards ORR. The overview of those catalysts and their constraints in MFCs are discussed in sections below.

6.1.1.2 Chemical Catalysts (Abiotic Catalysts)

The chemical catalysts are classified into many groups: (i) metal-based catalysts (Pt and Pt free), (ii) metal oxide-based catalysts, (iii) carbon-based catalysts (with and without hetero doping), (iv) metal-nitrogen-carbon (M-N-C) and/or metal-carbon catalysts (v) carbides, (vi) nitrides and (vi) their composites (with and without polymers).

6.1.2 Metal-based Catalysts in MFCs

Metal catalysts exhibit strong electrocatalytic activity towards ORR based on their d-band center and vacancies. The above metal catalysts can be categorized into Pt- and Pt-free catalysts. Many researchers have been working on cathode catalysts to minimize or replace Pt with other earth-abundant metals or carbon-based materials. Pt-free abiotic catalysts can be categorized into noble/transition metal and carbon-based materials (Gouse et al., 2020; Boukoureshtlieva et al., 2021; Dange et al., 2022) and composites. The Pt-free metal-based catalysts are the best alternative to lower the cost and high activity towards ORR. Various reports exist in the literature based on metal-based catalysts in MFC. The Pt-free metal catalyst can be divided into two categories: (a) noble metal-based catalysts and (b) transition metal-based catalysts.

6.1.2.1   Pt-based Catalysts

The Pt-based materials are extensively used as ORR catalysts in fuel cells due to their high electrocatalytic activity (Yan et al., 2014; Ren et al., 2020). Also, the Pt/C is the most typical commercial ORR catalyst due to its ability to reduce required critical oxygen concentration and activation energy. Plotting the correlation between the experimentally determined activities of the surfaces of single crystals of Au(111), Ag(111), Pd(111), Rh(111), Ir(111), and Ru(0001), Pt monolayers deposited on their surfaces, and also nanoparticles of these metals dispersed on high-surface-area carbon with the calculated metal d-band center energies, εd, revealed a volcano-type dependence (Lima et al., 2007). They found that Pt(111), Pt/C, and Pt/Pd(111) activity is higher than other metals. The above factors made Pt and Pt-based catalysts the most popular and extensively used catalysts in electrochemical technology in the form of single catalyst, alloy, bimetallic, and carbon matrix supported Pt (Wang et al., 2014; Mashkour et al., 2016). They have been extensively used in MFC during the initial stages of development because Pt-based catalysts exhibit high catalytic activity towards ORR and follow 4e$^-$ reduction of O_2 to H_2O. The Pt-based catalysts exhibit high power density, but high cost and low durability limit their use in MFC applications (Wang et al., 2014; Gouse et al., 2020). It can be easily poisoned with anions Cl$^-$, S^{2-}, and SO$_4^{2-}$ naturally present in wastewater. Also, Pt is easily covered by adsorbed oxygenated species, such as OH$^-$.

Pt-based alloys Pt–M (M = Co, Fe, Cu, Pd, and Ni, etc.) (Yan et al., 2012, 2014; Kim et al., 2020) and low amounts of Pt-loaded carbon electrodes are used to lower the cost of the catalysts (Greeley et al., 2009; Sathish Kumar, et al., 2012b; Reyes-Rodríguez, et al., 2015). For example, Pt-Ni alloy nanoparticles were dispersed on carboxyl multi-wall carbon nanotubes (MWNTs–COOH) by the NaBH$_4$ reduction method. The resulting Pt-Ni/MWNTs are used as air cathodes in SCMFC and produce a P$_{max}$ of 1.22 W/m^2, close to that of a Pt/C (20 wt% Pt) cathode (1.40 W/m^2) (Yan et al., 2012). High-entropy alloys (HEAs) with built-in stability through their low free-energy phases provide a promising route to prepare low content Pt-based alloy catalysts with less than 50 at % Pt while maintaining high ORR activity and stability in various environments, as shown in Fig. 6.4A (Li et al., 2020). A lower amount of Pt-coated carbon hybrids is impressive due to the low cost and high catalytic activity. The lower amount of Pt NPs can be mixed with carbon materials to make Pt-C, but it's challenging to attain homogenous dispersion. The Pt NPs can be grown on carbon surfaces by *in-situ* or electrodeposition methods. Size and morphology play a major role, and sometimes size is non-uniform. Recently, a conductive pyridine-Fe gel-template strategy has been proposed for fabricating highly-dispersed and low-loading Pt catalysts on carbon hybrids (Wang et al., 2020a). The resulting N_3/Fe/C-Pt catalyst exhibited high electrocatalytic activity and generated the P_{max} power density 504 mWm^{-2} for 400 h when MFC was fed with a culture medium containing $5gL^{-1}$ sucrose in the phosphate buffer solution. However, the poisoning of Pt by the anions deactivates the Pt, and the carbon participates in ORR, which leads to a decrease in the power density of MFC, production of reactive oxygen species via $2e^-$ reduction of O_2, and deactivation of bioanode by developing oxidative stress in microbes. Pt is the earliest adopted catalyst and still prevailing ORR electrocatalyst even though Pt's cost and poisoning effects limit its use in MFC.

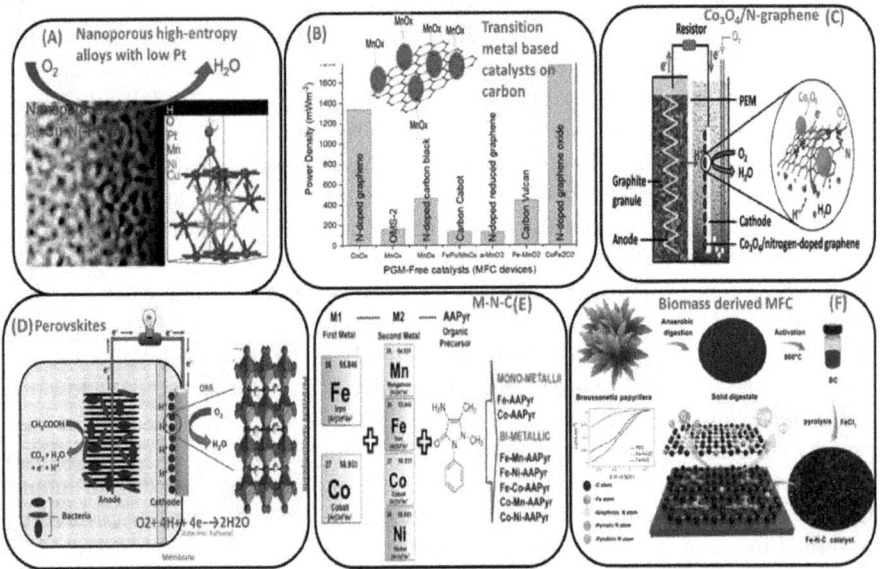

Figure 6.4   (A) Nanoporous high-entropy alloys with low Pt (Li et al., 2020). (B) Transition metal oxides on carbon (Oliveira et al., 2020). (C) Co_3O_4/N-graphene cathode in MFC (reprinted with permission from [(Su et al., 2013], copyright 2022; American Chemical Society). (D) MFC with the perovskite-based cathode (Nandikes, Peera, and Singh, 2022). (E) Bimetallic M-N-C based catalysts derived from heating of metal macrocycles (Kim et al., 2020). (F) Biomass-derived M-N-C, Reprinted with permission from (Yang et al., 2020a) (Copyright 2022; American Chemical Society).

6.1.2.2   Noble Metal-based Catalysts

The noble metals Pd, Au, Ag, Ru, and Ir-based materials exhibit efficient ORR activity, but the catalysts' cost is high, restricting the commercialization and large-scale applications (Sun et al., 2015; Nosheen et al., 2019; Han et al., 2020). Also, the electrolyte's pH, charged species, and microbes affect the Ru and Ir catalysts' stability and durability. A higher amount of reactive oxygen species are produced by Pd, Au, and Ag than Pt. Recently, the ultra-small gold nanoparticles (<2 nm diameter, 36 atoms of Au) have been found to exhibit a 4-electron reduction of ORR activity which may be due to the high surface density of Au with a low coordination number (Sumner et al., 2018). The silver-deposited activated carbon was used as an air cathode and generated the P_{max} of 1080 ± 60 mW/m^2 (Pu et al., 2014). Also, the Ag inhibits the growth of the biofilm on the cathodes. The bimetallic core-shell Au-Pd nanoparticles served as a cathode in membraneless SCMFC. This MFC delivered a P_{max} of 16.0 W/m^3 and remained stable for over 150 days (Yang et al., 2016). The Au, Ag, and Pd-based catalysts are also proven to be promising electrocatalysts for ORR but utilizing these metals is restricted due to the cleavage of O–O whilch is thermodynamically unfavorable and produces reactive oxygen species (Nosheen et al., 2019).

6.1.2.3 Transition Metal-based Catalysts

The transition metals, Cu, Fe, Ni, Co, Mn, and Ti-based catalysts have been developed. They can be subdivided into different categories of metal oxides (spinels to perovskites), hydroxides, selenides, nanoparticles, metals (Fe, Co, Ni) coordinated with macromolecules, metals (Fe, Co, Ni) coordinated with macromolecules pyrolyzed at high temperature to yield metal-nitrogen-carbon (M-N-C) catalysts and composites of above with carbon and/or polymer (Dange et al., 2022; Gouse et al., 2020; Wang et al., 2014). The metal oxides, such as PbO_2, perovskite oxides, V_2O_5, ZnO, TiO_2, Co_3O_4, MnOx, NiO, CuO, etc., are used as cathode catalysts in MFCs. Most metal oxides are active towards ORR in acid and alkali solutions. The manganese oxides (MnOx)-based catalysts have been extensively investigated because they are active in neutral medium, inexpensive, abundant, and possess a wide range of oxidation states. The ORR activity of MnOx follows the order: Mn_5O_8 < Mn_3O_4 < Mn_2O_3 < MnOOH, ~MnO_2. The different forms of manganese oxides α-MnO_2, β-MnO_2, and γ-MnO_2 were developed by hydrothermal method and tested as alternative catalysts to Pt (Zhang et al., 2009a). The β-MnO_2 shows high electrocatalytic activity and is used as a cathode catalyst in both cube and tube air-cathode MFCs using *Klebsiella pneumoniae* biofilm as biocatalyst, utilizing glucose as a substrate in the anode chamber. They found that the tube MFC produced higher output power, with the maximum volumetric power density of 3773 ± 347 mW/m^3 than cube MFC. The CNT-supported MnO_2 was also used as an air cathode in MFC and found that the β-MnO_2-based MFC yielded a power density of 97.8 mWm^{-2} which was 64.1% that of the Pt-based MFC. The COD removal was 84.8%, slightly higher than Pt-based MFC (tested as a control in the same conditions) (Lu et al., 2011).

Tofighi et al. used a carbon cloth coated with a composite of α-MnO_2, graphene oxide (GO), and activated carbon (MnO_2/GO/AC) as a cathode. The composite with 10% GO achieved a maximum power density of 148.4 mW m^{-2}, 280 fold higher than the AC (Tofighi et al., 2019). Figure 6.4(B) shows the MnO$_x$-based catalysts supported on carbon and power density of different types of transition metal oxide catalysts in air-cathode MFC devices (Oliveira et al., 2020). A PANI/MnO_2 nanocomposite coated carbon paper was developed (Ansari et al., 2016), which showed significant capacitance (525 F/g) with

P_{max} 37.6 mW/m^2. The MnO$_2$-modified carbon fabric was used as an air cathode in sediment microbial fuel cells to recover energy from river sediment (Salgado-dávalos et al., 2021). The TiO$_2$-modified graphite paste (GP) electrode was developed and used as a cathode in MFC. The GP-TiO$_2$ exhibits a higher power density (80 mW/m^2) than the bare GP (30 mW/m^2) (Mashkour et al., 2017). NiCo$_2$O$_4$ nanoplatelets were electrodeposited on carbon cloth for air cathode, showing a 12.9% higher maximum power density (P_{max}) than Pt/C catalyst (Cao et al., 2017). The MFC with NaCo$_2$O$_4$ cathode generated a power density of 0.6 W/m^2 (ref). The hybrid Co$_3$O$_4$ micro-particles directly grown on stainless steel mesh were used as an air cathode in batch-fed dual-chamber MFCs, and they generated 96% power density using commercial Pt/C electrodes (Gong et al., 2014).

The metal-based catalysts supported on N-doped carbon exhibit higher activity towards ORR. The Co$_3$O$_4$ supported on nitrogen-doped carbon nanotubes (NCNT) generated the P_{max} of 17 mW/m^2, which was 5.3 times higher than the NCNT (Song et al., 2015). The facile hydrothermal method was adopted to prepare a nanocomposite of cobaltosic oxide and nitrogen-doped graphene (Co$_3$O$_4$/N-G) for MFC applications, as shown in Fig. 6.4(C). The obtained P_{max} was 1340 ± 10 mW/m^2, almost four times higher than the plain cathode (340 ± 10 mW/m^2) and slightly lower than a commercial Pt/C catalyst (1470 ± 10 mW/m^2) in a neutral medium (Su et al., 2013). The nitrogenous mesoporous carbon coated with Co and Cu nanoparticles generated the P_{max} of 2033 mW/m^2 (Liang et al., 2020a). A strongly ordered porous hydrangea-like Cu$_2$O on N-doped activated carbon was developed by a single-step electrodeposition process using polyvinylpyrrolidone as a surfactant (Yang et al., 2017). The optimized Cu$_2$O/AC generated the P_{max} of 1610 ± 30 mW/m^2. Layered double hydroxides (LDHs) are also used as air cathodes in MFC due to their nanostructure, ease of preparation, and layered structure of the LDH to provide more electrochemical active sites. The nickel-based double hydroxide Ni$_3$Al-LDH generated the P_{max} of 3.2 µW/cm^2 (Djellali et al., 2021). The NiFe-LDHs were directly grown on Fe$_3$O$_4$ by the hydrothermal method to form Fe$_3$O$_4$@NiFe-LDHs (Jiang et al., 2020a). The resulting catalyst was loaded on stainless steel electrode and used as an air cathode in SCMFCs. The MFC with Fe$_3$O$_4$@NiFe-LDHs/SS cathode delivered a P_{max} of 211.40 ± 2.27 mW/m^2. Jiang et al. developed NiFe-LDH@Co$_3$O$_4$ core-shell structure and generated P_{max} of 467.35 ± 8.27 mW/m^2. The voltage output of NiFe-LDH@Co$_3$O$_4$-MFC was maintained at about 0.43 V, with few variations over eight days, indicating that the NiFe-LDH@Co$_3$O$_4$ catalyst has remarkable stability and durability (Jiang et al., 2020b).

A hybrid of CNTs decorated with copper selenide nanoparticles (CuSe-CNTs) also served as a cathode in MFC, and generated a P_{max} of 425.9 ± 5 mW/m^2 when the 1 : 1 ratio of CuSe and CNTs was used. The P_{max} is 1.90 and 1.60 times higher than the CNTs (244 ± 4 mW/m^2) and CuSe (258.8 ± 6 mW/m^2) (Tan et al., 2016). The RuCoSe air cathode and geobacter sulfurreducens-fed anode were used in microbial electrolysis cells to convert acetate into electricity (Rozenfeld et al., 2017). The RuCo$_2$Se exhibited improved tolerance than Pt at acetate concentration ≥500 mM and generated a higher P_{max} of 750 mW/m^2, comparable with Pt 900 mW/m^2. Recently, the structured NiO-coated stainless steel (NiO/SS) has been used as an air cathode with reduced graphene oxide coated SS to develop an MFC prototype to treat domestic sewage wastewater. They show that the methodologies developed for the coating of the electrodes aid in improving the performance of the MFC in delivering potential, current density, and power density up to 220%, 140%, and 700%, respectively, compared to the blank stainless steel electrodes (González Vázquez et al., 2022). The mixed metal oxides exhibit higher electrocatalytic activity due to breaking the O=O bond of molecular O_2 and then the migration of adsorbed oxygen to another metal, where electro-reduction occurs (Lima et al., 2006). Recently, the NiO-CuO supported on graphene (NiO-CuO/G) was used as an air cathode in MFC. The NiO-CuO/G cathode exhibits enhanced electrocatalytic activity toward ORR at a neutral pH and generate the P_{max} of 21.25 mW/m^2 (Ansari et al., 2016). The pervoskites composites are also used as cathode in MFC, shown in Fig. 6.4(D) (Nandikes et al., 2022).

6.1.2.4 Transition Metal-macrocycles-derived Catalysts

The macrocyclic complexes containing nitrogen atoms are active toward ORR (Dange et al., 2022; Gao et al., 2020; Orellana, 2022; Santoro et al., 2018; Viera et al., 2020). The utilization of transition metal macrocycles, phthalocyanines, and porphyrins for ORR is not new. In the early 1960s, the Co phthalocyanines were used as an electrocatalyst for ORR cathode in fuel cells by Jasinski (Jasinski, 1964). Macrocyclic molecules, such as tetraphenylporphyrin (TPP), tetramethoxyphenylporphyrin (TMPP), and phthalocyanines (Pc) form complexes with transition metals and facilitate ORR (Jasinski, 1964; Wiesener et al., 1989; Zhang et al., 2006). Later, it was shown that various Fe and Co macrocyclic complexes, such as N_4^-, $N_2O_2^-$, $N_2S_2^-$, O_4^-, and S_4^- produced a certain catalytic activity level for ORR (Gao et al., 2020; Jasinski, 1964; Oldacre, Friedman, and Cook, 2017; Santoro et al., 2018; Viera et al., 2020). The Fe and Co-based MN_4 macrocyclic complexes are ORR's most famous active catalysts. The activity can be partially due to ligands' inductive and mesomeric effects on the central ion. Depending on the nature of the ligand and chemical environment of the central metal ion, it follows either 4e$^-$ or 2e$^-$ reduction of ORR. Most macrocyclic complexes follow two e$^-$ transfer processes of ORR, which generates peroxide intermediates. An attack of the macrocycle rings by peroxide intermediates was the main reason for the catalyst's poor stability. Also, the catalyst can decompose via hydrolysis in the electrolyte. The Fe-complexes of phthalocyanines/porphyrin can promote the 4e$^-$ pathway, but they are not stable. At the same time, Co-complexes are stable but follow the 2e$^-$ pathway. But still, the Co complexes are of interest because of their stability, and one can promote either the 2e$^-$ or 4e$^-$ transfer process by changing the chemical environment of the Co metal center (Viera et al., 2020).

Various strategies have been developed to improve the stability and activity of the MN_4 complexes: (i) the carbon-based nanomaterials are used as a support, (ii) tuning the chemical environment of the central metal ions by introducing electron-withdrawing pheripherical substituents, and (iii) pyrolysis of MN_4. The heat treatment of MN_4 complexes results in good electrocatalysts with stability and high electrocatalytic activity (Kim et al., 2011; Kodali et al., 2017a;

Liu et al., 2014; Orellana et al., 2022; Santoro et al., 2018; Zhang et al., 2006). The Co-napthalocyanine carbon (CoNPc/C) was prepared by heating 5, 9, 14, 18, 23, 27, 32, 36-octabutoxy-2, 3-naphthalocyanine with carbon black at 700°C in the N_2 atmosphere. The resulting CoNPc/C was used as an air cathode in H-type MFC and generated Pmax of 64.7 mW/m^2 at 0.25 mA compared with Pt/C (81.3 mW/m^2) (Kim et al., 2011). The pyrolysis of CoTTP ([meso-tetrakis(2-thienyl)porphyrinato]Co(II)) with Vulcan XC-72 carbon (weight ratio of 1:9) at 800°C results in CoTTP/C catalyst. The CoTTP/C was treated wastewater containing 0.04 M methanol in SCMFC over 900 h. The CoTTP/C cathodes remained stable even at higher methanol concentrations than Pt (Liu et al., 2014). The aminoantipyrine (AAPyr) based Fe, Co, Ni, and Mn complexes were pyrolyzed at 900°C to form metal-AAPyr/C. The AAPyr act as both carbon and nitrogen precursor. The resulted catalysts are used as air-breathing cathode in MFC and delivered a P_{max} of 251 ± 2.3 μW/cm^2 (Fe-AAPyr/C), 196 ± 1.5 μW/cm^2 (Co-AAPyr/C), 171 ± 3.6 μW/cm^2 (Ni-AAPyr/C) and 160 ± 2.8 μW/cm^2 (Mn-AAPyr/C) (Kodali et al., 2017a). The heat treatment and pyrolysis of macrocyclic complexes at high temperatures have been reported to be one of the best methods to develop non-precious metal-nitrogen-carbon (M-N-C) based catalysts for ORR. M-N-C based catalysts are discussed in the section 'M-N-C-based catalyst in MFC' below.

6.1.2.5 Metal-nitrogen-carbon (M-N-C)-based Catalysts

Metal-nitrogen-carbon (M-N-C) based catalysts have emerged as a new class of active catalysts (Wang et al., 2014; Yuan et al., 2016; Ma et al., 2019a; Gao et al., 2020; Gouse et al., 2020; Dange et al., 2022). As a general strategy, the M-N-C catalysts can be synthesized by pyrolyzing metal precursors with suitable carbon and nitrogen precursors at various temperatures (from 700 to 1000°C). First, the M-N-C catalysts are produced by pyrolyzing metal-macrocycle complexes with a carbon mixture and utilized for ORR (Liu et al., 2014; Orellana, 2022; Viera et al., 2020). The metal, nitrogen, and carbon are covalently linked in the M-N-C catalysts. Researchers found that the direct involvement of M-N-C sites in the adsorption/desorption of oxygen and the high catalytic activity of the catalyst is directly dependent on the number of metals bonded to nitrogen atoms. The metal macrocycle-derived M-N-C catalysts have been studied for several years and used as air cathodes in MFC. Synthesizing an ORR catalyst for MFC by heat treatment of Iron(II)phthalocyanine and CoTMP (Zhao et al., 2005) opened several M-N-C-based catalysts by pyrolysis of the macrocyclic metal complex. Developing M-N-C catalysts thus promoted an entirely new era of research and found to exhibit excellent ORR activity in both acidic and alkaline medium. Various C and N precursors are utilized to develop M-N-C catalysts in past decades. Based on the precursor used to prepare M-N-C catalysts, the strategies are categorized into six different approaches: (1) pyrolysis of N-containing metal macrocyclic complexes with carbon materials include CNTs, CNFs, graphene oxide, graphene, activated carbon, Vulcan carbon, acetylene black, ketjen black, (2) pyrolysis of mixture of N-containing ligands, carbon, and metal precursor, (3) pyrolysis of the mixture of polymers and/or hydrogels containing N and C with metal precursors, (4) pyrolysis of the mixture of metal precursor and monomers with a hard template, i.e. porous silica template, (5) pyrolysis of metal-organic framework (MOFs) and (6) pyrolysis of biomass/biomass derived carbon with metal precursor.

Iron-chelated electrocatalysts were prepared by pyrolyzing sodium ferric ethylenediamine-N, N′-bis(2-hydroxyphenylacetic acid) (FeE) or sodium ferric diethylene triamine pentaacetic acid (FeD) supported on carbon Vulcan XC-72R carbon black and multi-walled carbon nanotubes (CNTs) at 800°C in a ceramic vessel under argon atmosphere. The FeE/CNT was used as an air cathode in MFC and generated a P_{max} of 127 ± 0.9 (mW/m^2) (Nguyen et al., 2014). The binuclear-cobalt-phthalocyanine (Bi-CoPc) pyrolyzed at different temperatures (300–1000°C) to form Bi-CoPc. The power density of the Bi-CoPc increased with increasing pyrolysis temperature and the order is Bi-CoPc/C-800 > Bi-CoPc/C-1000 > Bi-CoPc/C-600 > Bi-CoPc/C-300 > Bi-CoPc/C. The Bi-CoPc/C-800 generated P_{max} of 604 mW/m^2 in SCMFC (Li et al., 2015). The transition metal (Mn, Fe, Co, and Ni) nitrates and C and N precursor aminoantipyrine (AAPyr) was pyrolyzed at 900°C with a silica template to prepare M-N-C catalysts. Fe-AAPyr, Co-AAPyr, Mn-AAPyr, and Ni-AAPyr showed good ORR activity in neutral media (Rojas-Carbonell et al., 2017). But, the Fe-AAPyr had an electron transfer involving 4e$^-$ with lower than 5% peroxide. The Co-AAPyr, Mn-AAPyr, and Ni-AAPyr follow a $2 \times 2e^-$ mechanism with peroxide formation during the intermediate step. The M-N-C catalysts derived from metal-macrocycles exhibited good electrocatalytic activity with reasonable stability, but metal macrocycle-derived M-N-C catalysts are not feasible due to their high cost. Also, the activity depends on pyrolyzed temperature, orientation, the coordination number of center metal, bond formed between metal-complex and carbon support.

Different carbon materials are used to support the M-N-C catalysts to improve durability and power density. The Fe-AAPyr coated on activated carbon showed higher power density (1.3 W/m^2) with excellent durability (350 days) (Gajda et al., 2018). The bimetallic M-N-C catalysts are also developed by pyrolyzing AAPyr with two metal precursors, as shown in Fig. 6.4(E). The Fe-Mn-AAPyr catalyst recorded the highest performance in this study, with a maximum power density of 221.8 ± 6.6 μW/cm^2 (Kodali et al., 2017b). The M-N-C catalysts can also be derived from the metal-organic framework (MOFs) for ORR. MOFs are fascinating materials, and various shapes, such as nanocubes, nanowires, and nanorods of MOFs with porous structures, can be prepared by choosing suitable ligands, metal precursors, and reaction conditions (Du et al., 2020). The zeolitic imidazolate framework-8 (ZIF-8) was heated to 140°C, and the resulting ZIF-8 catalyst with Zn-N-C generated a P_{max} of 2103.4 mW/m^2 (Xue et al., 2019). Pyrolyzing MOFs developed the Fe, N, and S co-doped Fe/S@N/C containing Fe^{3+}, Zn^{2+}, and 2-methylimidazole (Luo et al., 2019). The Fe/S@N/C generated a P_{max} of 1196 mW/m^2. The bimetal Cu/Co/N-C was also derived from ZIF-67 (Wang et al., 2020b), used air cathode in MFC, and delivered a P_{max} of 1008 mW/m^2. The Fe-N-C was derived by pyrolyzing coordinated polymer (chitosan-Fe^{3+} hydrogel) with activated carbon (AC) at 800°C and resulted in a porous Fe-N-C catalyst on AC with a uniform distribution of Fe active sites (Yang et al., 2020b).

The low-cost chitosan acts as a nitrogen precursor, which can significantly reduce the cost of the cathode catalysts over other expensive N-precursors, which are generally used in synthesizing M-N-C catalysts. The Co/N-C materials were prepared by a one-step pyrolysis method in which the dried mixture of chitosan and Co precursor was pyrolyzed at different temperatures of 700°, 800°, 900°, and 1000°C. The Co/N-C prepared at 800°C shows good activity and 1738 ± 40 mW/m^2, which is 44.5% higher than that of Pt/C (1203 ± 18 mW/m^2) (Liang et al., 2020b). Fe-N-C can be derived from the *Broussonetia papyrifera* plant by solid digestion to produce bacterial cellulose, followed by pyrolysis with FeCl$_3$, as shown in Fig. 6.4(F). The MFCs with this Fe-N-C could have a maximum power density of up to 1308 mW m^{-2}, much higher than that of the N-C catalyst (638 mW m^{-2}). The activity of M-N-C catalysts depends on heteroatom, metal center, the coordination number of central metal, source of N and C, and loading of the catalysts. The high loading M-N-C catalysts result in high activity but impose oxygen transport issues due to the increased thickness of the catalysts. Also, one should consider the cost of the nitrogen and carbon precursor.

6.1.2.6 Carbon-based Catalysts in MFCs

Carbon-based materials are the cheapest, with various interesting physical properties, such as high surface area, porosity, electronic conductivity, and stability in different pHs. The rapid advancement in carbon materials leads to different methodologies to develop various carbon nanomaterials, such as carbon block (CB), activated carbon (AC), Vulcan carbon (VC), graphite, carbon nanotubes (CNTs), carbon nanofibers (CNFs), graphene, MOFs derived carbon, biomass-derived carbon, etc. Now, we have different carbon materials with different dimensions (0D, 1D, 2D, 3D, and bulk materials) and forms, as shown in Fig. 6.5. Mostly, the carbon cloth, carbon paper, carbon felt, carbon brush, and graphite-based air diffusion electrodes act as cathodes either as such or after loading the desired catalyst. The carbon-based materials can be used as such or in combination with metal/metal oxides (Gouse et al., 2020; Mashkour et al., 2021; Boukoureshtlieva et al., 2021; Dange et al., 2022). Recently, the heteroatom (N, S, P, F, etc.) doped carbon materials exhibit significant activity toward ORR, also known as metal-free electrocatalysts. The carbon materials show poor ORR performance due to insufficient active sites to reduce O$_2$. The carbon materials are chemically activated by treatment in harsh acid conditions to introduce oxygen functional groups to improve the activity of the carbon materials. Also, the metals, metal-oxides, and heteroatoms are introduced/anchored/doped on carbon materials to exhibit significant activity towards ORR to use as a cathode in MFC.

(i) Graphite

Graphite, a layered sp^2 carbon and a naturally occurring form of crystalline carbon, consists of stacked graphene layers. It was used as a catalyst for ORR due to its chemical stability and high conductivity. The graphite rod, plate, and brush are widely used in MFC as anode, cathode, and air diffusion electrodes due to their specific surface area, biocompatibility, chemical stability, and low cost

(Rahimnejad et al., 2015; Xing et al., 2017; Gouse et al., 2020). The performance of graphite as a cathode is significantly less due to the insufficient active sites to reduce oxygen. The graphite granules are treated with nitric acid, followed by thermal activation, and used as a cathode in DCMFC (Erable et al., 2009). The activated granules result in high open circuit voltage (OCP-1050 mV) than untreated graphite granules (660 mV), which might be due to the increased chemical functionalities (oxygen and nitrogen) on graphite. In another study, the graphite is chemically treated by just immersing the graphite in a series of chemical agents' H_3PO_4, HNO_3, $ZnCl_2$, urea, and melamine. The graphite treated with H_3PO_4 and HNO_3 shows improved MFC performance (Zhang et al., 2016).

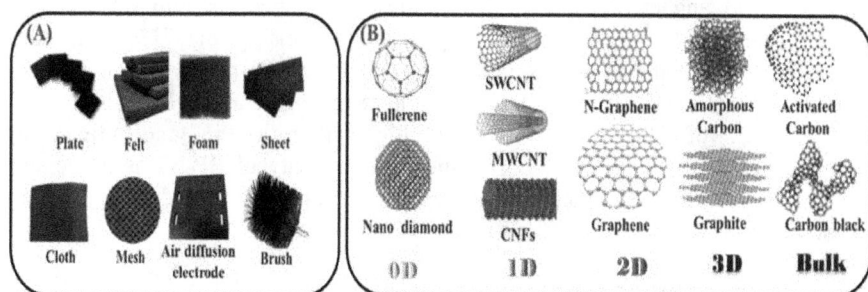

Figure 6.5　(A) Different forms of carbon electrodes used for MFC and (B) carbon-based materials with different dimensions.

It was further modified with metal oxide and sulfides (MnO_2, CuO/ZnO, MoS_2) (Jiang et al., 2016, 2017; Tajdid Khajeh et al., 2020) to improve the activity of graphite. The graphite biochar (BCw) was derived by a high-temperature gasification process using wood-biomass as a carbon source. The MnO_2/BCw/Nafion ink was prepared and used as an ORR cathode in SCMFCs inoculated with anaerobic sludge (Huggins et al., 2015). In another study, the graphite-MnO_2 and graphite-MoS_2 were prepared by dispersing metal oxide and metal sulfide in graphite colloidal solution (Jiang et al., 2017). The graphite MnO_2 showed a power density of 100 mW/m^2 in real wastewater, while the graphite-MoS_2 showed lesser power density with excellent stability over time. The 1 : 1 ratio of MnO_2 and MoS_2-graphite showed considerable power density and stability. The graphite cathode is modified by electrophoretic deposition of CuO/ZnO to improve the activity of the graphite and is found to have a 32% higher power density at CuO/ZnO/graphite than the graphite (Tajdid Khajeh et al., 2020).

(ii) Carbon black

Carbon black (CB) is a polycrystalline carbon with a high surface-area-to-volume ratio and possesses electrocatalytic properties. The CB is produced by the incomplete combustion of hydrocarbons such as coal, coal tar, vegetable matter, petroleum products, etc. Based on the source and heating methods, there are other subtypes of CBs: acetylene black, channel black, furnace black, lamp black, thermal black, and Vulcan carbon black. It was oxidized by an acid oxidation process using HNO_3, H_2SO_4, or a mixture of both to improve the electrochemical properties of the CB. The Vulcan XC-72R CB is modified by treating with nitric acid and ammonia and found to have chemical functionalities (oxygen and nitrogen-based) on the carbon surface. The Vulcan carbon treated with HNO_3 catalyst exhibited the highest catalytic performance in the ORR (1788 mW/m^3) than the ammonia-treated carbon (Duteanu et al., 2010). This study also found that introducing oxygen and nitrogen moieties to carbon black enhanced the ORR activity.

Recently, the acid-treated Vulcan XC-72R CB and pyrazinamide in a mass ratio of 1 : 30 were ball-milled for 1 h, followed by pyrolyzed in a tube furnace at different temperatures ($700°$, $800°$, $900°$, and $1000°C$) for 2 h in N_2 atmosphere (Wang et al., 2020c). The resulting N-doped CB-800 was used as a metal-free ORR catalyst in MFC and generated the P_{max} of 371 ± 3 mW/m^2. The N and F-doped CB (BP-NF) was also prepared by pyrolyzing a mixture of PTFE and CB in an ammonium atmosphere (Meng et al., 2015). The BP-NF showed a P_{max} of 676 mW/m^2 in SCMFC, higher than the Pt/C and individually doped carbon (BP-N and BP-F). The increased activity of BP-NF is associated with the C-N and C-F, which can synergistically polarize the adjacent carbons to facilitate ORR.

(iii) Activated carbon

Activated carbon (AC) is an inexpensive form of carbon with high porosity. It is also known as activated charcoal. AC is usually derived from waste products, such as coconut husks and waste from paper mills. First, the bulk sources are converted into charcoal; then, it is activated by the thermal method (Sathish Kumar et al., 2012c). AC is produced by carbon's thermal activation, which enhances the specific surface area and porosity. The AC has a surface area of more than 3000 m^2/g due to a high degree of microporosity. AC-based materials are used in MFC applications as a support and cathode catalyst. The carbon black is mixed with either activated carbon block (AC) or Vulcan carbon (VC) to improve the conductivity of AC (Yuan et al., 2014; Zhang et al., 2017). In this study, the resulting AC-CB possesses a higher conductivity of 0.53 S/cm than CB (0.40 S/cm) and AC (0.37 S/cm). The AC–CB generated the P_{max} of 1900 mW/cm^2, and the cost analysis shows that only \$0.3/m^2 cost is required to develop an AC-CB cathode (Zhang et al., 2017). The AC was cold pressed with PTFE on Ni mesh and used as a cathode in SCMFC. The AC cathode produced a P_{max} of 1220 ± 46 mW/m^2 with PTFE and 1150 ± 57 mW/m^2 without PTFE (Zhang et al., 2009b). Also, Pt/Pd was added to the AC-PTFE/Ni cathode to improve the ORR activity and generated the P_{max} of 1415 mW/m^2. In another study, the AC (supercapacitor activated carbon and nano AC), PTFE, and conductive carbon (Nano CC, 3000 mesh CC, XC-72, and F900-CC) based cathode on Ni mesh was developed and studied the effect of type and loading of CC and AC. The S-AC, PTFE, and F900-CC-based catalyst layers on Ni mesh produced a power density of 1190 ± 50 mW/m^2 comparable to a typical Pt-carbon cloth cathode (1320 mW/m^2). The above Ni foam cathode cost is \$50 m^{-2}, which is 1/30 of that of a Pt carbon cloth cathode (\$1500 m^{-2}) (Cheng and Wu, 2013).

In another study, conductive carbon (CB, mesoporous carbon (MC), CNTs), and metal particles (Cu and Au) were added to the AC to increase the performance of SCMFC (Zhang et al., 2017). They found that adding CB, MC and CNTs enhanced the MFC power density by 6–14% compared to AC. The heat-treated AC with CB reached the P_{max} of 1900 ± 76 mW/m^2, which is 41% higher than the Pt/C and 18% higher than plain AC. The enhanced performance is due to the increase in conductivity and hydrophobicity. At the same time, the addition of Cu and Au reduced the performance of SCMFC, which might be due to toxicity effects on the anode bacteria. The heteroatoms (N, P, O) were also introduced into the carbon structures of AC to enhance the ORR activity (Lv et al., 2019; Tian et al., 2018). The AC derived from silk fibroin was found to have nitrogen in its structure and exhibits enhanced ORR activity. After that, a lot of research is done to incorporate heteroatoms on AC for MFC applications. The AC is treated with the NH$_3$ gas, and the resulting N-AC contains N-functionalities. The N-AC generated a power density of 240 mW/m^2.

The nitrogen-doped AC is also synthesized by a pyrolyzing nitrogen-containing precursor (melamine, urea, cyanamide, phthalocyanine) with AC at high temperatures (Tian et al., 2018; Zhang et al., 2014). The AC pyrolyzed with melamine at 900°C (AC-N10-900) catalyst showed the highest BET surface area of 1911 m^2/g^{-1} and produced a power density of 1042 mW/cm^2. The N-AC derived from pyrolyzing phthalocyanine with AC at 1000°C generated the P_{max} of 1026 mW/m^2. The N and P-doped AC (NPC@AC) was prepared by pyrolyzing phytic acid doped polyaniline with AC (Lv, Zhang, and Chen, 2018). The resulting NPC@AC delivered a power density of 1233 mW/m^2. Fe and Co metals are combined with NAC to enhance the activity by generating M-N-C active sites (Kodali et al., 2017a; Yang et al., 2017). The Fe-N/AC (Liu et al., 2019a) delivered a P_{max} of 1092 mW/m^2, and the Fe-N-C/AC catalyst developed by pyrolyzing a Fe(III)-chitosan hydrogel generated a power density of 2.4 W/m^2 (Kodali et al., 2020b). The Fe-N-C/AC derived from MOF delivered the highest power density, 4.7 W/m^2, similar to a Pt-based catalyst (4.3 W/m^2) (Liu et al., 2012b).

(iv) Graphene and heteroatom-doped graphene-based materials

Graphene (Gr) is an allotrope of carbon consisting of a single layer of atoms arranged in a two-dimensional honeycomb lattice nanostructure (Geim and Novoselov, 2007; Vusa et al., 2014). It possesses excellent electronic conductivity, large surface area, high graphitic character, and ease of chemical and physical functionalization. Like other carbons, the pristine graphene is not active towards ORR due to a lack of defects and activity centers. The graphene with defects is found to have enhanced electrocatalytic activity. The Gr is doped with heteroatoms (N, B, S, P, and F), decorated with metal-based catalysts and/or M-N-C-based catalysts to exhibit good activity towards ORR (Qu et al., 2010; Feng et al., 2011a; Jeon et al., 2013; Kang et al., 2017; Zhao et al., 2019). The doping of heteroatom in sp^2 carbon perturbs the π electrons and changes the spin and charge densities of adjacent carbons atoms in the graphene matrix, creating a potential active site for O$_2$ adoption, charge transfer, and reduction process. Various forms of graphene are utilized as ORR catalysts, such as crumpled graphene, reduced graphene oxide (r-GO), monolayer and multilayered graphene sheets, graphene nanoribbons, and graphene quantum dots. Like M-N-C catalysts, generally, the heteroatom doped graphene can be prepared by (i) pyrolysis of organic molecules containing nitrogen with Gr, (ii) pyrolysis of organic frameworks without metal, (iii) pyrolysis of polymers, and (iv) pyrolysis of biomass.

The N-Gr was prepared by pyrolyzing the morphology-preserved thermal transformation of rod-shaped MOFs. The MOFs resulted in non-hollow (solid) carbon nanorods, which are further transformed into two- to six-layered graphene nanoribbons through sonochemical treatment followed by chemical activation (Pachfule et al., 2016). Feng et al. synthesized N-Gr on a gram-scale using a detonation technique with cyanuric chloride and trinitrophenol as reactants, which are low cost and easily accessible. NG delivered a P_{max} of 1350 mW/m^2 like that of Pt/C (1420 mW/m^2) (Feng et al., 2011a). The S and N doped Gr (N/S-Gr) was developed by pyrolyzing graphene oxide and poly dopamine-cysteine mixture at 900°C under an Ar atmosphere in a tubular furnace. The N/S-Gr exhibited ORR activity in a neutral medium and produced a P_{max} of 1368 mW/m^2, relatively higher than that of Pt/C (Zhao et al., 2019). In another study, the N-Gr on Ni mesh was prepared directly *in situ,* grown on the Ni surface, and used as a binder-free air cathode in MFC. The maximum power density of MFC based on NG can be boosted up to 1470 ± 80 mW/m^2, which is 32% higher than the conventional Pt/C air cathode (Wang et al., 2016). The electrochemical methods are also adopted to introduce heteroatoms on grapheme (Han et al., 2016; Manju et al., 2017). The graphite was anodically oxidized in an electrolytic solution (Na$_2$S$_2$O$_3$ + H$_2$SO$_4$ at a 3:1 ratio) to exfoliate the Gr, resulting S doped Gr (S-Gr). The S-Gr cathode produced a maximum power density of 51.22 ± 6.01 mWm^2, 1.92 ± 0.34 times higher than Pt/C (Han et al., 2016). The metals, metal-oxides, and M-N-C-based catalysts supported on graphene are also used as air cathodes in MFC (Song et al., 2015; Sun et al., 2015; Mashkour et al., 2017; Khater et al., 2021; González Vázquez et al., 2022) (Sun et al., 2015; Tofighi et al., 2019; Gao et al., 2020; Orellana, 2022).

(v) Carbon nanotubes and nanofibers

Carbon nanotubes (CNTs) and nanofibers (CNFs) are also used as cathode catalysts and support in MFC. CNTs are one or multiple graphene sheets wrapped concentrically to form a well-defined cylindrical hollow structure. There are different types of CNTs (SWCNTs, MWNTs, and DWCNTs). Various strategies are developed, such as acid or alkaline activation, heteroatom doping on CNTs, and mixing CNTs with other catalysts to enhance the activity of ORR. The N-doped CNTs (NCNTs) were grown, using the chemical vapour deposition technique (CVD) for ORR in MFC to produce electricity efficiently and durably (Feng et al., 2011b). The NCNTs exhibited a 4e$^-$ reduction of O$_2$ in neutral PBS and produced a P_{max} of 1600 ± 50 mW/m^2, which was higher than Pt/C (1393 ± 35 mW/m^2). In another study (He et al., 2016), microvillus-like N-CNTs are grown on carbon cloth by a CVD method, and N-CNT/CC exhibits an output power density of 542 mW/m^3. The PANI/MWNTs cathode generated a P_{max} of

476 mW/m^2 (Jiang et al., 2014) The MoS$_2$/CNTs-based cathode generated the P$_{max}$ of 53.0 mW/m^2, which is much higher than those MFCs with pure CNTs (21.4 mW/m^2) or solely MoS$_2$ (14.4 mW/m^2) cathode (Xu et al., 2019). The Co/N-CNT was prepared by pyrolysis of carbon nitride and cobalt acetate and achieved a maximum power density of 1260 mW/m^2, which was 16.6% higher than that of Pt/C catalyst (1080 mW/m^2). The Fe-N-CNT was also prepared (Lu et al., 2017; Türk et al., 2018) and delivered a P$_{max}$ of 6 W/m^3.

CNFs are cylindrical nanostructures with graphene layers arranged as stacked cones. The CNFs show high electrical conductivity and high BET surface area. The NCNFs are prepared to exhibit enhanced ORR activity. The NCNFs on SSM delivered a power density of 0.096 W/cm^2. The 3D N-doped CNFs are developed by heating clip-like polypyrrole (Yang et al., 2014). The 3D arrangement of CNFs produces a porous hierarchical fiber structure essential for better mass transport of oxygen. The NCF-900 catalyst showed 4e$^-$ reduction of O$_2$ in neutral media and generated a P$_{max}$ of 1377 mW/cm^2 in H-type MFCs. The N-CNFs were prepared by pyrolyzing the electrospun poly (acrylonitrile) at 900°C, generating a power density of 1.15 ± 0.04 W/g (Massaglia et al., 2019).

6.1.3   Others

Other catalysts, such as biomass-derived carbón, composites, hybrids, carbón nitrides, metal carbides, metal-carbon composites, and conductive polymer-carbon composites are used as air cathodes in MFCs. A variety of biomass materials, such as yeast cells, mushrooms, plant biomass, silk-cotton, cornstalk, egg white, coconut shell, chitin, pectin, human hair, etc. are used as C, N, O, S, P, and F source to prepare carbon-based materials by pyrolyzing at high temperature (Borghei et al., 2018; Li et al., 2021; Wang et al., 2020d; Yang and Chen, 2020; Zhang et al., 2022). For example, the activated carbon derived from ground nutshells delivered a P$_{max}$ of 521 mW/m^2 in SCMFC (Karthick et al., 2022). The pectin biomass derived hierarchical porous S, N doped graphitized carbon produced 1161 mW/m^2 due to enhanced ORR activity (Ma et al., 2019). The graphitized carbon doped with N and P derived from chitosan delivered 1603 mW/m^2, which is five-fold higher than the control (Liang et al., 2019). The N, P-doped graphitized carbon derived from cornstalk produced 1122 mW/m^2 power density with only a 10.2% loss over 80 days of operation (Sun et al., 2016). Heteroatom-doped lamellar-structured carbon with a high surface area is synthesized from alfalfa leaves and activated using KOH (Deng et al., 2017). The activated alfalfa leaves derived carbon produced a P$_{max}$ of 1328.9 mW/m^2. The biochar microspheres (BCMs) are derived from pomelo peels and activated by thermal methods to produce activated BCMs (a-BCMs) (Zhang et al., 2020). The a-BCMs cathode in MFC produced 907.2 mW/m^2 and continuously ran for 90 days. The efficiency, activity, and cost of the biomass-derived carbons depend on biomass, pyrolysis method, activation method, number of active sites, etc.

Composites/hybrids are also prepared by combining carbon, polymer, metal oxides, metal nitrides, and carbides. For example, a silver-tungsten carbide-Vulcan carbon hybrid (Ag-WC/C was developed, which exhibits a 4e$^-$ reduction of ORR in a neutral electrolyte in MFC (Gong et al., 2013). The porous carbon with Co_3O_4 (Co/N-C-800) generated the P_{max} of 1738 ± 40 mW/m^2, which is 44.5% higher than that of Pt/C (1203 ± 18 mW/m^2) (Liang et al., 2020b). The 3D N-graphene and Co_3O_4 (Co_3O_4/N-Gr) (Tan et al., 2018) produced a P_{max} of 578 ± 10 mW/m^2. The perovskite composite was also used for MFC applications (Nandikes et al., 2022). Ternary nanotube a-MnO_2/GO/AC (Tofighi et al., 2019) composite with 10% GO produced 148.4 mW/m^2. The carbon-based hybrids are also prepared to enhance activity and conductivity. Porous N-Gr/CB composites produced a P_{max} of 936 mW/m^2, which was 26% higher than that of NG (Liu et al., 2019b). The PANI@GO hybrids were prepared via *in-situ* polymerization of aniline monomers on graphene oxide (GO) surface and carbonized at 1600°C (Kang et al., 2017). The PANI/MWNTs composite cathode produced a P_{max} of 4760 mW/m^2 (Jiang et al., 2014). Fibrous PANI–MnO_2 nanocomposites were prepared using a one-step and scalable *in-situ* chemical-oxidative polymerization method. The fibrous PANI–MnO_2 cathode catalyst showed an improved power density of 0.0588 W/m^2, higher than that of pure PANI and carbon paper, respectively (Ansari et al., 2016).

6.2 BIOTIC CATALYSTS

The biotic catalysts are the enzymes or living microorganisms that can be used as cathodic ORR catalysts in MFC (Erable et al., 2012; Gao et al., 2020; Santoro et al., 2022). The enzymes, particularly the multicopper oxidase, are responsible for biocatalysis reduction of oxygen. The most known multicopper oxidase enzymes are ascorbate oxidase, laccase, and bilirubin oxidase (BOx). The BOx has higher ORR kinetic and durability than laccase and ascorbate oxidase. Microbial catalysis is based on the microbial film, which can reduce oxygen. The biotic catalysts are two types: (1) enzyme-based and (2) microbial-based catalysts. A hybrid lactate/air biofuel cell has been created using a microbial anode with *Shewanella MR1* integrated into a chitosan–carbon nanotube porous matrix and a DET-based laccase air-breathing cathode (Higgins et al., 2011). This MFC produced a power density of 26 W/m^3. The laccase from *T. versicolor* was used as a cathode with ABTS (2,20 -azino-bis(3-ethylbenzo-thiazoline-6-sulfonic acid) diammonium salt) in MFC (Schaetzle et al., 2009). The MFC with the laccase-bound cathode showed a low power output in the absence of the redox mediator (6.5 mW/m^2 at 0.23 V), indicating that direct electron transfer from the enzyme hydrogel was absent or occurred at a very low rate. An increased power density was observed in the presence of ABTS (37 mW/m^2 at 0.47 V). The BOx-based air-breathing cathode was used to treat activated sludge in MFC, and OCP was higher than 500 mV *vs.* Ag/AgCl over 32 days (Santoro et al., 2016).

The yeast cells are used as air-breathing cathodes, and the enzymes (Lacasse and BOx) present on the surface of yeast cells are responsible for surface ORR activity (Szczupak et al., 2012). Live microalgal cells modified by L-cys/Au@ carbon dots/bilirubin oxidase (BOD) layers-based cathode produced an ORR current density of 655.2 µA/cm^2 with fast ORR kinetics, which is 2.68 times higher than that of a BOD cathode fed with pure O_2. The resulting power density is 2.39 times higher than that of a BOD cathode biofuel cell in an O_2-saturated solution (Qing et al., 2022). The enzymes are the best catalysts to use in neutral media and are sensitive to any inhibition, requiring a particular method to immobilize on to the electrode surface. Also, enzymes are deactivated with complex media and temperature.

6.2.1 Chemical Fouling and Biofouling

Microbe-based catalysts' activity depends on parameters of biofilm formation, surface enzymes, and complex media of the MFC. Chemical fouling and biofouling are the primary concern of biocatalysts.

6.3 CONSTRAINTS OF ORR CATALYSTS IN MFCS

(a) We have a series of catalysts from single atoms to bulk materials, including noble metal, transition metal, carbon, heteroatom-doped carbon, metal-carbon, metal-nitrogen-carbon enzyme, and microbes. Coating catalysts on carbon materials, such as carbon cloth, brush, paper, felt, plate, foam, and air-diffusion electrodes (Fig. 6.5(A)) is a prerequisite to use them in MFC and to improve their performance. It also requires a specific approach in the preparation of the MFC electrodes. However, all those electrodes have stability, water permeability, and fouling limitations due to either radical and/or bacteria. All the above odds add additional challenges in developing successful MFC for large-scale commercialization. In addition to the carbon-based support, metal support, such as Pt mesh, stainless steel mesh, and nickel foam, are also used in some instances. However, care must be taken in restricting the support's unwanted reactions in the MFCs.

(b) Even though the metal-based catalysts are cost-effective and active towards ORR, only a few are suitable for neutral media and complex wastewaters. Most of the electrodes are active only for a few days (less than a month). Regeneration of the cathode is another challenge; the possible leaching of metals into treated water from the cathode may lead to toxic effects. It becomes another effluent to treat, increasing the operational complexity and price. MnO_2 is active towards ORR in a neutral medium in different forms. However, there is an ambiguity in whether the α-form or β-form is better in this case. One must explore the relationship between the activity and oxidation state, composition, phase, size, and morphology of the catalyst in situ to avoid the above or similar ambiguities.

(c) Most metal-coordinated macrocycles promote the 2-electron reduction of O_2 to peroxide, whereas the 4-electron reduction of O_2 to H_2O is preferred for fuel cell applications. However, one can change the O_2 reduction path to H_2O by tuning the chemical environment of the metal centers of metal-coordinated macrocycles.

(d) CNTs and CNFs-based catalysts have gained considerable attention, but the synthesis of CNTs and CNFs by a facile and cheap method is still in its infancy. AC-based catalysts also show good activity, but the conductivity decreases after activation. So, we need conductive materials and AC to use as an electrode.

(e) Graphene-based materials are of great interest, but the activity depends on the dopant and metal center. Few graphene-based materials show $4e^-$ reduction activity while the others, either $2e^-$ or mixed electron transfer. Also, the activity of the carbon-based materials depends on the electrolyte's pH.

(f) The commercial Pt/C has mostly been used as a cathode to study the newly developed anode and/or membrane performance in MFCs. Cathode efficiency is crucial in the production of electricity in MFCs. Most of the cathodes in MFC, such as air cathodes, underwater aqueous cathodes, and biocathodes, usually use carbon paper, graphite chips, felt, and fabric. Consequently, industrial use of such non-catalytic materials is restricted due to their sluggish cathode reduction kinetics. Also, continuous reduction of oxygen on carbon surface results in a decrease in performance. The 4-electron pathway is predominant on noble-metal electrocatalysts, whereas the peroxide pathway is on graphite, gold, oxide-covered metals, and most carbon materials.

(g) Chemical fouling and biofouling of the catalysts are two major concerns in developing biotic and abiotic cathodes. One should know electrochemistry, chemistry, microbiology, biology, ecology, material sciences, and engineering to design the best cathode catalyst.

(h) Considering power density alone is not enough to compare the performance and efficiency of an MFC. The direct comparison of the performance of an MFC with different catalysts is difficult because it varies with various parameters, such as type of MFC, organic substrate, anode, cathode, cocatalyst, wastewater, nutrients, and temperature. One of the major aspects is an energy balance in MFC137, i.e. the consumption of energy to develop MFC should be less than the energy produced by the same MFC.

6.4   CONCLUSION

MFC-based technologies are considered important electrochemical devices for future energy generation devices, wastewater treatment, and fuel production from organic waste. One should either know or have collaborators from electrochemistry, chemistry, microbiology, biology, ecology, material sciences, and engineering to design the best MFC. ORR cathode with $4e^-$ oxygen reduction to water in neutral media. Carbon-based M-N-C materials are better ORR catalysts to use as cathodes in MFC. But identifying the low-cost precursor, cheap method to produce M-N-C, and compatible way to load M-N-C on cathode support appear challenging. Recently, the low loading of Pt on ternary-based metals surface and/ or M-N-C is another better ORR catalyst to use in MFC. On the other hand, heteroatom-doped carbons are also gaining importance as metal-free catalysts in neutral media, while biotic catalysts are also used due to their compatibility in complex media. This chapter presents an overview of cathode catalysts used for MFC and their constraints.

ACKNOWLEDGEMENTS

We would like to acknowledge the Vellore Institiute of Technology, Vellore, Tamil Nadu, India for providing Post-Doctoral Fellowship, Indian Institute of Technology Kanpur, Uttar Pradesh, India and Chung-Ang University, Anseong, The Republic of Korea for the financial and technical support.

REFERENCES

Abd-Elrahman, N.K., N. Al-Harbi, N.M. Basfer, Y. Al-Hadeethi, A. Umar and S. Akbar. 2022. Applications of nanomaterials in microbial fuel cells: A review. Molecules. 27: 7483.

Allen, R.M. and H.P. Bennetto. 1993. Microbial fuel-cells. Appl. Biochem. Biotechnol. 39: 27–40.

Ansari, S.A., N. Parveen, T.H. Han, M.O. Ansari and M.H. Cho. 2016. Fibrous Polyaniline@ manganese oxide nanocomposites as supercapacitor electrode materials and cathode catalysts for improved power production in microbial fuel cells. Phys. Chem. Chem. Phys. 18: 9053–9060.

Apollon, W., S.K. Kamaraj, H. Silos-Espino, C. Perales-Segovia, L.L. Valera-Montero, V.A. Maldonado-Ruelas, et al., 2020. Impact of opuntia species plant bio-battery in a semi-arid environment: demonstration of their applications. Appl. Energy. 279: 115788.

Apollon, W., A.I. Luna-Maldonado, S.K. Kamaraj, J.A. Vidales-Contreras, H. Rodríguez-Fuentes, J.F. Gómez-Leyva, et al., 2021. Progress and recent trends in photosynthetic assisted microbial fuel cells: a review. Biomass Bioenergy. 148: 106028.

Apollon, W., I. Rusyn, N. González-gamboa, T. Kuleshova, A.I. Luna-maldonado, J.A. Vidales-contreras, et al., 2022a. Improvement of zero waste sustainable recovery using microbial energy generation systems: a comprehensive review. Sci. Total Environ. 817.

Apollon, W., L.L. Valera-Montero, C. Perales-Segovia, V.A. Maldonado-Ruelas, R.A. Ortiz-Medina, J.F. Gómez-Leyva, et al., 2022b. Effect of ammonium nitrate on novel cactus pear genotypes aided by biobattery in a semi-arid ecosystem. Sustainable Energy Technol. Assess. 49: 101730.

Boas, J.V., V.B. Oliveira, M. Simoes and A.M.F. Pinto. 2022. Review on microbial fuel cells applications, developments and costs. J. Environ. Manage. 307: 114525.

Borghei, M., J. Lehtonen, L. Liu and O.J. Rojas. 2018. Advanced biomass-derived electrocatalysts for the oxygen reduction reaction. Adv. Mater. 30: 1 703691.

Boukoureshtlieva, R., T. Stankulov and A. Momchilov. 2021. Carbon-based cathode catalyts used in microbial fuel cell for wastewater treatment and energy recovery. Ecol. Eng. Environ. Prot. 24–33.

Cao, C., L. Wei, G. Wang and J. Shen. 2017. *In-situ* growing $NiCo_2O_4$ nanoplatelets on carbon cloth as binder-free catalyst air-cathode for high-performance microbial fuel cells. Electrochim. Acta. 231: 609–616.

Cheng, S. and J. Wu. 2013. Air-cathode preparation with activated carbon as catalyst, PTFE as binder and nickel foam as current collector for microbial fuel cells. Bioelectrochemistry. 92: 22–26.

Choudhury, P., U.S.P. Uday, T.K. Bandyopadhyay, R.N. Ray and B. Bhunia. 2017. Performance improvement of microbial fuel cell (MFC) using suitable electrode and bioengineered organisms: a review. Bioengineered. 8: 471–487.

Cohen, B. 1931. The bacterial culture as an electrical half-cell. J. Bacteriol. 21: 18–19.

Damjanovic, A., M.A. Genshaw and J.O'M. Bockris. 1967. The Mechanism of oxygen reduction at platinum in alkaline solutions with special reference to H_2O_2. J. Electrochem. Soc. 114: 1107.

Damjanovic, A., D.B. Sepa and M.V. Vojnovic. 1979. New evidence supports the proposed mechanism for O_2 reduction at oxide-free platinum electrodes. Electrochim. Acta. 24: 887–889.

Dange, P., N. Savla, S. Pandit, R. Bobba and S.P. Jung. 2022. A comprehensive review on oxygen reduction reaction in microbial fuel cells. J. Renewable Mater. 10: 665–697.

Deng, L., Y. Yuan, Y. Zhang, Y. Wang, Y. Chen, H. Yuan, et al., 2017. Alfalfa leaf-derived porous heteroatom-doped carbon materials as efficient cathodic catalysts in microbial fuel cells. ACS Sustainable Chem. Eng. 5: 9766–9773.

Djellali, M., M. Kameche, H. Kebaili, M.M. Bouhent and A. Benhamou. 2021. Synthesis of nickel-based layered double hydroxide (LDH) and their adsorption on carbon felt fibres: application as low cost cathode catalyst in microbial fuel cell (MFC) Environ. Technol. 42: 492–504.

Du, Lei, L. Xing, G. Zhang and S. Sun. 2020. Metal-organic framework derived carbon materials for electrocatalytic oxygen reactions: recent progress and future perspectives. Carbon. 156: 77–92.

Duteanu, N., B. Erable, S.M. Senthil Kumar, M.M. Ghangrekar and K. Scott. 2010. Effect of chemically modified vulcan XC-72R on the performance of air-breathing cathode in a single-chamber microbial fuel cell. Bioresour. Technol. 101: 5250–5255.

Erable, B., N. Duteanu, S.M. Senthil Kumar, Y. Feng, M.M. Ghangrekar and K. Scott. 2009. Nitric acid activation of graphite granules to increase the performance of the non-catalyzed oxygen reduction reaction (ORR) for MFC applications. Electrochem. Commun. 11: 1547–1549.

Erable, B., D. Føron and A. Bergel. 2012. Microbial catalysis of the oxygen reduction reaction for microbial fuel cells: a review. Chem. Sus. Chem. 5: 975–87.

Fan, Y., S.-K. Han and H. Liu. 2012. Improved performance of CEA microbial fuel cells with increased reactor size. Energy Environ. Sci. 5: 8273–8280.

Feng, L., Y. Chen and L. Chen. 2011a. Easy-to-operate and low-temperature synthesis of gram-scale nitrogen-doped graphene and its application as cathode catalyst in microbial fuel cells. ACS Nano. 5: 9611–9618.

Feng, L., Y. Yan, Y. Chen and L. Wang. 2011b. Nitrogen-doped carbon nanotubes as efficient and durable metal-free cathodic catalysts for oxygen reduction in microbial fuel cells. Energy Environ. Sci. 4: 1892–1899.

Gaixiu, Y.Z.Y., L. Pengmei, X. Kong, L. Li, G. Chen, T. Lu, et al., 2014. Application of surface-modified carbon powder in microbial fuel cells. Chin. J. Catal. 35: 770–775.

Gajda, I., J. Greenman, C. Santoro, A. Serov, C. Melhuish, P. Atanassov, et al., 2018. Improved power and long-term performance of microbial fuel cell with Fe-N-C catalyst in air-breathing cathode. Energy. 144: 1073–1079.

Gao, M., J.Y. Lu and W.W. Li. 2020. Oxygen reduction reaction electrocatalysts for microbial fuel cells. pp. 73-96. *In*: L. Singh, D.M. Mahapatra and H. Liu (eds). Novel Catalyst Materials for Bioelectrochemical Systems: Fundamentals and Applications. ACS Symposium Series, Vol. 1342. American Chemical Society.

Geim, A.K. and K.S. Novoselov. 2007. The rise of graphene. Nat. Mater. 6: 183–191.

Gong, X.-B., S.-J. You, X.-H. Wang, Y. Gan, R.-N. Zhang and N.-Q. Ren. 2013. Silver–tungsten carbide nanohybrid for efficient electrocatalysis of oxygen reduction reaction in microbial fuel cell. J. Power Sources. 225: 330–337.

Gong, X.-B., S.-J. You, X.-H. Wang, J.-N. Zhang, Y. Gan and N.-Q. Ren. 2014. A novel stainless steel mesh/cobalt oxide hybrid electrode for efficient catalysis of oxygen reduction in a microbial fuel cell. Biosens. Bioelectron. 55: 237–241.

González V., O. Francisco, C.F. Reyes, M.O. Morales, S.K. Kamaraj, M.D.R.M. Virgen, et al., 2022. Facile scalable manufacture of improved electrodes using structured surface coatings of nickel oxide as cathode and reduced graphene oxide as anode for evaluation in a prototype development on microbial fuel cells. Int. J. Hydrogen Energy. 47(70): 30248–30261.

Gouse, S., T. Maiyalagan and C. Liu. 2020. A review on carbon and non-precious metal based cathode catalysts in microbial fuel cells. Int. J. Hydrogen Energy. 46: 3056–3089.

Greeley, J., I.E.L. Stephens, A.S. Bondarenko, T.P. Johansson, H.A. Hansen, T.F. Jaramillo, et al., 2009. Alloys of platinum and early transition metals as oxygen reduction electrocatalysts. Nat. Chem. 1: 552–556.

Han, T.H., N. Parveen, S.A. Ansari, J.H. Shim, A.T.N. Nguyen and M.H. Cho. 2016. Electrochemically synthesized sulfur-doped graphene as a superior metal-free cathodic catalyst for oxygen reduction reaction in microbial fuel cells. RSC Adv. 6: 103446–103454.

Han, J., J. Bian and C. Sun. 2020. Recent advances in single-atom electrocatalysts for oxygen reduction reaction. Research (Washington, D.C.). 2020: 9512763.

He, Y.-R., F. Du, Y.-X. Huang, L.-M. Dai, W.-W. Li and H.-Q. Yu. 2016. Preparation of microvillus-like nitrogen-doped carbon nanotubes as the cathode of a microbial fuel cell. J. Mater. Chem. A. 4: 1632–1636.

Higgins, S.R., C. Lau, P. Atanassov, S.D. Minteer and M.J. Cooney. 2011. Hybrid biofuel cell: microbial fuel cell with an enzymatic air-breathing cathode. ACS Catal. 1: 994–97.

Huggins, T.M., J.J. Pietron, H. Wang, Z.J. Ren and J.C. Biffinger. 2015. Graphitic biochar as a cathode electrocatalyst support for microbial fuel cells. Bioresour. Technol. 195: 147–153.

Jasinski, R. 1964. A new fuel cell cathode catalyst. Nature. 201: 1212–1213.

Jeon, I.-Y., S. Zhang, L. Zhang, H.-J. Choi, J.-M. Seo, Z. Xia, et al., 2013. Edge-selectively sulfurized graphene nanoplatelets as efficient metal-free electrocatalysts for oxygen reduction reaction: the electron spin effect. Adv. Mater. 25: 6138–6145.

Jiang, Y., Y. Xu, Q. Yang, Y. Chen, S. Zhu and S. Shen. 2014. Power generation using polyaniline/multi-walled carbon nanotubes as an alternative cathode catalyst in microbial fuel cells. Int. J. Energy Res. 38: 1416–1423.

Jiang, B., T. Muddemann, U. Kunz, H. Bormann, M. Niedermeiser, D. Haupt, et al., 2016. Evaluation of microbial fuel cells with graphite plus MnO_2 and MoS_2 paints as oxygen reduction cathode catalyst. J. Electrochem. Soc. 164: H3083–H3090.

Jiang, B., T. Muddemann, U. Kunz, L.G. Silva e Silva, H. Bormann, M. Niedermeiser, et al., 2017. Graphite/MnO_2 and MoS_2 composites used as catalysts in the oxygen reduction cathode of microbial fuel cells. J. Electrochem. Soc. 164: E519–E524.

Jiang, L., J. Chen, Y. An, D. Han, S. Chang, Y. Liu, et al., 2020a. Enhanced electrochemical performance by nickel-iron layered double hydroxides (LDH) coated on Fe_3O_4 as a cathode catalyst for single-chamber microbial fuel cells. Sci. Total Environ. 745: 141163.

Jiang, L., J. Chen, D. Han, S. Chang, R. Yang, Y. An, et al., 2020b. Potential of core-shell NiFe layered double hydroxide@Co_3O_4 nanostructures as cathode catalysts for oxygen reduction reaction in microbial fuel cells. J. Power Sources. 453: 227877.

Kamaraj, S.K., A.E. Rivera, S. Murugesan, J. García-Mena, O. Maya, C. Frausto-Reyes, et al., 2019. Electricity generation from nopal biogas effluent using a surface modified clay cup (Cantarito) microbial fuel cell. Heliyon. 5: e01506.

Kang, Z., K. Jiao, X. Xu, R. Peng, S. Jiao and Z. Hu. 2017. Graphene oxide-supported carbon nanofiber-like network derived from polyaniline: a novel composite for enhanced glucose oxidase bioelectrode performance. Biosens. Bioelectron. 96: 367–372.

Karthick, S., S. Vishnuprasad, K. Haribabu and N.J. Manju. 2022. Activated carbon derived from ground nutshell as a metal-free oxygen reduction catalyst for air cathode in single chamber microbial fuel cell. Biomass Convers. Biorefin. 12: 1729–1736.

Karube, I., T. Matsunaga, S. Tsuru and S. Suzuki. 1976. Continous hydrogen production by immobilized whole cells of clostridium butyricum. Biochim. Biophys. Acta—Gen. Subj. 444: 338–343.

Khater, D.Z., R.S. Amin, M.O. Zhran, Z.K. Abd El-Aziz, M. Mahmoud, H.M. Hassan, et al., 2021. Enhancement of microbial fuel cell performance by anodic communities adaptation and cathodic mixed nickel copper oxides (NiO-CuO) on graphene electrocatalyst, 1–18.

Kim, J.R., J.-Y. Kim, S.-B. Han, K.-W. Park, G.D. Saratale and S.-E. Oh. 2011. Application of Co-Naphthalocyanine (CoNPc) as alternative cathode catalyst and support structure for microbial fuel cells. Bioresour. Technol. 102: 342–347.

Kim, J., Y. Hong, K. Lee and J.Y. Kim. 2020. Highly stable Pt-Based ternary systems for oxygen reduction reaction in acidic electrolytes. Adv. Energy Mater. 10: 2002049.

Kodali, M., C. Santoro, A. Serov, S. Kabir, K. Artyushkova, I. Matanovic, et al., 2017a. Air breathing cathodes for microbial fuel cell using Mn^-, Fe^-, Co- and Ni-containing platinum group metal-free catalysts. Electrochim. Acta. 31: 115–124.

Kodali, M., C. Santoro, S. Herrera, A. Serov and P. Atanassov. 2017b. Bimetallic platinum group metal-free catalysts for high power generating microbial fuel cells. J. Power Sources. 366: 18–26.

Kumar, R., L. Singh, A.W. Zularisam and F.I. Hai. 2018. Microbial fuel cell is emerging as a versatile technology: a review on its possible applications, challenges and strategies to improve the performances. Int. J. Energy Res. 42: 369–394.

Li, B., M. Wang, X. Zhou, X. Wang, B. Liu and B. Li. 2015. Pyrolyzed binuclear-cobalt-phthalocyanine as electrocatalyst for oxygen reduction reaction in microbial fuel cells. Bioresour. Technol. 193: 545–548.

Li, S., X. Tang, H. Jia, H. Li, G. Xie, X. Liu, et al., 2020. Nanoporous high-entropy alloys with low pt loadings for high-performance electrochemical oxygen reduction. J. Catal. 383: 164–171.

Li, S., S.-H. Ho, T. Hua, Q. Zhou, F. Li and J. Tang. 2021. Sustainable biochar as an electrocatalysts for the oxygen reduction reaction in microbial fuel cells. Green Energy Environ. 6: 644–659.

Liang, B., K. Li, Y. Liu and X. Kang. 2019. Nitrogen and phosphorus dual-doped carbon derived from chitosan: an excellent cathode catalyst in microbial fuel cell. Chem. Eng. J. 358: 1002–1011.

Liang, B., C. Ren, Y. Zhao, K. Li and C. Lv. 2020a. Nitrogenous mesoporous carbon coated with Co/Cu nanoparticles modified activated carbon as air cathode catalyst for microbial fuel cell. J. Electroanal. Chem. 860: 113904.

Liang, B., Y. Zhao, K. Li and C. Lv. 2020b. Porous carbon Co-doped with inherent nitrogen and externally embedded cobalt nanoparticles as a high-performance cathode catalyst for microbial fuel cells. Appl. Surf. Sci. 505: 144547.

Lima, F.H.B., J.F.R. De Castro and E.A. Ticianelli. 2006. Silver-cobalt bimetallic particles for oxygen reduction in alkaline media. J. Power Sources. 161: 806–812.

Lima, F.H.B., J. Zhang, M.H. Shao, K. Sasaki, M.B. Vukmirovic, E.A. Ticianelli, et al., 2007. Catalytic activity−d-band center correlation for the O_2 reduction reaction on platinum in alkaline solutions. J. Phys. Chem. C. 111: 404–410.

Liu, B., C. Brückner, Y. Lei, Y. Cheng, C. Santoro and B. Li. 2014. Cobalt porphyrin-based material as methanol tolerant cathode in single chamber microbial fuel cells (SCMFCs). J. Power Sources. 257: 246–253.

Liu, Y., Y.-S. Fan and Z.-M. Liu. 2019a. Pyrolysis of iron phthalocyanine on activated carbon as highly efficient non-noble metal oxygen reduction catalyst in microbial fuel cells. Chem. Eng. J. 361: 416–427.

Liu, Y., Z. Liu, H. Liu and M. Liao. 2019b. Novel porous nitrogen-doped graphene/carbon black composites as efficient oxygen reduction reaction electrocatalyst for power generation in microbial fuel cell. Nanomaterials. 9: 836.

Lu, M., S. Kharkwal, H. Yong, S. Fong and Y. Li. 2011. Carbon nanotube supported MnO_2 catalysts for oxygen reduction reaction and their applications in microbial fuel cells. Biosens. Bioelectron. 26: 4728–4732.

Lu, B., T.J. Smart, D. Qin, J.E. Lu, N. Wang, L. Chen, et al., 2017. Nitrogen and Iron-Co-doped carbon hollow nanotubules as high-performance catalysts toward oxygen reduction reaction: a combined experimental and theoretical study. Chem. Mater. 29: 5617–5628.

Luo, X., W. Han, H. Ren and Q. Zhuang. 2019. Metallic organic framework-derived Fe, N, S Co-doped carbon as a robust catalyst for the oxygen reduction reaction in microbial fuel cells. Energies. 12: 3846.

Lv, K., H. Zhang and S. Chen. 2018. Nitrogen and phosphorus Co-soped carbon modified activated carbon as an efficient oxygen reduction catalyst for microbial fuel cells. RSC Adv. 8: 848–855.

Lv, C., B. Liang, M. Zhong, K. Li and Y. Qi. 2019. Activated carbon-supported multi-doped graphene as high-efficient catalyst to modify air cathode in microbial fuel cells. Electrochim. Acta. 304: 360–369.

Ma, R., G. Lin, Y. Zhou, Q. Liu, T. Zhang, G. Shan, et al., 2019a. A review of oxygen reduction mechanisms for metal-free carbon-based electrocatalysts. Npj Comput. Mater. 5: 78.

Ma, Y., S. You, B. Jing, Z. Xing, H. Chen, Y. Dai, et al., 2019b. Biomass pectin-derived N, S-Enriched carbon with hierarchical porous structure as a metal-free catalyst for enhancing bio-electricity generation. Int. J. Hydrogen Energy. 44: 16624–16638.

Manju, V., CS.R. Vusa, P. Arumugam and S. Berchmans. 2017. A facile and versatile electrochemical tuning of graphene for oxygen reduction reaction in acidic, neutral and alkali media. ChemistrySelect. 2: 8541–8552.

Mashkour, M., M. Rahimnejad and M. Mashkour. 2016. Bacterial cellulose-polyaniline nano-biocomposite: a porous media hydrogel bioanode enhancing the performance of microbial fuel cell. J. Power Sources. 325: 322–328.

Mashkour, M., M. Rahimnejad, S.M. Pourali, H. Ezoji, A. El-Mekawy and D. Pant. 2017. Catalytic performance of nano-hybrid graphene and titanium dioxide modified cathodes fabricated with facile and green technique in microbial fuel cell. Prog. Nat. Sci.: Mater. Int. 27: 647–651.

Mashkour, M., M. Rahimnejad, F. Raouf and N. Navidjouy. 2021. A review on the application of nanomaterials in improving microbial fuel cells. Biofuel Res. J. 30: 1400–1416.

Massaglia, G., V. Margaria, A. Sacco, M. Castellino, A. Chiodoni, F.C. Pirri, et al., 2019. N-doped carbon nanofibers as catalyst layer at cathode in single chamber microbial fuel cells. Int. J. Hydrogen Energy. 44: 4442–4449.

Meng, K., Q. Liu, Y. Huang and Y. Wang. 2015. Facile synthesis of nitrogen and fluorine Co-doped carbon materials as efficient electrocatalysts for oxygen reduction reactions in air-cathode microbial fuel cells. J. Mater. Chem. A. 3: 6873–6877.

Nandikes, G., S.G. Peera and L. Singh. 2022. Perovskite-based nanocomposite electrocatalysts: an alternative to platinum ORR catalyst in microbial fuel cell cathodes. Energies. 15: 272.

Nguyen, M.-T., B. Mecheri, A. D'Epifanio, T. Sciarria, F. Adani and S. Licoccia. 2014. Iron chelates as low-cost and effective electrocatalyst for oxygen reduction reaction in microbial fuel cells. Int. J. Hydrogen Energy. 39: 6462–6469.

Nørskov, J.K., J. Rossmeisl, A. Logadottir, L. Lindqvist, J.R. Kitchin, T. Bligaard, et al., 2004. Origin of the overpotential for oxygen reduction at a fuel-cell cathode. J. Phys. Chem. B. 108: 17886–17892.

Nosheen, F., T. Anwar, A. Siddique and N. Hussain. 2019. Noble metal based alloy nanoframes: syntheses and applications in fuel cells. Front. Chem. 7: 456.

Oldacre, A.N., A.E. Friedman and T.R. Cook. 2017. A self-assembled cofacial cobalt porphyrin prism for oxygen reduction catalysis. J. Am. Chem. Soc. 139: 1424–1427.

Costa de Oliveira, M.A., A. D'Epifanio, O. Hitoshi and B. Mecheri. 2020. Platinum group metal-free catalysts for oxygen reduction reaction: applications in microbial fuel cells. Catal. 10: 475.

Orellana, W. 2022. Evidence of carbon-supported porphyrins pyrolyzed for the oxygen reduction reaction keeping integrity. Sci. Rep. 1–11.

Orellana, W., C.Z. Loyola, J.F. Marco and F. Tasca. 2022. Evidence of carbon-supported porphyrins pyrolyzed for the oxygen reduction reaction keeping integrity. Sci. Rep. 12: 8072.

Pachfule, P., D. Shinde, M. Majumder and Q. Xu. 2016. Fabrication of carbon nanorods and graphene nanoribbons from a metal–organic framework. Nat. Chem. 8: 718–724.

Potter, M.C. and A.D. Waller. 1911. Electrical effects accompanying the decomposition of organic compounds. Proc. R. Soc. Lond. B. 84: 260–276.

Pu, L., K. Li, Z. Chen, P. Zhang, X. Zhang and Z. Fu. 2014. Silver electrodeposition on the activated carbon air cathode for performance improvement in microbial fuel cells. J. Power Sources. 268: 476–481.

Qing, S., L.-L. Wang, L.-P. Jiang, X. Wu and J.-J. Zhu. 2022. Live microalgal cells modified by L-Cys/Au@carbon dots/bilirubin oxidase layers for enhanced oxygen reduction in a membrane-less biofuel cell. Smart Mat. 3: 298–310.

Qu, L., Y. Liu, J.-B. Baek and L. Dai. 2010. Nitrogen-doped graphene as efficient metal-free electrocatalyst for oxygen reduction in fuel cells. ACS Nano. 4: 1321–1326.

Rahimnejad, M., A. Adhami, S. Darvari, A. Zirepour and S.-E. Oh. 2015. Microbial fuel cell as new technology for bioelectricity generation: a review. Alexandria Eng. J. 54: 745–756.

Ren, X., Q. Lv, L. Liu, B. Liu, Y. Wang, A. Liu, et al., 2020. Current progress of Pt and Pt-based electrocatalysts used for fuel cells. Sustainable Energy Fuels. 4: 15–30.

Reyes-Rodríguez, J.L., S.K. Kamraj and O. Solorza-Feria. 2015. Synthesis and function-alization of green carbon as a Pt catalyst support for the oxygen reduction reaction. Int. J. Hydrogen Energy. 40: 17253–17263.

Rojas-Carbonell, S., C. Santoro, A. Serov and P. Atanassov. 2017. Transition metal-nitrogen-carbon catalysts for oxygen reduction reaction in neutral electrolyte. Electrochem. Commun. 75: 38–42.

Rozenfeld, S., M. Schechter, H. Teller, R. Cahan and A. Schechter. 2017. Novel RuCoSe as non-platinum catalysts for oxygen reduction reaction in microbial fuel cells. J. Power Sources. 362: 140–146.

Salgado-dávalos, V., S. Osorio-avilés, S.K. Kamraj, L. Vega-alvarado, K. Juárez, S. Silva-martínez, et al., 2021. Sediment microbial fuel cell power boosted by natural chitin degradation and oxygen reduction electrocatalysts. Clean Soil, Air, Water. 49: 2000465.

Santoro, C., S. Babanova, Be. Erable, A. Schuler and P. Atanassov. 2016. Bilirubin oxidase-based enzymatic air-breathing cathode: operation under pristine and contaminated conditions. Bioelectrochemistry. 108: 1–7.

Santoro, C., M. Kodali, S. Herrera, A. Serov and I. Ieropoulos. 2018. Power generation in microbial fuel cells using platinum group metal-free cathode catalyst: effect of the catalyst loading on performance and costs. J. Power Sources. 378: 169–175.

Santoro, C., P. Bollella, B. Erable, P. Atanassov and D. Pant. 2022. Oxygen reduction reaction electrocatalysis in neutral media for bioelectrochemical systems. Nat. Catal. 5: 473–484.

Sathish Kumar, K., O. Solorza-Feria, G. Vázquez-Huerta and J.P. Luna-Arias. 2012a. Electrical stress-directed evolution of biocatalysts community sampled from a sodic-saline soil for microbial fuel cells. J. New Mater. Electrochem. Syst. 181–186.

Sathish Kumar, K., O. Solorza-Feria, R. Hernández-Vera and G.Vazquez-Huerta. 2012b. Comparison of various techniques to characterize a single chamber microbial fuel cell loaded with sulfate reducing biocatalysts. J. New Mater. Electrochem. Syst. 195–201.

Sathish Kumar, K., G. Vázquez-Huerta, A. Rodríguez-Castellanos and O. Solorza-Feria. 2012c. Microwave assisted synthesis and characterizations of decorated activated carbon. Int. J. Electrochem. Sci. 7: 5484.

Schaetzle, O., F. Barrière and U. Schröder. 2009. An improved microbial fuel cell with laccase as the oxygen reduction catalyst. Energy Environ. Sci. 2: 96–99.

Song, T.-S., D.-B. Wang, H. Wang, X. Li, Y. Liang and J. Xie. 2015. Cobalt oxide/nanocarbon hybrid materials as alternative cathode catalyst for oxygen reduction in microbial fuel cell. Int. J. Hydrogen Energy. 40: 3868–3874.

Su, Y., Y. Zhu, X. Yang, J. Shen, J. Lu, X. Zhang, et al., 2013. A highly efficient catalyst toward oxygen reduction reaction in neutral media for microbial fuel cells. Ind. Eng. Chem. Res. 52: 6076–6082.

Sumner, L., N.A. Sakthivel, H. Schrock, K. Artyushkova, A. Dass and S. Chakraborty. 2018. Electrocatalytic oxygen reduction activities of thiol-protected nanomolecules ranging in size from Au28(SR)20 to Au279(SR)84. Ind. Eng. Chem. Res. C. 122: 24809–24817.

Sun, M., H. Liu, Y. Liu, J. Qu and J. Li. 2015. Graphene-based transition metal oxide nanocomposites for the oxygen reduction reaction. Nanoscale. 7: 1250–1269.

Sun, Y., Y. Duan, L. Hao, Z. Xing, Y. Dai, R. Li, et al., 2016. Cornstalk-derived nitrogen-soped partly graphitized carbon as efficient metal-free catalyst for oxygen reduction reaction in microbial fuel cells. ACS Appl. Mater. Interfaces. 8: 25923–25932.

Szczupak, A., D. Kol-Kalman and L. Alfonta. 2012. A hybrid biocathode: surface display of O_2-reducing enzymes for microbial fuel cell applications. Chem. Commun. 48: 49–51.

Tajdid Khajeh, R., S. Aber and K. Nofouzi. 2020. Efficient improvement of microbial fuel cell performance by the modification of graphite cathode via electrophoretic deposition of CuO/ZnO. Mater. Chem. Phys. 240: 122208.

Tan, L., Z.-Q. Liu, N. Li, J.-Y. Zhang, L. Zhang and S. Chen. 2016. CuSe decorated carbon nanotubes as a high performance cathode catalyst for microbial fuel cells. Electrochim. Acta. 213: 283–290.

Tan, L., S.-J. Li, X.-T. Wu, N. Li and Z.-Q. Liu. 2018. Porous Co_3O_4 decorated nitrogen-doped graphene electrocatalysts for efficient bioelectricity generation in MFCs. Int. J. Hydrogen Energy. 43: 10311–10321.

Tian, X., M. Zhou, M. Li, C. Tan, L. Liang and P. Su. 2018. Nitrogen-doped activated carbon as metal-free oxygen reduction catalyst for cost-effective rolling-pressed air-cathode in microbial fuel cells. Fuel. 223: 422–430.

Tofighi, A., M. Rahimnejad and M. Ghorbani. 2019. Ternary nanotube A-MnO_2/GO/AC as an excellent alternative composite modifier for cathode electrode of microbial fuel cell. J. Therm. Anal. Calorim. 135: 1667–1675.

Türk, K.K., I. Kruusenberg, E. Kibena-Põldsepp, G.D. Bhowmick, M. Kook, K. Tammeveski, et al., 2018. Novel multi-walled carbon nanotube based nitrogen impregnated Co and Fe cathode catalysts for improved microbial fuel cell performance. Int. J. Hydrogen Energy. 43: 23027–23035.

Ucar, D., Y. Zhang, I. Angelidaki and A.E. Franks. 2017. An overview of electron acceptors in microbial fuel cells. Front. Microbiol. 8: 1–14.

Viera, M., J. Riquelme, C. Aliaga, J.F. Marco, W. Orellana, J.H. Zagal, et al., 2020. Oxygen reduction reaction at penta-coordinated Co phthalocyanines. Front. Chem. 8: 1–12.

Vusa, C.S.R., S. Berchmans and S. Alwarappan. 2014. Facile and green synthesis of graphene. RSC Adv. 4: 22470–22475.

Wang, Z., C. Cao, Y. Zheng, S. Chen and F. Zhao. 2014. Abiotic oxygen reduction reaction catalysts used in microbial fuel cells. Chem. Electro. Chem. 1: 1813–1821.

Wang, Q., X. Zhang, R. Lv, X. Chen, B. Xue, P. Liang, et al., 2016. Binder-free nitrogen-doped graphene catalyst air-cathodes for microbial fuel cells. J. Mater. Chem. A. 4: 12387–12391.

Wang, H., L. Wei, C. Yang, J. Liu and J. Shen. 2020a. A Pyridine-Fe gel with an ultralow-loading Pt derivative as ORR catalyst in microbial fuel cells with long-term stability and high output voltage. Bioelectrochemistry. 131: 107370.

Wang, H., L. Wei, J. Liu and J. Shen. 2020b. Hollow bimetal ZIFs derived Cu/Co/N Co-coordinated ORR electrocatalyst for microbial fuel cells. Int. J. Hydrogen Energy. 45: 4481–4489.

Wang, X., C. Yuan, C. Shao, S. Zhuang, J. Ye and B. Li. 2020c. Enhancing oxygen reduction reaction by using metal-free nitrogen-doped carbon black as cathode catalysts in microbial fuel cells treating wastewater. Environ. Res. 182: 109011.

Wang, M., S. Wang, H. Yang, W. Ku, S. Yang, Z. Liu, et al., 2020d. Carbon-based electrocatalysts derived from biomass for oxygen reduction reaction: a minireview. Front. Chem. 8: 116.

Wiesener, K., D. Ohms, V. Neumann and R. Franke. 1989. N₄ Macrocycles as electrocatalysts for the cathodic reduction of oxygen. Mater. Chem. Phys. 22: 457–475.

Wroblowa, H.S., Yen-Chi-Pan and G. Razumney. 1976. Electroreduction of oxygen: a new mechanistic criterion. J. Electroanal. Chem. Interfacial Electrochem. 69: 195–201.

Xing, Z., N. Gao, Y. Qi, X. Ji and H. Liu. 2017. Influence of enhanced carbon crystallinity of nanoporous graphite on the cathode performance of microbial fuel cells. Carbon. 115: 271–278.

Xu, Y., S. Zhou and M. Li. 2019. Enhanced bioelectricity generation and cathodic oxygen reduction of air breathing microbial fuel cells based on MoS_2 decorated carbon nanotube. Int. J. Hydrogen Energy. 44: 13875–13884.

Xue, W., Q. Zhou, F. Li and B.S. Ondon. 2019. Zeolitic imidazolate framework-8 (ZIF-8) as robust catalyst for oxygen reduction reaction in microbial fuel cells. J. Power Sources. 423: 9–17.

Yan, Z., M. Wang, B. Huang, J. Zhao and R. Liu. 2012. Carboxyl multi-wall carbon nanotubes supported Pt-Ni alloy nanoparticles as cathode catalyst for microbial fuel cells. Int. J. Electrochem. Sci. 7: 10825–10834.

Yan, Z., M. Wang, Y. Lu, R. Liu and J. Zhao. 2014. Ethylene glycol stabilized $NaBH_4$ reduction for preparation carbon-supported Pt–Co alloy nanoparticles used as oxygen reduction electrocatalysts for microbial fuel cells. J. Solid State Electrochem. 18: 1087–1097.

Yang, X., W. Zou, Y. Su, Y. Zhu, H. Jiang, J. Shen, et al., 2014. Activated nitrogen-doped carbon nanofibers with hierarchical pore as efficient oxygen reduction reaction catalyst for microbial fuel cells. J. Power Sources. 266: 36–42.

Yang, G., D. Chen, P. Lv, X. Kong and Y. Sun. 2016. Core-shell Au-Pd nanoparticles as cathode catalysts for microbial fuel cell applications. Nature Publishing Group. 1–9.

Yang, T., K. Li, Z. Liu, L. Pu and X. Zhang. 2017. One-step synthesis of hydrangea-like Cu_2O@N-doped activated carbon as air cathode catalyst in microbial fuel cell. J. Electrochem. Soc. 164: F270–F275.

Yang, W. and S. Chen. 2020. Biomass-derived carbon for electrode fabrication in microbial fuel cells: a review. Ind. Eng. Chem. Res. 59: 6391–6404.

Yang, G., Z. Zhang, X. Kang, L. Li, Y. Li and Y. Sun. 2020a. Fe–N–C composite catalyst derived from solid digestate for the oxygen reduction reaction in microbial fuel cells. ACS Appl. Energy Mater. 3: 11929–11938.

Yang, W., X. Wang, R. Rossi and B.E. Logan. 2020b. Low-cost Fe–N–C catalyst derived from Fe (III)-chitosan hydrogel to enhance power production in microbial fuel cells. Chem. Eng. J. 380: 122522.

Yuan, H., Y. Hou, I.M. Abu-Reesh, J. Chen and Z. He. 2016. Oxygen reduction reaction catalysts used in microbial fuel cells for energy-efficient wastewater treatment: a review. Materials Horizons. 3: 382–401.

Zhang, L., J. Zhang, D.P. Wilkinson and H. Wang. 2006. Progress in preparation of non-noble electrocatalysts for pem fuel cell reactions. J. Power Sources. 156: 171–182.

Zhang, L., C. Liu, L. Zhuang, W. Li, S. Zhou and J. Zhang. 2009a. Manganese dioxide as an alternative cathodic catalyst to platinum in microbial fuel cells. Biosens. Bioelectron. 24: 2825–2829.

Zhang, F., S. Cheng, D. Pant, G.V. Bogaert and B.E. Logan. 2009b. Power generation using an activated carbon and metal mesh cathode in a microbial fuel cell. Electrochem. Commun. 11: 2177–2179.

Zhang, B., Z. Wen, S. Ci, S. Mao, J. Chen and Z. He. 2014. Synthesizing nitrogen-doped activated carbon and probing its active sites for oxygen reduction reaction in microbial fuel cells. ACS Appl. Mater. Interfaces. 6: 7464–7470.

Zhang, L., Z. Lu, D.M. Li, J. Ma, P. Song, G. Huang, et al., 2016. Chemically-activated graphite enhanced oxygen reduction and power output in catalyst-free microbial fuel cells. J. Cleaner Prod. 115: 332–336.

Zhang, X., Q. Wang, X. Xia, W. He, X. Huang and B.E. Logan. 2017. Addition of conductive particles to improve the performance of activated carbon air-cathodes in microbial fuel cells. Environ. Sci. Water Res. Technol. 3: 806–810.

Zhang, Y., L. Deng, H. Hu, Y. Qiao, H. Yuan, D. Chen, et al., 2020. Pomelo peel-derived, N-doped biochar microspheres as an efficient and durable metal-free ORR catalyst in microbial fuel cells. Sustainable Energy Fuels. 4: 1642–1653.

Zhang, G., X. Liu, L. Wang and H. Fu. 2022. Recent advances of biomass derived carbon-based materials for efficient electrochemical energy devices. J. Mater. Chem. A. 10: 9277–9307.

Zhao, F., F. Harnisch, U. Schröder, F. Scholz, P. Bogdanoff and I. Herrmann. 2005. Application of pyrolysed Iron(II) phthalocyanine and CoTMPP-based oxygen reduction catalysts as cathode materials in microbial fuel cells. Electrochem. Commun. 7: 1405–1410.

Zhao, C., J. Li, Y. Chen and J. Chen. 2019. Nitrogen and sulfur dual-doped graphene as an efficient metal-free electrocatalyst for the oxygen reduction reaction in microbial fuel cells. New J. Chem. 43: 9389–9395.

Chapter 7

Power Electronics Interfaces for Portable and Vehicle PEMFC Systems

José L. Díaz-Bernabé*, A. Rodríguez-Castellanos,
Sebastián Citalán-Cigarroa and Omar Solorza-Feria

Chemistry Department, CINVESTAV-IPN, Av. IPN 2508,
Sn. Pedro Zacatenco, 073060, CDMX, Mexico.

7.1 INTRODUCTION

The polymer electrolyte membrane fuel cell (PEMFC) has long been recognized as a promising zero-emission, suitable power source for a wide range of applications, like portable electronic telecommunications and hybrid electric vehicle (HEV). PEMFC devices efficiently convert diverse chemical fuels into electricity, helping toward a sustainable clean energy economy. Hydrogen, a clean energy source with the highest specific energy density, represents the best alternative to fossil fuels used in conventional energy production (Wu and Yang, 2013).

The PEMFC is an unregulated energy source depending on cell temperature, humidity, and fuel supply availability, so it needs a power converter unit (PCU) to connect with loads (Gou et al., 2010). A direct-current to direct-current (DC-DC) converter works upon the electric power from the PEMFC source to provide tracking of the PEMFC operating point, to fulfill the nominal voltage for load demand under normal conditions, and during PEMFC transients. When the

*For Correspondence: Email: jldiaz@cinvestav.mx

PEMFC source output voltage is greater than that of device to connect, a PCU interface is not necessary, but this omission may lead to a sudden malfunction by stressed internal electronics, and to an unsuccessful management of PEMFC clean energy. Conversely, other connected loads as inverters, motor-drives, LED controllers, energy generators, drones, etc. rigorously require a PCU to guarantee a proper work, see Figure 7.1.

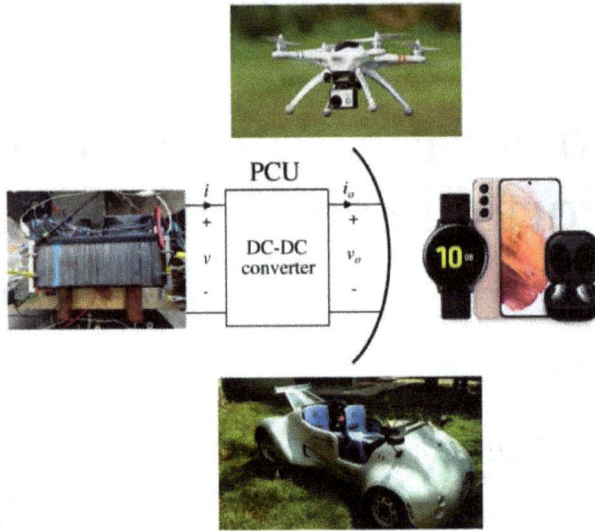

Figure 7.1   Overview of typical PEMFC applications.

7.2 PORTABLE PEMFC SYSTEMS

A fuel cell is classified as 'portable' when it is built into or powers a non-stationary device. Portable fuel cell applications (Bagotsky et al., 2015) are in portable military equipment (e.g. personal power supply for soldiers and skid-mounted fuel cell generators), auxiliary power units or APUs (e.g. in travel and trucking industries), drones, and electronic devices for daily life as mobile phones, digital still video-cameras, personal navigation system, internet of things (IoT), medical monitoring, etc. Fuel cell systems availability is according to their power output, with systems capable of operating at <5 W and up to several kilowatts. Standard output classifications are < 2, 10–50, and 100–250 W (Revankar and Majumdar 2014).

Commonly, portable devices involve an external alternating-current (AC) power adaptor to properly provide the direct-current (DC) voltage for internal electronics. In portable PEMFC systems, the PEMFC-PCU pair must properly provide the required regulated DC voltage to the connected devices with a clean energy source. To accomplish its task, the PCU requires a set of well-qualified voltage and current transducers for PEMFC and load, and high-performance devices as field programmable gate array (FPGA), microcontrollers, or suitable monolithic integrated circuits to implement control algorithms and monitoring functions.

Several DC-DC topologies for fuel cell applications have been reviewed in literature (Zhang et al., 2012; Kolli et al., 2015; Sivakumar et al., 2016). Usually the power conversion is carried out by means of one or more switches working under high-frequency pulse-width modulation (PWM) technique. The steady-state conversion ratio $G(u)$ relates the output voltage in respect to input voltage as a function of the duty-cycle u. The non-isolated DC-DC converters provide an ohmic path for the electric current between the PEMFC module and the load. The isolated DC-DC converters are another kind of PCU for portable systems which involve a high-frequency transformer providing a high conversion ratio and galvanic isolation for the PEMFC. Galvanic isolation keeps the PEMFC source from the sudden power surges of the load transients. For alternating current (AC) requirements, a DC-DC converter might be employed as a former stage to regulate the input voltage of DC-AC inverters. In high-power grid-connected PEMFC applications, the galvanic isolation may be a fundamental constriction.

Figure 7.2 shows diagrams of some non-isolated and isolated DC-DC topologies, where main switches are represented by a MOSFET transistor and a diode. In these ideal circuits, L and C are reactive components, so the equivalent dynamic system will have two state variables. Having four or more reactive components, advanced converter circuits yield a much more complex dynamic system but a high voltage-conversion ratio (García–Vite et al., 2017; Rosas-Caro et al., 2017; Diaz-Saldierna et al., 2021). The controller, a high-performance microcontroller or digital signal processor (DSP), continuously samples the state variables and then modifies the control variable u to maintain regulation of output voltage or current or PEMFC operating point. However, with an analog solution, a single-chip integrated circuit works with external components to implement the required control loop (Segura et al., 2011).

The Buck converter is shown in Fig. 7.2(a). It is suitable when the load voltage v_o, is less than the PEMFC voltage, v, at the worst operating point. The L and C parameters, forming a low-pass filter, produce a low-ripple output voltage. Because of its linear conversion ratio $G(u)$ (Table 7.1), it is a preferred topology for three-phase H-bridge motor-drives and multilevel voltage source inverters.

Figure 7.2(b) shows the Boost topology which it is suitable when load voltage is always greater than input voltage, as in macrobiotic PEMFC cells development. Perfectly, the voltage gain could increase as the duty-cycle u, but remains constricted to a narrow band (doubtless below $u < 0.65$) because of transistors-diode stress, and inductor magnetic flux saturation. Serialized converters—sequential connected power conversion units—and interleaved converters—paralleling power conversion units with phase-shifted control signals, are other choices to increase the voltage or current conversion ratio. In an optimum design, the inductor L helps to reduce ripple current at the PEMFC module terminals.

Figure 7.2(c) shows the well-known Buck-Boost. It is suitable whenever basic buck or boost functions cannot regulate the load voltage because transients on the PEMFC operating point yield a large deviation of the voltage v. The polarity of the voltage is reversed in respect to the input voltage polarity, and either increases as $u > 0.5$ or decreases as $u < 0.5$. Figure 7.2(d) shows the no-inverting Buck-Boost circuit, which makes the buck and boost functions by means of two diodes and two transistors, and the polarity of the output voltage is not reversed in respect to that of the input voltage. Figure 7.2(e) shows the Interleaved Boost converter with phases number $n = 2$. In the circuit, two power units are stacked, and control signals may be equal, but phase-shifted by T/n (T means the inverse of switching frequency). It is demonstrated that when duty-cycle u is close to n/T, the current ripple is successfully reduced so that it does not affect the PEMFC system. Figure 7.2(f) shows the Interleaved Buck-Boost converter with phases number $n = 2$, which reverses the output voltage with respect to the input voltage v.

a) Buck

b) Boost

c) Buck-Boost

d) No-Inverting Buck-Boost

e) Interleaved Boost (IBC)

f) Interleaved Buck-Boost (IBBC)

g) Non-isolated Half-Bridge

h) Isolated Half-Bridge

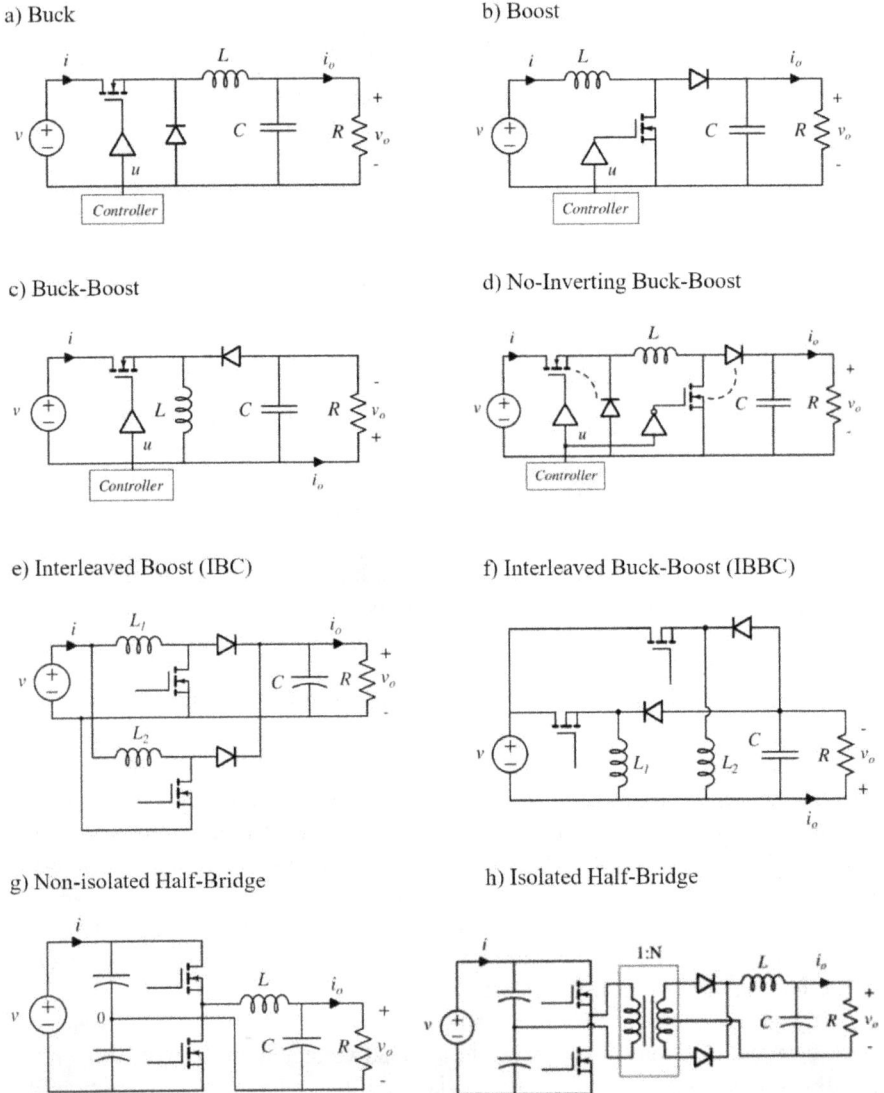

Figure 7.2 Basic non-isolated DC-DC converters (Reprinted with permission from B.W. Erickson, Fundamentals of Power Electronics [Chicago, CRC Press, 2003], pp. 112–113).

Figure 7.2(g) shows the non-isolated Half-bridge converter, where output voltage scales down and reverses in respect to the input voltage for $u < 0.5$. On the other hand, the output voltage increases and does not reverse in respect to the input voltage for $u > 0.5$. At the output, L and C form a low-pass filter to reduce voltage ripple at the load R. Like Buck converter, non-isolated Half-bridge is a useful structure for DC motor-drives and the building block for complex voltage source inverters (VSI). Finally, Fig. 7.2(h) is the isolated Half-bridge converter. In this circuit, the transformer winding turns ratio N allowing it to easily increase or decrease output voltage. Like Buck converter and Half-bridge, L and C form a low-pass filter to reduce voltage ripple at the load R. The parameter N directly shapes the steady-state conversion ratio.

Lossless auxiliary snubber sub-circuits that decrease stress on transistors and diodes, or techniques as zero voltage switching (ZVS), zero current switching (ZCS), or synchronous switching scheme make the basic topologies to reach power transformation efficiencies of 90%. However, advanced resonant converter topologies can reach a maximum power conversion efficiency above of 97% but the cost and design complexity are higher than the previous mentioned topologies.

Table 7.1   Steady-state voltage conversion factor for CCM DC-DC converters

Converter	Circuit	$G(u) = v_o/v$
Buck	a	u
Boost	b	$\dfrac{1}{1-u}$
Buck-Boost	c	$-\dfrac{u}{1-u}$
No-inverting Buck-Boost	d	$\dfrac{u}{1-u}$
Interleaved Boost	e	$\left(\dfrac{1}{1-u}\right)$
Interleaved Buck-Boost	f	$\left(\dfrac{u}{1-u}\right)$
Non-isolated Half-Bridge	g	$\dfrac{2u-1}{2}$
Isolated Half-Bridge	h	$N\left(\dfrac{u}{2}\right)$

As mentioned, the basic non-isolated topologies can be implemented for portable PEMFC systems below 250 W. The advantages of interleaved topologies are power rate below 15 kW, higher power density, reduced current ripple for PEMFC source, less semiconductor devices stress, reduced inductors size, and current sharing strategy. However, the complexity increases because n-phases require n transistor drive circuit, n current transducers and n phase-shifted control signals. Table 7.1 shows the steady-state continuous current mode (CCM) conversion ratio for depicted DC-DC converters. As shown, conversion factor of interleaved converters is equal to that of basic topologies, but the resulting input current ripple is reduced by either increasing the phase number n or with a duty-cycle nearing $u = 2\pi T/n$. Initial survey and analysis of these topologies can be found in Erickson and Maksimovic (2003) and Mohan et al. (2003).

7.3 VEHICLE PEMFC SYSTEMS

PEMFC source is quote suitable for merely hybrid electric vehicle (HEV) applications because a PEMFC has low operation temperature, small design, no undermining difficulties under suitable care, and improves the battery capacity by enlarging the recharging cycle. Usually, the architecture of fuel cell HEV (FCHEV) powertrain involves a PEMFC stack with its DC-DC converter and a Li-ion battery module (Bayindir et al., 2011; Wu et al., 2015; Sanli and Gunlu, 2016; Andaloro et al., 2017; Napoli et al., 2017; Lü et al., 2018, Zhuang et al., 2020; Di Trolio et al., 2020). The Energy Management Strategy (EMS) algorithm schedules the energy sources according to established rules for battery state of charge (SOC), PEMFC fuel consuming constriction, or vehicle autonomy under a specific drive-cycle (Allaoua et al., 2017; Martinez et al., 2017; Krithika and Subramani, 2018; Soumeur, et al., 2020; Teng et al., 2020). In Fig. 7.3, the PEMFC stack should supply either the motor-drive power during the cruiser speed periods (PEMFC mode) or the charging current for a section of the battery module (charging mode) enlarging the vehicle autonomy (Fernandez et al., 2016).

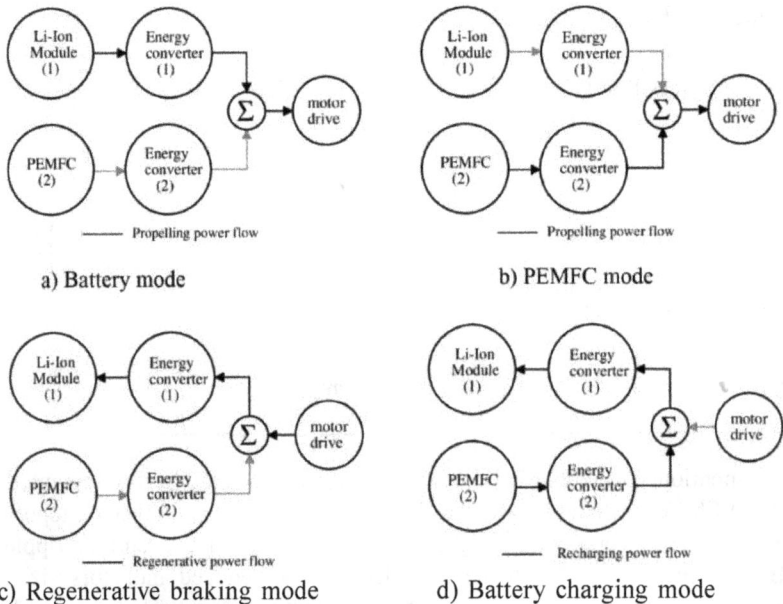

a) Battery mode

b) PEMFC mode

c) Regenerative braking mode

d) Battery charging mode

Figure 7.3 Hybrid operating modes for FCHEV (Reprinted with permission from M. Ehsani et al., 2018, Hybrid Electric Vehicles, in Modern Electric, Hybrid Electric and Fuel Cell Vehicles [CRC Press], pp. 114).

PEMFC vehicle developments require a meticulous investigation of the dynamic response of the powertrain and the power limitations of the PEMFC system under critical conditions, so all the required components of the powertrain should be analyzed, modeled, and defined before being built. The results allow us to attune important parameters of the PEMFC system and to fulfill the expected performance. Authors (Daud et al., 2017; Di Wu and Tang, 2020) have explained the control requirements of auxiliary component and the fuels requirements to guarantee the response of PEMFC source in HEV powertrain. Several authors (Pires et al., 2014; Zhang et al., 2016; Das H.S. et al., 2017; Wang et al., 2019; Das et al., 2019; Bairabathina and Balamurugan, 2020; Kumar et al., 2020; Affam et al., 2021) have reviewed useful converter topologies and architectures to tie the PEMFC and battery sources to powertrain of HEV. Table 7.2 shows a set of early reviewed DC-DC converter topologies for purely battery electric vehicle (BEV) and hybrid electric vehicles (HEV) (Chakraborty, et al., 2019). As shown, interleaved step-up converters are useful topologies because PEMFC output voltage needs to be properly lifted for a high-voltage powertrain (Gao et al., 2016; Forouzesh et al., 2017; Pires et al., 2019; Diaz-Saldierna et al., 2021). Bidirectional DC-DC converters help to retrieve the regenerative energy braking produced in the deceleration periods of a drive-cycle. However, according to the driving pattern, only about 15% of this energy should be successfully retrieved to the battery module.

Table 7.2   Overview of DC-DC converters for HEV

	BEV & HEV ≥ 10 kW	BEV & HEV < 10 kW
DC-DC converter topology	Interleaved *n-phase* boost converter (IBC) Full-bridge boost converter (FBC) Multi-device interleaved bidirectional boost converter (MDIBBC)	Boost converter (BC) Boost converter with resonant Circuit (BCRC) Interleaved *n-phase* boost converter (IBC) Buck-Boost converter (BBC) Isolated ZVS converter (ZVSC) Z-source converter

Electrochemical and suitable empirical models allow us to analyze and simulate a whole PEMFC resource in power systems environments (Bagotsky, 2009; Kulikovsky, 2009; Nalbant et al., 2018) since they do not consider in-depth single-cell chemistry but describe comprehensive properties of a PEMFC stack.

The advantages, modeling approaches, and common issues of battery modules based on Li-ion cell technology have been reviewed in references (Kroeze and Krein, 2008; Kulkarni and Agrawal, 2010; Seaman et al., 2014; Tomasov et al., 2019; Tamilselvi et al., 2021; Alsharif et al., 2022). By expanding the electrical model of a single cell proposed by Chen (Chen and Rincon-Mora, 2006), the performance of a Li-ion battery module is investigated. Parameters related to the motor-drives, BLDC motors, and vehicle dynamics under a specific drive-cycle give a framework to analyze the fuel cell and PCU behavior.

7.3.1   Powertrain Analysis

The generalized structure of a series powertrain is satisfactory for early FCHEV powertrain evaluation (Fig. 7.4a). Key components as the PEMFC source, a power converter unit (PCU), battery module, motor-drive, BLDC motors, and vehicle body dynamics are explained in the following analysis. The two-port PCU converter connects the PEMFC source to motor-drive as the battery module without power converter ties directly to powertrain. The pair motor-drive and BLDC motor provide the necessary electromagnetic torque τ_m to overcome both static and rolling torque τ_L as the vehicle body develops the linear velocity V. In the motor-drive, the speed reference ω^* modifies the motor winding voltages v_t that comes from a practical drive-cycle pattern.

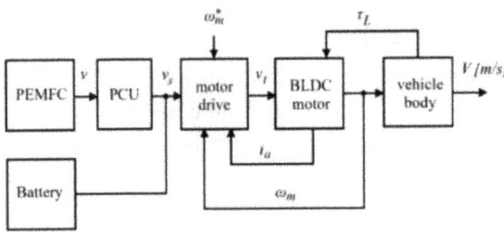

a) Series powertrain   b) FCHEV Sicarú

Figure 7.4   FCHEV development. (a) Simplified HEV series powertrain; (b) FCHEV prototype named Sicaru.

Figure 7.5 shows the comprehensive behavior of a PEMFC source by means of its polarization curve of the voltage as a function of the current density. The voltage varies by three kind of major losses regions as the current density value increases. The activation losses take part at the left side whereby the current density is lower, the ohmic losses are sizeable at the center of the curve, and the mass transport losses (concentration losses) are more significant at the right side.

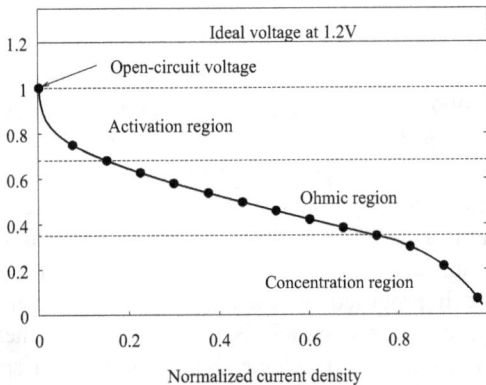

Figure 7.5   Polarization curve of a PEMFC source (Reprinted with permission from Gou et al., Fuel Cells: Modeling, Control, and Applications [CRC Press], 2010).

A simplified electrochemical model is widely used in analysis of power systems because it disregards the kinetics of charged particles inside the membrane exchange assembly (MEA) or the thermodynamics of the gasses in the plate-channels, and only ideal conditions for the fuels, materials, and structures are sustained. The open-circuit voltage continuously varies due to the internal partial pressures of hydrogen, oxygen, and water saturation:

$$pp_H = 0.5\left(P_H \exp\left(-\frac{1.653J}{T_K^{1.334}} \right) - P_{H_2O} \right)$$

$$pp_{O_2} = P_{air} \exp\left(-\frac{4.192J}{T_K^{1.334}} \right) - P_{H_2O} \tag{1}$$

$$\log(P_{H_2O}) = 1.4454 \times 10^{-7} T_C^3 - 9.1837 \times 10^{-5} T_C^2 + 2.953 \times 10^{-2} T_C - 2.1794$$

where J is the current density, T_C is single assembly plate temperature, P_{H_2O} is the pressure for water saturation, P_H is the pressure for hydrogen, P_{air} is the pressure for the air blowing into the cells, and F is the Faraday constant. The open-circuit voltage should correspond to the Nernst voltage:

$$E_0 = -\frac{G_{f,liq}}{2F} - \frac{RT_k}{2F} \log\left(\frac{P_{H_2O}}{pp_H \sqrt{pp_{O_2}}} \right) \tag{2}$$

where $G_{f,liq}$ is the Gibss constant in liquids, and R is the ideal gas coefficient.

The activation losses and the mass transport losses (concentration losses) are numerically estimated by:

$$V_{act} = -\underbrace{\frac{RT}{2\alpha F}}_{B} \log\left(\frac{J}{J_0} \right)$$

$$V_{cond} = \begin{cases} 0, & \left(1 - \dfrac{J}{J_L}\right) \le 0 \\[3mm] -\alpha_1 J^k \log\left(1 - \dfrac{J}{J_L}\right), & \text{otherway} \end{cases} \tag{3}$$

where J_0 is the exchange density current with unit of A/cm², J_L is the current density where the curve turns away from linear region in A/cm², B is Tafel slope, k is the mass transport coefficient, and α_1 is the amplification constant.

Considering that the internal ohmic losses due to a resistance density R_S produces a potential drop, the instantaneous voltage of a PEMFC single cell would be:

$$v_c = E_O + V_{cond} + V_{act} - JR_S \tag{4}$$

Finally, given the MEA effective area A (cm²) and the required number of series cell N_S, the expected stack voltage v and output current i are close to:

$$v = N_S v_c$$

$$i = AJ \tag{5}$$

This equation ideally considers homogeneous voltage allocation among cells number and the same effective area for every single cell.

Assume that the Buck-Boost converter in Fig. 7.2(c) regulates the PEMFC output voltage as the motor-drive voltage v_s requires it. The average behavior of the Buck-Boost converter for CCM is estimated by:

$$\dot{x}_1 = \acute{u}x_2 + uv$$

$$\dot{x}_2 = \acute{u}x_1 - \frac{1}{RC}x_2$$

$$i = ux_1$$

$$u = [0,1]$$

(6)

where $x_1 = i_L$ is the current flowing through L, and $x_2 = -v_o$ is the voltage at capacitor C, R is the equivalent load related to motor-drive power requirements, and $\acute{u} = (1 - u)$ is the complementary control signal.

Battery module is analyzed by means of the single Li-ion cell equivalent circuit described by Chen and Rincon-Mora (2006) where a large capacitor C stands for the charge storage capacity and its voltage indicates the state-of-charge (SOC). The battery SOC value corresponds to the time integration of cell current $i_b(t)$ (Piller et al. 2001):

$$SOC(t) = SOC(t_0) + \frac{1}{C}\int_0^\tau i_b(\tau)d\tau$$

(7)

where $SOC(t_0)$ is the SOC initial condition. The open-circuit voltage v_{oc} is given by a third order polynomial equation as a function of the instantaneous SOC:

$$v_{oc} = a_1 SOC^3 + a_2 SOC^2 + a_3 SOC^1 + a_4 \exp(a_4 SOC) + a_5$$

(8)

where a_1 to a_5 are constant terms defined in Chen, and Rincon-Mora (2006). Then, the cell output-voltage is approximated by:

$$v_{cell} = v_{oc} - R_s i_b(t)$$

(9)

Finally, a lattice of N_p-N_s cells determines the increased discharging current, the nominal charge storage capacity, the open-circuit voltage, and the series resistances of the battery module.

The motor-drive implements a three-phase H-bridge converter and a modern control algorithm to provide reliable speed tracking or enough electromagnetic torque. The behavioral model for the motor-drive is:

$$v_t = u\left(\frac{v_s}{2}\right)$$

$$i_s \approx \hat{i}_a$$

(10)

where v_t is the phase voltage, v_s is the DC-Link voltage, i_s represents either the battery current or PEMFC current depending on the EMS mode, and u is the control variable. The motor-drive average input-current i_s is close to the maximum phase-current i_a, for either trapezoidal-wave or sinusoidal-wave BLDC motors (Mohan et al., 2003). On other hand, the controller continuously modifies the phase voltage v_t or phase-current i_a by means of duty-cycle u to regulate the angular speed ω_m or the electromagnetic torque τ_m, respectively.

The averaged behavior of a BLDC motor (per-phase approach) is depicted by Mohan et al. (2003):

$$\frac{d}{dt}i_a = \frac{1}{L_a}(v_t - i_a R_a - K_E \omega_m)$$
$$\frac{d}{dt}\omega_m = \frac{1}{J}(\tau_m - K_E \omega_m - \tau_L) \tag{11}$$

where J_m is the inertia, B is the friction coefficient, K_T is the torque constant, K_E is the electrical constant, R_a is the per-phase winding resistance, L_a is the per-phase inductance, and $\tau_m = K_T i_a$ is the electromagnetic torque.

The developed linear velocity on the wheel's axle corresponds to the angular speed on the motor-shaft (Larminie and Lowry, 2012):

$$V = \left(\frac{R_d}{G}\right)\omega_m \quad \text{[m/s]} \tag{12}$$

where R_d (m) is the radius of wheels, and G is the gear-box ratio from the motor-shaft to the wheel's axle. Conversely, the equivalent instantaneous load torque τ_L on the motor-shaft is:

$$\tau_L = \left(\frac{R_d}{G}\right)\left(Mgf_r \cos\alpha - \frac{1}{2}\rho_a C_d A_f V^2 + M\delta\frac{dV}{dt}\right)[nm] \tag{13}$$

where M is the vehicle mass (kg), f_r is the rolling coefficient, $\cos\alpha$ is the slope of the road, C_d is the drag resistance coefficient, A_f (m²) is front area of the vehicle, ρ_a is the air density, and V (m/s) is the instantaneous speed.

7.3.2 FCHEV Performance

Figure 7.3(b) shows a front view of the developed small FCHEV named Sicarú and Table 7.3 gives its important characteristics. As shown, a 36 V powertrain involves four wheel-drive (4WD) layout, so one BLDC motor is tied to each wheel, 3.6 kW Li-ion battery module, and 2 kW PEMFC system. Table 7.4 displays the major parameters of the PEMFC stack, and Table 7.5 shows the configuration of the PCU used to tie the PEMFC source to the powertrain.

Table 7.3 Important specifications of FCHEV-Sicarú

C_d	f_r	A_f(m²)	Mass (kg)	R_d (m)	Battery (kW)	PEMFC (kW)	Motors
0.3	0.01	1.5	300	0.27	3.6	2	4–1 kW, 36 V

Table 7.4   Parameters of the PEMFC stack

Type	Ns	A (cm^2)	J (A/cm^2)	Power (kW)	Material	Reactants	Reactant Pressure (atm)
PEM	90	500	150	2	Pt, 20%	H$_2$, air	1 atm

Table 7.5   Requirements of the PCU for the FCHEV

Topology	F_s (kHz)	L_a (H)	C (F)	Q1, Q2
Buck-Boost converter	250	220E-6	100E-6	IRF254

Figure 7.6 shows the expected performance of the FCHEV powertrain under suitable J227b drive-cycle. The single cycle lasts 100s, the maximum velocity of 32 km/h is reached at t = 24 s and maintained by 26 s. At t = 50 s the vehicle decelerates over 15 s, and then at t = 65 s it develops a zero velocity which is held up by 35 s until the end of a cycle. Figure 7.6(a) shows the motor-drive current drawn under four cycles of this speed pattern. At the beginning of each cycle, a 150 A peak motor-current is demanded because motor torque must beat the static rolling resistance. However, in an unprotected current framework, the maximum motor-drive current shall easily reach 200 A. At the cruiser speed interval, the motor-drive draws a constant current of 30 A. During the deceleration period, the motor-drive returns a −50 A regenerative current, but only a maximum of 15% might recharge the battery module. The PEMFC or Li-ion battery supplies this current i_s according to the proposed EMS strategy.

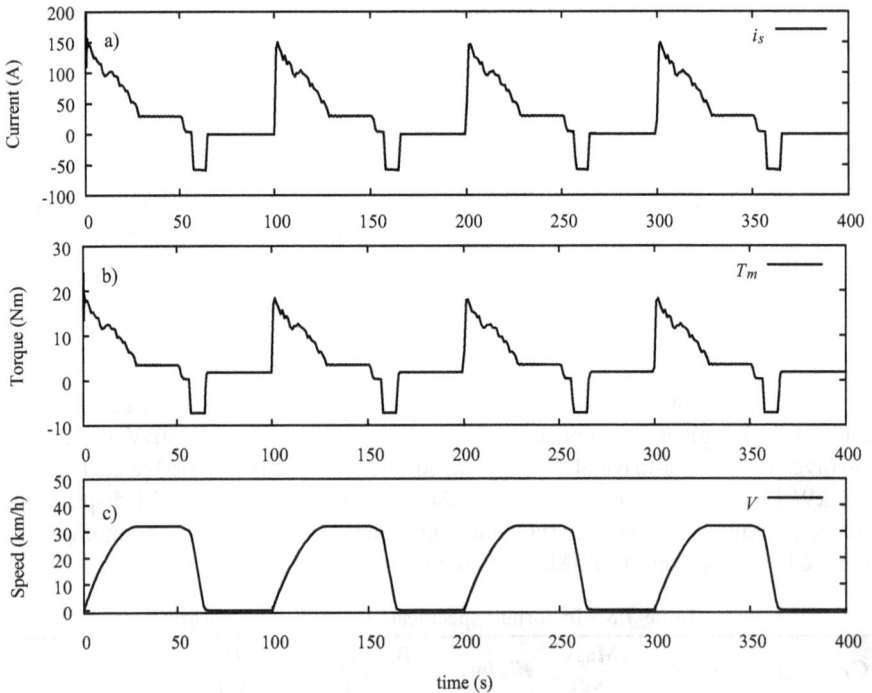

Figure 7.6   FCHEV powertrain: (a) Motor-drive current, (b) Electromagnetic torque, (c) Developed velocity.

Figure 7.6(b) shows the variations of the electromagnetic torque. During cruiser speed period, the torque remains close to 4 nm. At zero speed lapse, the load torque is not zero because it corresponds to the vehicle rolling resistance as shown by first term on right-side of Equation (12). Consequently, at the beginning of the acceleration period, an electromagnetic torque of 24 nm provides the tractive effort on wheels and turns the vehicle into motion. Finally, by deceleration lapse the torque goes to –7 nm. Figure 7.6(c) shows the developed linear velocity of J227b drive-cycle. As shown in Figure 7.6, the torque and speed traces should correspond to typical torque-speed characteristics of an equivalent electric motor of FCHEV.

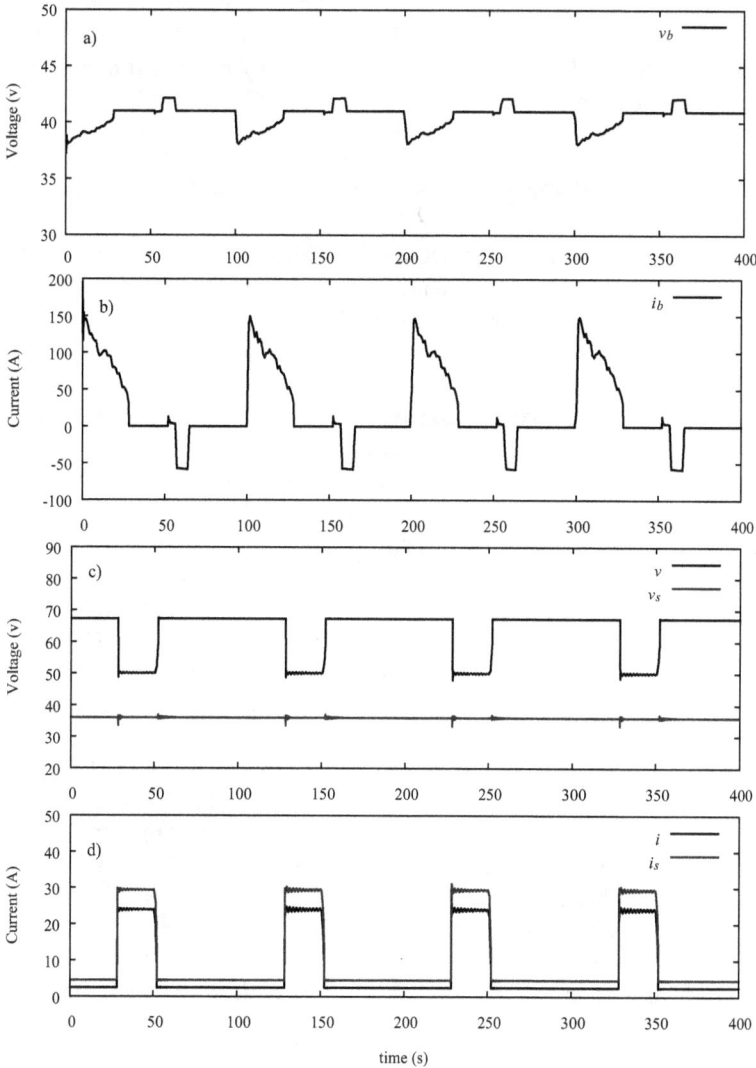

Figure 7.7 Energy sources behavior under the proposed scheduling. a) Battery voltage, b) Battery current; c) PEMFC voltage variations (black trace) and PCU regulated output voltage (gray trace); d) PEMFC (black trace) and motor-drive (gray trace) currents.

In Fig. 7.7(a), the battery voltage decreases for the acceleration period but increases by the deceleration lapse, indicating a recharging mode and agreeing with the shape of the motor-drive current. Figure 7.7(b) shows the battery current response and how it holds up the positive and negative current peaks in the acceleration and deceleration modes. Positive value refers to discharging and negative value refers to charging battery currents, respectively.

According to the described EMS, the PEMFC mode supplies the motor-drive current during the continuous-current stage. Figure 7.7(c) shows how the PEMFC voltage diminishes when it supplies the motor-drive current, and the regulated output voltage v_s seen at terminals of PCU. Figure 7.7(d) shows the transients of the PEMFC and motor-drive currents. It is noticed that PEMFC does not remain in the open-circuit mode, so a small current is drawn whenever it does not provide the motor-drive current.

7.4 CONCLUSION

In this chapter, an overview of DC-DC converter interfaces for portable and vehicular PEMFC systems is presented. Portable PEMFC systems might cover power ratings below 250 W or a bit more. Since its low output-voltage, a PEMFC PCU usually requires a step-up high-conversion ratio DC-DC converter as non-isolated Boost, Buck-Boost or its derivations. Serialized or cascaded power units allow improvement in the voltage conversion ratio and interleaved topologies allow increase in both the voltage and current conversion ratios.

The simplified serial powertrain for HEV was presented to analyze the power requirements of PEMFC source. All components of FCHEV play an important role in the evaluation of the PEMFC stack. Several architectures and topologies for DC-DC converter had been developed to integrate electrochemical storage sources as PEMFC, batteries and supercapacitors to FCHEV and BEV applications. For high-voltage powertrains, a step-up high-conversion ratio DC-DC converter is necessary as isolated Full-bridge converter, Boost Converter, Boost with resonant circuit converter, multi-phase interleaved Boost converter, multiphase Interleaved Buck-Boost converter among others.

Figures 7.6 and 7.7 show the expected performance of a FCHEV-Sicaru by means of real-time emulation of the described powertrain. The J227b drive-cycle is suitable for low-speed vehicles, and it serves as a velocity reference framework in this accomplished analysis. The dynamic behavior of the powertrain allows verification of the behavior for a 2 kW PEMFC stack with non-inverting Buck-Boost converter as alternative energy source by the maximum velocity stage of the proposed drive-cycle. A well-attuned control algorithm for the Buck-Boost converter yields a good voltage regulation for the motor-drive under the operating mode transients.

REFERENCES

Affam, A., Y.M. Buswig, Al-Khalid B.H. Othman, N.B. Julai and O. Qays. 2021. A review of multiple input DC-DC converter topologies linked with hybrid electric vehicles and renewable energy systems. Renewable Sustainable Energy Rev. 135: 110186. https://doi.org/10.1016/j.rser.2020.110186

Allaoua, B., K. Asnoune and B. Mebarki. 2017. Energy management of PEM fuel cell/supercapacitor hybrid power sources for an electric vehicle. Int. J. Hydrogen Energy. 42: 21158. https://doi.org/10.1016/j.ijhydene.2017.06.209

Alsharif, K.I., A.H. Pesch, V. Borra, F.X. Li, P. Cortes, E. Macdonald, et al., 2022. A novel modal representation of battery dynamics. IEEE Access. 10: 16793–16806. https://doi.org/10.1109/ACCESS.2022.3149617

Andaloro, L., A. Arista, G. Agnello, G. Napoli, F. Sergi and V. Antonucci. 2017. Study and design of a hybrid electric vehicle (lithium batteries-PEMFC). Int. J. Hydrogen Energy. 42: 3166–3184. https://doi.org/10.1016/j.ijhydene.2016.12.082

Bagotsky, V.S.. 2009. Mathematical modeling of fuel cells. pp. 255–261. *In*: V.S. Bagotsky (ed.). Fuel Cells Problems and Solutions. John Wiley & Sons. Hoboken NJ. USA.

Bagotsky, V.S., A.M. Skundin and Y.M. Volfkovich. 2015. Electrochemical power sources: batteries, fuel cells, and supercapacitors. John Wiley & Sons. Hoboken. NJ. USA.

Bairabathina, S. and S. Balamurugan. 2020. Review on non-isolated multi-input step-up converters for grid-independent hybrid electric vehicles. Int. J. Hydrogen Energy. 45: 21687–21713. https://doi.org/10.1016/j.ijhydene.2020.05.277

Bayindir, K.Ç., M.A. Gözüküçük and A. Teke. 2011. A comprehensive overview of hybrid electric vehicle: Powertrain configurations, powertrain control techniques and electronic control units. Energy Convers. Manage. 52: 1305–1313. https://doi.org/10.1016/j.enconman.2010.09.028

Chakraborty, S., V. Hai-Nam, M.M. Hasan, T. Dai-Duong, M. El Baghdadi and O. Hegazy. 2019. DC-DC converter topologies for electric vehicles, plug-in hybrid electric vehicles and fast charging stations: state of the art and future trends. Energies. 12: 1569. https://doi.org/10.3390/en12081569

Chen, M. and G.A. Rincon-Mora. 2006. Accurate electrical battery model capable of predicting runtime and I-V performance. IEEE Trans. Energy Convers. 21: 504–511. http://doi.org/10.1109/TEC.2006.874229

Das, H.S., C.W. Tan and A.H.M. Yatim. 2017. Fuel cell hybrid electric vehicles: a review on power conditioning units and topologies. Renewable Sustainable Energy Rev. 76: 268–291. https://doi.org/10.1016/j.rser.2017.03.056

Das, V., S. Padmanaban, K. Venkitusamy, R. Selvamuthukumaran, F. Blaabjerg and P. Siano. 2019. Recent advances and challenges of fuel cell-based power system architectures and control—a review. Renewable Sustainable Energy Rev. 73: 10–18. https://doi.org/10.1016/j.rser.2017.01.148

Daud, W.R.W., R.E. Rosli, E.H. Majlan, S.A.A. Hamid, R. Mohamed and T. Husaini. 2017. PEM fuel cell system control: a review. Renewable Energy. 113: 620–638. https://doi.org/10.1016/j.renene.2017.06.027

Di Trolio, P., P. Di Giorgio, M. Genovese, E. Frasci and M. Minutillo. 2020. A hybrid power-unit based on a passive fuel cell/battery system for lightweight vehicles. App. Energy. 279: 432–442. https://doi.org/10.1016/j.apenergy.2020.115734

Di Wu and H. Tang. 2020. Review of system integration and control of proton exchange membrane fuel cells. Electrochem. Energy Rev. 51: 1–40. https://doi.org/10.1007/s41918-020-00068-1

Diaz-Saldierna, L.H., J. Leyva-Ramos, D. Langarica-Cordoba and M.G. Ortiz-Lopez. 2021. Energy processing from fuel-cell systems using a high-gain power DC-DC converter: analysis, design, and implementation. Int. J. Hydrogen Energy. 46: 25264–25276. https://doi.org/10.1016/j.ijhydene.2021.05.046

Erickson, R.W. and D. Maksimovic. 2003. Fundamentals of Power Electronics. Kluwer Academic Press, USA.

Fernandez, R.A., F.B. Cilleruelo and I.V. Martínez. 2016. A new approach to battery powered electric vehicles: a hydrogen fuel-cell-based range extender system. Int. J. Hydrogen Energy. 41: 4808–4819. https://doi.org/10.1016/j.ijhydene.2016.01.035

Forouzesh, M., Y.P. Siwakoti, S.A. Gorji, F. Blaabjerg and B. Lehman. 2017. Step-up DC-DC converters: a comprehensive review of voltage-boosting techniques, topologies, and applications. IEEE Trans. Power Electron. 32: 9143–9178. https://doi.org/10.1109/TPEL.2017.2652318

Gao, D., Z. Jin, J. Liu and M. Ouyang. 2016. An interleaved step-up/step-down converter for fuel cell vehicle applications. Int. J. Hydrogen Energy. 41: 22422–22432. https://doi.org/10.1016/j.ijhydene.2016.09.171

García–Vite, P.M., C.A. Soriano–Rangel, J.C. Rosas–Caro and F. Mancilla–David. 2017. A DC-DC converter with quadratic gain and input current ripple cancelation at a selectable duty cycle. Renewable Energy. 101: 431–436. https://doi.org/10.1016/j.renene.2016.09.010

Gou, B., W. Na and B. Diong. 2010. Fuel Cells: Modeling, Control, and Applications. CRC Press, Boca Raton, FL. USA.

Kolli, A., A. Gaillard, A. De Bernardinis, O. Bethoux, D. Hissel and Z. Khatir. 2015. A review on DC/DC converter architectures for power fuel cell applications. Energy Convers. Manage. 105: 716–730. https://doi.org/10.1016/j.enconman.2015.07.060

Krithika, V. and C. Subramani. 2018. A comprehensive review on choice of hybrid vehicles and power converters, control strategies for hybrid electric vehicles. Int. J. Energy Res. 42: 1789–1812. https://doi.org/10.1002/er.3952

Kroeze, R.C. and P.T. Krein. 2008. Electrical battery model for use in dynamic electric vehicle simulations. Proc. IEEE Power Electron. Spec. Conf. 1336–1342. https://doi.org/10.1109/PESC.2008.4592119

Kulikovsky, A.A. 2009. Analytical models of a polymer electrolyte fuel cell. pp. 199–252. *In*: S.J. Paddison and Keith S. Promislow (eds). Device and Materials Modeling in PEM Fuel Cells. Springer New York, New York. doi:10.1007/978-0-387-78691-9_7

Kulkarni, M. and V.D. Agrawal. 2010. A tutorial on battery simulation-matching power source to electronic system. Proc.: 14th IEEE/VSI VLSI Design and Test Symposium, Chandigarh. 11 pages.

Kumar, D., R.K. Nema and S. Gupta. 2020. A comparative review on power conversion topologies and energy storage system for electric vehicles. Int. J. Energy Res. 44: 7863–7885. https://doi.org/10.1002/er.5353

Larminie, J. and J. Lowry. 2012. Electric Vehicle Technology Explained. John Wiley & Sons, UK.

Lü, X., Y. Qu, Y. Wang, C. Qin and G. Liu. 2018. A comprehensive review on hybrid power system for pemfc-hev: issues and strategies. Energy Convers. Manage. 171: 1273–1291. https://doi.org/10.1016/j.enconman.2018.06.065

Martinez, C.M., X. Hu, D. Cao, E. Velenis, B. Gao and M. Wellers. 2017. Energy management in plug-in hybrid electric vehicles: recent progress and a connected vehicles perspective. IEEE Trans. Veh. Technol. 66: 4534–4549. https://doi.org/10.1109/TVT.2016.2582721

Mohan, N., T.M. Undeland and W.P. Robbins. 2003. Power Electronics: Converters, Applications, and Design. John Wiley & Sons. Hoboken. NJ. USA.

Nalbant, Y., C.O. Colpan and Y. Devrim. 2018. Development of a one-dimensional and semi-empirical model for a high temperature proton exchange membrane fuel cell. Int. J. Hydrogen Energy. 43: 5939–5950. https://doi.org/10.1016/j.ijhydene.2017.10.148

Napoli, G., S. Micari, G. Dispenza, S. Di Novo, V. Antonucci and L. Andaloro. 2017. Development of a fuel cell hybrid electric powertrain: a real case study on a minibus application. Int. J. Hydrogen Energy. 42: 28034–28047. https://doi.org/10.1016/j.ijhydene.2017.07.239

Piller, S., M. Perrin and A. Jossen. 2001. Methods for state-of-charge determination and their applications. J. Power Sources. Elsevier. 96: 113–120. https://doi.org/10.1016/S0378-7753(01)00560-2

Pires, V.F., E. Romero-Cadaval, D. Vinnikov, I. Roasto and J.F. Martins. 2014. Power converter interfaces for electrochemical energy storage systems—a review. Energy Convers. Manage. 86: 453–475. https://doi.org/10.1016/j.enconman.2014.05.003

Pires, V.F., A. Cordeiro, D. Foito and J.F. Silva. 2019. High step-up DC-DC converter for fuel cell vehicles based on merged quadratic boost–ćuk. IEEE Trans. Veh. Technol. 68: 7521–7530. https://doi.org/10.1109/TVT.2019.2921851

Revankar, S.T. and P. Majumdar. 2014. Fuel Cells: Principles, Design, and Analysis. CRC Press. Boca Raton, FL. USA.

Rosas-Caro, J.C., V.M. Sanchez, J.E. Valdez-Resendiz, J.C. Mayo-Maldonado, F. Beltran-Carbajal and A. Valderrabano-Gonzalez. 2017. Quadratic buck-boost converter with positive output voltage and continuous input current for PEMFC systems. Int. J. Hydrogen Energy. 42: 30400–30406. https://doi.org/10.1016/j.ijhydene.2017.10.079

Sanli, A.E. and G. Gunlu. 2016. Investigation of the vehicle application of fuel cell-battery hybrid systems. pp. 61–94. *In*: T.H. Karakoc, M.B. Ozerdem, M.Z. Sogut, C.O. Colpan, O. Altuntas and E. Açıkkalp (eds). Sustainable Aviation: Energy and Environmental Issues. Cham: Springer International Publishing. https://doi.org/10.1007/978-3-319-34181-1_8

Seaman, A., T.-S. Dao and J. McPhee. 2014. A survey of mathematics-based equivalent-circuit and electrochemical battery models for hybrid and electric vehicle simulation. J. Power Sources. Elsevier. 256: 410–423. https://doi.org/10.1016/j.jpowsour.2014.01.057

Segura, F., J.M. Andujar and E. Duran. 2011. Analog current control techniques for power control in PEM fuel-cell hybrid systems: a critical review and a practical application. IEEE Trans. Ind. Electron. 58: 1171–1184. https://doi.org/10.1109/TIE.2010.2049710

Sivakumar, S., M. Jagabar Sathik, P.S. Manoj and G. Sundararajan. 2016. An assessment on performance of DC-DC converters for renewable energy applications. Renewable Sustainable Energy Rev. 58: 1475–1485. https://doi.org/10.1016/j.rser.2015.12.057

Soumeur, M.A., B. Gasbaoui, O. Abdelkhalek, J. Ghouili, T. Toumi and A. Chakar. 2020. Comparative study of energy management strategies for hybrid proton exchange membrane fuel cell four wheel drive electric vehicle. J. Power Sources. 462: 228167. https://doi.org/10.1016/j.jpowsour.2020.228167

Tamilselvi, S., S. Gunasundari, N. Karuppiah, A. Razak, S. Madhusudan, V.M. Nagarajan, et al., 2021. A review on battery modelling techniques. Sustainability. 13(18): 10042. https://doi.org/10.3390/su131810042

Teng, T., X. Zhang, H. Dong and Q. Xue. 2020. A comprehensive review of energy management optimization strategies for fuel cell passenger vehicle. Int. J. Hydrogen Energy. 45: 20293–20303. https://doi.org/10.1016/j.ijhydene.2019.12.202

Tomasov, M., M. Kajanova, P. Bracinik and D. Motyka. 2019. Overview of battery models for sustainable power and transport applications. Transp. Res. Procedia. Elsevier. 40: 548–555. https://doi.org/10.1109/TIE.2010.2049710

Wang, H., A. Gaillard and D. Hissel. 2019. A review of DC/DC converter-based electrochemical impedance spectroscopy for fuel cell electric vehicles. Renewable Energy. 141: 124–138. https://doi.org/10.1016/j.renene.2019.03.130

Wu, J. and H. Yang. 2013. Platinum-based oxygen reduction electrocatalysts. Acc. Chem. Res. 46: 1848–1857. https://doi.org/10.1021/ar300359w

Wu, G., X. Zhang and Z. Dong. 2015. Powertrain architectures of electrified vehicles: review, classification and comparison. J. Franklin Inst. Elsevier. 352: 425–448. https://doi.org/10.1016/j.jfranklin.2014.04.018

Zhang, Z., R. Pittini, M.A.E. Andersen and O.C. Thomsena. 2012. A review and design of power electronics converters for fuel cell hybrid system applications. Energy Procedia. 20: 301–310. https://doi.org/10.1016/j.egypro.2012.03.030

Zhang, N., D. Sutanto and K.M. Muttaqi. 2016. A review of topologies of three-port DC-DC converters for the integration of renewable energy and energy storage system. Renewable Sustainable Energy Rev. 56: 388–401. https://doi.org/10.1016/j.rser.2015.11.079

Zhuang, W., Shengbo Li (Eben), X. Zhang, D. Kum, Z. Song, G. Yin, et al., 2020. A survey of powertrain configuration studies on hybrid electric vehicles. App. Energy. 262: 114553. https://doi.org/10.1016/j.apenergy.2020.114553

Chapter **8**

Biomass-derived Carbon Electrode Materials for Fuel Cells

Diana C. Martínez-Casillas

Escuela Nacional de Estudios Superiores Unidad Juriquilla,
Universidad Nacional Autónoma de México, Campus UNAM-Juriquilla,
Boulevard Juriquilla 3001, CP 76230 Santiago de Querétaro, Qro, Mexico
Email: d.martinez@unam.mx

8.1   INTRODUCTION

Continuous high-energy consumption and depletion of fossil fuel reserves lead societies to increase green and clean energy generation from renewable sources, such as geothermal heat, solar radiation and heat, water flow, and wind. Although there are advantages of renewable energy sources, their use is limited due to intermittency in the production. Thus, combination of various energy conversion and storage devices is needed for an efficient exploitation of renewable energy sources. Recently, electrochemical devices, like batteries, fuel cells (FC), and supercapacitors (SC) have emerged as very attractive technologies and ongoing research is focusing on improving the performance of these devices.

FC is one of the most promising electrochemical devices in energy conversion, due to its low temperature performance without pollutant emissions, and diversity in applications, such as providing energy in space stations, portable power generation devices, and a wide variety of vehicles (Srinivasan, 2006). Electrode material determines the performance and cost of this energy conversion system;

therefore, electrode materials are a major issue for this electrochemical device. In this regard, due to its unique properties, such as large surface area, porosity, high conductivity, good stability, and relatively low cost, carbonaceous materials (such as activated carbon, carbon black, carbon nanotubes, graphene, nanocarbons, etc.) have been widely used as electrode materials in energy conversion and storage devices (Beguin and Frackowiak, 2009; Escobar et al., 2021). Furthermore, carbon materials have shown electrocatalytic activity for the oxygen reduction reaction (ORR) in alkaline and microbial fuel cells (Li et al., 2017; An et al., 2022).

Among the many carbonaceous materials investigated, the recent biomass-derived carbons, also known as biochars, are the most promising candidates because of their high earth abundance, low-cost, heteroatom content (such as nitrogen, oxygen, sulfur, and phosphorous), and environmentally friendly nature. Biochar is a carbon-rich material generated after thermochemical conversion with or without oxygen of biomass between 150°–1000°C (Lehmann and Joseph, 2015; Ok et al., 2019). Generally, biochar contains 50–90% fixed carbon, 0–40% volatiles, 1–15% moisture, and 0.5–5% ash (Brewer and Brown, 2012). The physical and electrochemical properties of biochars are determined by the type of biomass used, the thermochemical process, parameters applied, and post-treatments in the carbonization process.

Figure 8.1 presents the general process for the obtention of biomass-derived carbon electrode materials (BCEM). The first step is the selection of biomass, whatever natural organic material can be easily transformed into biochar. The second stage is biomass conditioning where the biomass is hardly used as obtained; mainly it is usually cleaned, dried, and grinded. Then the biomass gets transformed into biochar which can be pre-treated and/or post-treated for activation and/or doping, which is a key factor in the process. Finally, the obtained material is tested as electrode in fuel cell. In this chapter, recent developments and ongoing trends in research of biomass-derived carbon electrode material are reviewed.

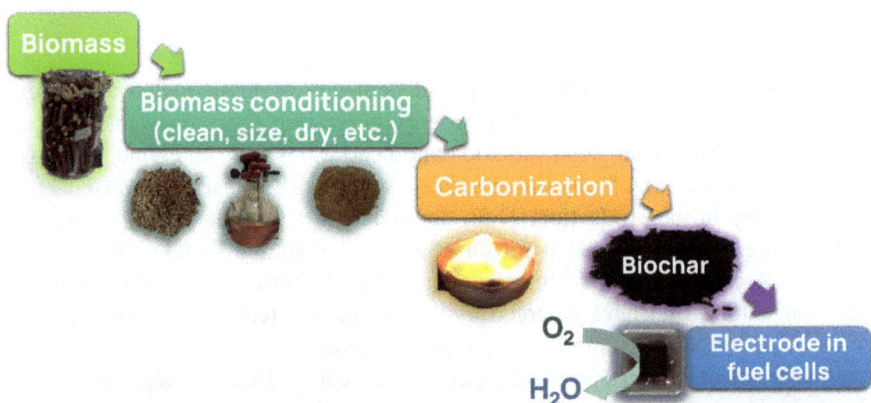

Figure 8.1   Schematic process of the preparation of BCEM for fuel cell applications.

8.2 CARBON ELECTRODE MATERIALS FROM BIOMASS

All renewable organic material derived from plants and animals, and wastes generated from their transformations, are considered biomass. Chemically, it is a biopolymer made up of proteins, carbohydrates, cellulose, hemicellulose, lignin, lipids, starches, extractives, water, ash, and other trace components (Vassilev et al., 2010), i.e. biomass is rich in carbon and heteroatoms, such as N, O, P, and S. Moreover, biomass can be classified in two groups, depending on its composition as is presented in Fig. 8.2. On the one hand, lignocellulosic biomass, for example wood, plants, agricultural crops residues, forestry residues have as their main components—carbohydrate polymers (cellulose and hemicellulose), an aromatic polymer (lignin), and small amounts of simple sugar, protein, starches, and lipids (Lee et al., 2007; Vassilev et al., 2010). On the other hand, non-lignocellulosic biomass, like algae, livestock wastes, food supply chain residues, are composed of proteins, lipids saccharides, inorganics, and heteroatoms (Vassilev et al., 2010; Li and Jiang, 2017).

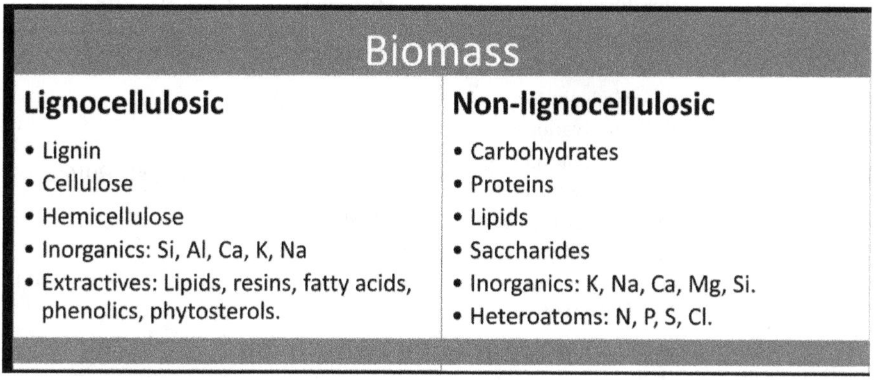

Biomass	
Lignocellulosic	**Non-lignocellulosic**
• Lignin	• Carbohydrates
• Cellulose	• Proteins
• Hemicellulose	• Lipids
• Inorganics: Si, Al, Ca, K, Na	• Saccharides
• Extractives: Lipids, resins, fatty acids, phenolics, phytosterols.	• Inorganics: K, Na, Ca, Mg, Si.
	• Heteroatoms: N, P, S, Cl.

Figure 8.2 Composition of lignocellulosic and non-lignocellulosic biomasses.

Considering that carbon is the main element of biomass, it can be chemically and/or thermally treated in a wide range of temperatures to produced biochars which can be used in a variety of applications including, electrode materials for energy conversion devices. As can be appreciated in Table 8.1, many lignocellulosic and non-lignocellulosic biomasses, such as agricultural waste, aquatic plants, consumable products, and residues have been used as precursors to produce carbon electrode materials for alkaline and microbial fuel cells. Furthermore, many preparation processes have been used to produce such BCEM.

Table 8.1 Biomass-derived carbon electrode materials for fuel cells

Biomass	Preparation process	Fuel cell type	References
Agave waste	Solar pyrolysis	Alkaline	Campos Roldán et al., 2021
Algae: *Sargassum* spp.	KOH activation, pyrolysis	Alkaline	Pérez-Salcedo et al., 2019
Amaranthus waste	Pyrolysis	Alkaline	Gao et al., 2015
Banana	Hydrothermal carbonization, pyrolysis	Microbial	Yuan et al., 2014
Banana peel	Pre-carbonization, co-pyrolysis for KOH activation, NH_3 treatment	Alkaline	Zhang et al., 2017
Basswood	Delignifcation, pyrolysis, NH_3 treatment	Alkaline	Tang et al., 2018
Cattle bones	Pre-carbonization, co-pyrolysis for KOH activation	Alkaline	Zan et al., 2017
Chicken feathers	Co-pyrolysis for KOH activation	Alkaline	Tyagi et al., 2020
Corn cob	Pyrolysis	Microbial	Li et al., 2018
Corn straw	Co-pyrolysis for KOH activation	Microbial	Wang et al., 2017
Corn stalk	Co-pyrolysis for KOH activation	Alkaline	Cao et al., 2019
Eggplant	KOH activation, pyrolysis with NH_3	Alkaline	Zhou et al., 2016a
Fresh egg white	Molten salt method	Alkaline	Chen et al., 2018
Human hair	Pre-carbonization, co-pyrolysis for NaOH activation	Alkaline	Chaudhari et al., 2014
Leather waste	Pyrolysis, KOH activation, Hydrothermal N doping	Alkaline	Alonso-Lemus et al., 2016; Lardizábal-Gutiérrez et al., 2016
Lotus stem	Co-pyrolysis for N doping	Alkaline	Weththasinha et al., 2017
Moss	Hydrothermal carbonization, pyrolysis	Alkaline/ Microbial	Zhou et al., 2016d
Onion	Hydrothermal carbonization, co-pyrolysis for $ZnCl_2$ activation	Alkaline	Yang et al., 2018
Pomelo peel	Pre-treatment for doping, pyrolysis, activation	Alkaline	Wang et al., 2018
Poplar catkins	$ZnCl_2$ activation, pyrolysis	Alkaline	Gao et al., 2017
Seaweed: *Ascophyllum nodosum*	KOH activation, pyrolysis	Alkaline	Perez-Salcedo et al., 2020
Seaweed: Enteromorpha	Pyrolysis	Alkaline	Zhang et al., 2019

(Contd.)

Biomass	Preparation process	Fuel cell type	References
Sheep bones	Pre-carbonization, co-pyrolysis for KOH activation	Alkaline	Li et al., 2017
Soybean shells	Pre-carbonization, co-pyrolysis for KOH activation, H_2SO_4 treatment, co-pyrolysis for NH_3 activation	Alkaline	Zhou et al., 2016b
Soybean straw	Pyrolysis	Alkaline	Liu et al., 2020
Spent coffee grounds	Co-pyrolysis for N and Fe doping Pyrolysis, HNO_3 treatment, co-pyrolysis for N and P doping	Alkaline	Srinu et al., 2018
Spider silk	$ZnCl_2$ activation, pyrolysis	Alkaline/ Microbial	Zhou et al., 2016c
Tea residue	Pyrolysis	Alkaline	Wu et al., 2018
Walnut green peels	Hydrothermal carbonization, pyrolysis	Alkaline	Zhou et al., 2022
Water hyacinth	$ZnCl_2$ activation, pyrolysis	Alkaline	Liu et al., 2015
Yuba	$ZnCl_2$ activation, pyrolysis	Alkaline	Zhang et al., 2021

Among the diversity of biomasses reported, in general the authors select biomass without considering availability, carbonization yield, elemental composition, and other factors that are related to the BCEM performance and costs. In this sense, some of the biomasses in the table must be avoided; for example, edible products, such as banana (Yuan et al., 2014), fresh egg white (Chen et al., 2018), onion (Yang et al., 2018), and yuba (Zhang et al., 2021). Human hair (Chaudhari et al., 2014) and spider silk (Zhou et al., 2016a) are other biomasses that should be avoided because of diversity and scarcity, respectively. On the contrary, the use of highly abundant biomasses, such as organic residues from agriculture, farming, and human activities are a feasible alternative to produce BCEM. The obtained materials have been investigated as oxygen reduction reaction (ORR) electrocatalysts in alkaline fuel cells with good performances. Results showed onset potential values > 0.8 V vs RHE, and electron transfer numbers estimated from K–L slopes are close to 4, indicating that BCEM follow a direct pathway for the ORR. Thus, the biomass-derived carbon electrode material could replace expensive Pt-based electrocatalyst for fuel cells.

8.3 PRODUCTION OF BIOMASS-DERIVED CARBON MATERIALS

The development of effective strategies for transformation of biomass to carbon materials is a topic of significant interest, since properties and characteristics of biochar depend on the production process used. During this transformation, moisture and volatiles of biomass are removed, while its structural components go through depolymerization, crosslinking, and fragmentation. Therefore, it is necessary to find the most suitable biomass processing strategy, which could include pre-treatments and/or post-modifications to active and/or dope the biochar as is presented in Fig. 8.3.

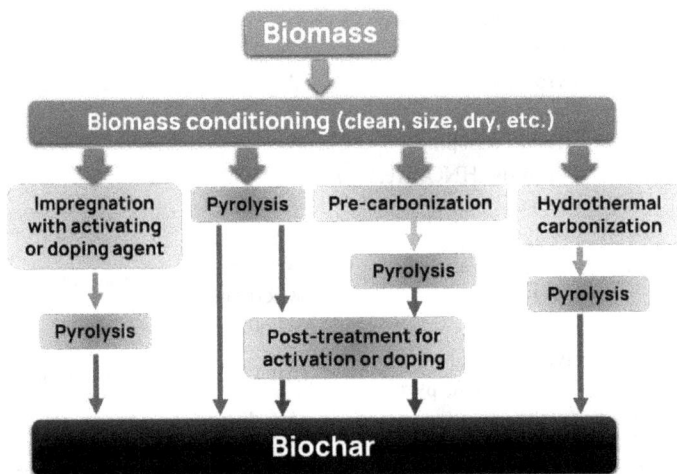

Figure 8.3   Process strategies to produce biochars.

Some treatments are physical/chemical activation and surface functionalization, which are made to increase the specific surface area (SSA) and porosity, or to form functional groups on the biochar. Many strategies have been investigated to produce BCEM for fuel cells with activity towards ORR (Table 8.1). Pyrolysis is the oldest and the most used direct thermochemical process for biochar production. This method proceeds in a wide range of temperatures under an oxygen-free atmosphere. Most approaches focused on the temperature, time, chemical activation, heteroatom doping and hydrothermal treatments for the transformation of biomass into BCEM with enhanced performance as discussed in the following paragraphs.

8.3.1   Non-activation

Gao et al. (2015) investigated the ORR activity of BCEM produced from the pyrolysis of amaranthus waste at 600°C, 700°C, 800°C, and 900°C under a N_2 atmosphere with resident time of 120 minutes and washed with 2 M HCl. The generated materials exhibit SSA (determined from N adsorption-desorption isotherms and BET equation) in the range of 663–1008 m^2 g^{-1}. A higher pyrolysis temperature leads to a higher specific surface area and pore volume. Also, nitrogen was detected on the BCEM, with the N contents being in the range of 1.42–2.47 at% for all the carbons, with the maximum content found in the sample pyrolyzed at 700°C, showing that biomass is an excellent precursor to produce nitrogen self-doped biochars. It should be noted that N content and chemical bonding state in the biochar played a key role in catalytic performance of N-doped catalysts for ORR (Zhang et al., 2021). The BCEM reported onset potentials (E_{onset}) in the range of 0.23–0.27 V vs Hg/HgO close to the commercial Pt/C catalyst ($E_{onset} = 0.3$ V *vs.* Hg/HgO) in alkaline medium. The best self-doped carbon material, i.e. NDC-L-800 was also tested for methanol oxidation reaction, with the results indicating high selectivity and stability.

Another example of self-doped biomass-derived carbon electrode material is presented by Wu et al. (2018). They pyrolyzed tea residue at 1000°C for 2 h under N_2 atmosphere, then washed with 2 M HCl and deionized water. The obtained material exhibited an SSA and total pore volume of 856 $m^2 g^{-1}$ and 0.65 $cm^3 g^{-1}$, respectively. The atomic percentages of N and F were calculated to be 2.8% and 2.2%, respectively by X-ray photoelectron spectroscopy (XPS). The BCEM showed an E_{onset} of 0.81 V vs RHE with an average number of electron transfer (n) around 3.8, indicating that this doped material is a good ORR electrocatalyst alternative for alkaline fuel cell.

Li et al. (2018) reported pyrolysis under N_2 during 2 h at six different temperatures of corn cob. The electrochemical active areas were 68 $m^2 g^{-1}$, 105 $m^2 g^{-1}$, 155 $m^2 g^{-1}$, 366 $m^2 g^{-1}$, 656 $m^2 g^{-1}$, and 446 $m^2 g^{-1}$ at 250°C, 350°C, 450°C, 550°C, 650°C, and 750°C, respectively, indicating that a higher pyrolysis temperature leads to a higher electrochemical active area. The obtained BCEM were composed by C, O, K and N as indicated by XPS measurements. Moreover, rising pyrolysis temperature diminishes the atomic contents of O, K, and N, whereas the atomic C content increases in the materials. All BCEM were tested as cathode electrode in a single chamber microbial fuel cell, the material produced at 650°C presenting the best performance with an open voltage of 0.459 V and a maximum power density (P_{max}) of 458.85 mW m^{-3}. This result could be attributed to the diversity and abundance of active sites evidenced by the high electrochemical active area given by the self-doping.

The pyrolysis of *Enteromorpha* was investigated by Zhang et al. (2019) at three different temperatures for 3 h under argon atmosphere. After pyrolysis, the carbonized materials were washed with 5 M HCl. The SSA were 604 $m^2 g^{-1}$, 778 $m^2 g^{-1}$, and 768 $m^2 g^{-1}$ at 800°C, 1000°C, and 1100°C, respectively. Surface chemical compositions determined by XPS show that N and S atoms were incorporated in the carbon matrix. The obtained materials exhibit catalytic activity for the ORR in alkaline medium (0.1 M KOH). The BCEM produced at 1000°C presented E_{onset} of 0.95 V *vs.* RHE, which is just about 10 mV below to that of the 20% commercial Pt/C, and an average electron transfer number of 3.8. This behavior is related by the co-doping effect of N and S.

Campos-Roldán et al. (2021) studied the solar pyrolysis of agave waste at 500°C, 700°C, and 900°C under Ar atmosphere for 1 h with a heating rate of 30°C min^{-1}. These materials obtained low specific surface areas in the range of 9.5–48.8 $m^2 g^{-1}$ and a higher pyrolysis temperature leading to a lower SSA. This effect is related to the high heating rate in the solar pyrolysis. The BCEM obtained at 500°C exhibiedt the highest electrocatalytic activity in 0.1 M KOH with $E_{onset} = 0.87$ V *vs.* RHE. Although its performance is lower than the commercial

Pt electrocatalyst, which is higher than the N self-doped BCEM obtained at 1000°C reported earlier (Wu et al., 2018).

Furthermore, Zhou et al. (2016) reported the preparation of N self-doped BCEM from moss, using hydrothermal carbonization (HTC) and subsequent pyrolysis process. First the biomass was carbonized by HTC at 180°C for 24 h using deionized water as solvent; then, the obtained hydrochar was pyrolyzed at 900°C under nitrogen atmosphere for 2 h with a yield of 52%. The resulting BCEM obtained SSA and E_{onset} in alkaline medium of 409 m^2g^{-1} and 0.935 V vs RHE, respectively. This material was also evaluated as cathodic catalyst in an air-single chamber MFC, achieving a P_{max} of 703 ± 16 mW m^2, which was slightly higher to that of the MFC using commercial Pt/C.

8.3.2 Chemical Activation

Activation is the most applied method to increase SSA and adjust pore structure of carbonaceous materials by opening the occlusion holes, expanding pores, and/or forming new ones. Additionally, activation also could modify the functional groups on the carbon surface. Depending on the activation mechanism, activation methods can be divided into chemical and physical activation. In physical activation, the biochar is exposed to the flow of an oxidizing gas at above 700°C, when oxidizing agents like CO_2, H_2O steam or a mixture of them are used. On the other hand, chemical activation involves the addition of activators, such as NaOH, KOH, K_2CO_3, H_3PO_4, H_2O_2, and $ZnCl_2$ (Anto et al., 2021). Figure 8.4 shows the effect of different chemical activations in the SSA to produce biomass-derived Carbon electrodes from different biomasses.

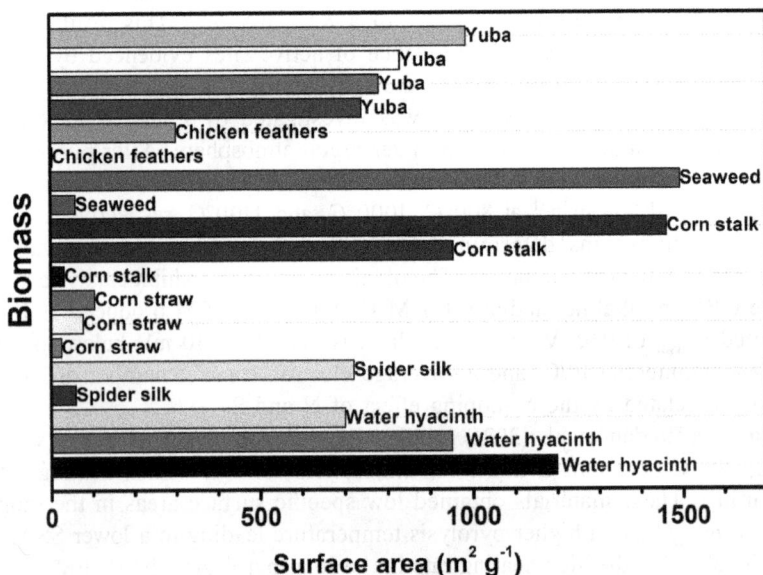

Figure 8.4 SSA of biomass-derived carbon electrodes obtained by different chemical activations.

Chemical activation can be done by two techniques. The first one, is a one-step method, also known as co-pyrolysis, which is a direct carbonization/ activation process. In this method, the raw biomass is previously impregnated with the activating agent and then pyrolyzed under inert atmosphere (N_2 or Ar) at high temperature. The second one is a two-step method, where the biomass is thermally treated (carbonized under an inert gas flow or by hydrothermal process with solvent) and then activated, using chemical activating agent. Regardless of the biomass, activating agent, and activation process used, the specific surface area changes with activation. It is for this reason that chemical activation is performed to increase pore fraction and SSA of carbon materials.

8.3.2.1 Co-pyrolysis

$ZnCl_2$ is commonly used as a dehydrating chemical activating agent with a good performance. Liu et al. (2015) reported the production of nitrogen self-doped biomass-derived carbon electrode from water hyacinth by co-pyrolysis with $ZnCl_2$ at 600°C, 700°C, and 800°C. The produced materials exhibit SSA in the range of 700–1199 m^2 g^{-1} as can be observed in Fig. 8.4, a higher pyrolysis temperature leading to a lower pore volume and specific surface area. Nitrogen and oxygen were detected on the BCEM by XPS measurements; the N contents were in the range of 2.80–5.02 at% for the three carbonaceous materials, with the maximum content found in the sample pyrolyzed at 600°C. However, the BCEM obtained at 700°C presented the highest E_{onset} of 0.98 V *vs.* RHE, which are even more positive than that of commercial Pt/C, indicating that the activity towards ORR is not related to the SSA or N content.

Another example of self-doped biomass-derived carbon electrode material chemically activated with $ZnCl_2$ is presented by Zhou et al. (2016a). They produced carbon nanofibers from spider silk; first, the obtained BCEM were evaluated for ORR in alkaline medium and then tested in a microbial fuel cell (MFC). Chemical activation increases SSA by 12 times in comparison with the material without activation (Fig. 8.4). As expected, activated carbon showed higher catalytic activity than non-activated BCEM in 0.1 M KOH. Moreover, in an MFC, the activated BCEM exhibited a P_{max} of 1800 ± 82 mW m^{-2}, which was superior to that of a MFC with Pt/C as cathodic electocatalyst (1152 ± 73 mW m^{-2}).

The co-pyrolysis with $ZnCl_2$ under N_2 atmosphere for 2 h of yuba was investigated by Zhang et al. (2021) at four different temperatures (750°C, 800°C, 850°C, and 900°C). In this case, the specific surface area gradually increased with pyrolysis temperature increasing, whereas N content diminishes. All the BCEM reported E_{onset} values above 0.9 V vs RHE, suggesting that ORR activity is related to the type of nitrogen present in the activated carbons, i.e. pyridinic-N, pyrrolic-N, and graphitic-N. Nonetheless, similar E_{onset}, BCEM proceed by different pathways as evidenced by the electron transfer numbers. Sample produced at 750°C with E_{onset} = 0.93 V vs RHE follows a 2e^- pathway, whereas the material obtained at 850°C presents an average electron transfer number of 3.81, indicating a 4e^- pathway with a E_{onset} = 0.97 V vs RHE.

Another commonly used activating agent is KOH, which usually increases SSA by developing porosity. This effect was demonstrated by Cao et al. (2019), who studied the ORR activity of N self-doped BCEM produced from co-pyrolysis at 850°C under N_2 flow of corn stalk with KOH. They applied two pyrolysis times: 30 (BC-K-0.5) and 60 minutes (BC-K-1); also prepared a biochar without activation (BC) for comparison. Chemical activation with KOH increases the SSA from 33 m^2g^{-1} to 955 m^2g^{-1} and 1461 m^2g^{-1} for the materials obtained at 60 and 30 minutes, respectively. Elemental analysis demonstrated nitrogen doping, that N content was in the range of 0.77–2.13 at%, the material without activation presented the highest N content. However, from XPS measurements BC-K-0.5 showed N content of 4.21 at%; this could be related to the higher SSA of this biochar. BC-K-0.5 also obtained the best performance with E_{onset} (–0.06 V *vs.* Hg/HgO) and stability for ORR in alkaline medium.

One more study of self-doped BCEM chemically activated with KOH is reported by Pérez-Salcedo et al. (2020). They prepared two biochars—the first one by just pyrolyzing the *Ascophyllum nodosum* at 700°C under N_2 atmosphere for 2 h; the second one was prepared by co-pyrolysis with KOH under the same conditions. The elemental analyses indicate that the produced materials are N and S doped. Chemical activation increases SSA and pore volume by almost 25 and 17 times, respectively, indicating that KOH develops the porosity of the material. The effect of chemical activation also can be observed in the ORR activity, achieving E_{onset} and n of 0.878 V vs RHE and 4, respectively, which are higher than that obtained for the biochar without activation.

Tyagi et al. (2020) investigated the effect of the activating agent and pyrolysis temperature of BCEM derived from chicken feathers for the ORR activity in alkaline medium. Among the three activating agents (KOH, $ZnCl_2$, H_3PO_4), KOH was found to be the most effective, with E_{onset} of the KOH activated biochars more positive as compared with $ZnCl_2$ and H_3PO_4 activated materials. Specifically, the sample obtained at 900°C and activated with KOH presented the highest performance as electrocatalyst towards ORR.

8.3.2.2   Pre-treatment and Co-pyrolysis

In addition to chemical activation, some researchers performed a carbonization at lower temperature in an inert atmosphere before pyrolysis. For example, Li et al. (2017) reported the production of N-doped BCEM from sheep bones. First, biomass was converted into biochar, executing a thermal treatment at 450°C under N_2 atmosphere for 2 h; then the activation was achieved by mixing the biochar with KOH and pyrolyzing it under N_2 atmosphere at 850°C. The obtained material exhibited high specific and micropore surface areas of 1961 $m^2\,g^{-1}$ and 1166 m^2g^{-1}, respectively. By XPS measurements it was determined that the main species of N present on the BCEM were pyridinic-N and quaternary-N. This material presents an n of 3.85, indicating a 4e⁻ pathway with E_{onset} of 0.97 V *vs.* RHE in alkaline electrolyte. The results indicated that the obtained BCEM is an excellent N-doped carbon electrocatalyst for the ORR.

Another example of N self-doped biomass-derived carbon electrode material prepared by a two-step process and chemically activated with KOH is presented by Wang et al. (2017). Initially a biochar was produced by pyrolyzing corn straw under inert atmosphere at 500°C; then the obtained material was modified with KOH and pyrolyzed at 900°C. The resulting BCEM obtained SSA of 105 m^2g^{-1} and P_{max} of 8.36 W m^{-3} used as anodic and cathodic material in a MFC.

As mentioned earlier, another two-step activation process used is an HTC followed by a co-pyrolysis with the activating agent. In this sense, Yuan et al. (2014) investigated the ORR activity of BCEM produced using banana as precursor. First, the biomass was carbonized by HTC at 180°C for 12 h, using deionized water as solvent; then, the obtained hydrochar was impregnated with KOH and pyrolyzed at two different temperatures (500°C and 900°C). The material activated at 900°C exhibited the highest activity towards ORR with $E_{onset} = -0.28$ V *vs.* SCE. Moreover, this material was evaluated as cathode catalyst in an air-single chamber MFC, achieving a stable voltage of 0.47 V and P_{max} of 528.2 mW m^2.

The production of N, S self-doped BCEM by a hydrothermal and subsequent pyrolysis process from onion was studied by Yang et al. (2018). Biomass was carbonized by HTC with an NH_4HCO_3 solution at 200°C for 12 h; after that, the obtained product was mixed with $ZnCl_2$ and pyrolyzed at 800°C under N_2 atmosphere. Activation increases SSA by 1200 m^2g^{-1} in comparison to the material synthesized without chemical activation. The doping was confirmed by XPS and the atomic percentages were 6.23 at% and 0.36 at%, for N and S, respectively. Electrochemical characterization in 0.1 M KOH of the BCEM showed $E_{onset} = 0.88$ V *vs.* RHE, a $n = 3.8$, and long stability.

8.3.3 Doping Treatments

Some investigations have been performed to increase the natural heteroatom content in the biomass. In this regard, Alonso-Lemus et al. (2016) produced BCEM from leather waste by pyrolysis at four different temperatures under N_2 atmosphere, followed by a chemical activation with KOH at 750°C and a final hydrothermal treatment with hydrazine at 180°C for 24 h ofr nitrogen doping. The generated materials exhibit SSA and N content in the range of 819–2100 m^2g^{-1} and 3.39–7.65 at%, respectively; a higher pyrolysis temperature leads to a lower SSA and N content. Materials present higher N content than that nitrogen self-doped, indicating a successful doping process. The E_{onset} values are 0.905 V, 0.905 V, 0.930 V, and 0.920 V *vs.* RHE at 700°C, 800°C, 900°C, and 1000°C, respectively.

Zan et al. (2017) investigated the ORR activity of nitrogen and phosphorus co-doped produced from cattle bones. First, the biomass was pre-carbonized under Ar atmosphere at 400°C for 3 h, then the obtained biochar was activating by co-pyrolysis with KOH at 800°C for 1 h in Ar atmosphere; finally, this activated carbon was doped by another co-pyrolysis with phytic acid and dicyandiamide at 900°C for 3 h under Ar atmosphere. The produced material showed a SSA of 1516 m^2g^{-1}, and presence of N (3.2 at%) and P (4.0 at%) confirming co-doping. The electrochemical characterization in 0.1 M KOH electrolyte indicated outstanding ORR activity, great durability, and superior methanol tolerance.

Another example of nitrogen co-doped BCEM from *Sargassum spp.* by a two-step process is presented by Pérez-Salcedo et al. (2019). The first stage was a co-pyrolysis with KOH at three different temperatures (700°C, 750°C, and 800°C); followed by a hydrothermal treatment with pyridine at 180°C for 24 h. The produced materials exhibited SSA in the range of 2513–2675 m^2g^{-1} and specific surface area increase with increasing pyrolysis temperature. Doping was evidenced by elemental analysis and XPS. These measurements detected the presence of N on the produced biochars. The BCEM showed E_{onset} values of 0.812, 0.811, and 0.870 V *vs.* RHE for the materials produced at 700°C, 750°C, and 800°C, respectively. However, the BCEM obtained at 750°C showed the best performance when evaluated as cathode electrode in PEM fuel cell, achieving the highest maximum power density (12.72 mW cm^{-2}). Therefore, it is important to prepare membrane electrode assembly (MEA) to evaluate the materials in a real PEMFC.

Liu et al. (2020) reported the co-pyrolysis of soybean straw with melamine, MgO, and $Fe(NO_3)_3$ under N_2 atmosphere at 800°C for 2 h. The resulting material exhibited an SSA of 520 m^2 g^{-1} and excellent ORR performance with *n* of 3.97 and E_{onset} = 0.989 V *vs.* RHE in alkaline conditions (0.1 M KOH). The BCEM was also characterized in acidic medium (0.1 M $HClO_4$) achieving E_{onset} of 0.88 V *vs.* RHE and *n* = 3.89, indicating that this material could be used as a cathode electrocatalyst for fuel cell.

8.4   OUTLOOK

Biomass-derived carbon materials are promising candidates to replace precious metals as cathodic electrode materials in fuel cells. Figure 8.5 shows the SSA vs onset potential relation. Generally high specific surface area provides more active sites for catalytic reactions; thus higher onset potentials. Nevertheless, the BCEM with a moderate specific surface area of 520.9 m^2g^{-1} achieved the highest E_{onset} which is attributed to a combination of porosity and doping in the material (Liu et al., 2020).

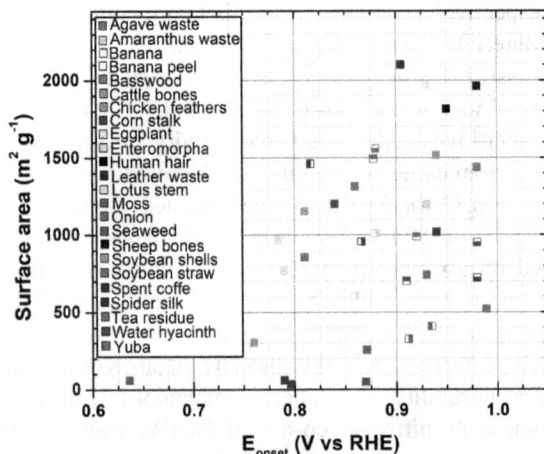

Figure 8.5   SSA *vs.* onset potential of BCEM for ORR in alkaline medium.

The studies described in this chapter show that the electrocatalytic activity towards ORR is not directly dependent on the specific surface area or heteroatom content. There should exist a balance between carbon defects, graphitic domains, heteroatoms content, and SSA induced by pyrolysis to boost the ORR activity. Therefore, some strategies to achieve an enhanced electrocatalytic activity include chemical activation, pre-treatments such as hydrothermal or carbonization and doping along with biomass pyrolysis. Thus, the proper production method of biomass-derived carbon electrode materials plays a main role in their performance towards ORR. Notwithstanding all the encouraging results reported for BCEM, some of the BCEM presented in Table 8.1 and Fig. 8.5 have use as electrode materials for real practical and commercial FC applications due to their limited availability, other uses (like food), and low yield production. Forthcoming investigations should be carried out in developing methods to produce consistent carbonaceous materials from biomass on a large scale, in order to consider the BCEM for commercial applications.

Additionally, fundamental (theoretical and experimental) studies are essential to comprehend the properties of BCEM that are produced through different methods and precursors. For example, the role of heteroatoms, particularly the nitrogen species, and textural properties on the electrocatalytic performance of BCEM, should be investigated. Moreover, to determine the viability of BCEM, it is important to study the material's performance under real operating conditions in a fuel cell.

REFERENCES

Alonso-Lemus, I.L., F.J. Rodríguez-Varela, M.Z. Figueroa-Torres, M.E. Sánchez-Castro, A. Hernández-Ramírez, D. Lardizábal-Gutiérrez, et al., 2016. Novel self-nitrogen-doped porous carbon from waste leather as highly active metal-free electrocatalyst for the ORR. Int. J. Hydrogen Energy. 41(48): 23409–23416. https://doi.org/10.1016/j.ijhydene.2016.09.033

An, F., X. Bao, X. Deng, Z. Ma and X. Wang. 2022. Carbon-based metal-free oxygen reduction reaction electrocatalysts: Past, present and future. New Carbon Mater. 37(2): 338–354. https://doi.org/10.1016/S1872-5805(22)60590-0

Anto, S., M.P. Sudhakar, T. Shan Ahamed, M.S. Samuel, T. Mathimani, K. Brindhadevi, et al., 2021. Activation strategies for biochar to use as an efficient catalyst in various applications. Fuel. 285: 119205. https://doi.org/10.1016/j.fuel.2020.119205

Beguin, F. and E. Frackowiak (eds). 2009. Carbons for Electrochemical Energy Storage and Conversion Systems. CRC Press. https://doi.org/10.1201/9781420055405

Brewer, C.E. and R.C. Brown. 2012. Biochar. A Comprehensive Renewable Energy. 357–384. Elsevier. https://doi.org/10.1016/B978-0-08-087872-0.00524-2

Campos Roldán, C.A., A. Ayala-Cortés, R.G. González-Huerta, H.I. Villafán-Vidales, C.A. Arancibia-Bulnes, A.K. Cuentas-Gallegos, et al., 2021. Metal-free electrocatalysts obtained from agave waste by solar pyrolysis for oxygen reduction reaction. Int. J. Hydrogen Energy. 46(51): 26101–26109. https://doi.org/10.1016/j.ijhydene.2020.12.095

Cao, W., B. Wang, Y. Xia, W. Zhou, R. Wen, Y. Jia, et al., 2019. Preparation of highly-active oxygen reduction reaction catalyst by direct co-pyrolysis of biomass with KOH. Int. J. Electrochem. Sci. 250–261. https://doi.org/10.20964/2019.01.33

Chaudhari, K.N., M.Y. Song and J.-S. Yu. 2014. Transforming hair into heteroatom-doped carbon with high surface area. Small. 10(13): 2625–2636. https://doi.org/10.1002/smll.201303831

Chen, Y., S. Ji, H. Wang, V. Linkov and R. Wang. 2018. Synthesis of porous nitrogen and sulfur co-doped carbon beehive in a high-melting-point molten salt medium for improved catalytic activity toward oxygen reduction reaction. Int. J. Hydrogen Energy. 43(10): 5124–5132. https://doi.org/10.1016/j.ijhydene.2018.01.095

Escobar, B., D.C. Martínez-Casillas, K.Y. Pérez-Salcedo, D. Rosas, L. Morales, S.J. Liao, et al., 2021. Research progress on biomass-derived carbon electrode materials for electrochemical energy storage and conversion technologies. Int. J. Hydrogen Energy. 46(51): 26053–26073. https://doi.org/10.1016/j.ijhydene.2021.02.017

Gao, S., K. Geng, H. Liu, X. Wei, M. Zhang, P. Wang, et al., 2015. Transforming organic-rich amaranthus waste into nitrogen-doped carbon with superior performance of the oxygen reduction reaction. Energy Environ. Sci. 8(1): 221–229. https://doi.org/10.1039/C4EE02087A

Gao, S., X. Li, L. Li and X. Wei. 2017. A versatile biomass-derived carbon material for oxygen reduction reaction, supercapacitors and oil/water separation. Nano Energy. 33: 334–342. https://doi.org/10.1016/j.nanoen.2017.01.045

Lardizábal-Gutiérrez, D., D. González-Quijano, P. Bartolo-Pérez, B. Escobar-Morales, F.J. Rodríguez-Varela and I.L. Alonso-Lemus. 2016. Communication—synthesis of self-doped metal-free electrocatalysts from waste leather with high ORR activity. J. Electrochem. Soc. 163(2): H15–H17. https://doi.org/10.1149/2.0191602jes

Lee, D.K., V.N. Owens, A. Boe and P. Jeranyama. 2007. Composition of Herbaceous Biomass Feedstocks. Brookings. SD. USA. South Dakota State University.

Lehmann, J. and S. Joseph (eds). 2015. Biochar for Environmental Management: Science, Technology and Implementation, 2nd Ed. Routledge, Taylor & Francis Group.

Li, D.-C. and H. Jiang. 2017. The thermochemical conversion of non-lignocellulosic biomass to form biochar: a review on characterizations and mechanism elucidation, Bioresour. Technol. 246: 57–68. https://doi.org/10.1016/j.biortech.2017.07.029

Li, S., C. Cheng and A. Thomas. 2017. Carbon-based microbial-fuel-cell electrodes: from conductive supports to active catalysts. Adv. Mater. 29(8): 1602547. https://doi.org/10.1002/adma.201602547

Li, S., R. Xu, H. Wang, D.J.L. Brett, S. Ji, B.G. Pollet, et al., 2017. Ultra-high surface area and mesoporous N-doped carbon derived from sheep bones with high electrocatalytic performance toward the oxygen reduction reaction. J. Solid State Electrochem. 21(10): 2947–2954. https://doi.org/10.1007/s10008-017-3630-3

Li, M., H. Zhang, T. Xiao, S. Wang, B. Zhang, D. Chen, et al., 2018. Low-cost biochar derived from corncob as oxygen reduction catalyst in air cathode microbial fuel cells. Electrochim. Acta. 283: 780–788. https://doi.org/10.1016/j.electacta.2018.07.010

Liu, X., Y. Zhou, W. Zhou, L. Li, S. Huang and S. Chen. 2015. Biomass-derived nitrogen self-doped porous carbon as effective metal-free catalysts for oxygen reduction reaction. Nanoscale. 7(14): 6136–6142. https://doi.org/10.1039/C5NR00013K

Liu, Y., M. Su, D. Li, S. Li, X. Li, J. Zhao, et al., 2020. Soybean straw biomass-derived Fe–N co-doped porous carbon as an efficient electrocatalyst for oxygen reduction in both alkaline and acidic media. RSC Adv. 10(12): 6763–6771. https://doi.org/10.1039/C9RA07539A

Ok, Y., D.C.W. Tsang, N. Bolan and J.M. Novak (eds). 2019. Biochar from Biomass and Waste: Fundamentals and Applications. Elsevier.

Pérez-Salcedo, K., X. Shi, A. Kannan, R. Barbosa, P. Quintana and B. Escobar. 2019. N-doped porous carbon from sargassum spp. as efficient metal-free electrocatalysts for O_2 reduction in alkaline fuel cells. Energies. 12(3): 346. https://doi.org/10.3390/en12030346

Pérez-Salcedo, K.Y., S. Ruan, J. Su, X. Shi, A.M. Kannan and B. Escobar. 2020. Seaweed-derived KOH activated biocarbon for electrocatalytic oxygen reduction and supercapacitor applications. J. Porous Mater. 27(4): 959–969. https://doi.org/10.1007/s10934-020-00871-7

Srinivasan, S. 2006. Fuel Cells: From fundamentals to Applications. Boston. MA. Springer. Science+Business Media. LLC Springer, e-books.

Srinu, A., S.G. Peera, V. Parthiban, B. Bhuvaneshwari and A.K. Sahu. 2018. Heteroatom engineering and co-doping of N and P to porous carbon derived from spent coffee grounds as an efficient electrocatalyst for oxygen reduction reactions in alkaline medium. Chemistry Select. 3(2): 690–702. https://doi.org/10.1002/slct.201702042

Tang, Z., Z. Pei, Z. Wang, H. Li, J. Zeng, Z. Ruan, et al. 2018. Highly anisotropic, multichannel wood carbon with optimized heteroatom doping for supercapacitor and oxygen reduction reaction. Carbon. 130: 532–543. https://doi.org/10.1016/j.carbon.2018.01.055

Tyagi, A., S. Banerjee, S. Singh and K.K. Kar. 2020. Biowaste derived activated carbon electrocatalyst for oxygen reduction reaction: effect of chemical activation, Int. J. Hydrogen Energy. 45(34): 16930–16943. https://doi.org/10.1016/j.ijhydene.2019.06.195

Vassilev, S.V., D. Baxter, L.K. Andersen and C.G. Vassileva. 2010. An overview of the chemical composition of biomass. Fuel. 89(5): 913–933. https://doi.org/10.1016/j.fuel.2009.10.022

Wang, B., Z. Wang, Y. Jiang, G. Tan, N. Xu and Y. Xu. 2017. Enhanced power generation and wastewater treatment in sustainable biochar electrodes-based bioelectrochemical system. Bioresour. Technol. 241: 841–848. https://doi.org/10.1016/j.biortech.2017.05.155

Wang, N., T. Li, Y. Song, J. Liu and F. Wang. 2018. Metal-free nitrogen-doped porous carbons derived from pomelo peel treated by hypersaline environments for oxygen reduction reaction. Carbon. 130: 692–700. https://doi.org/10.1016/j.carbon.2018.01.068

Weththasinha, H.A.B.M.D., Z. Yan, L. Gao, Y. Li, D. Pan, M. Zhang, et al., 2017. Nitrogen doped lotus stem carbon as electrocatalyst comparable to Pt/C for oxygen reduction reaction in alkaline media. Int. J. Hydrogen Energy. 42(32): 20560–20567. https://doi.org/10.1016/j.ijhydene.2017.06.011

Wu, D., Y. Shi, H. Jing, X. Wang, X. Song, D. Si, et al., 2018. Tea-leaf-residual derived electrocatalyst: Hierarchical pore structure and self, nitrogen and fluorine co-doping for efficient oxygen reduction reaction. Int. J. Hydrogen Energy. 43(42): 19492-19499. https://doi.org/10.1016/j.ijhydene.2018.08.201

Yang, S., X. Mao, Z. Cao, Y. Yin, Z. Wang, M. Shi, et al., 2018. Onion-derived N, S self-doped carbon materials as highly efficient metal-free electrocatalysts for the oxygen reduction reaction. Appl. Surf. Sci. 427: 626–634. https://doi.org/10.1016/j.apsusc.2017.08.222

Yuan, H., L. Deng, Y. Qi, N. Kobayashi and J. Tang. 2014. Nonactivated and activated biochar derived from bananas as alternative cathode catalyst in microbial fuel cells. Sci. World J. 1–8. https://doi.org/10.1155/2014/832850

Zan, Y., Z. Zhang, H. Liu, M. Dou and F. Wang. 2017. Nitrogen and phosphorus co-doped hierarchically porous carbons derived from cattle bones as efficient metal-free electrocatalysts for the oxygen reduction reaction. J. Mater. Chem. A. 5(46): 24329–24334. https://doi.org/10.1039/C7TA07746G

Zhang, J., C. Zhang, Y. Zhao, I.S. Amiinu, H. Zhou, X. Liu, et al., 2017. Three dimensional few-layer porous carbon nanosheets towards oxygen reduction. Appl. Catal. B. 211: 148–156. https://doi.org/10.1016/j.apcatb.2017.04.038

Zhang, F., J. Miao, W. Liu, D. Xu and X. Li. 2019. Heteroatom embedded graphene-like structure anchored on porous biochar as efficient metal-free catalyst for ORR. Int. J. Hydrogen Energy. 44(59): 30986–30998. https://doi.org/10.1016/j.ijhydene.2019.09.239

Zhang, J.J., Y. Sun, L.K. Guo, X.N. Sun and N.B. Huang. 2021. Ball-milling effect on biomass-derived nanocarbon catalysts for the oxygen reduction reaction. ChemistrySelect. 6(24): 6019–6028. https://doi.org/10.1002/slct.202100752

Zhang, Jian, Jingjing Zhang, F. He, Y. Chen, J. Zhu, D. Wang, et al., 2021. Defect and doping co-engineered non-metal nanocarbon ORR electrocatalyst. Nano Micro Lett. 13(1): 65. https://doi.org/10.1007/s40820-020-00579-y

Zhou, H., J. Zhang, J. Zhu, Z. Liu, C. Zhang and S. Mu. 2016a. A self-template and KOH activation co-coupling strategy to synthesize ultrahigh surface area nitrogen-doped porous graphene for oxygen reduction. RSC Adv. 6(77): 73292–73300. https://doi.org/10.1039/C6RA16703A

Zhou, H., J. Zhang, I.S. Amiinu, C. Zhang, X. Liu, W. Tu, et al., 2016b. Transforming waste biomass with an intrinsically porous network structure into porous nitrogen-doped graphene for highly efficient oxygen reduction. Phys. Chem. Chem. Phys. 18(15): 10392–10399. https://doi.org/10.1039/C6CP00174B

Zhou, L., P. Fu, X. Cai, S. Zhou and Y. Yuan. 2016c. Naturally derived carbon nanofibers as sustainable electrocatalysts for microbial energy harvesting: a new application of spider silk. Appl. Catal. B. Env. 188: 31–38. https://doi.org/10.1016/j.apcatb.2016.01.063

Zhou, L., P. Fu, D. Wen, Y. Yuan and S. Zhou. 2016d. Self-constructed carbon nanoparticles-coated porous biocarbon from plant moss as advanced oxygen reduction catalysts. Appl. Catal. B. Env. 181: 635–643. https://doi.org/10.1016/j.apcatb.2015.08.035

Zhou, Y., L. Yan and J. Hou. 2022. Nanosheets with high-performance electrochemical oxygen reduction reaction revived from green walnut peel. Molecules. 27(1): 328. https://doi.org/10.3390/molecules27010328

Chapter 9

Core-Shell Catalysts for Oxygen Reduction Reaction in Acidic Medium

Ildefonso Esteban Pech Pech[1] and Andrés Godínez García[2]*

[1]Centro de investigación en Corrosión,
Universidad Autónoma de Campeche.

[2]Tecnológico Nacional de México Campus Tlalnepantla.
Departamento de Ciencias Básicas,
División de Estudios de Posgrado e Investigación.

9.1 INTRODUCTION

Greenhouse gases have increased since the mid-20th century due to the heavy reliance on fossil fuels for power generation. Cars use petroleum derivatives, and electronic devices mostly run on energy from thermoelectric plants that use natural gas or coal. Therefore, the amount of CO_2 in the atmosphere continues to increase; as a consequence, according to NASA data, in 2020, the planet's average temperature was 1.2°C. The 2015 Paris agreement establishes that the temperature must be kept below 1.5°C to avoid a catastrophe for the global ecosystem, with the disappearance of many species of plants and animals. The amount of CO_2 must be reduced and energy obtained from clean sources to prevent the global temperature from exceeding 1.5°C, as well as improve the technologies that can

*For Correspondence: Email: andres.gg@tlalnepantla.tecnm.mx

use this energy in automobiles and electronic devices. In the continuing search for greener energy sources, lithium-ion batteries and hydrogen fuel cells are two technologies of public interest to achieve this goal.

A key driver for interest in H_2 is its use as an energy source and storage medium, which also finds uses in transportation. Different types of fuel cells differ in the electrolyte they use. Table 9.1 shows the characteristics that distinguish them. Each type of fuel cell tends to be more appropriate for certain applications. For example, polymeric electrical membrane fuel cells have proven to be suitable for automobiles, while molten carbonate fuel cells appear more suitable for gas turbines.

Hydrogen, when used in light-duty fuel cell electric vehicles (FCEVs) is a zero CO_2 emission alternative fuel that will play a vital role in the future of the automotive industry. Fuel cells, particularly PEMFCs, are expected to provide a viable long-term solution to improve energy efficiency in automobiles, decrease environmental pollution and contribute to a more sustainable civilization. Therefore, extensive efforts have been put in researching and developing PEMFC technology. Although, this last technology has achieved significant progress, some challenges still hinder its practical use and commercialization. One of them is to improve all the components, mainly the catalyst, made of platinum, which is one of the scarcest metals in nature.

Both PEMFC electrodes, anode and cathode, use Pt as electrocatalyst. From them, the cathode is the one that presents the most significant challenge due to the slow kinetics of oxygen reduction reaction (ORR). The ORR is the most significant limiting factor for this device's performance due to the electrocatalyst's low stability and catalytic activity of existing materials. Therefore, the development of new catalysts is vital to improving the performance of such electrochemical energy devices. Core-shell nanostructured electrocatalysts were recently developed (Gawande et al., 2015) and applied in various electrochemical energy devices (Li et al., 2012; Wang et al., 2017a; Jun et al., 2017; Jiang et al., 2018). However, in this chapter, the central focus is on the current state of core-shell nanostructured electrocatalysts for the ORR in PEM fuel cells.

The following subsections describe the main characteristics of PEMFC and how the catalyst plays a fundamental role in future commercialization.

9.2 PEMFC

A PEMFC consists of two electrodes separated by an electrolyte. Oxygen passes over one electrode and hydrogen over the other. When hydrogen is ionized, it loses an electron, and when this happens, both (hydrogen and electron) take different paths toward the second electrode. The hydrogen migrates to the other electrode through the electrolyte while the electron migrates through a conductive material. This process produces water, electric current, and useful heat (Fig. 9.1). Fuel cells are 'stacked' into a multi-layered ensemble to generate usable amounts of current.

Table 9.1 Fuel Cells classification (Boudghene Stambouli and Traversa, 2002; Kamarudin et al., 2009; Kirubakaran et al., 2009)

Fuel Cells	Temperature	Electrolyte	Electrodes	Catalyst	Reaction at the anode	Reaction at the cathode
Proton exchange membrane or polymer electrolyte membrane fuel cell (PEMFC)	Its operative temperature is below 100°C	Uses a water-based, acidic polymer membrane as the electrolyte	Carbon	Platinum	$H_2 \rightarrow 2H^+ + 2e^-$	$O_2 + 4H^+ + 4e^- \rightarrow 2H_2O$
Direct methanol fuel cell (DMFC)	Its operative temperature is below 100°C	Uses a polymer membrane	Carbon	Platinum–ruthenium catalyst on its anode	$CH_3OH + H_2O \rightarrow CO_2 + 6H^+ + 6e^-$	$3/2O_2 + 6e^- + 6H^+ \rightarrow 3H_2O$
Alkaline fuel cell (AFC)	65–220°C	Mobilized or immobilized potassium hydroxide in asbestos matrix	Platinum	Platinum	$H_2 + 2OH^- \rightarrow 2H_2O + 2e^-$	$O_2 + 2H_2O + 4e^- \rightarrow 4OH^-$
High temperature PEMFC (HT-PEMFC)	It operates up to 200°C	A mineral acid-based system	Carbon	Platinum	$H_2 \rightarrow 2H^+ + 2e^-$	$O_2 + 4H^+ + 4e^- \rightarrow 2H_2O$
Phosphoric acid fuel cell (PAFC)	205°C	Immobilized liquid phosphoric acid in SiC carbon	Consists of an anode and a cathode made of finely dispersed platinum	Platinum	$H_2 \rightarrow 2H^+ + 2e^-$	$O_2 + 4H^+ + 4e^- \rightarrow 2H_2O$
Molten carbonate fuel cell (MCFC)	Operating at temperatures of about 650°C	Immobilized liquid molten carbonate in LiAlO$_2$	Nickel and nickel oxide	Electrode material	$H_2 + CO_3^{2-} \rightarrow H_2O + CO_2 + 2e^-$ $CO + CO_3^{2-} \rightarrow 2CO_2 + 2e^-$	$O_2 + CO_2 + 4e^- \rightarrow 2CO_3^{2-}$
Solid oxide fuel cell (SOFC)	600–1000°C	Uses a solid ceramic such as perovskites	Perovskite and perovskite/metal cermet	Electrode material	$H_2 + O^{2-} \rightarrow H_2O + 2e^-$ $CO + O^{2-} \rightarrow CO_2 + 2e^-$ $CH_4 + 4O^{2-} \rightarrow 2H_2O + CO_2 + 8e^-$	$O_2 + 4e^- \rightarrow 2O^{2-}$

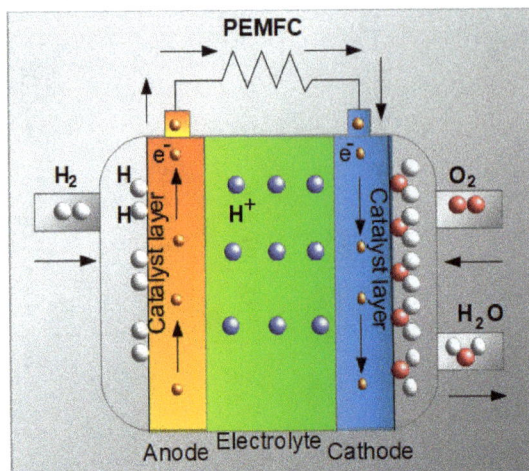

Figure 9.1 Operation of proton exchange membrane or polymer electrolyte membrane fuel cell (PEMFC).

PEM fuel cells are devices that operate with high energy efficiencies compared to internal combustion engines and generate relatively high energy densities. If hydrogen (H_2) is used as fuel, PEMFCs generate zero CO_2 emissions. They can operate at low temperatures, below 80°C, unlike an internal combustion engine with quick starts and quick responses to load changes. PEMFCs have already shown their effectiveness in almost all technological applications, including buses, automobiles, ships, locomotives, underwater vehicles, cogeneration generation systems, portable power systems, and backup power systems.

The current state of fuel cell systems indicates that durability and cost remain the main challenges to overcome. The US Department of Energy's 2020 goal for systems using fuel cells is \$40/kW cost with 65% efficiency at peak power and 12.5 g Pt. Cost analysis of PEMFC cells identifies that the catalyst contributes 41% to the total cost compared to bipolar plates, membrane, gas diffusion layer, electrodes, and gaskets. Therefore, current research focuses on improving the cathode's catalyst layer. The ORR reaction is slow and Pt/C loadings typically required are 0.35 mg Pt/cm^2. Reducing the total Pt loading in PEMFCs is a research field. Here novel catalyst structures and alloys that can enhance the catalytic activity of ORR are studied to reduce the total Pt loading in PEMFCs, thus reducing the total PEMFC cost (Schmidt et al., 2017). The electrocatalyst activity can be improved by increasing the surface area to favor the accessibility of reagents to the active sites to obtain better mass activity (mA/mg Pt). In addition, the catalyst is subject to charging cycles between 0.60 V and 0 V, water generation as a function of current density, variable relative humidity, and voltage losses due to startup and shutdown. Each of these conditions degrades the performance of the catalyst layer over time. Despite this, it is expected to find electrocatalytic materials using loadings of less than 0.1 mg Pt per kW by 2050. This last would put PEMFCs on par with current Pt loadings used in catalytic converters (approx. 1 g for gasoline and between 8 and 10 g for diesel (Pollet et al., 2019).

9.2.1 Limitations

There are three main limitations of PEMFCs, the cost to obtain hydrogen, the storage of hydrogen that must be greater than 81 g L^{-1} (the density of liquid hydrogen is 70.8 g L^{-1}), and the high cost of the Pt PEMFC catalysts. The catalysts must have high activity and stability to carry out the hydrogen oxidation reaction (HOR) at the anode and ORR at the cathode. In addition, they must be tolerant to small contaminating impurities in the feed gases. Until today, commercialized catalysts are made with Pt, which is scarce and expensive and requires improved performance in terms of activity and stability. In addition to replacing Pt with other lower-cost metals, extensive research is being carried out to reduce the amount of Pt in catalysts and improve their performance. The main strategies to reduce Pt load consist of synthesizing catalysts with conductive carbon support, creating nanostructured materials and alloys of Pt with other cheaper metals, and creating core-shell nanostructures. Where the core is made up of more abundant materials and Pt in the shell, these strategies are effective and have been able to reduce the cost of the catalyst and improve its performance. However, there is still much room for improvement to reduce the cost of PEMFCs so they can be massively commercialized (Banham et al., 2015; Shao et al., 2016).

Therefore, finding a catalyst with best stability, high electrical conductivity, moderate surface adsorption, and catalytic activity is the prime objective. To achieve this, it has been observed that the composition, morphology, and structure substantially affect the characteristics sought.

Stability: The catalyst environment in PEMFCs is aggressive due to its acidity and high temperature, as well as the presence of water and oxygen, which cause corrosion of the catalyst supports and partial dissolution of the catalysts. One more peculiarity is the high cathodic potential in the presence of O_2 and H_2O, which also favors oxidation and dissolution. Therefore, a better catalyst must have better corrosion and chemical resistance than commercial catalysts.

High electrical conductivity: The catalyst material must have meager electrical resistance, and high electrical conductivity since fuel-cell reactions involve electron transfer processes. The support materials must have a high electrical conductivity to avoid energy losses due to heat release.

Moderate surface adsorption: If the adsorption forces of the reactants on the active sites of the catalyst are too strong, it won't be easy to release the intermediate products, which become final products, leading to a slow reaction rate. On the contrary, if they are too weak, only small amounts of reactants will be absorbed, thus making the reaction rate slow. Therefore, the surface adsorption energy of reagents on active sites is an essential property for electrocatalysts to be selective and have high catalytic activity. Again, the adsorption capacity of the reactive will be significantly affected by the catalyst's morphology, composition, and structure, which cause differences in reaction mechanisms and rates.

Catalytic activity: A high-performance catalyst must have many active sites per unit of volume with adsorption energy conducive to carrying out a reaction. A catalyst with high catalytic activity is because it has many of these sites for the reactants and intermediates to be converted into products. Shell and core nanostructured catalysts typically possess high active site densities, resulting in higher catalytic activities.

The catalyst morphology, composition, and structure affect all of these characteristics, thus, they are the parameters that need to be optimized to get a catalyst with maximum performance.

9.2.2 Oxygen Reduction Reaction

The paths followed by ORR involve a transfer of two or four electrons associated with elementary steps (Fig. 9.2) (Wroblowa et al., 1976).

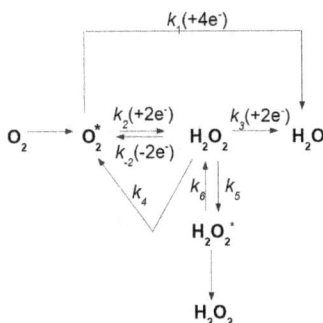

Figure 9.2 Possible paths followed by the ORR.

The four-electron pathway in acidic media is as follows (Eq. 1):

$$O_2 + 4H^+ + 4e^- \rightarrow 2H_2O \text{ (acidic media)} \tag{1}$$

The two-electron pathway can proceed in two different ways (Eq. 2 and Eq. 3):

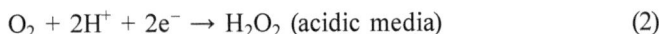

$$O_2 + 2H^+ + 2e^- \rightarrow H_2O_2 \text{ (acidic media)} \tag{2}$$

Or

$$H_2O_2 + 2H^+ + 2e^- \rightarrow 2H_2O \text{ (acidic media)} \tag{3}$$

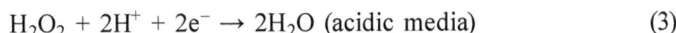

In theoretical studies based on density functional theory (DFT), it was found that molecular oxygen can follow a dissociative mechanism that generates the O* intermediate or an associative mechanism that generates the OOH* intermediate species, which subsequently reacts to produce water. Electrocatalysts sometimes follow the 'indirect' two-electron transfer pathway, which generates hydrogen peroxide $(H_2O_2)^*$, an intermediate species that dissociates into 2OH* and reacts to form H_2O. Hydrogen peroxide species reduce the efficiency of the reaction and increase the rate at which catalyst degradation occurs (Janik et al., 2009). In some studies, the rate-limiting step was proposed to be the breaking of the O–O bond in the oxygen molecule following a dissociative mechanism, but this may depend on the type of element present on the surface. However, the free energy diagram obtained by Norskov shows that in the direct pathway, the associative mechanism dominates, and the barrier where the protonation of O* or OH* occurs has lower energy, which is the rate-limiting step (Nørskov et al., 2004; Wu et al., 2020).

From DFT theoretical studies, it has been discovered that the center of the d-band of transition metals is a parameter that can be correlated with the adsorption energy of the species produced during the reaction. (Hammer and Norskov, 2000). Some intermediate species generated exhibit a linear correlation with the d-band center parameter (Hammer and Norskov, 2000). A volcano-like correlation has been found between the center of the d-band and ORR activity (Stamenkovic et al., 2006). This last study leads us to the fact that the activity of electrocatalysts can be improved ORR by optimizing the electronic structure, that is, the d-band center parameter of the catalyst materials. In the case of Pt, it has been observed that this parameter is altered by the particles' structure, size, composition, and morphology (Shao et al., 2011; Huang et al., 2015). In fact, during the past two decades, intense research efforts have been carried out to synthesize nanometric Pt alloys by controlling these material parameters and studying the effects on ORR electrocatalysis (Wu et al., 2020).

9.2.3 Traditional Catalysts

To date, the electrocatalysts with the highest performance for ORR continue to be Pt-composite nanoparticles. However, due to the scarcity of Pt, the cost of the catalyst is very high, which means that the PEMFC's massive commercialization has not yet been carried out (Wu and Yang, 2013; Shao et al., 2016).

A wide variety of other materials have been investigated to meet this challenge, including metal chalcogenides (Morawa Eblagon and Brandaõ, 2015), oxynitrides (Chisaka et al., 2017), non-noble metal-based materials (Chao et al., 2013) and the alloys of Pd (Fouda-Onana et al., 2009), among others materials. However, the performance of Pt-based alloys stands out among all these materials, and they remain the most promising candidates for commercial applications.

In this chapter, we focus on mainly summarizing electrocatalysts with core-shell type structures under acidic conditions, as they are the most promising to achieve the goal set by the Department of Energy of reducing the amount of platinum to less than 0.1 mg Pt per kW by 2050 (Wu et al., 2020).

9.2.4 Core-Shell Catalysts

A significant challenge often faced by chemists working on new electrocatalysts for ORR is the development of robust, selective, active, low-cost, and environmentally friendly catalytic materials. Core-shell nanostructured materials supported on a porous and conductive material seem to be the most suitable to face this challenge (Jiang et al., 2018).

The development of electrocatalysts with a core-shell type structure has the following advantages: (1) by using a core as a support, the surface structure will be tuned according to the specific characteristics that are desired in terms of surface area, porosity, and surface energy which gives as a result in catalysts with better performance, (2) synergistic effects are frequently generated between the shell and the core, thus achieving more excellent selectivity, yield, and efficiency in catalytic applications, (3) the properties of the core and the shell can be combined to obtain the most suitable properties according to the application.

In core-shell catalysts, the catalytic activity of an electrocatalyst depends on synergistic interactions between the core and shell. There are three main effects that play essential roles in the final catalytic activity of a specific material: (a) group effects, which originate from different atomic groups on the surface and determine the adsorption of the material; (b) effect of the ligand, which is due to the interaction between the cover and the core that affects the band-structures and the charge transfer between the components, and finally, (c) geometric effects that originate from the different reactivity of the atoms found on the surface and the three-dimensional structural constraints of the material (Gawande et al., 2015).

The activity of these materials for a reaction can be controlled by making adjustments to their structure, size, and composition (Peng and Yang, 2009). Besides the ORR, core-shell catalysts have been used in alcohol oxidation and oxygen evolution reactions (OER) (Choi et al., 2013; Hwang et al., 2013).

The most studied core-shell type nanostructures are those where nanoparticles of non-precious metals, such as Ni, Fe, or Co are coated with a layer of precious metals, such as Au, Pt, and Pd. However, methods have also been developed to coat non-precious metal alloys and also generate precious metal alloys on the surface (Gawande et al., 2015).

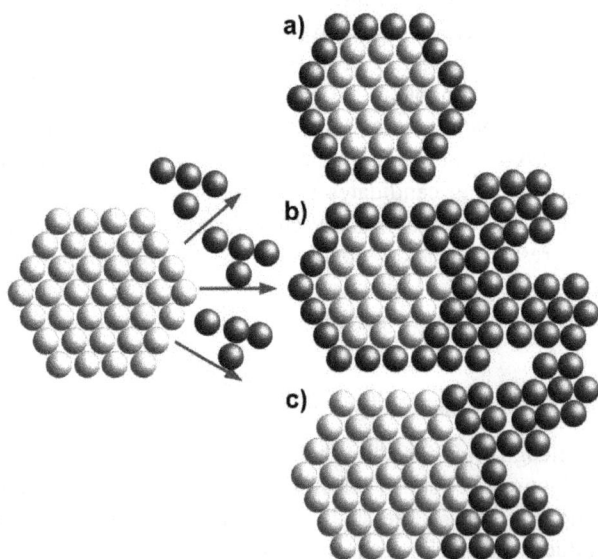

Figure 9.3 Ways in which shells on nanoparticles can grow (Yang, 2011).

9.2.4.1 Strategies for Synthesizing Core-Shell Catalysts

The catalyst's structure, morphology, and composition have been shown to affect its performance. The synthesis methods involve the characteristics mentioned above. Therefore, they are vital to obtaining an optimal structure, morphology, and composition to improve the stability and catalytic activity in the PEMFC reactions.

In this section, a summary of the methods for the synthesis of Pt-based core-shell nanostructures is presented since they are the most promising to be used as highly active electrocatalysts for ORR in PEMFCs (Yang, 2011; Oezaslan et al., 2013; Jiang et al., 2018).

To date, information in the literature on the synthesis of well-formed multi-metallic core-shell nanostructures with sizes less than 10 nm is scarce. In principle, the synthesis of core-shell multi-metallic nanoparticles is thermodynamically favored. Synthesis is possible since heterogeneous nucleation of the second metal component on the nanoparticle core has a lower critical energy barrier than homogeneous nucleation. Three different main types of nanostructures are formed that depend on the overall excess energy of the system and are closely related to the terms of surface energy and strain energy due to lattice mismatch at the interface. The ways in which the nanostructures can grow are: (a) layer by layer, (b) island on the shell, and finally, (c) in the form of an island without forming a homogeneous shell (Fig. 9.3) (Yang, 2011).

9.2.5 Chemical Methods

9.2.5.1 Deposition of a Metal on Preformed Colloidal Nanoparticles

A widely used method to synthesize bimetallic nanoparticles with a core-shell structure is to deposit a metal on preformed suspended nanoparticles. Suppose the nanoparticles in colloidal suspension do not react with the shell precursor cations, and the deposited metal atoms can remain stable on the surface of the core. In that case, a metal layer can be deposited on metal nanoparticles (1) by reducing metal cations (2) on the nanoparticles in colloidal suspension. Assuming that the suspended nanoparticles do not have potent ligands on their surface that prevent the formation of the shell. The problem that Yuan Wang and Naoki Toshima solve with their method is to avoid the redox reaction between the metallic cations that will form the cover and the metallic core if the metallic cations have a higher redox potential than the core. By overcoming this issue, it is possible to have more control over the characteristics of the formed shell. Since the surface of a small particle is composed of metal atoms that occupy a large proportion of the total metal atoms in the particle, the fate of the metal cations formed by the oxidation of the metal nucleus through the redox reaction mentioned above cannot be predicted (Wang and Toshima, 1997).

9.2.5.2 Hydrogen-Sacrificial Protective Method

In order to avoid this problem and provide a reliable preparation method to establish a controllable shell on the core, Yuan Wang and Naoki Toshima proposed a hydrogen-sacrificial protective strategy. The strategy is based on the properties of noble metals, such as Pd and Pt, among others, to adsorb hydrogen and split it to form a metal-H bond on the surface. Hydrogen atoms adsorbed on noble metals have a powerful reducing capacity due to their low redox potential. Therefore, the adsorbed hydrogen is oxidized, and its electrons reduce the noble metal cations.

That is to say, when metal cations have a higher redox potential like Pt^{2+} or Pt^{4+} in contact with metal nuclei like Pd with adsorbed hydrogen, it can be expected that it is the hydrogen atom adsorbed on the surface that provides its electron and leaves the particle, as a proton, and although palladium has a lower redox potential than Pt, it will not dissolve to leave the nucleus as a cation (Wang and Toshima, 1997). Besides, the interactions between the protective agent chain, which can be PVP, and the suspended nanoparticles are weak interactions, which do not prevent the metal atom formed in the solution from covering the surface of the nanoparticles because it is not difficult to move its protective layer. Thus, the hydrogen-sacrificial protective method is reliable for establishing a controllable metal skin on PVP-protected homogeneously dispersed small metal particles for various systems (Wang and Toshima, 1997).

9.2.5.3 The Three most Used Methods for the Growth of Core-Shell Nanostructures

Figure 9.4 shows the three most used methods for the growth of core-shell nanostructures. In the first, using the underpotential deposition (UPD) technique, a Cu monolayer is formed on the core (Fig. 9.4a). This Cu layer is sacrificed; by galvanic replacement, Cu atoms give up their electrons to Pt ions dissolved in an electrolytic solution. The Pt atoms formed are deposited on different metallic nanoparticles forming monolayers of Pt on their surface. In the second method, nanostructures of two or more elements, including Pt, are formed; the most straightforward atoms to oxidize are electrochemically removed by applying a potential difference. Thus, as Pt is one of the most stable metals, the nanoparticles will contain a surface rich in Pt (Fig. 9.4b).

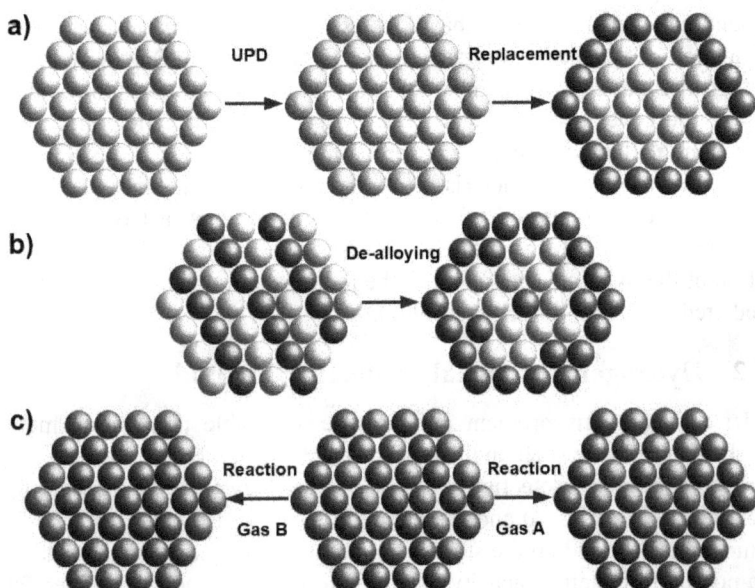

Figure 9.4 Methods for the synthesis of core-shell nanoparticles (Yang, 2011).

The method can be applied to a wide variety of multi-metallic systems since the preferential dissolution of a metal is based on the redox potentials of the metals.

In the third method, Fig. 9.4c, reactive gases such as NO, CO, H_2, and O_2 are used. Gases, due to their electronic properties, will adsorb with greater intensity on some metals than others; this helps pull them to the surface. Therefore, the surface layer of alloy nanoparticles will be richer in some of the metals that compose it (Jiang et al., 2018).

For example, for a nanostructure composed of a Pd-Rh alloy, exposure to an oxidizing gas of NO (at ambient pressures) results in an Rh-rich nanoparticle on the surface (Tao et al., 2008). However, if the alloy is exposed to CO, the Rh atoms can return to the nanoparticle's core. Therefore, using reactive gases is another alternative to designing core-shell type nanostructures (Yang, 2011).

9.2.5.4   Modification of the Traditional Underpotential Deposition (UPD) Method

Using core-shell nanoparticles composed of a metallic core (can be Au, Pd, Ru, among others) and a Pt monolayer on nanometric materials is the best option for each Pt atom to participate in the ORR catalytic process. Synthesis methods to obtain core-shell nanoparticles must be simple to be used in practical applications.

Wang et al. proposed a modification of the traditional underpotential deposition (UPD) method to make it simpler (Wang et al., 2020). Traditionally, Pt-shell nanoparticles are obtained by the underpotential deposition (UPD) method with Cu as mediator and displacement of the Cu mediator by Pt (Fig. 9.4a). The disadvantages of this method are:

- High-precision potential control is required for Cu deposition through a potentiostat.
- Can only be used for gram-scale production of nanostructures.
- The fixed potential to carry out the underpotential deposition of copper does not consider the influence of the nanoparticle characteristics that form the core, such as the particle size distribution and the difference in surface structure, so it is unable to achieve a homogeneous monolayer of Cu. Therefore also non-homogeneous Pt monolayers are obtained (Shao et al., 2016).

Several strategies have been investigated to address these issues, but the use of these methods in a practical way is still limited by the requirements of specific substrates, electrochemical equipment, and reducing and stabilizing agents (Lai et al., 2012; Wu and Yang, 2013; Wang et al., 2020).

The synthesis of Pd@Pt nanoestructures exemplifies Wang's method with Pd as the core and Pt as the shell. The steps are the growth of the Cu mediator on the Pd nanoparticles, and subsequently, the Cu is galvanically displaced by Pt.

Continuously stirring the suspended Pd/C particles will come into contact with the Cu wire. A galvanic cell will be established, where the cathode is the Pd nanoparticles and the nearby Cu atoms (monolayer), the anode is the Cu wire, and the electrolyte is a mixture of H_2SO_4 (0.5 M) and $CuSO_4$ (0.01 M). At that moment of contact, the Pd nanoparticles obtain a weak current from the oxidation of the Cu wire to Cu^{2+} (anodic reaction). In turn, this weak current will lead to Cu^{2+} deposition, forming a Cu monolayer on the surface of the labeled Pd nanoparticles (cathodic reaction). Due to the difference in redox potentials occurring in adjacent regions, the dissolution of Cu with low redox potential only drives the deposition of the Cu monolayer with higher redox potential. A disadvantage of the method is that it requires tens of hours of stirring to form Cu monolayers on Pd nanoparticles. Next, the Cu wire is removed, and a 10.0 mM K_2PtCl_4 solution previously saturated with N_2 is slowly added to the previous suspension. This step requires only a few minutes, and the suspended nanoparticles are filtered and dried. After this process, the core-shell nanostructures with Pt as the shell can be obtained (Wang et al., 2020).

9.2.5.5   The Electrochemical Deposition Method

The method generally consists of depositing nanoparticles supported on some conductive carbon, such as Vulcan carbon, and the powder obtained is, in turn, deposited on a conductor, such as Toray paper. The Toray paper with the deposited nanoparticles is introduced into an electrochemical cell and works as a working electrode while a Pt mesh is used as a counter electrode. A potential difference is applied between both electrodes so that the anodic dissolution of Pt occurs. The cations of Pt will be transferred from the platinum counter electrode to the working electrode containing the metal nanoparticles where the shell will be formed.

The Aucore-Ptshell/C system exemplifies this method. Toray paper is impregnated with Au/C (gold nanoparticles supported on Vulcan carbon XC72 dispersed in a mixture of 96% ethanol and Teflon in a ratio of 2.5:1) and immersed in an electrochemical cell containing an electrolyte with a pH of ~0.3–0.4, the reference electrode which is a saturated calomel electrode (SCE) and the platinum mesh. The potential of the working electrode, which is the Toray paper impregnated with Au/C, is subjected to a potential difference more significant than the corrosion potential of Pt (~+1.18 V against NHE) using the cyclic voltammetry technique at a rate of 50 mV/s. The potential applied to the working electrode must be between −0.35 and +1.0 V with reference electrode SCE. The deposition occurs in 24 hours. A change in Au coloration to gray should be observed at this time (Pitois et al., 2010).

Zhang et al. used the method to form Pt monolayers on Au/Ni and Pd/Co cores. The synthesis involved impregnating the Vulcan X-72C carbon with mixed solutions of noble and non-noble metal salts. A suspension formed by carbon and metal salts ($NiCl_2$, $AuCl_3$, K_2PtCl_4, $PdCl_2$, and $CoCl_2$) was reduced with H_2, using a temperature in the 600–850°C range, which depends on the type of metal to be treated. The heat treatment time varied between one and two hours to segregate the core alloy's noble metal to the surface, either Au or Pd. After treatment, a monolayer of Cu was deposited by the UPD method and replaced by Pt using the galvanic replacement method. In this way, the Pt core-shell nanostructures were formed; Pt as shell and the Au/Ni and Pd/Co alloys as a core (Zhang et al., 2005).

By repeating the same process of Cu monolayer deposition and galvanic replacement by Pt, the number of Pt monolayers can be controlled.

Although it is a simple method, the metals that constitute the nucleus must be thoroughly investigated since their segregation properties, and electronic effects induce deformations on the Pt monolayer.

Synthesizing this type of structure has the advantage that if the noble metal cover is selected correctly, the activity of a Pt monolayer can be increased through electronic and/or geometric effects that induce tension on the Pt monolayer.

9.2.5.6 General Considerations to Apply the Methods of De-alloying, Segregation, and Galvanic Replacement

Although current methods allow us to have reasonable control over the structure, there is still a wide field to investigate in this area since better synthesis methods are required that facilitate control over nanoparticle structure to tune the electronic and geometric factors and enhance catalytic activity.

To correctly apply the synthesis methods in the design of nanometric structures with optimized surface properties, the following general characteristics of each method must be taken into account:

- The de-alloying method requires selectively dissolving of the component with the lowest reduction potential of Pt alloys using electrochemical corrosion, so that a metal with a lower reduction potential than Pt must always be present (Gong et al., 2020). Shell thickness and porosity can be optimized by changing the size and composition of the initial alloy precursors, as well as the de-alloying conditions, such as solution acidity, temperature, and applied potential (Mani et al., 2008; Li et al., 2014; Strasser and Kühl, 2016).

- Surface segregation method: This method uses bimetallic nuclei containing Pt and some other metal. Through thermal treatment in a controlled atmosphere, using gases like H_2 and CO (Mayrhofer et al., 2009; Wang et al., 2010), a driving force is generated that causes the segregation of Pt to the surface. This driving force is due to the system's surface energy reduction (Ma and Balbuena, 2008). When such an alloy is heated, the element with the lowest surface energy tends to diffuse towards the nanoparticle's surface (Sasaki et al., 2011).

- The galvanic replacement method: As in the de-alloying method, it starts from the reduction potentials difference of the metals that form the nanostructure. Noble metals, such as Pt, have a higher reduction potential than non-noble metals, such as Ni, Co, and Cr, among others. Therefore, if non-noble metal nuclei and Pt cations are suspended in an aqueous solution, the nuclei will transfer electrons to the Pt cations. Thus, some nuclei atoms oxidize and dissolve while the cations Pt are reduced to form the shell (Xia et al., 2013a). A disadvantage of this method is oxides that form on the core surface when the process is carried out under ambient conditions. The presence of oxides causes the formation of independent nanoparticles of the metals that are supposed to form the shell. While if the process is carried out in an inert atmosphere at all times since the formation of the core, then it is more likely that the formation of the uniform shell will occur. Zhang et al. reported the observations mentioned above in studies carried out with the system of Ni nanoparticles covered with a FePt alloy as the shell (Zhang et al., 2014).

The galvanic replacement method is simple and versatile for synthesizing core-shell nanostructures, but dissolving part of the core atoms often leads to a deteriorated core-shell interface. Another disadvantage occurs when the difference in reduction potentials between the two metals is too significant, which is common when the metals that make up the nanostructure are a combination of noble and non-noble metals. If such a system is present, the galvanic replacement will control growth. The problem is that overgrowth does not always lead to core and shell structures because it can also follow the islands' growth mode (Alinezhad et al., 2019; Gong et al., 2020).

According to studies by Xia and colleagues, the island-like growth may be due to high interfacial tension or slow surface diffusion. These studies showed that the surface diffusion rate of Pt atoms is essential in determining the type of growth (Xia et al., 2013b). The experiment was performed using Pd nanocubes. They found that Pt islands were formed on Pd nanocubes at a relatively low reaction temperature, and as the temperature increased, the coating improved to a uniform shell at high temperatures. The island-like structure is the path followed to avoid an increase in interfacial energy due to lattice mismatch between the two metals (Gong et al., 2020).

A second element is introduced in the core to reduce the lattice mismatch, reduce the interfacial strain, and favor the core-shell growth mode. In order to prove this, Lentijo-Mozo et al., 2015 synthesized a Co rod and tried to grow an Au film on it, but gold islands were formed instead. In a second attempt, the rod was subjected to a treatment where an Sn buffer film was grown, and the gold film was successfully grown on it (Gong et al., 2020).

This way, a new possibility is opened to obtain cheaper catalysts using non-precious metal cores. Until now, one of the best catalysts that have been obtained with core-shell structures is the Pd catalyst with an ultra-low Pt film (0.025 mg Pt/cm^2). The Pd@Pt core-shell system was presented at high current densities with low voltage loss. The amount of Pt used is so low that only 2 g would be required for an 80 kW light vehicle, which is equivalent to the amount of catalyst used in a catalytic converter of combustion engine vehicles (Kongkanand et al., 2016). However, the core Pd cost cannot be ignored, as the stability of the Pt monolayer is as essential as the catalytic activity. Finding cheaper and more stable core materials that increase the stability of the Pt film is one of the leading research objectives in this field.

9.2.5.7 The Pulse Electrochemical Deposition (PED)

Tian et al. tried to reduce the cost of the materials that form the core, and under the principle that each Pt atom participates in the reaction proposed TiNiN nanostructures with a Pt shell (Tian et al., 2016). Transition metal nitrides were investigated to replace the carbon-based supports of noble metal catalysts in PEMFCs. These materials have the characteristics that they are thermally stable, good electrical conductors as they are metallic, have high hardness as they have covalent bonds, and are resistant to corrosion under fuel cell operating conditions. Another feature is that Pt nanoparticles adhere more strongly than to carbon support, which increases stability and facilitates electron transfer when ORR is catalyzed. Therefore, a Pt shell on these materials is expected to be more stable under PEMFC operating conditions.

The formation of the TiNiN@Pt core-shell catalyst was carried out by the pulsed electrochemical deposition (PED) method. The procedure consisted of nitriding an ammonia complex of Ti and Ni in an NH_3 atmosphere at a temperature of 700°C to obtain TiNiN nanoparticles (NPs). Several Pt monolayers were deposited on the nanoparticles of material obtained by the PED method. The method consisted of placing 5 mg of TiNiN NPs in 1 ml of Nafion/ethanol solution (0.25 wt% Nafion) to form a solution using ultrasound, and 4 µl of this suspension was taken and deposited on a vitreous carbon electrode (0.196 cm^{-2}). This electrode was immersed in a polyvinylpyrrolidone solution (PVP, 50 mmol), $H_2PtCl_6 \cdot 6H_2O$ (5 mmol), 0.4 M H_2SO_4, and 0.1 M Na_2SO_4 solution, and pulses were applied. The connection time was 0.003 s and the disconnection time was 0.03 s between each pulse (Tian et al., 2016).

The efficacy of the PED method was also demonstrated on other non-PGM substrates (Kongkanand et al., 2016; Gong et al., 2020).

9.2.6 Physical Methods

Physical methods are less widely used than chemical methods to form core-shell nanostructures. However, they can be an alternative to trying new ways of obtaining this type of material. Among the most common methods are physical vapor deposition (PVD), atomic layer deposition (ALD), and chemical vapor pulse deposition (CVPD). These methods feature that the substrate does not have to be immersed in an aqueous solution.

An overview of these methods follows.

9.2.6.1 PVD Method

The method is generally used to cover flat substrates. Esposito et al. used it to generate Pt-coated WC and W_2C films. The methodology they followed was first to coat a polycrystalline tungsten sheet by sputtering. Deposition was carried out using a heated Pt source in an XPS chamber. The coated substrate was placed in front of the Pt source for some time (Esposito et al., 2012), though with this method, it is challenging to obtain spherical core-shell nanoparticles.

9.2.6.2 ALD Method

Unlike the PVD method, the ALD is capable of three-dimensional coating substrates uniformly. With this method, it is possible to have reasonable control over the thickness of the film, and ultra-thin films can be produced on different types of substrates, such as microspheres and carbon nanotubes. Consequently,

WC@Pt core-shell particles were synthesized, extending the thin-film system to powder nanoparticles (Hsu et al., 2012).

9.2.6.3 CVPD Method

The chemical pulse vapor deposition (CVPD) method is exemplified in the study conducted by Seo et al. (2014). In this study, a layer of Pt is deposited using the CVPD method, which allows the selective deposition of Pt on the surface of Co/C particles. Vulcan carbon was impregnated with Co to obtain Co/C. Pt was subsequently deposited on the Co/C surface, using the CVPD method. The amount of Pt deposited on Co was controlled by the number of applied pulses. For this, $Pt(PF3)4$ vapor was introduced into a reactor containing Co/C at room temperature. Then, the material was thermally treated at 200°C for 1 h, in a reactor containing Co/C, using a flow of hydrogen-nitrogen to reduce the Pt precursor and deposit on the Co surface (Seo et al., 2014).

9.3 CORE-SHELL CATALYSTS FOR ORR

Different parameters are used to describe stability and catalytic efficiency in catalysts. The electrochemical active surface area (ECSA, $m^2/g_{catalyst}$), the catalytic activity in terms of mass (Mass activity or MA, $A/g_{catalyst}$), and the catalytic activity in terms of surface area (specific activity or SA A/m^2) are three key parameters typically used to describe the electrocatalytic characteristics in ORR catalysts. The ECSA allows to determine the number of catalytic sites available to carry out the oxygen reduction reaction. This first parameter is estimated from the hydrogen absorption/desorption region (0–0.3 V vs RHE) by cyclic voltammetry (CV) (Pech-Pech et al., 2018). On the other hand, the mass activity is usually estimated by linear sweep voltammetry (LSV) from the normalized current against the Pt loading used in the catalyst through the Koutecky-Levich equation, while the specific activity is calculated in the same way but is normalized against active area found in the ECSA (Pech-Pech et al., 2018; Shi et al., 2021). In addition, the stability of catalysts is usually estimated by accelerated durability test (ADT), which consists of measuring the variations of ECSA and MA after a given number of CV cycles (Wang et al., 2015; Zhang et al., 2021).

The design of nanoparticles with core-shell structures is one of the promising alternatives to reduce the cost of Pt-based catalysts used in the PEMFCs because these materials have shown higher activity and stability for the ORR in comparison with traditional Pt catalysts (Oezaslan et al., 2013; Sasaki et al., 2020; Shi et al., 2021). The high catalytic activity that exhibits this type of nanostructures has been mainly associated with the strain variations observed from the changes in the lattice constants (geometric or strain effect) and to the ligand effect associated with the transfer of electronic charge between the shell surface atoms and the inner core atoms (Hammer and Nørskov, 2000; Kitchin et al., 2004). In the next sections, we will discuss these effects in terms of Pd-based and Ag-based catalysts.

9.3.1 Pd@Pt Core-Shell Catalysts

Palladium-based core-shell nanostructures have been widely explored to catalyze the ORR due to the lower cost of Pd in comparison with Pt (Table 9.2); also, this type of nanostructures is demonstrated to improve the catalytic activity compared to monometallic Pt catalysts (Wang et al., 2009b; Xie et al., 2014; Xiang et al., 2018; He et al., 2021). In this regard, Xie, Shuifen et al. synthesized Pd nanocubes coated with a shell from one to six atomic layers of Pt and compared their electrocatalytic properties with those of Pt with the same morphology (Xie et al., 2014). The authors found that the key to controlling the atomic layers of Pt in the synthesis of these nanostructures is the slow injection (4.0 mL/h) of Pt precursor at high temperature (200°C) on these cubic Pd structures used as substrates for the surface deposition of Pt atoms. The importance of this reported methodology lies in the possibility of controlling the thickness of Pt in the core-shell structures, thus increasing the utilization efficiency of this metal, and consequently reducing the cost of the catalyst. Other strategies have been developed to achieve the goal of controlled deposition of Pt atoms in this type of nanostructures (Zhao et al., 2018; Xie et al., 2019). On the other hand, the electrochemical results revealed that among the synthesized catalysts by Xie Shuifen et al., the cubic nanostructure containing two to three atomic layers of Pt ($Pd@Pt_{2-3L}/C$) was the one that showed the highest current per unit of area or specific activity (1.6 times higher than that of cubic Pt structures); however, the highest mass activity (current per unit of mass) was exhibited for the catalyst with one layer of Pt ($Pd@Pt_{1L}/C$). The enhanced specific activity was ascribed to a combination of ligand and strain effects between the atoms of Pd and Pt, while the highest mass activity was attributed to the balance between specific activity and the Pt dispersion. They also studied the stability of the $Pd@Pt_{1L}/C$ and $Pd@Pt_{2-3L}/C$ nanostructures. Their results showed that the Pd cores dissolve and the Pt atoms migrate to the corners and edges to form a Pt-rich concave surface. However, they observed a lower dissolution of these cores when the catalysts were synthesized with thicker Pt overlayers ($Pd@Pt_{4L}/C$ and $Pd@Pt_{6L}/C$).

Recently, Zhang Yafeng et al. developed a one-pot solvo-thermal method for the synthesis of Pd@Pt core-shell nanoparticles. They found that the thickness of Pt shells (Pt_{nL}) on the Pd core can be tuned by modifying the reaction temperature (Zhang et al., 2021). The Pd@Pt core-shell nanostructures with a particle size of 15.1 nm and a shell thickness of 0.78 nm were obtained at 140°C, which is equivalent to 3.4 atomic layers of Pt deposited on the Pd core ($Pd@Pt_{3.4}$); also, they modified the Pt monolayers by increasing the reaction temperature in a ratio of three monolayers per 10°C. Their results showed that when increasing the shell thickness (from 3.4 to 13.9 monolayers) the surface compressive strain decreases (from -1.85 to -0.18%). In this regard, all synthesized catalysts were more stable and showed higher specific activity and mass activity compared to a Pt/C catalyst. However, the catalyst with a surface compression of 1.85% ($Pd@Pt_{3.4}/C$) showed the highest mass activity. The catalytic activities revealed for these catalysts were associated with the oxygen surface adsorption energy. According to this, they found that the $Pd@Pt_{3.4}/C$ catalyst showed an adsorption energy (0.15 eV) closer to the reported optimal value (0.2 eV) (Greeley et al., 2009); in addition, their results showed that thicker Pt shells lead to oxygen adsorption energy more negative. Therefore, they concluded that the catalyst with the highest surface strain (−1.85%) has the highest mass activity due to its optimal oxygen adsorption energy on the surface Pt shell.

As mentioned above, the combination of ligand and strain effects influences oxygen reduction catalysis. Particle size control is a physical alternative that allows modifying of these two effects in core-shell structures. In this sense, Wang et al. evaluated the effect of a variety of Pd cores with different sizes (3.5, 3.9, 5.7 nm) on the lattice strain of surface Pt atoms to establish a correlation between the strain effect and the catalytic activity for the oxygen reduction reaction (Wang et al., 2013). They found that the Pt atoms suffer from the highest compressive strain when these atoms are deposited on the Pd atoms with the smallest particle size and the highest surface roughness (Wang et al., 2013). Based on other works (Jiang et al., 2001; Huang et al., 2008; Wang et al., 2009a), the authors associated the changes in lattice strain of smaller particle size with a larger surface curvature; also, surface roughness was associated with the radial contraction due to the presence of more surface atoms with low coordination number. On the other hand, the catalytic activity results revealed an increase in specific activity for the ORR when the Pt monolayer is compressed. Therefore, these results indicate that the particle size of core atoms influences the structural control of atoms in the shell, and therefore the strain effect, and consequently the catalytic activity. Chen Liu et al. also found that Pd core sizes influence catalytic activity in core Pd – shell Pt catalysts (Liu et al., 2020). In this regard, all synthesized catalysts with Pd core sizes of 2.3, 4.3, and 8 nm exhibited a higher activity than Pt/C at 25°C. This behavior was associated with the compression effect from Pd cores (strain effect), which is in accord with the results of Wang. However, unlike the Wang results, the results obtained at 25°C by Chen Liu et al. revealed an increase in specific activity when core size increases. Shao et al. found that the specific activity significantly increases as the nanoparticle size of platinum grows from 1.3 nm to 2.2 nm and slightly increases for larger particle sizes (2.2–5 nm). Also, they reported that {111} facets are the most catalytically active faces and that the particles about 2.1 nm have the weakest interaction with oxygen (Shao et al., 2011). Batyr Garlyyev et al. theoretically found that different Pt sizes (1.1, 2.07, and 2.87 nm) can produce optimal activities (Garlyyev et al., 2019); in addition, their results revealed that size distributions significantly influence catalytic activities due to surface rearrangements that result in the exposure of different active sites or surface facets. On the other hand, Xu Zhang and Gang Lu found that the oxygen surface adsorption energy for different Pd cores sizes (3.2, 4.8, 6.4, and 8 nm) and a Pt shell thickness of 0.6 nm are slightly closer to optimal oxygen surface adsorption energy for smaller cores (Zhang and Lu, 2014). Wei Wang and co-workers used room-temperature electron reduction for the preparation of Pd@Pt core-shell nanostructures (Wang et al., 2017b). These nanostructures showed higher catalytic activity and better stability in comparison with a commercial Pt/C catalyst. The highest activity revealed for the Pd@Pt core-shell catalyst was attributed to its particle size (2.6 nm) and the {111} surface facet it exhibited would improve the oxygen surface adsorption energy, as well as the breaking of O–O bond and OH* desorption in the oxygen reduction reaction. On the other hand, Xue Wang et al. synthesized Pd@Pt core-shell concave decahedra (Wang et al., 2015). Platinum nanostructures with high-index facets (Miller indices with at least one index larger than one) on their surface revealed higher catalytic activity compared to nanostructures with low-index facets, such as {100} and {111} facets (Yu et al., 2011; Ma et al., 2015). Therefore, the enhancement in specific activity observed in these Pd@Pt core-shell concave decahedra was mainly attributed to two factors: (i) the combination of ligand and strain effects, and (ii) the high-index facet associated with the concave surface. In addition, the results showed that the enhanced specific activity together with the high electrochemical active surface area (ECSA) led to an increase in mass activity in these nanostructures. On the other hand, their results revealed that the Pd cores in these nanostructures are dissolved during the durability tests; consequently, these core-shell structures are transformed into cage-like structures made of Pt. The loss of specific activity that occurred after durability tests was associated with a decrease in the ligand and strain effects caused by Pd dissolution in the structure. Despite this, these nanostructures still showed a higher mass activity than the Pt catalyst (Table 9.2).

Table 9.2 Electrochemical parameters for Pd-based core-shell catalysts reported in some works

Catalyst	Activity Mass (A/mg_{Pt})	Specific Activity (mA/cm^2)	ECSA initial (m^2/g_{Pt})	ADT-Cycles	Activity Mass after ADT (A/mg_{Pt})	References
Pd@Pt$_{1L}$	0.34	0.32	103.54	5,000	0.18	
Pd@Pt$_{2-3L}$	0.23	0.49	47.50	10,000	0.19	
Pd@Pt$_{4L}$	0.15	0.36	39.58	10,000	0.14	Xie et al., 2014
Pd@Pt$_{6L}$	0.07	0.31	24.79	10,000	0.06	
Pd@Pt$_{3.4}$	0.95	0.50	188.35	8,000	0.60	
Pd@Pt$_{5.3}$	0.89	1.45	61.04	8,000	0.87	
Pd@Pt$_{8.2}$	0.71	1.68	42.17	8,000	0.59	Zhang et al., 2021
Pd@Pt$_{11}$	0.61	1.77	34.14	8,000	0.39	
Pd@Pt$_{13.9}$	0.42	1.32	31.73	8,000	0.26	
Pt/Pd$_{(core=2.3nm)}$	0.47	0.39	–	–	–	
Pt/Pd$_{(core=4.3nm)}$	0.58	0.48	–	–	–	Liu et al., 2020
Pt/Pd$_{(core=8.0nm)}$	0.69	0.58	–	–	–	
Pt$_1$Pd$_1$	0.49	–	–	10,000	0.47	
Pt$_1$Pd$_2$	0.34	–	–	10,000	0.32	Wang et al., 2017b
Pt$_1$Pd$_4$	0.23	–	–	10,000	0.22	
Pd@Pt$_{(Pt=29.5\%wt)}$	1.59	1.67	95.90	10,000	0.67	Wang et al., 2015

Some values reported in this Table were estimated from bar graph using Image J. The values were adjusted to two significant numbers.

These works demonstrate that the highest catalytic activity in Pd-based core-shell catalysts can be achieved by tuning their core sizes, shell thicknesses, and surface facets in order to optimize the oxygen surface adsorption energy, and therefore improve the catalysis of the oxygen reduction reaction.

9.3.2 Ag@Pt Core-Shell Catalysts

The silver-based core-shell catalysts for the ORR in an acidic medium have been little investigated. However, this type of nanostructures has demonstrated to be a promising alternative to catalyze this reaction (Table 9.3). In this regard, Feng Yuan Yuan et al. studied the catalytic behavior of a series of core-shell catalysts with different atomic Pt/Ag ratios (Feng et al., 2011). These catalysts were prepared by reducing $PtCl_6^{2-}$ ions on Ag nanoparticles of 6.3 nm in an aqueous solution. During the synthesis, they found that the nanostructures are transformed from a Ag@PtAg core-shell structure to a hollow PtAg alloy structure when the Pt/Ag ratio increases. In addition, the results obtained by XPS showed that the electronic structure of Pt atoms is influenced by Ag atoms, and consequently, this structure can be modified by adjusting the Pt/Ag ratio. On the other hand, their electrochemical results revealed that both the mass activity and specific activity increase for the Pt/Ag ratios from 0.1 to 0.5; but these activities decrease slightly for the Pt/Ag ratio of 0.6. In this regard, the ECSA (except for a Pt/Ag ratio of 0.1) and mass activities of all those synthesized catalysts were lower than those of the Pt catalyst, which shows that the dispersion of catalysts on support plays a key role in this property. On the contrary, the specific activities for Pt/Ag ratios of 0.4, 0.5, and 0.6 were higher in comparison with the Pt catalyst. This behavior was mainly related to the ligand effect associated with the interaction of Ag with Pt; also, the lower activities observed for Pt/Ag ratios lower than 0.4 were associated with expansive strain (strain effect) caused by the higher lattice parameter that Ag has compared to Pt. The strain and ligand effects of Ag on Pt in this type of structure have also been reported by Rui Zhang et al. (Zhang et al., 2018), who found that the electron transfer from Ag to Pt easily occurs due to the higher electronegativity of Pt (ligand or electronic effect) and that Ag induces an expansive strain on the surface Pt due to its larger lattice parameter (strain effect). Interestingly, they found that the strain effect is negligible when the ligand effect is considered. On the other hand, Liu Min et al. synthesized Ag@Pt core-shell nanostructures through a galvanic replacement reaction by incorporating chloroplatinic acid (H_2PtCl_6) into an aqueous solution with Ag nanoparticles (Liu et al., 2016). These nanostructures were tuned from a core-shell to alloy structure and then to a hollow structure by increasing the concentration of H_2PtCl_6 (from 0.55 mM to 0.7 and 0.83 mM, respectively) in the aqueous solution. The particle sizes of these synthesized nanostructures were similar (between 12 and 14 nm); however, the Ag@Pt core-shell nanostructure revealed the highest mass activity and the highest specific activity for oxygen reduction. In this regard, the catalytic activities increased in the following order: Ag@Pt core-shell > hollow alloy > solid alloy > commercial Pt/C. In this study, the origin of the enhancement in the catalytic activity in the ORR was again associated with the ligand effect of the Ag atoms on the Pt atoms. Shutang Chen et al. synthesized

Ag@Pt core-shell nanostructures with different sizes (22.4, 12.5, and 5.8 nm); then, these nanostructures were transformed into hollow nanostructures with different shell thickness in the presence of acetic acid (Chen et al., 2017). In the synthesis methodology, they found that the presence of oxygen and the thickness of the Pt shell play an important role in the formation of hollow nanostructures with a Pt–Ag alloy shell. In addition, the Ag content in the hollow Pt–Ag nanostructure increased when the nanostructures sizes decreased. On the other hand, despite having different Ag contents, the hollow nanostructure of 12.5 and 22.4 nm showed identical specific activities; also, the highest specific activity was revealed for the hollow nanostructure of 5.8 nm. Shutang Chen et al. concluded in their work that the unfavorable lattice expansion effect associated with adding silver into Pt–Ag alloys is mitigated by the hollow morphology and that the ligand effect associated with Ag dopants into Pt–Ag alloys could be contributing to improving the catalytic activity in these nanostructures.

It is clear that the high catalytic activity for the oxygen reduction reaction is mainly associated with the ligand and strain effects. However, these studies suggest that the ligand effect seems to be the most important factor for enhancing the catalytic activity in Ag-based core-shell catalysts.

9.3.3   Ag@PtM Core@Shell Catalysts

Our work group reported the synthesis of Ag@PtPd core-shell type nanostructures. These nanostructures with an average particle size of 9.1 nm were obtained by reducing Pt and Pd ions on the surface of previously synthesized Ag nanoparticles using sodium borohydride and sonication of high-intensity (Pech-Pech et al., 2015a). These Ag@PtPd nanostructures showed an expansion strain on the lattice constant for Pt and Pd monometallic nanoparticles and Pt-Pd alloy nanoparticles of smaller sizes (around 2–3 nm); however, although the Ag@PtPd nanostructures showed the largest lattice constant as well as the largest particle size, these nanostructures revealed higher mass activity and specific activity in comparison with the synthesized monometallic and bimetallic catalysts, as well as a specific activity similar to that of a commercial Pt catalyst. We found these results contradictory to other studies that have shown that compressive strain tends to weaken the interaction between oxygen atoms and the surface metal, which contributes to increasing the kinetics of the ORR, whereas expansive strain shows the opposite effect. Therefore, as in Ag@Pt core-shell catalysts, in this study, the highest activities found in the catalyst were mainly associated with the ligand effect of Ag on the PtPd surface metals.

In a second work, the effect of the PtPd content on catalytic activity in this type of Ag@PtPd catalysts was reported (Pech-Pech et al., 2015b). The electrochemical results showed that when the Pt/Pd ratio increases on the Ag surface, then the bulk features of Pt predominate in the cyclic voltammogram signal, and when this Pt/Pd ratio decreases, then the bulk features of Pd are predominant in this signal. On the other hand, the catalyst with the highest content of Pd atoms in the shell ($Ag@Pt_{0.1}Pd_{0.5}$ with 50% of Pd and 10% of Pt) showed the highest ECSA, which was associated with a better dispersion; in addition, the catalyst with the highest Pt atomic content in the shell ($Ag@Pt_{0.1}Pd_{0.5}$ with 10% of Pd and 50% of Pt) revealed the lowest expansive strain. However, the catalyst with 30% of Pd and 30% of Pt ($Ag@Pt_{0.3}Pd_{0.3}$) exhibited both the highest mass activity and the highest specific activity compared to the other synthesized catalysts and the commercial Pt catalyst. Based on these results, this work concluded that among the ligand, strain and dispersion effects, the ligand effect has a greater influence on the ORR; as a consequence, this optimal ligand effect may be producing the highest catalytic activity on $Ag@Pt_{0.1}Pd_{0.5}$ nanostructures. These results show that the optimization of the atomic ratio of the shell elements has a fundamental role in the electronic and structural properties and hence in the catalytic activity for oxygen reduction.

In a third work, the effect of metal loading on catalytic activity was reported. For this work, the studied material was the $Ag@Pt_{0.1}Pd_{0.1}$ catalyst with different metal loadings (5, 10, 20, and 30 wt% of total metals on carbon) (Pech-Pech et al., 2018). The results showed that all the synthesized $Ag@Pt_{0.1}Pd_{0.1}$ catalysts have similar morphologies, sizes, and tensile strains. However, the higher metal loading (percentage of $Ag@Pt_{0.1}Pd_{0.1}$ deposited on carbon) generated a slight increase in the agglomeration. On the other hand, the electrochemical results showed that the ECSA increases as the percentage of $Ag@Pt_{0.1}Pd_{0.1}$ deposited on carbon increases from 10 to 30%. This increase in the ECSA was associated with the increase in the number of active sites due to the higher content of metals ($Ag@Pt_{0.1}Pd_{0.1}$) in the catalysts. Despite this, the catalyst with the lowest metal loading (5% of $Ag@Pt_{0.1}Pd_{0.1}$ on carbon) gave the highest ECSA. Since all $Ag@Pt_{0.1}Pd_{0.1}$ nanostructures showed similar morphologies and sizes, the highest ECSA found in the catalyst with the lowest metal loading was associated with the lowest agglomeration (hence to the highest dispersion) of its $Ag@Pt_{0.1}Pd_{0.1}$ nanoparticles on carbon. Therefore, this study demonstrates that the ECSA can be increased by increasing the catalyst loading or by increasing the dispersion of the catalyst on the support. Interestingly, among all synthesized catalysts, the catalyst with a metal loading of 5% showed the highest mass and the highest specific activity. The high catalytic activity exhibited for this catalyst with 5% of $Ag@Pt_{0.1}Pd_{0.1}$ nanoparticles was associated with its higher number of active sites available for catalyzing the ORR. In addition, this catalyst with 5% of metals (equivalent to 1% of Pt) showed a lower ECSA but a mass activity higher than a commercial catalyst with 20% of platinum. In this regard, the commercial catalyst showed a particle size around four times smaller than the $Ag@Pt_{0.1}Pd_{0.1}$ catalyst; therefore, the lower ECSA shown by the $Ag@Pt_{0.1}Pd_{0.1}$ catalyst was associated with its larger particle size, and its enhanced mass activity was mainly associated with the high dispersion of $Ag@Pt_{0.1}Pd_{0.1}$ nanoparticles and ligand effect of Ag on Pt and Pd atoms.

The high utilization of Pt in Ag-based catalysts for oxygen reduction reaction in acidic electrolytes was also addressed by Shuping Yu et al. (Yu et al., 2016). In their research work, they synthesized Ag@Pt core-shell nanostructures, which were doped with different contents of phosphotungstic acid (HPW from 0 to 50%) and dispersed on multi-walled carbon nanotubes (MWCNs). Ag atoms were found to expand the Pt lattice (strain effect) in the Ag@Pt nanostructures; also, the Pt lattice expansion was slightly higher in the HPW-doped nanostructures. On the other hand, it was found that the incorporation of HPW increases the particle size in these Ag@Pt nanostructures. The electrochemical results showed that the catalyst with 25% HPW content has the highest ECSA, as well as the highest mass activity and highest specific activity. This enhanced catalytic activity was attributed to the high utilization of Pt and faster electron transfer rate associated with the synergic effect between phosphotungstic acid and the Ag@Pt structure.

Table 9.3 Electrochemical parameters for Ag-based core-shell catalysts reported in some works

Catalyst	Activity Mass (A/mg_{Pt})	Specific Activity (mA/cm^2)	ECSA initial (m^2/g_{Pt})	References
$Pt_{0.1}Ag$	0.043	0.028	155	
$Pt_{0.2}Ag$	0.036	0.052	70	
$Pt_{0.4}Ag$	0.050	0.124	40	Feng et al., 2011
$Pt_{0.5}Ag$	0.085	0.146	58	
$Pt_{0.6}Ag$	0.079	0.132	60	
$Ag@Pt_{core-shell}$	0.051	0.120	45.03	
$Ag@Pt_{Alloy}$	0.016	0.065	24.89	Liu et al., 2016
$Ag@Pt_{Hollow}$	0.200	0.100	19.36	
$Ag@Pt_{(Particle\ size=5.8nm)}$	—	1.111	—	
$Ag@Pt_{(Particle\ size=12.5nm)}$	—	0.863	—	Chen et al., 2017
$Ag@Pt_{(Particle\ size=22.4nm)}$	—	0.853	—	
$Ag@Pt_{0.1}Pd_{0.1}$	0.021	0.034	37.48	
$Ag@Pt_{0.1}Pd_{0.3}$	0.025	0.024	39.08	
$Ag@Pt_{0.1}Pd_{0.5}$	0.024	0.011	56.10	Pech-Pech et al., 2015b
$Ag@Pt_{0.3}Pd_{0.1}$	0.023	0.053	36.32	
$Ag@Pt_{0.5}Pd_{0.1}$	0.019	0.052	34.34	
$Ag@Pt_{0.2}Pd_{0.2}$	0.030	0.038	53.42	
$Ag@Pt_{0.3}Pd_{0.3}$	0.044	0.069	41.24	
$(Ag@Pt_{0.1}Pd_{0.1})_x\ (x=5\%wt)$	0.029	0.038	46.15	
$(Ag@Pt_{0.1}Pd_{0.1})_x\ (x=10\%wt)$	0.020	0.034	37.89	Pech-Pech et al., 2018
$(Ag@Pt_{0.1}Pd_{0.1})_x\ (x=20\%wt)$	0.019	0.032	38.65	
$(Ag@Pt_{0.1}Pd_{0.1})_x\ (x=30\%wt)$	0.023	0.033	45.49	
$Ag@PtCo$	0.450	0.630	132.06	Deng et al., 2021

Some values reported in this Table were estimated from bar graph using Image J. The values were adjusted to three significant numbers.

Xiaoting Deng et al. synthesized Ag@PtCo core-shell mesoporous nanoflowers with a diameter of around 40 nm (Deng et al., 2021). These Ag@PtCo nanostructures exhibited the highest ECSA compared to Pt and PtCo nanostructures, which were synthesized for comparative purposes in this work. In addition, the authors found that the ECSA of the Ag@PtCo nanostructures is higher than that of a commercial Pt catalyst. The highest number of active sites found in the Ag@PtCo nanostructures was attributed to its mesoporous nanoflower structure. On the other hand, the Ag@PtCo nanostructure exhibited the highest catalytic activities compared to the synthesized catalysts and the commercial catalyst. In addition, the Ag@PtCo catalyst showed the best oxidation resistance (or stability) during the durability tests. The best catalytic properties of Ag@PtCo were associated with the ligand effect of Ag on Pt. In this regard, the XPS results of the Ag@PtCo catalyst showed that electrons are transferred from Co and Ag atoms to Pt atoms due to electronegativity differences; also, the highest Pt(0)/Pt(II) ratio found in Ag@PtCo was associated with the highest percentage of surface Pt in the metallic state (related to more active sites). On the other hand, their DFT studies showed that Ag@PtCo has a smaller O–O bond-breaking energy in comparison with Pt and PtCo. Therefore, based on these results, the authors of this work concluded that the excellent performance exhibited for Ag@PtCo nanostructure can be attributed to the Ag effect in PtCo alloy.

These results suggest that the ligand effect seems to be the most important factor for enhancing the catalytic activity in Ag-based core-shell catalysts, even if the shell in these catalysts is composed of more than one metal.

9.3.4 Recent Advances in Core-Shell Catalyst for the ORR

Currently, different strategies are still being developed to improve the catalytic parameters and the durability of core-shell catalysts for ORR. One of these strategies is the synthesis of multi-metallic alloys in the shell with a concave structure, which have been shown to increase the amount of low coordination atoms, as well as the amount of step atoms and the number of defects on the surface of core-shell catalysts. This strategy has allowed the obtention of enhanced catalytic activities and a higher stability in this type of structures (Chen et al., 2022). Core-shell structures with multi-metallic cores also have demonstrated to be an excellent strategy to increase ORR catalysis; in this sense, the composition of the core and the thickness of the shell have been two key parameters to achieve this goal (Shi et al., 2022). On the other hand, the incorporation of interlayers between the atoms of the shell and those of the core has demonstrated to be an alternative to significantly improve the stability of the core and the ligand effect; consequently, a very high catalytic activity has been obtained together with a surprising durability for this type catalysts (He et al., 2021). Another alternative to produce highly durable catalysts is the incorporation of thin protective layers of carbon with hereteoatoms on the surface of the core-shell catalysts. In this regard, the interaction of the heteroatoms with Pt atoms has shown a synergistic effect on the ligand effect, which in turn has had a positive impact on ORR catalysis (Yan et al., 2022). The synthesis of a shell with an ultrathin thickness deposited on a core with a different morphology of a sphere was a promising strategy in the development of core-shell catalysts for ORR (Chen et al., 2022). The above-mentioned research works demonstrate the extensive efforts to optimize the strain and ligand effects, and consequently, produce highly durable materials for the efficient catalysis of the oxygen reduction reaction (Table 9.4).

Table 9.4 Electrochemical parameters for some recent advances in catalysts for ORR

Catalyst	Activity Mass (A/mg$_{Pt}$)	Specific Activity (mA/cm²)	ECSA initial (m²/g$_{Pt}$)	ADT-Cycles	Activity Mass after ADT (A/mg$_{Pt}$)	References
Au@PtNiAu-COCS$_{(concave\ octahedral\ core-shell)}$	2.02	3.84	63.48	10,000 50,000	1.84 1.15	Feng et al., 2022
Pt$_1$Cu$_1$@Pt	0.45	1.16	38.80	30,000	0.34	Shi et al., 2022
Pt$_2$Cu$_1$@Pt	0.68	1.92	35.26	30,000	0.60	
Pt$_3$Cu$_1$@Pt	0.34	1.09	30.67	30,000	0.22	
Pd@a-Pd-P@PtSML-cube$_{(Pd@a-Pd-P\ nanocubes\ as\ seeds)}$	3.35	2.91	115.10	50,000	3.06	He et al., 2021
Pd@a-Pd-P@PtSML-octa$_{(Pd@a-Pd-P\ octahedra\ as\ seeds)}$	4.99	5.42	92.10	50,000	4.58	
N–C/PtNi	0.963	—	48.91	120,000	0.27	Yan et al., 2022
Pd@PtNi NSs-1$_{(Shell\ thickness=1.25)}$	2.637	0.254	185.35	10,000	1.56	Chen et al., 2022
Pd@PtNi NSs-2$_{(Shell\ thickness=2.41)}$	2.019	1.251	249.88	10,000 80,000	1.35 1.02	
Pd@PtNi NSs-3$_{(Shell\ thickness=3.39)}$	0.746	0.547	240.25	10,000	0.47	

Some values reported in this Table were estimated from bar graph using Image J. The values were adjusted to two significant numbers.

9.4   CONCLUSIONS AND PERSPECTIVES

The high catalytic activity that exhibits the core-shell nanostructures can be mainly associated with two factors, the strain effect and the ligand effect (Fig. 9.5). The ligand effect is associated with the transfer of electronic charge between the shell surface atoms and the inner core atoms; the strain effect is associated with the strain variations in the lattice constants. In this work, we have explored some parameters, such as the morphology, particle size, core composition, shell composition, surface facets and shell thickness, which have demonstrated to be an alternative to tune these effects in order to increase the catalytic activity for the ORR. The number of active sites associated with the dispersion and the porous structures demonstrated a significant influence on electrochemical active surface area (ECSA) and therefore on catalytic activity. In addition, the ligand effect has shown a predominant role in catalytic activity for nanostructures that have an expansion effect on the surface Pt lattice. The theoretical studies together with the control on synthesis methodology will be the key factors to achieving the highest catalytic activity with the lowest cost in this type of core-shell nanostructures.

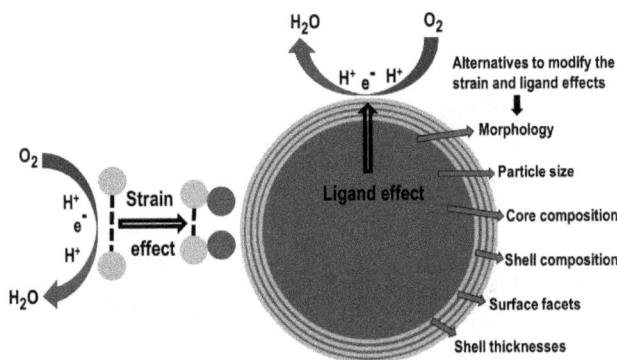

Figure 9.5   Alternatives to improve the catalysis of the oxygen reduction reaction related to ligand and strain effects.

ACKNOWLEDGMENTS

One of the authors, AGG, acknowledges the National Technological Institute of Mexico (TecNM) and its program, Call 2021 Scientific Research Projects, for the financial support.

REFERENCES

Alinezhad, A., L. Gloag, T.M. Benedetti, S. Cheong, R.F. Webster, M. Roelsgaard, et al., 2019. Direct growth of highly strained Pt islands on branched Ni nanoparticles for improved hydrogen evolution reaction activity. J. Am. Chem. Soc. 141(41): 16202–16207. https://doi.org/10.1021/jacs.9b07659

Banham, D., S. Ye, K. Pei, J.I. Ozaki, T. Kishimoto and Y. Imashiro. 2015. A review of the stability and durability of non-precious metal catalysts for the oxygen reduction reaction in proton exchange membrane fuel cells. J. Power Sources [Internet]. 285: 334–348. https://doi.org/10.1016/j.jpowsour.2015.03.047

Boudghene Stambouli, A. and E. Traversa. 2002. Fuel cells, an alternative to standard sources of energy. Renew. Sustain. Energy Rev. 6(3): 295–304. https://doi.org/10.1016/S1364-0321(01)00015-6

Chao, Y.S., D.S. Tsai, A.P. Wu, L.W. Tseng and Y.S. Huang. 2013. Cobalt selenide electrocatalyst supported by nitrogen-doped carbon and its stable activity toward oxygen reduction reaction. Int. J. Hydrogen Energy [Internet]. 38(14): 5655–5664. https://doi.org/10.1016/j.ijhydene.2013.03.006

Chen, S., S. Thota, G. Singh, T.J. Aimola, C. Koenigsmann and J. Zhao. 2017. Synthesis of hollow Pt–Ag nanoparticles by oxygen-assisted acid etching as electrocatalysts for the oxygen reduction reaction. RSC Adv. 7(74): 46916–46924.

Chen, Q., Z. Chen, A. Ali, Y. Luo, H. Feng, Yuanyan Luo, et al., 2022. Shell-thickness-dependent Pd@PtNi core–shell nanosheets for efficient oxygen reduction reaction. Chem. Eng. J. 427: 131565. https://doi.org/10.1016/j.cej.2021.131565

Chisaka, M., A. Ishihara, H. Morioka, T. Nagai, S. Yin, Y. Ohgi, et al., 2017. Zirconium oxynitride-catalyzed oxygen reduction reaction at polymer electrolyte fuel cell cathodes. ACS Omega. 2(2): 678–684. https://doi.org/10.1021/acsomega.6b00555

Choi, R., S.H. Choi, C.H. Choi, K.M. Nam, S.I. Woo, J.T. Park, et al., 2013. Designed synthesis of well-defined Pd@Pt core-shell nanoparticles with controlled shell thickness as efficient oxygen reduction electrocatalysts. Chem. A Eur. J. 19(25): 8190–8198. https://doi.org/10.1002/chem.201203834

Deng, X., S. Yin, Z. Xie, F. Gao, S. Jiang and X. Zhou. 2021. Synthesis of silver@ platinum-cobalt nanoflower on reduced graphene oxide as an efficient catalyst for oxygen reduction reaction. Int. J. Hydrogen Energy. 46(34): 17731–17740.

Esposito, D.V., S.T. Hunt, Y.C. Kimmel and J.G. Chen. 2012. A new class of electrocatalysts for hydrogen production from water electrolysis: Metal monolayers supported on low-cost transition metal carbides. J. Am. Chem. Soc. 134(6): 3025–3033. https://doi.org/10.1021/ja208656v

Feng, Y.-Y., G.-R. Zhang, J.-H. Ma, G. Liu and B.-Q. Xu. 2011. Carbon-supported Pt_x Ag nanostructures as cathode catalysts for oxygen reduction reactionm. Phys. Chem. Chem. Phys. 13(9): 3863–3872.

Feng, H., Y. Luo, B. Yan, H. Guo, L. He, Z.Q. Tian, et al., 2022. Highly stable cathodes for proton exchange membrane fuel cells: novel carbon supported Au@PtNiAu concave octahedral core-shell nanocatalyst. J. Colloid Interface Sci. 626: 1040–1050.

Fouda-Onana, F., S. Bah and O. Savadogo. 2009. Palladium-copper alloys as catalysts for the oxygen reduction reaction in an acidic media I: Correlation between the ORR kinetic parameters and intrinsic physical properties of the alloys. J. Electroanal. Chem. [Internet]. 636(1–2): 1–9. https://doi.org/10.1016/j.jelechem.2009.06.023

Garlyyev, B., K. Kratzl, M. Rück, J. Michalička, J. Fichtner, J.M. Macak, et al., 2019. Optimizing the size of platinum nanoparticles for enhanced mass activity in the electrochemical oxygen reduction reaction. Angew. Chemie. Int. Ed. 58(28): 9596–9600.

Gawande, M.B., A. Goswami, T. Asefa, H. Guo, A.V. Biradar, D.L. Peng, et al., 2015. Core-shell nanoparticles: synthesis and applications in catalysis and electrocatalysis. Chem. Soc. Rev. [Internet]. [accessed 2022, Jul. 7]. 44(21): 7540–7590. https://doi.org/10.1039/C5CS00343A

Gong, S., Y.X. Zhang and Z. Niu. 2020. Recent advances in earth—abundant core/noble-metal shell nanoparticles for electrocatalysis. ACS Catal. 10(19). https://doi.org/10.1021/acscatal.0c02587

Greeley, J., I.E.L. Stephens, A.S. Bondarenko, T.P. Johansson, H.A. Hansen, T.F. Jaramillo, et al., 2009. Alloys of platinum and early transition metals as oxygen reduction electrocatalysts. Nat. Chem. 1(7): 552–556.

Hammer, B. and J.K. Nørskov. 2000. Theoretical surface science and catalysis—calculations and concepts. Adv. Catal. 45: 71–129. https://doi.org/10.1016/S0360-0564(02)45013-4

He, T., W. Wang, X. Yang, F. Shi, Z. Ye, Y. Zheng, et al., 2021. Deposition of atomically thin Pt shells on amorphous palladium phosphide cores for enhancing the electrocatalytic durability. ACS Nano. 15(4): 7348–7356.

Hsu, I.J., Y.C. Kimmel, X. Jiang, B.G. Willis and J.G. Chen. 2012. Atomic layer deposition synthesis of platinum–tungsten carbide core–shell catalysts for the hydrogen evolution reaction. Chem. Commun. 48(7): 1063–1065. https://doi.org/10.1039/c1cc15812k

Huang, W.J., R. Sun, J. Tao, L.D. Menard, R,G. Nuzzo and J.M. Zuo. 2008. Coordination-dependent surface atomic contraction in nanocrystals revealed by coherent diffraction. Nat. Mater. 7(4): 308–313.

Hwang, S.J., S.J. Yoo, J. Shin, Y.H. Cho, J.H. Jang, E. Cho, et al., 2013. Supported core at shell electrocatalysts for fuel cells: Close encounter with reality. Sci. Rep. 3(Ml): 1–7. https://doi.org/10.1038/srep01309

Huang, X., Z. Zhao, L. Cao, Y. Chen, E. Zhu, Z. Lin, et al., 2015. High-performance transition metal–doped Pt3Ni octahedra for oxygen reduction reaction. Science. 348(6240): 1230–1234.

Janik, M.J., C.D. Taylor and M. Neurock. 2009. First-principles analysis of the initial electroreduction steps of oxygen over Pt(111). J. Electrochem. Soc. 156(1): B126–B135.

Jiang, Q., L.H. Liang and D.S. Zhao. 2001. Lattice contraction and surface stress of fcc nanocrystals. J. Phys. Chem. B. 105(27): 6275–6277.

Jiang, R., S. Tung, Z. Tung, L. Li, L. Ding, X. Xi, et al., 2018. A review of core-shell nanostructured electrocatalysts for oxygen reduction reaction. Energy Storage Mater. [Internet]. 12: 260–276. https://doi.org/10.1016/j.ensm.2017.11.005

Jun, D.W., C.S. Yoon, U.H. Kim and Y.K. Sun. 2017. High-energy density core-shell structured Li[Ni$_{0.95}$Co$_{0.025}$Mn$_{0.025}$]O$_2$ cathode for lithium-ion batteries. Chem. Mater. [Internet]. [accessed 2022, Jul. 7]. 29(12): 5048–5052. https://doi.org/10.1021/ACS. CHEMMATER.7B01425/SUPPL_FILE/CM7B01425_SI_001.PDF

Kamarudin, S.K., F. Achmad and W.R.W. Daud. 2009. Overview on the application of direct methanol fuel cell (DMFC) for portable electronic devices. Int. J. Hydrogen Energy. 34(16): 6902–6916. https://doi.org/10.1016/J.IJHYDENE.2009.06.013

Kirubakaran, A., S. Jain and R.K. Nema. 2009. A review on fuel cell technologies and power electronic interface. Renew. Sustain Energy Rev. 13(9): 2430–2440. https://doi.org/10.1016/J.RSER.2009.04.004

Kitchin, J.R., J.K. Nørskov, M.A. Barteau and J.G. Chen. 2004. Modification of the surface electronic and chemical properties of Pt (111) by subsurface 3d transition metals. J. Chem. Phys. 120(21): 10240–10246.

Kongkanand, A., N.P. Subramanian, Y. Yu, Z. Liu, H. Igarashi and D.A. Muller. 2016. Achieving high-power PEM fuel cell performance with an ultralow-Pt-content core-shell catalyst. ACS Catal. 6(3): 1578–1583. https://doi.org/10.1021/acscatal.5b02819

Lai, L., J.R. Potts, D. Zhan, L. Wang, C.K. Poh, C. Tang, et al., 2012. Exploration of the active center structure of nitrogen-doped graphene-based catalysts for oxygen reduction reaction. Energy Environ. Sci. 5(7): 7936–7942. https://doi.org/10.1039/c2ee21802j

Lentijo-Mozo, S., R.P. Tan, C. Garcia-Marcelot, T. Altantzis, P.-F. Fazzini, T. Hungria, et al., 2015. Air- and water-resistant noble metal coated ferromagnetic cobalt nanorods. ACS Nano. 9(3): 2792–2804.

Li, X., A. Dhanabalan, L. Gu and C. Wang. 2012. Three-dimensional porous core-shell Sn@Carbon composite anodes for high-performance lithium-ion battery applications. Adv. Energy Mater. [Internet]. [accessed 2022, Jul. 7], 2(2): 238–244. https://doi.org/10.1002/AENM.201100380

Li, X., Q. Chen, I. McCue, J. Snyder, P. Crozier, J. Erlebacher, et al., 2014. Dealloying of noble-metal alloy nanoparticles. Nano Lett. 14(5): 2569–2577. https://doi.org/10.1021/nl500377g

Liu, M., F. Chi, J. Liu, Y. Song and F. Wang. 2016. A novel strategy to synthesize bimetallic Pt-Ag particles with tunable nanostructures and their superior electrocatalytic activities toward the oxygen reduction reaction. RSC Adv. 6(67): 62327–62335.

Liu, C., T. Uchiyama, K. Yamamoto, T. Watanabe, X. Gao, H. Imai, et al., 2020. Effect of temperature on oxygen reduction reaction kinetics for Pd core—Pt shell catalyst with different core size. ACS Appl. Energy Mater. 4(1): 810–818.

Ma, Y. and P.B. Balbuena. 2008. Pt surface segregation in bimetallic Pt_3M alloys: a density functional theory study. Surf. Sci. 602(1): 107–113. https://doi.org/10.1016/j.susc.2007.09.052

Ma, L., C. Wang, B.Y. Xia, K. Mao, J. He, X. Wu, et al., 2015. Platinum multicubes prepared by Ni_{2+}-mediated shape evolution exhibit high electrocatalytic activity for oxygen reduction. Angew Chemie. Int. Ed. 54(19): 5666–5671.

Mani, P., R. Srivastava and P. Strasser. 2008. Dealloyed Pt-Cu core-shell nanoparticle electrocatalysts for use in PEM fuel cell cathodes. J. Phys. Chem. C. 112(7): 2770–2778. https://doi.org/10.1021/jp0776412

Mayrhofer, K.J.J., V. Juhart, K. Hartl, M. Hanzlik and M. Arenz. 2009. Adsorbate-induced surface segregation for core-shell nanocatalysts. Angew Chemie—Int. Ed. 48(19): 3529–3531. https://doi.org/10.1002/anie.200806209

Morawa Eblagon, K. and L. Brandaõ. 2015. RuSe electrocatalysts and single wall carbon nanohorns supports for the oxygen reduction reaction. J. Fuel Cell Sci. Technol. 12(2): 1–8. https://doi.org/10.1115/1.4029422

Nørskov, J.K., J. Rossmeisl, A. Logadottir, L. Lindqvist, J.R. Kitchin, T. Bligaard, et al., 2004. Origin of the overpotential for oxygen reduction at a fuel-cell cathode. J. Phys. Chem. B. 108(46): 17886–17892. https://doi.org/10.1021/jp047349j

Oezaslan, M., F. Hasche and P. Strasser. 2013. Pt-based core--shell catalyst architectures for oxygen fuel cell electrodes. J. Phys. Chem. Lett. 4(19): 3273–3291.

Pech-Pech, I.E., D.F. Gervasio, A. Godínez-Garcia, O. Solorza-Feria and J.F. Pérez-Robles. 2015a. Nanoparticles of Ag with a Pt and Pd rich surface supported on carbon as a new catalyst for the oxygen electroreduction reaction (ORR) in acid electrolytes: Part 1. J. Power Sources. 276. https://doi.org/10.1016/j.jpowsour.2014.09.112

Pech-Pech, I.E., D.F. Gervasio and J.F. Pérez-Robles. 2015b. Nanoparticles of Ag with a Pt and Pd rich surface supported on carbon as a new catalyst for the oxygen electroreduction reaction (ORR) in acid electrolytes: Part 2. J. Power Sources. 276: 374–381.

Pech-Pech, I.E., D. Gervasio, S.A. Aguila and J.F. Perez-Robles. 2018. Electrocatalysis of oxygen reduction when varying the mass ratio of metal nanoparticles to carbon support for catalysts with a 10 to 10 to 80 mol\% of Pt and Pd on Ag. Int. J. Hydrogen Energy. 43(32): 15205–15216.

Peng, Z. and H. Yang. 2009. Designer platinum nanoparticles: control of shape, composition in alloy, nanostructure and electrocatalytic property. Nano Today. https://doi. org/10.1016/j.nantod.2008.10.010

Pitois, A., A. Pilenga, A. Pfrang, G. Tsotridis, B.L. Abrams and I. Chorkendorff. 2010. Temperature dependence of CO desorption kinetics at a novel Pt-on-Au/C PEM fuel cell anode. Chem. Eng. J. [Internet]. 162(1): 314–321. https://doi.org/10.1016/j. cej.2010.05.002

Pollet, B.G., S.S. Kocha and I. Staffell. 2019. Current status of automotive fuel cells for sustainable transport. Curr. Opin. Electrochem. [Internet]. 16: 90–95. https://doi. org/10.1016/j.coelec.2019.04.021

Sasaki, K., K.A. Kuttiyiel, L. Barrio, D. Su, A.I. Frenkel, N. Marinkovic, et al., 2011. Carbon-supported IrNi core-shell nanoparticles: Synthesis, characterization, and catalytic activity. J. Phys. Chem. C. 115(20): 9894–9902. https://doi.org/10.1021/jp200746j

Sasaki, K., K.A. Kuttiyiel and R.R. Adzic. 2020. Designing high performance Pt monolayer core--shell electrocatalysts for fuel cells. Curr. Opin. Electrochem. 21: 368–375.

Schmidt, O., A. Hawkes, A. Gambhir and I. Staffell. 2017. The future cost of electrical energy storage based on experience rates. Nat. Energy. 2(8): 1–8. https://doi.org/10.1038/ nenergy.2017.110

Seo, S.-J., H.-K. Chung, J.-B. Yoo, H. Chae, S.-W. Seo and S. Min Cho. 2014. Co-Pt core-shell nanostructured catalyst prepared by selective chemical vapor pulse deposition of Pt on Co as a cathode in polymer electrolyte fuel cells. J. Vac. Sci. Technol. A. Vacuum. Surfaces Film. 32(1): 01A129. https://doi.org/10.1116/1.4853135

Shao, M., A. Peles and K. Shoemaker. 2011. Electrocatalysis on platinum nanoparticles: Particle size effect on oxygen reduction reaction activity. Nano. Lett. 11(9): 3714–3719. https://doi.org/10.1021/nl2017459

Shao, M., Q. Chang, J.P. Dodelet and R. Chenitz. 2016. Recent advances in electrocatalysts for oxygen reduction reaction. Chem. Rev. 116(6): 3594–3657. https://doi.org/10.1021/ acs.chemrev.5b00462

Shi, F., J. Peng, F. Li, N. Qian, H. Shan, P. Tao, et al., 2021. Design of highly durable core-shell catalysts by controlling shell distribution guided by *in-situ* corrosion study. Adv. Mater. 33(38): 2101511.

Shi, W., A.-H. Park and Y.-U. Kwon. 2022. Scalable synthesis of (Pd, Cu)@ Pt core-shell catalyst with high ORR activity and durability. J. Electroanal. Chem. 116451.

Stamenkovic, V., B.S. Mun, K.J.J. Mayrhofer, P.N. Ross, N.M. Markovic, J. Rossmeisl, et al., 2006. Changing the activity of electrocatalysts for oxygen reduction by tuning the surface electronic structure. Angew. Chem. Int. Ed. 45(18): 2897–2901.

Strasser, P. and S. Kühl. 2016. Dealloyed Pt-based core-shell oxygen reduction electrocatalysts. Nano Energy [Internet]. 29: 166–177. https://doi.org/10.1016/j.nanoen.2016.04.047

Tao, F., M.E. Grass, Y. Zhang, D.R. Butcher, J.R. Renzas, Z. Liu, et al., 2008. Reaction-driven restructuring of Rh-Pd and Pt-Pd core-shell nanoparticles. Science. 322(5903): 932–934.

Tian, X., J. Luo, H. Nan, H. Zou, R. Chen, T. Shu, et al., 2016. Transition metal nitride coated with atomic layers of Pt as a low-cost, highly stable electrocatalyst for the oxygen reduction reaction. J. Am. Chem. Soc. 138(5): 1575–1583. https://doi.org/10.1021/jacs. 5b11364

Wang, Y. and N. Toshima. 1997. Preparation of Pd–Pt bimetallic colloids with controllable core/shell structures. J. Phys. Chem. B. [Internet]. 101(27): 5301–5306. https://doi.org/10.1021/jp9704224

Wang, L., A. Roudgar and M. Eikerling. 2009a. Ab initio study of stability and site-specific oxygen adsorption energies of Pt nanoparticles. J. Phys. Chem. C. 113(42): 17989–17996.

Wang, J.X., H. Inada, L. Wu, Y. Zhu, Y. Choi, P. Liu, et al., 2009b. Oxygen reduction on well-defined core-shell nanocatalysts: particle size, facet, and Pt shell thickness effects. J. Am. Chem. Soc. 131(47): 17298–17302.

Wang, D., H.L. Xin, Y. Yu, H. Wang, E. Rus, D.A. Muller, et al., 2010. Pt-decorated PdCo@Pd/C core-shell nanoparticles with enhanced stability and electrocatalytic activity for the oxygen reduction reaction. J. Am. Chem. Soc. 132(50): 17664–17666. https://doi.org/10.1021/ja107874u

Wang, X., Y. Orikasa, Y. Takesue, H. Inoue, M. Nakamura, T. Minato, et al., 2013. Quantitating the lattice strain dependence of monolayer Pt shell activity toward oxygen reduction. J. Am. Chem. Soc. 135(16): 5938–5941.

Wang, X., M. Vara, M. Luo, H. Huang, A. Ruditskiy, J. Park, et al., 2015. Pd@Pt core-shell concave decahedra: a class of catalysts for the oxygen reduction reaction with enhanced activity and durability. J. Am. Chem. Soc. 137(47): 15036–15042.

Wang, K.C., H.C. Huang and C.H. Wang. 2017a. Synthesis of Pd@Pt$_3$Co/C core–shell structure as catalyst for oxygen reduction reaction in proton exchange membrane fuel cell. Int. J. Hydrogen Energy. 42(16): 11771–11778. https://doi.org/10.1016/J.IJHYDENE.2017.03.084

Wang, W., Z. Wang, J. Wang, C.-J. Zhong and C.-J. Liu. 2017b. Highly active and stable Pt-Pd alloy catalysts synthesized by room-temperature electron reduction for oxygen reduction reaction. Adv. Sci. 4(4): 1600486.

Wang, X., Y. Orikasa, M. Inaba and Y. Uchimoto. 2020. Reviving galvanic cells to synthesize core-shell nanoparticles with a quasi-monolayer Pt shell for electrocatalytic oxygen reduction. ACS Catal. 10(1): 430–434. https://doi.org/10.1021/acscatal.9b03672

Wroblowa, H.S., Y.-C. Pan and G. Razumney. 1976. Electroreduction of oxygen: a new mechanistic criterion. J Electroanal Chem. Interfacial Electrochem. 69(2): 195–201. https://doi.org/10.1016/s0022-0728(76)80250-1

Wu, J. and H. Yang. 2013. Platinum-based oxygen reduction electrocatalysts. Acc. Chem. Res. 46(8): 1848–1857. https://doi.org/10.1021/ar300359w

Wu, D., X. Shen, Y. Pan, L. Yao and Z. Peng. 2020. Platinum alloy catalysts for oxygen reduction reaction: advances, challenges and perspectives. Chem. Nano. Mat. 6(1): 32–41. https://doi.org/10.1002/cnma.201900319

Xia, X., Y. Wang, A. Ruditskiy and Y. Xia. 2013a. 25th anniversary article: galvanic replacement: a simple and versatile route to hollow nanostructures with tunable and well-controlled properties. Adv. Mater. 25(44): 6313–6333. https://doi.org/10.1002/adma.201302820

Xia, X., S. Xie, M. Liu, H.C. Peng, N. Lu, J. Wang, et al., 2013b. On the role of surface diffusion in determining the shape or morphology of noble-metal nanocrystals. Proc. Natl. Acad. Sci. USA. 110(17): 6669–6673. https://doi.org/10.1073/pnas.1222109110

Xiang, T., L. Fang, J. Wan, L. Liu, J.J. Gao, H.T. Xu, et al. 2018. Thickness-tunable core-shell Co@Pt nanoparticles encapsulated in sandwich-like carbon sheets as an enhanced electrocatalyst for the oxygen reduction reaction. J. Mater. Chem. A. 6(43): 21396–21403.

Xie, S., S.-I. Choi, N. Lu, L.T. Roling, J.A. Herron, L. Zhang, et al., 2014. Atomic layer-by-layer deposition of Pt on Pd nanocubes for catalysts with enhanced activity and durability toward oxygen reduction. Nano Lett. 14(6): 3570–3576.

Xie, C., Z. Niu, D. Kim, M. Li and P. Yang. 2019. Surface and interface control in nanoparticle catalysis. Chem. Rev. 120(2): 1184–1249.

Yan, Z., Y. Zhang, C. Dai, Z. Zhang, M. Zhang, W. Wei, et al., 2022. Porous, thick nitrogen-doped carbon encapsulated large PtNi core-shell nanoparticles for oxygen reduction reaction with extreme stability and activity. Carbon. N.Y. 186: 36–45.

Yang, H. 2011. Platinum-based electrocatalysts with core-shell nanostructures, Angew. Chemie.—Int. Ed. 50(12): 2674–2676. https://doi.org/10.1002/anie.201005868

Yu, T., D.Y. Kim, H. Zhang and Y. Xia. 2011. Platinum concave nanocubes with high-index facets and their enhanced activity for oxygen reduction reaction. Angew Chemie. 123(12): 2825–2829.

Yu, S., Y. Wang, H. Zhu, Z. Wang and K. Han. 2016. Synthesis and electrocatalytic performance of phosphotungstic acid-modified Ag@ Pt/MWCNTs catalysts for oxygen reduction reaction. J. Appl. Electrochem. 46(9): 917–928.

Zhang, J., F.H.B. Lima, M.H. Shao, K. Sasaki, J.X. Wang, J. Hanson, et al., 2005. Platinum monolayer on nonnoble metal-noble metal core-shell nanoparticle electrocatalysts for O_2 reduction. J. Phys. Chem. B. [Internet]. 109(48): 22701–22704. https://doi.org/10.1021/jp055634c

Zhang, X. and G. Lu. 2014. Computational design of core/shell nanoparticles for oxygen reduction reactions. J. Phys. Chem. Lett. 5(2): 292–297.

Zhang, S., Y. Hao, D. Su, V.V.T. Doan-Nguyen, Y. Wu, J. Li, et al., 2014. Monodisperse core/shell Ni/FePt nanoparticles and their conversion to Ni/Pt to catalyze oxygen reduction. J. Am. Chem. Soc. 136(45): 15921–15924. https://doi.org/10.1021/ja5099066

Zhang, R., W. Xia, W. Kang, R. Li, K. Qu, Y. Zhang, et al., 2018. Methanol oxidation reaction performance on graphene-supported PtAg alloy nanocatalyst: contrastive study of electronic and geometric effects induced from Ag doping. ChemistrySelect. 3(13): 3615–3620.

Zhang, Y., J. Qin, D. Leng, Q. Liu, X. Xu, B. Yang, et al., 2021. Tunable strain drives the activity enhancement for oxygen reduction reaction on Pd@Pt core-shell electrocatalysts. J. Power Sources. 485: 229340.

Zhao, M., X. Wang, X. Yang, K.D. Gilroy, D. Qin and Y. Xia. 2018. Hollow metal nanocrystals with ultrathin, porous walls and well-controlled surface structures. Adv. Mater. 30(48): 1801956.

Chapter **10**

Non-Noble Metal Catalysts in Oxygen Reduction Reaction

Rajendran Rajaram[1,5*], V. Maruthapandian[2], Divya Velpula[3],
K. Naga Mahesh[4] and Sankararao Mutyala[4*]

[1]Corrosion Engineering and Materials Electrochemistry Lab,
Department of Metallurgical and Materials Engineering,
Indian Institute of Technology Madras, Chennai–600036, India.

[2]Clean Energy Lab, Department of Chemistry, Indian Institute of Technology,
Madras, Chennai-600036, India.

[3]Center for Nano Science and Technology, Institute of Science and Technology,
JNTU Hyderabad, India-500085.

[4]Nanosol Energy Pvt. Ltd., Hyderabad, Telangana, India-502 032.

[5]Department of Chemistry, Madanapalle Institute of Technology and Science,
Angallu (V), Madanapalle, Andhrapradesh, India-517325.

10.1   INTRODUCTION

Reduction of oxygen plays a crucial role in energy-converting systems, like fuel cells, metal-air batteries, etc. Based on the nature of reaction medium, ORR mechanism follows two pathways: (i) direct 4e⁻ transfer (ii) 2e⁻ transfer via formation of H_2O_2. In addition to these two, there is a possibility of 1e⁻ transfer pathway from O_2 to superoxide (O_2^-) in non-aqueous medium.

$$O_2 + 4H^+ + 2e^- \rightarrow H_2O_2 \tag{1}$$

$$O_2 + 4H^+ + 4e^- \rightarrow 2H_2O \tag{2}$$

*For Correspondence: Rajendran Rajaram (madurairajaramac@gmail.com);
Sankararao Mutyala (mutyala.sankararao@gmail.com)

Among them, the reaction which follows 4e⁻ pathway produces higher
energy due to the instance 4e⁻ transfer process rather than the others. In Proton
Exchange Membrane Fuel Cells (PEMFCs), ORR occurs at cathode and its
kinetics is generally sluggish. To increase the rate of ORR, and to achieve the
practically usable level in fuel cells, an ORR catalyst is necessary. In the present
scenario, Pt and its related catalysts are found to be most practical materials
(Kozuch and Shaik, 2008, 2011; Skúlason et al., 2010). However, its high-cost
limits its application in viable fuel cells. In addition, it is not freely available in
nature. Hence, researchers across the globe have developed alternative non-noble
metal catalysts like carbon materials doped with hetero atoms, polymers, metal
chalcoginides, sulfides, oxides, etc. (Behret et al., 1981; Shi et al., 2007; Nilekar
et al., 2008).

10.2 MATERIALS IN ORR

Driven by the crucial importance of ORR in the energy conversion process, the
catalyst design should follow some principles, viz. (1) increasing intrinsic catalytic
activities, (2) exposing huge number of active sites and mass transport abilities,
(3) improving electron transfer abilities. Pt and other precious metals can enhance
the cost of fuel cells, hence a large number of non-noble catalysts including
porous carbon, carbon allotropes, metal alloys, metal derivatives, metal-organic
frameworks (MOFs), and conducting polymers, etc. were intensively investigated
as electrocatalysts for ORR.

Since the invention of the conducting polymer (CP) polyacetylne in 1977,
the role of CP is found inevitable in various electrochemical applications, like
organic electronics, electrochromic devices, sensors, actuators, and fuel cells
(Tsakova and Seeber, 2016; Le et al., 2017). It induces interest to produce a variety
of other polymers, like polyanilne (PANI), polypyrrole (PPy), Polythiophene
(PT), poly(3,4)ethylenedioxythiophene (PEDOT), poly(p-phenylene vinylene)
(PPV), etc. They are associated with a number of advantages, like chemical
diversity, flexibility, low density, easy-to-control shape, adjustable conductivity,
morphology, resistance against corrosion, reversible doping/dedoping process.
CP features unsaturation (alternative single and double bonds) which provides
optical, electrochemical, and electrical/electronic properties. Therefore, development
of CP/CP composites towards ORR has gained importance with researchers
focusing their attention on this aspect (Qiu et al., 2012; Liu and Tao, 2017).

10.3 PANI-BASED MATERIALS FOR ORR

PANI is chiefly present in two different forms: emeraldine (ES) and leucoemeraldine
(LS). Emeraldine is the oxidized form of PANI whereas leucoemeraldine is its
reduced form. The former is conductive and the latter is an insulator. PANI is
associated with high conductivity, exclusive redox tunability, better chemical
and thermal stability, adhesion, hydrophilic properties, and low production cost.

In addition, its increased surface area, porous, networked nanostructures assist the formation of wide morphological features, like nanoparticles, nanofibers, nanotubes, nanobelts, and nanoflakes. They can be established by altering the solvents, the oxidants during the polymerization process. Further, PANI can act as a source of carbon which is used in batteries, capacitors, etc. Hence, it is identified as a vital catalyst for various electrochemical applications, like sensors, batteries, Oxygen Evolution Reaction (OER), ORR, and electronic technologies, etc. In the meantime, the activity of the polymer gets inhibited when the pH rises beyond 6. In addition, its low solubility, and meager cycle time limit its applications (Zhong et al., 2012; Stamenović et al., 2018; Chen et al., 2020). In 1994, Cui et al. examined ORR activity of PANI modified glassy carbon electrode (PANI/GCE) in a neutral solution. It is documented that the activity of electrode is poor due to pH rise at the interface which reduces conductivity of PANI. The ORR curve starts to rise at −0.15 V, which is not a notable deviation from the bare glassy carbon electrode. The authors further addressed that raw PANI does not have active sites for ORR in neutral solution. However, upon incorporation of Pt nanoparticles in the matrix, its catalytic activity increases (Cui and Lee, 1994). Therefore, functionalization of PANI has gained enormous attention in order to harvest its electrocatalytic activity. The lone pair of electrons present in N center of PANI act as a coordinate site for transition metals. Hence, a large number of reports were directed to the development of various catalyst using PANI doped with transition metals, like Pt, Pd, Ag, Co, Fe, etc., for ORR applications. In addition to the metals, non-metals like CNTs, Prussian blue (PB) and its analogues have also been doped with PANI. Further, PANI-derived carbon materials are also explored for its electrocatalytic activity towards ORR. The efficiencies of the catalysts are described below.

10.3.1 PANI-transition Metal Composite Materials for ORR

Lai *et. al.* prepared PtNPs decorated PANI–Nafion composite via electrochemical as well as chemical route and examined its ORR activity (Lai et al., 1999). Coutanceau et al. briefly studied ORR activity of PtNPs decorated PANI. In this work, the authors optimized the thickness of the polymer as 0.5 μm. Subsequently, the loading of PtNPs was carried out from 11–600 μg cm^{-2} and it was reported that after 200 μg cm^{-2} of Pt loading, the observed response resembled bulk Pt. The kinetic parameters, like exchange current density, Tafel slope, and number of electrons transferred at the interface at each loading were calculated using Koutecky–Levich (K-L) equation (Coutanceau et al., 2000). In addition to this, Ag NPs have also played a crucial role in ORR along with PANI. In this regard, Stamenovic et al. prepared a catalyst by combining Ag-PANI-polyvinylpyrrollidine (PVP) towards the efficient ORR. They prepared three different combinations of the composite by adjusting the pH, and the concentration of PVP, and studied its ORR activity in alkaline and acidic media. Among them, the one with the lowest Ag content was highly selective to O_2 and was able to produce excellent ORR activity.

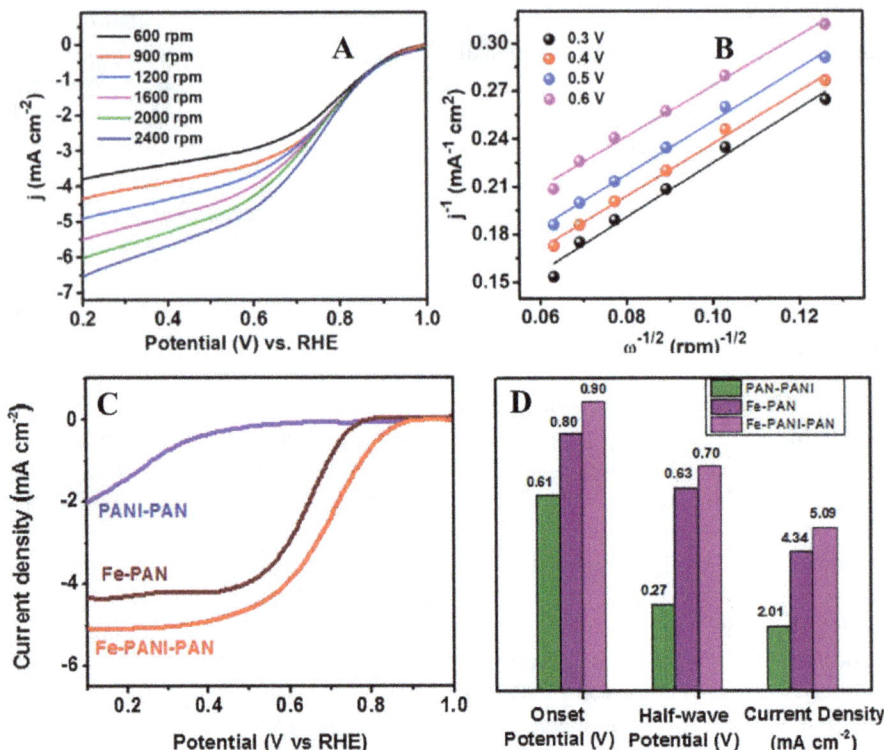

Figure 10.1 (A) Linear sweep voltammetry (LSV) curves of Cu_xO-C/PANI at various rotations range of 600–2400 rpm at 10 mV s^{-1}, (B) its corresponding K-L plots at different potentials (Reproduced from [Chhetri et al., 2021]), (C). LSV of Fe-PAN, Fe-PANI-PAN and PANI-PAN in the electrolyte, 0.1 M $HClO_4$ at 1600 rpm, and (D) evaluation of various electrochemical parameters (Reproduced from [Zamani et al., 2014]).

It is associated with high electronic conductivity as well as dispersion of Ag nanoparticles in PANI (Stamenović et al., 2018). Similarly, Pd also attracts considerable attention towards ORR along with PANI. In this regard, Wang et al. functionalized Pd with the polymer, PANI and utilized it for ORR in alkaline medium. Further, Co-PANI hybrids, like PANI-Co-CNT and Cobalt-Porphyrin/PANI nanocomposites, were also developed to understand their ORR activity in an acidic medium. These reports confirmed that the reaction with these catalysts was via 4e$^-$ pathway (Yin et al., 2014; Zhou et al., 2008). Zamani et al. synthesized a low-cost iron–PANI–polyacrylonitrile (PAN) nanofibers by electrospun to replace Pt. PANI into Fe-PAN helps to improve the E_{onset} from 0.8–0.9 V, $E_{1/2}$ is 0.63–0.70 V *vs.* RHE and thus the role of PANI towards ORR is addressed and shown in Fig. 10.1 (Zamani et al., 2014). In addition, PANI was doped with Cu and the electrocatalytic activity of the composite was examined using ORR as exhibited in Fig. 10.1. In this Cu–MOF was prepared and carbonized at 600°C (2°C min^{-1}) for 4 h. The carbonized material, Cu_xO–C was added with PANI in an acidic medium along with the reducing agent, ammonium persulfate. The resultant product was

Cu_xO–C/PANI. During ORR, it is able to deliver the signal at the E_{onset} of 0.94 V with the $E_{1/2}$ of 0.76 V (Chhetri et al., 2021).

10.3.2 PANI-Non-metal Composite Materials for ORR

Manesh et al. deposited PANI grafted MWCNT on GCE by electrochemical route. While examining the electrocatalytic activity of electrode via ORR in the acidic medium, it produces the signal at positive applied overpotential (0.61 V). Using chronoamperometric technique, the rate constant obtained was 7.92×10^2 M^{-1} s^{-1} (Manesh et al., 2006). Like this, Prussian blue-PANI-based catalysts are also developed for ORR applications (Fu et al., 2011; Wang et al., 2016) and synthesized by hydrothermal route to increase the ORR activity. The resultant materials showed finite activity in both alkaline and acidic conditions. During the alkaline electrolysis, the $E_{1/2}$ was found as 0.85 V, which is close to commercial platinum catalyst, whereas in the acidic medium, the half-wave potential is more negative with respect to the commercial catalyst, Pt/C. In both the media, the reaction follows the $4e^-$ transfer pathway. This report clearly stated that the performance of the catalyst was better with other non-pyrolysed Prussian blue (PB)/PBA based catalysts (Wang et al., 2016). Further, Fu et al., developed another PB-anchored PANI catalyst towards ORR. PANI was deposited over the surface of spectrographic pure graphite rod (SPG), followed by the conversion of emeraldine base (EB) PANI into emeraldine salt (ES) PANI using 5% ammonia. Then this electrode was dipped for 12 h into the solution of 0.1 M $Fe_2(SO_4)_3$ to obtain $(FeSO_4)_4^{2-}$ anchored PANI. Finally, this electrode was dipped in 0.1 M H_2SO_4 having 0.05 M K_3[Fe(CN)$_6$] for 12 h to get the PB/PANI modified electrode. This electrode was found successful towards ORR in acidic medium (Fu et al., 2011). Besides these catalysts, nitrogen-doped carbon matrices, which are derived from pyrolysis of PANI, have also shown excellent electrocatalytic activity. Apart from PANI, nitrogen can be obtained from other organic compounds, like pyridine, pyrrole and graphitic nitrogen. Zhong et al. synthesized polymerized aniline, using PVP as a stabilizer and it was heated to produce nitrogen-doped carbon (NDC). This material was used for ORR applications in alkaline medium (Zhong et al., 2012).

10.4 POLYINDOLE-BASED CATALYST MATERIALS FOR ORR

Polyindole (PIN) has also received considerable attention in electrochemical applications. PIN is the polymer of indole, consisting of 5-membered indole and 6-membered benzene rings. It shows the properties of polyphenylene and polypyrrole together. When compared with PANI, PIN shows slow hydrolytic degradation and better thermal stability. Also its redox potential is comparable with PPy. Like PANI, PIN does not form salts like leucoemeraldine and pernigraniline during the charge-discharge phase. Its improved internal conductivity is very helpful for energy storage applications. Because of the poor polymerization ability and conductivity, PIN has not been focused like other polymers. However, introduction of various PIN-based composites/copolymers/derivatives leads to considerable attention on the development of PIN-based catalysts. Magdalena et al. deposited PIN on Au substrate in aqueous acidic electrolyte and studied its electroctalytic activity. The report confirms that the polymer is effective for ORR to produce hydrogen peroxide, i.e. $2e^-$ transfer mechanism in Na_2SO_4, showing finite activity and high faradaic efficiency at low pH value (Magdalena et al., 2020). Verma et al. synthesized V_2O_5 hydrothermally and it was integrated with nanosized Au encapsulated PIN to prepare Au-V_2O_5/PIN. It was characterized by using various spectroscopy and microscopy techniques. The catalyst is able to increase the current against ORR in the electrolyte of 0.1 M KOH + 0.1 M C_2H_5OH due to synergism between V_2O_5 NPs and Au-embedded PIN (Verma et al., 2020). Another report combines Fe, PIN and CNT/C to prepare an ORR catalyst for microbial fuel cell (MFC) applications. In this work, the authors prepared two catalysts, namely, FePc/CNTs and FePc/PID/CNTs. Both of them showed that the ORR waves at more positive potential and higher current density to Pt/C relatively. Among these two materials, FePc/PID/CNTs exhibited the best performance. It helped to attain the extreme power density (799 mW m^{-2}) and current density (3480 mA m^{-2}) in MFCs. The values are found better than the effectiveness of Pt/C, which produces the power and current densities at 646 mW m^{-2} and 3011 mA m^{-2} respectively (Nguyen et al., 2016). Sun et al. prepared the N and Fe co-doped carbon nanospheres via the pyrolysis of poly(bis(N-indolyl)-1,2,4,5-tetrazine)-Fe(II) at 900°C and synthesized two catalysts, namely Fe/N-C-900-HT and Fe/N-C-900-MS. ORR response of the materials was evaluated in both acidic and basic media. It is reported that due to increased specific surface area, and higher N/Fe content, the response of Fe/N-C-900-HT is better than Fe/N-C-900-MS. In the acidic medium, Fe/N-C-900-HT showed slightly lower E° value along with higher current density and E$_{1/2}$. The values were 0.913 V, 0.786 V, and 5.88 mA cm^{-2} respectively. In case of Pt/C, the values were 0.91 V, 0.75 V and 5.34 mA cm^{-2} respectively. It is documented that Fe/N-C-900-HT is one of the finest materials for ORR. The performance of Fe/N-C-900-HT is comparable with Pt/C in H_2–O_2 fuel cells (Sun et al., 2019).

10.5 PPy-BASED CATALYST FOR ORR

Pyrrole is an easily oxidizable, commercially available, water-soluble organic compound. PPy exhibits a number of advantages, like high electrical conductivity, exceptional redox properties, good electron affinity, low oxidation potential, and superior environmental stability. Hence, its role is crucial in several sectors, like batteries, sensors and supercapacitors, etc. Because of its improved conductivity, detailed studies were carried out to know its catalytic properties. The physical properties of PPy, like rigidity, crystallinity, and brittle nature limit its applications. Therefore, it is necessary to improve the processibility of the polymer. Researchers found that the processibility can be improved via co-polymerization with different nanomaterials. So, several matrices have been incorporated with PPy in order to improve its electrocatalytic properties and their efficiencies as described in following sections.

10.5.1   PPy-transition Metal Composite Materials for ORR

Feng et al. synthesized pyrolyzed carbon-supported cobalt PPy without and with the dopants. They monitored the effect of the matrix in the presence as well absence of dopants in ORR. They concluded that the number of electrons transferred at electrode-electrolyte interface was 2.7 and 3.1 with respect to presence and absence of electrode of dopants. Among the developed catalysts, the one which is prepared by using the dopant BSNa produced better catalytic activity due to its greater surface area, micropores, nitrogen and Co contents (Feng et al., 2012). Yuan et al. prepared the catalyst, Co-PPy-TsOH/C by heating carbon-based Co-PPy at different temperatures for various time intervals. During pyrolysis, precursor cobalt acetate was converted into CoO and metallic Co by thermal decomposition and chemical reduction with increment in temperature and time. It was concluded that the one which is pyrolyzed at 800°C for 2 hours produces better ORR activity in the acidic medium and is highly selective to four electron transfer pathway (Yuan et al., 2014). A similar catalyst was tried by Li et al. using 0.1 M KOH. The authors monitored the effect of Co precursors during ORR. Several Co precursors, like Co-acetate, Co-sulfate, Co-nitrate and Co-oxalate were used to prepare the catalyst. The authors concluded that 800°C and Co-acetate are the finite optimized conditions for pyrolysis (Li et al., 2015). Chung et al. identified the catalyst Co-PPy-C and its ORR activity was carried out in 0.5 M H_2SO_4. It is able to show the activity with $E_{1/2}$ of 60 mV. The electrode was stable and showed better ORR performance (Chung et al., 2014). In another report, the authors loaded Co on polypyrrole-supported carbon and optimized the temperature as well as the amount of Co in the system towards ORR applications in alkaline medium. With the help of XPS analysis and RRDE, the structure to the property correlations was elucidated, expressing a dual site ORR mechanism. Co-N_x type site facilitates a 2e⁻ reduction, by which formation of HO_2^- takes place. In the second step, OH⁻ and O_2 are formed by either electrochemical reduction or chemical disproportionation. Using RRDE alone, they could not conclude which type of mechanism is taking place on Co_xO_y/Co surface (Olson et al., 2010). Similarly, Nguyen-thanh et al. developed an ORR catalyst using PPy and Co deposited carbon black (CB) at different ratios. The experiment was carried out in the electrolyte $HClO_4$ (pH = 1) using RRDE and concluded that the best combination of the catalyst has the weight ratio of CB:PPy as 2 and molar ratio of Pyrrole:Co as 4. Using the catalyst, the onset potential and mass activity were achieved as 0.785 V and 1 A/g $_{cata}$ respectively. Further, they observed that the number of electrons transferred was 3.5 with the formation of H_2O_2, whose yield was 28%. Hence, it was believed that the reaction follows 4-electron transfer mechanism, where H_2O_2 is found as the intermediate (Nguyen-thanh et al., 2011).

In addition to Co, people have combined PPy with Co and other transition metals like Ni and Cu. In this regard, Cu/Co containing PPy hydrogel was prepared, using supramolecular self-alloying strategy and showed its catalytic activity in Fig. 10.3. Its activity against ORR has been recorded in alkaline medium. Here, the authors prepared the materials, PPy/CuPcTs, PPy/CuPcTs/Co, which can retain 75% and 80% current density for 35000s. The scheme involved in the preparation of PPy/CuPcTs hydrogel is provided as Figure 10.2. Hence, it was concluded that it is highly stable during ORR (Meng et al., 2020). Li et al. prepared hydrogels, with Ni containing PPy (NiPcTs/PPy hydrogel) and Ni and Co containing PPy (NiPcTs/Co/PPy hydrogel). They have been converted into catalysts after pyrolyzing. Both of them were able to retain the current densities of 75% and 80% respectively for 35000s. The activity of the catalyst is similar to 20% Pt/C (Li et al., 2020).

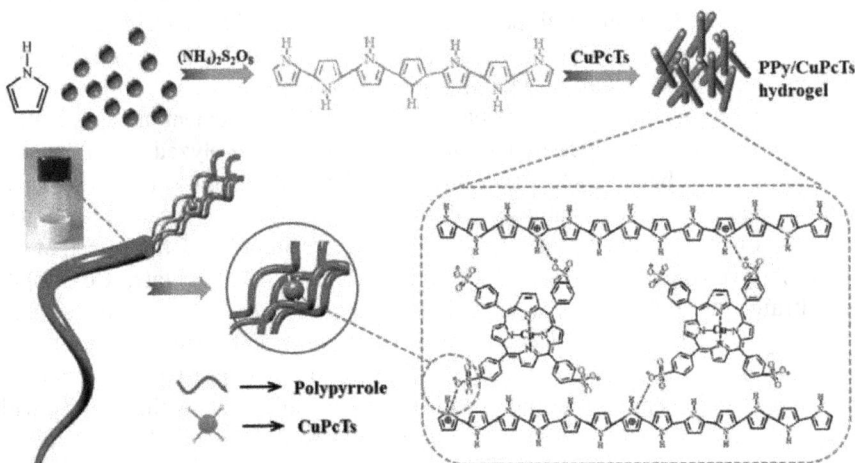

Figure 10.2 Schematic representation for the synthesis and assembly mechanism of PPy/CuPcTs hydrogel (Meng et al., 2020).

Nguyen-cong et al. have synthesized a spinel, $Ni_xCo_{3-x}O_4$ (x = 0.3 and 1) with Co-nitrate and Ni-nitrate. At first, glassy carbon electrode (GCE) was electrochemically modified by using PPy. The electrode was subsequently modified as GC/PPy/PPy(Ox)/PPy. With the help of the synthesized spinel, the authors obtained GC/PPy/PPy($Ni_xCo_{3-x}O_4$)/PPy composite electrode. The electrode was tried for ORR applications. The reaction took place at the oxide particles with the formation of H_2O_2 (Nguyen-cong et al., 2003). Bozzini et al., 2015 synthesized ternary manganese-based PPy composite materials, such as Mn–Co–Cu/PPy, Mn–Co–Mg/PPy, Mn–Ni–Mg/PPy and examined their electrochemical ORR activity in KOH. Among the prepared catalysts, Mn–Co–Cu/PPy produces better ORR activity via 4 electron reduction mechanism. $E_{1/2}$ and E_{onset} were obtained as –0.22 V and –0.122 V respectively. Further, Fe has also played a crucial role in ORR along with PPy. In this regard, Tran et al. synthesized Fe-treated nitrogen-doped carbon using pyrolysis–leaching–stabilization (PLS) sequence of PPy, where the polymer served as a source of carbon and nitrogen. The prepared material was found as an effective ORR catalyst in both acidic as well as alkaline medium. Pyrolysis is one of the stages for the synthesis of the catalyst. It was carried out at different temperatures, ranging from 800 to 1000°C. Among them, the catalyst which is prepared at 900°C produces better oxygen reduction reaction at E_{onset} 0.96 V (*vs.* RHE) and this activity is very near to commercial (Pt-based) catalyst

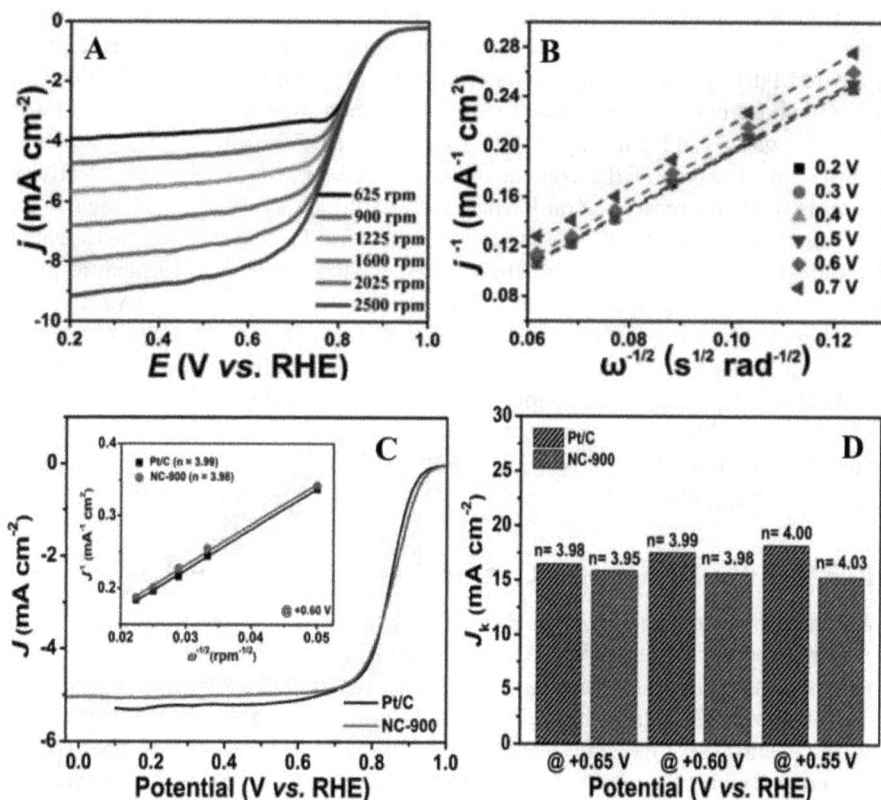

Figure 10.3 (A) LSV responses of CNTs@PPy-P800 at different rpm. (B) Its equivalent K-L plots (An et al., 2016). (C) Comparison of LSVs of NC-900 and Pt/C in basic medium at 1600 rpm, 10 mV s^{-1} scan rate and its corresponding K-L plots at 0.60 V (*inset*). (D) J_k and n of catalysts. NC-900 and Pt/C at 0.65, 0.60, and 0.55 V (*vs.* RHE) during ORR (Yang et al., 2016).

with $E_{1/2}$ of 0.877 V which was more positive to Pt/C. In the acidic condition also the performance of the material is comparable with Pt/C. In both cases, four electron transfer mechanism is followed. In addition, it is concluded that the catalyst is stable and possesses methanol-tolerance effect. The better catalytic activity is attributed to the new synthesis protocols adopted (Tran et al., 2016). Shang et al. derived another Fe-N-C catalyst using graphite, distilled pyrrole, and $FeCl_3 \cdot 6H_2O$. The catalyst having 4% iron has better ORR activity, where the mechanism follows 4-electron transfer (Shang et al., 2015). Osmieri et al. developed another Fe-N-C catalyst with the help of porous carbon, PPy and Iron acetate. They served as sources of carbon, nitrogen, and iron respectively. ORR studies are carried in basic and acidic solutions (Osmieri et al., 2016). Ren et al. found a hydrothermal route to synthesize Fe_2O_3/PPy/GO [GO – graphene oxide] and optimized various parameters, like Fe content in GO and heat treatment. It showed electrochemical (ORR) activity, where E_{onset} was observed as −0.1 V (*vs.* SCE) in alkaline medium. The stability of the material reduces only 5%

after 5000 cycles (Ren et al., 2015). Besides Co and Fe, other transition metals, like Pd, Mn, Cu also contributed in ORR along with PPy. Góral-kurbie et al. prepared Pd/PPy via varying the amount of Pd (2–20%) and utilized for ORR in H_2SO_4 whose concentration was 0.5 M. The activity was found good when Pd is in the range of 2–12% in PPy. In this range, the reaction proceeds via 4-electron transfer mechanism. If the content of Pd increases beyond 12%, the selectivity towards water decreases (Góral-kurbiel and Drelinkiewicz, 2013). Electrocatalytic ORR activity of the composite made by using PPy, Pt, and Cu was investigated by Saremi et al. They modified the catalyst on the surface of a graphitic plate via potential cycling and galvanostatic (GS) methods. Among these two, the one which is prepared by using GS method provides better activity via 4-electron mechanism due to its higher porosity (Saremi and Salehisaki, 2014). Further, KOH-activated method was employed to yield oxygen-doped carbonaceous PPy nanotubes (OCPN) and Ag nanoparticles were grown in it *in-situ*. This catalyst was able to show ORR activity, which is comparable to Pt/C in basic medium where the E_{onset} observed was –0.003 V (*vs.* SCE) (Xiao et al., 2018).

10.5.2   PPy-Non-metal Composite Materials for ORR

Wei et al. analyzed the synergetic effect between SWCNT and PPy. The authors prepared the composite PPy/SWCNT and analyzed it for its ORR activity. They concluded that the composite having weight ratio of 1:2 (PPy:CNT) is able to show better ORR activity and found the mixture of 2- and 4-electron pathway during electrolysis (Wei and Tanabe, 2011). Further, activated carbon fiber (ACF) was coated with PPy to obtain ACF-PPy composite. It was carbonized at various temperatures (500°C and 800°C) to obtain N-doped ACF, and it was checked for its electrochemical ORR activity in alkaline medium at these temperatures where this catalyst material showed finite activity at 800°C. The electrode follows 2-electron pathway (Ramírez-pérez et al., 2018). An et al. synthesized nitrogen-doped carbon nanotubes (NCNTs) by pyrolyzing CNT@PPy composite at 800°C. The E_{onset} and $E_{1/2}$ obtained during the measurement were 0.95 V and 0.81 V respectively. The limiting current density obtained with this catalyst was found superior than other N-CNT-based catalysts. The kinetic parameters are examined by using the K-L equation. K-L graphs in the range of 0.2–0.7 V showed a good linear range with consistent slope, which hints at the first order kinetics involved in ORR and showed 3.72–3.94e$^-$, suggesting the 4e$^-$ transfer process (An et al., 2016). Marjanovic et al. reported PPy-NTs along with ferric chloride and methyl orange. Upon carbonization, nanotubes were converted into N-containing, conducting, tubular carbon material (C-PPy-NT). This material produces significant ORR behavior in at alkaline medium. During this analysis, the E_{onset} was observed as –0.1 V (*vs.* SCE). The authors found that the value is comparable with carbon matrices derived from other polymers, like PANI (Ćirić-Marjanović et al., 2014). Yang et al. produced nitrogen-doped carbon (NC) by pyrolyzing the triazine-based PPy network (TPN) at 900°C. The material was employed as an electrocatalyst in alkaline medium, where it exhibited 0.97 V (*vs.* RHE), 0.85 V of E_{onset} and

$E_{1/2}$ consequently. They are comparable with the values received from Pt/C catalyst (Yang et al., 2016). Lin et al. prepared nitrogen-doped graphene (NG) via pyrolyzing GO-PPy composite at higher temperature for 30 min. in an inert atmosphere. It catalyzes ORR in 0.1 M KOH by following 4-electron transfer mechanism. It is highly stable and produces methanol tolerance effect (Lin et al., 2013).

10.6 CARBON-BASED MATERIALS FOR ORR

Carbonaceous materials were cheaper and promising to replace Pt-based catalysts for fuel cell applications because they possess high surface area, tunable surface morphology, conductivity, O_2 adsorption, and ease of heteroatom doping, etc. In these aspects, numerous heteroatom-doped carbon materials were studied for ORR characteristics. Several metals in various forms were anchored with carbon materials to derive their electrocatalytic activity against ORR. Dai's group first showed carbon materials as promising candidates for ORR applications in fuel cell in 2009 (Gong et al., 2009). This study showed carbon-based materials for replacing the costly noble metals, and opened a new area of heteroatom functionalized carbon as an efficient catalyst for the ORR. In this respect, B, S, P and I-doped CNTs and graphene and edge-halogenated (Cl, Br or I) graphene nano materials were emerged (Yang et al., 2011; Liu et al., 2011; Sheng et al., 2012; Yang et al., 2012; Yao et al., 2012; Jeon et al., 2013a; Jeon et al., 2013b; Ma et al., 2014; Wang et al., 2015). In connection with mono hetero atom, binary and ternary heteroatom doping also increase the ORR activity. In this respect, B and N-doped carbon nanotubes and graphene (Wang et al., 2012; Xue et al., 2013), N and S-doped graphene (Xu et al., 2013; Wang et al., 2014); B, N and P-doped graphene (Choi et al., 2013; Ma et al., 2016); were identified as better electrocatalysts for ORR activity. Yang et al. identified B-doped CNTs as electrocatalysts, where the peak current rises to 8.0 mA mg^{-1} with respect to loading of B in the catalyst. Further, peak movement is also observed from 0.43 to 0.35 V *vs.* SCE as displayed in Fig. 10.4 (A and B). CO-tolerance characteristics of the catalyst were also evaluated. This clearly signifies B doping turn carbon nanotubes into finite ORR catalyst with enhanced peak current, high stability and less carbon monoxide poisoning (Yang et al., 2011). Sheng et al. prepared Boron doped Graphene (BG) towards ORR in the same experimental conditions (Sheng et al., 2012). The electrode is able to show activity with E_{onset} of −0.34 V. The ORR of bare and pure graphene/GCE showed at −0.45V, −0.36V where the observed reduction peaks are as shown in Fig. 10.4 (C and D). The observations hint at ORR taking place via two-step processes with the production of HO_2^- intermediate. In connection with the B-doped carbon, Yang *et. al.* prepared S-doped graphene using a scalable synthetic approach to monitor the ORR in 0.1 M KOH (Yang et al., 2012). It showed higher catalytic activity towards O_2 with higher methanol tolerance ability than Pt/C. The authors found that S doping is the dominant factor for higher catalytic activity.

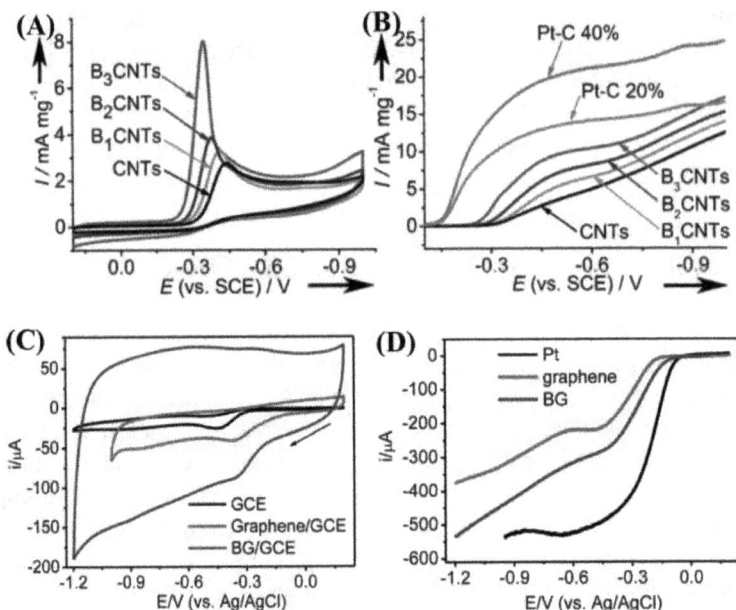

Figure 10.4 (A) Cyclic voltammetric responses of BCNT materials (50 mV s^{-1}). (B) Comparison of RDE responses of BCNT catalysts with Pt/C catalysts (2500 rpm and 10 mV s^{-1}) (Reproduced from [Yang et al., 2011]). (C) Comparison of Cyclic voltammograms and (D) LSV response of ORR of BG/GCE with bare GCE, and graphene/GCE in alkaline electrolyte (Reproduced from [Sheng et al., 2012]).

Similar to these catalysts, edge selective sulfurized graphene, and sulfur-doped graphene were also introduced by various researchers (Yang et al., 2012; Jeon et al., 2013b; Ma et al., 2014; Wang et al., 2015). In the one-step magnesiothermic reduction strategy, the CO_2 transforms into S-doped graphene in the presence of small amounts of Na_2SO_4 (S source) (Wang et al., 2015). At high temperature (800°C), Mg metal reduces carbon in CO_3^{2-} to graphene materials. Meantime, SO_4^{2-} is served as S source for formation of C–S bond in the carbon network with +6 oxidation state reduced into bottom oxidation by magnesium. Hence, it clearly signifies that the incorporated S in graphene greatly influences the morphological features of graphene sheets. The resultant materials exhibit finite activity in terms of onset potential, current density, charge transfer properties. Interestingly, towards the renewal of materials to valuable uses, Ma et al. also tried sulfur-incorporated graphene materials with ORR catalyst in an alkaline medium (Ma et al., 2014). The authors obtained S-doped graphene from graphene sulfur composite, which is cathode material in Li-S battery. After the charge-discharge process (0.2 C rate of 100 cycle), S incorporation of graphene was observed. Afterwards, electrocatalytic ORR was recorded in an alkaline condition and compared with Pt/C catalyst. Interestingly, S-doped graphene from the Li-S batteries showed E_{onset} and ORR potential at –0.15 and –0.34 V *vs.* SCE. The observed $E_{1/2}$ potential (–0.37 V *vs.* SCE) and limiting current are higher than that of pristine graphene. These results were closer to the commercial benchmark Pt/C catalyst. Liu et al. studied P-doped graphite layer as finite material for the ORR (Liu et al., 2011). The authors prepared P-incorporated graphite using theromolysis route and triphenylphosphine (TPP) using tubular furnace at 1000°C. The flow rate of Ar and toluene were maintained as 600 mL min^{-1} and 10 mL min^{-1} respectively with 2.5 wt.% TPP. Cyclic voltammogram showed two significant peaks and interestingly ORR activity of the catalyst material showed better response than bare glassy carbon electrode (GCE) and graphite/GCE. These results are clear indicative of P incorporation in carbon network, and the structural properties of resultant materials. Hence, P incorporated graphite/GCE exhibited a finite onset potential for the ORR than Pt–C/GCE. Yao et al. prepared the iodine-incorporated graphene (Yao et al., 2012) (I-graphene-900), which showed positive E_{onset} and improved current density closer to Pt/C and other counter samples. Through XPS analysis, the origin of I-graphene sample from the various temperatures (during preparation) of 500–900°C was studied where I_5^- transformed to I_3^- followed by decrease in its weight percentage from 1.21 to 1.05. This suggests that I_3^- has a higher ORR catalytic activity than I_5^-. Due to its more negative charge, more positive charge on carbon network is seen for better adsorption of O_2 followed by reduction to OH^-. LSV curves of few of the Carbon based catalysts towards ORR is shown in Fig. 10.5.

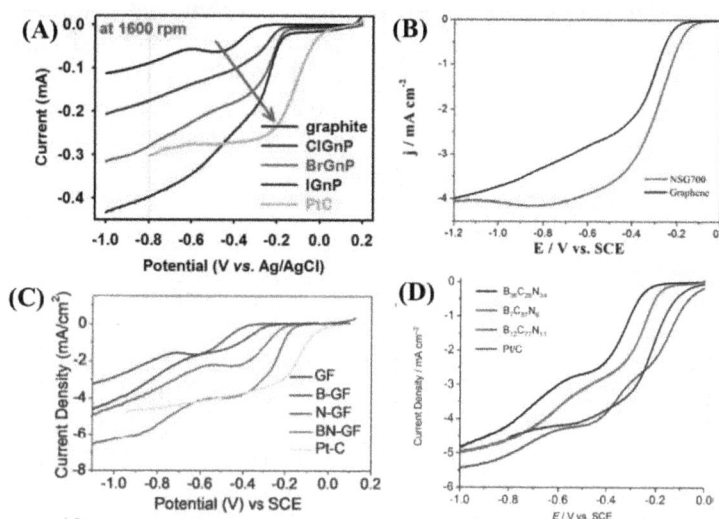

Figure 10.5 (A) LSVs of halogenated (X = Cl, Br, I) graphene nanoplatelets (XGnPs = ClGnP, BrGnP, IGnP) at 1600 rpm and 0.01 V s^{-1} (Reproduced from [Jeon et al., 2013a]). (B) Ring current density of NSG700 and graphene (Reproduced from [Wang et al., 2014]). (C) LSV of GF/GC, B-GF/GC, N-GF/GC, BN-GF/GC and Pt–C/GCE (Reproduced from [Xue, 2013]). (D) LSV curves of BCN graphene with different compositions with Pt/C catalyst (Reproduced from [Wang et al., 2012]).

In connection with the B, S and P-doped carbon materials, Jeon et al. synthesized halogen-incorporated graphene nanoplates by ball milling techniques in the presence of respective precursors for ORR in an alkaline medium (Jeon et al., 2013a). As shown in Fig. 10.6(A), ORR catalytic activity of the catalysts is in the following order: pristine graphite < ClGnP < BrGnP < IGnP < Pt/C in 0.1 M KOH. The ORR peak shift in the positive direction of ClGnP, BrGnP and IGnP are moved to -0.24 V, -0.22 V, -0.22 V respectively and the corresponding catalytic current values are to -0.39, -0.60 and -0.78 mA cm^{-2}. The results showed a finite activity in terms of current density, onset potentials, and ORR response compared to bare graphite materials (-0.28 mA cm^{-2}). To support these findings and to get insights of catalytic activities of XGnP, the authors carried out the DFT calculation studies further. XPS and other supporting spectroscopy studies revealed that incorporated halogen atoms have a strong bond formation with carbon network and synergistic effect between these two atoms are responsible for improved electrolcatalytic activity. Further, these halogen atoms have huge number of edge plane sites. With halogen atoms in the single sp^2 C–X dangling bond, binding affinity of O$_2$ was lesser. Where halogen was substituted in the C site of zigzag edges, the halogenated edges revealed decent binding affinity with O$_2$ and oxygen reduction reaction responses of halogen-incorporated materials as IGnP > BrGnP > ClGnP.

Xu et al. studied S and N-doped graphene layer (SNGL) as an effective catalyst for ORR applications (Xu et al., 2013). S and N-induced graphene was prepared using pyrimidine and thiophene sources. This dual S and N-doped pyrrolic/ graphitic N dominant structure exhibits higher catalytic activity than counter of mono N-doped graphene. The authors prepared S and N-doped graphene by the chemical vapour deposition (CVD) method and investigated the effect of precursor by changing thiophene volume ratio as 10, 20 and 33 vol.%. They found improvement in catalytic activity with the concentration of thiophene and K-L equations are approximately 4.8 and 10.1 mA cm^{-2} respectively for SNGL-10 and SNGL-33 respectively. Both the values are greater than that of non-doped CNT and graphene. Similarly in another study, N and S-doped graphene (NSG) was synthesized by using graphene oxide and thiourea by Wang et al. (Wang et al., 2014). The NSG showed 4e^{-} transfer efficient activity, higher E$_{onset}$ with higher current density in 0.1 M KOH and comparable to precious benchmark platinum catalyst as shown in Fig. 10.6(B). Xue et al. synthesized B and N-doped graphene in CVD, using methane along with NH$_3$ (BN-GFs) (Xue et al., 2013). The doping level of boron was 2.1 atom %, which was prepared by using toluene and triethyl borate. They are identified as the sources of C and B respectively. The authors compared ORR activity of BN-GF with undoped GF, B-GF, N-GF, Pt/C (Pt: 45 wt.%, Vulcan XC-72R) through cyclic voltammetry in basic medium at scan rate of 0.1 V/s its E$_{onset/peak}$ and the peak current, are shown in Fig. 10.6C. Significantly, the synergetic effect of B and N in the co-doped graphene showed highest ORR activity, which is slightly better than Pt/C. The N and B atoms are responsible for the active sites in connection with carbon.

Further, dual-doped (X = B, P and S) graphene (NXG) was prepared by Ma et al. via a facile solvothermal method using ionic liquids and used it as an electrocatalyst for ORR (Ma et al., 2016). The limiting of current density during ORR was observed at 4.4 mA cm^{-2} for NPG, which is higher than pristine graphene. The improved performance of Co-incorporated graphene is because of polarized heteroatom adjacent carbon for the adsorption and bonding of HO$_2$.

Further, addition of heteroatoms (B or P into N-doped graphene) results in asymmetry of charge delivery for the fast e⁻ movement at basal plane of graphene. Minimization of energy gap of HOMO and LUMO is observed. NPG showed closer catalytic activity of Pt/C.

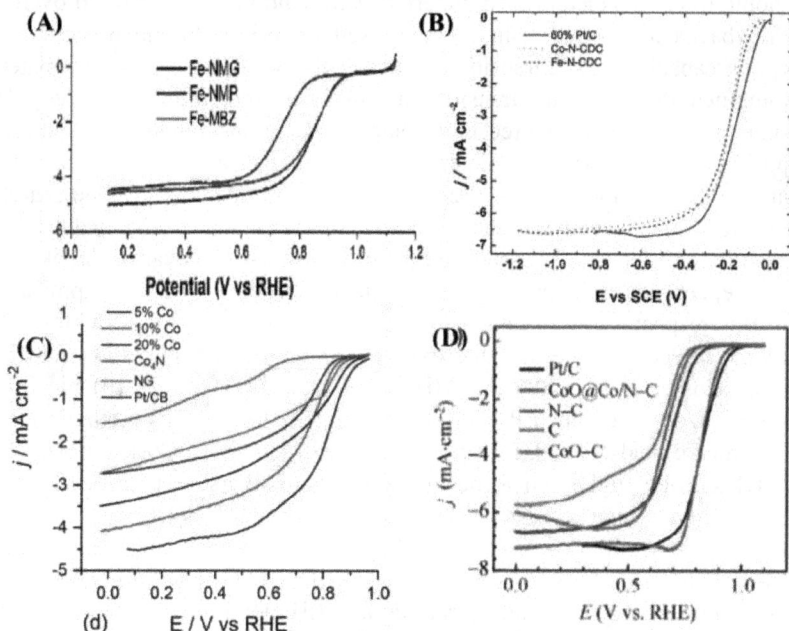

Figure 10.6 (A) RRDE polarization data for Fe-NMG, Fe-NMP and Fe-MBZ (Reproduced from [Hossen et al., 2017]). (B) A comparison of Fe-N-CDC, Co-N-CDC and platinum materials modified GCE (Reproduced from [Ratso et al., 2017]). (C) Co₄N/NG catalyst materials at various loading of cobalt and platinum on GCE (Reproduced from [Varga et al., 2018]) and (D) CoO@Co/N–C, N–C, C, CoO–C, and Pt/C (Reproduced from [Huang et al., 2014]).

To extend the heteroatoms doping effect, Wang et al. further studied the triple incorporation of B, C, N in the carbon towards improvement of ORR response (Wang et al., 2012). Herein, the BCN graphene was synthesized through facile approach with tunable B/N co-doping by thermal annealing of GO along with boric acid and NH_3. Further resultant BCN graphene showed ORR activity, which can be comparable to platinum catalyst in the alkaline medium, as shown in Fig. 10.6(D). To support these findings, the authors calculated the HOMO and LUMO energy gap of BCN and pristine graphene.

10.7 TRANSITION METAL-BASED CATALYSTS FOR ORR

Due to higher gas adsorption characteristics and heteroatom in the functionalized carbon, metal-free carbonaceous materials attracted interest in ORR. However, carbon corrosion leads to poor stability, which lowers its catalytic activity when compared to Pt/C. Since the carbon materials are less adhesive with gas diffusive layer, it can be peeled off quickly. Mechanical strength of the carbon materials is also less. Hence, the fuel cell components require a new kind of catalyst materials. Non-noble transition metal-based materials were paid more attention to overcome the drawbacks associated with carbon-based materials. In the meantime, the adsorption capacity of bare transition metals is also very low. So, carbon materials are combined along with the transition metals to enhance the ORR activity. Here we summarize the interesting recent developments of transition metal-based carbon composites for ORR. The combination shows better catalytic activity than those of parent transition metal-based (Fe, Co, Mn and Ni) materials. While searching Pt-free ORR electrocatalysts, the Fe-based materials gain a significant amount of attention. Because it is associated with a number of advantages, like large abundance, cost-effective, making easy commercialization of fuel cell applications.

In this respect, Fe-N-C catalyst was prepared by Hossen et al. with the help of organic molecules to assess the oxygen reduction reaction response in basic medium (Hossen et al., 2017), i.e. (i) mebendazole (Fe-NMG), (ii) nicarbazin, methylimidazole and glucoril (Fe-NMP) and (iii) nicarbazin, methylimidazole and pipemedic acid (Fe-MBZ) were used to extract the catalyst. Among them, Fe-NMG showed higher ORR performance than others in both rotating ring electrode and fuel cell test, as can be visualized in Fig. 10.6(A).

In connection with carbon and heteroatom-doped carbon materials, Ratso et al. identified two catalysts: (i) carbide-derived carbon (CDC) by selective removal of non-carbon atoms in the metal carbide and (ii) metal (Co or Fe) incorporated nitrogen-doped carbon, which was derived from carbide materials. They were investigated for their ORR behavior under alkaline conditions (Ratso et al., 2017). In 0.1 M KOH O_2 saturated solution, Fe incorporated nitrogen-doped carbide-derived carbon (Fe-N-CDC) showed the onset potential –0.1 V *vs.* SCE. Under identical conditions, Fe-N-CDC and Co-N-CDC showed catalytic activity closer to each other as well as Pt/C, as shown in Fig. 10.6(B). Varga et al. studied transition metal-nitrogen doped graphene materials, like Co_4N/NDG, a finite material towards ORR in basic solution (Varga et al., 2018). Cobalt-nitride (Co_4N) was decorated on nitrogen-doped graphene sheets. In the ORR catalytic activity measurements, E_{onset} was observed as 0.91 V (5% Co), which is in closer to platinum/CB (0.96 V). Further, the authors observed that E_{onset} decreases slightly while there is increase of Co content (0.88 V in case of 10 and 20% of prepared Co_4N/nitrogen doped graphene). Notably, the onset potential was only 0.69 V on Co_4N alone in the absence of graphene support, as shown in Fig. 10.6(C). Based on these studies, the authors believed that composite constituent of both N-doped graphene and metal are active sites in the ORR. Here electronegativity of nitrogen is higher than carbon and partially-charged carbon and nitrogen species were highly favorable for adsorption of O_2. In the N incorporated graphene and metal composite, the metal atoms are positively charged due to the electronegative nitrogen present in their neighboring side. Hence, it further favors the ORR.

In connection with previous studies, Huang et al. inspected the ORR response of CoO@Co/N–C (Huang et al., 2014). The material was prepared by mixing cobalt (II) chloride hexahydrate and phenanthroline at room temperature. Then carbon black powder was added with refluxion at 95°C for 4 h and then annealed by flowing Ar at 700°C for 1 h for ORR studies in basic medium. CoO@Co/N–C exhibited ORR, where E_{onset}, peak and limiting current densities at 0.99, 0.79 V (*vs.* RHE) 7.07 mA cm^{-2} respectively, as shown in Fig. 10.6(D). Based on a detailed analysis, higher catalytic activity of CoO@Co/N–C was due to synergism between C–N, Co–N–C, and CoO@Co moieties. In connection with this, Liang and co-workers revealed Co_3O_4 nanoparticles decorated with mild oxidized GO as ORR catalyst in alkaline medium (Liang et al., 2011). The hybrid Co_3O_4/ graphene oxide exhibits superior stability and comparable catalytic ORR response with respect to Pt. In the K-L plot analysis, Co_3O_4/GO showed 3.9 e$^-$ transfer @ 0.60V *vs.* RHE, which is closer to the benchmark catalyst Pt/C. Similar to this, Ania et al. studied the ORR activity of Cu/rGO, which was obtained by thermal decomposition of copper-based MOF with graphite oxide in alkaline medium (Ania et al., 2015). The Cu/rGO showed better catalytic activity, CH_3OH oxidation and long-term stability.

Yang et al. prepared rock-salt-type $MnCo_2O_3$/C catalyst to study ORR in alkaline medium (Yang et al., 2019). The rock-salt-type $MnCo_2O_3$/C and Co_3O_4/C were prepared by ammonia assisted complex formation followed by solvothermal process with carbon support ($MnCo_2O_3$ and carbon support as 1:1). The other counter of CoO/C and $MnCo_2O_3$/C were obtained by reacting Co_3O_4/C and $MnCo_2O_4$/C with ammonia around 290°C for 180 min. In 1 M KOH, $MnCo_2O_4$/C showed promising ORR half wave potential of 0.84 V than compared to other samples, as shown in Fig. 10.7(A) and that is closer to the performance of 20 wt.% Pt/C (0.89 V). In the K-L plot analysis, $MnCo_2O_3$/C reaches the diffusion-limited current density value at −3.6 mA cm^{-2} closer to Pt/C, which signifies the 4e$^-$ transfer process. Towards the development of non-precious and highly active ORR catalyst, Fu et al. studied nitrogen-doped graphene and transition metal-based composites (FeCo–N–rGO) and its counters, N–rGO, Fe–N–rGO and Co–N–rGO (Fu et al., 2013). The sample was prepared over pyrolysis of a mixture having Fe, Co salts alone and with polyaniline and rGO in controlled Ar atmosphere at 850°C. The ORR measurements suggest that Fe–N–rGO, Co–N–rGO and FeCo–N–rGO had positive E_{onset} as well as higher limiting current density compared to N–rGO (Fig. 10.7(B)). Further, FeCo–N–rGO has lower E_{onset} than Pt/C and positive shift of half-wave potential. Zeng et al. prepared and characterized carbon supported non-precious transition metal nitrides (Co_3N/C, MnN/C, Fe_3N/C, VN/C, and CrN/C) as efficient oxygen reduction electrocatalysts for alkaline medium (Zeng et al., 2022). The authors prepared by dispersion of metal precursor on carbon support at 80°C in oil-bath followed by annealing at higher temperatures, 300°C to 800°C in the presence of ammonia for deliberate incorporation of N on the metal carbide carbon composite. The ORR studies of synthesized materials are assessed by using alkaline solution and as shown in Fig. 10.7(C). From the analysis, the authors concluded the mass transport limiting current density value to be 3.7 mA cm^{-2}. Further, ORR activity of metal nitride electrocatalysts was found in the order of Co_3N/C > MnN/C > Fe_3N/C > VN/C ≈ Ni_3N/C > CrN/C > TiN/C.

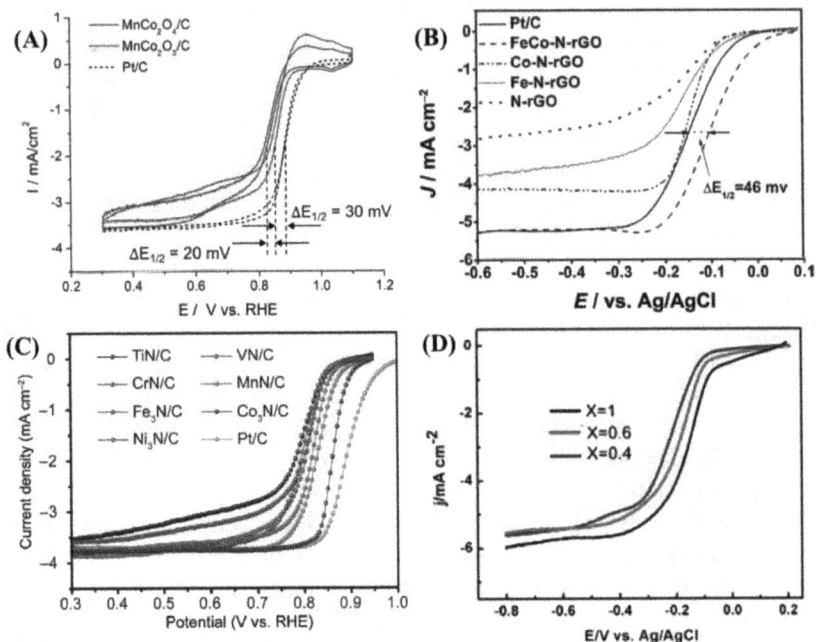

Figure 10.7 (A) Cyclic voltammetric CV curves of MnCo$_2$O$_4$/C, MnCo$_2$O$_3$/C, and Pt/C (Reproduced from [Yang et al., 2019]). LSV curves of (B) N–rGO, Fe–N–rGO, Co–N–rGO, FeCo–N–rGO and Pt/C (Reproduced from [Fu et al., 2012]), (C) MxN/C [20 wt. %] and Pt/C (20 wt. %) (Reproduced from (Zeng et al., 2022]). (D) C–Mn$_x$Fe$_{3-x}$O$_4$ NPs with different x (Reproduced from [Zhu et al., 2013]).

In connection with rock-salt-type MnCo$_2$O$_3$/C studies by Yang et al., Zhu et al. studied monodisperse M$_x$Fe$_{3-x}$O$_4$ (M = Fe, Cu, Co, Mn) nanoparticles on carbon support for ORR activity in alkaline medium (Zhu et al., 2013). The authors prepared M$_x$Fe$_{3-x}$O$_4$ NPs by thermal decomposition of precursors (various ratios of M(acac)$_2$/Fe(acac)$_3$ Fe(acac)$_3$ using Cu(acac)$_2$, or Co(acac)$_2$, or Mn(acac)$_2$) in benzyl ether and 1,2-tetradecanediol as a mild reducing agent to reduce Fe^{3+} to Fe^{2+} with oleylamine and oleic acid as a stabilizing agent. During the ORR activity measurements, a positive shift of the onset reduction potential was observed from C–Fe$_3$O$_4$ to C–MnFe$_2$O$_4$ (−0.340 V and −0.210 V *vs* Ag/AgCl). Other ferrite NPs appeared as −0.322 V for C–Cu$_{0.7}$Fe$_{2.3}$O$_4$, −0.302 V for C–Co$_{0.8}$Fe$_{2.2}$O$_4$ and −0.217 V for C–Mn$_{0.6}$Fe$_{2.4}$O$_4$ NPs. These studies suggest that O$_2$ is most easily reduced on Mn$_x$Fe$_{3-x}$O$_4$, in particular on MnFe$_2$O$_4$ NP as shown in Fig. 10.7(D). During the K–L plot analysis at 1600 rpm, E$_{onset}$ of C–Fe$_3$O$_4$, C–Cu$_{0.7}$Fe$_{2.3}$O$_4$, C–Co$_{0.8}$Fe$_{2.2}$O$_4$, C–Mn$_{0.6}$Fe$_{2.4}$O$_4$ NPs and C–MnFe$_2$O$_4$ NPs were found as −0.342, −0.319, −0.244, −0.186 V and −0.154 V (*vs.* Ag/AgCl) respectively. The number (n) of electrons transferred during ORR on Fe$_3$O$_4$, MnFe$_2$O$_4$ as well as on Cu- and Co-ferrite NPs were calculated from the K-L plot analysis and found to be ~3.83 and ~4.18, respectively (~3.84 for C–Cu$_{0.7}$Fe$_{2.3}$O$_4$, ~3.99 for C–Co$_{0.8}$Fe$_{2.2}$O$_4$, and ~4.12 for C–Mn$_{0.6}$Fe$_{2.4}$O$_4$ NPs). These suggest that 4e$^-$ transfer ORR from these NP catalysts which are closer to commercial benchmark Pt/C catalyst.

10.8 IMPORTANCE OF ORR IN FUEL CELLS

Since fuel cells are one of the possible economically viable power sources, they have attracted great attention on a number of energy-storing applications. They are identified as the more efficient devices among other energy conversion devices. In PEMFCs, oxidation of hydrogen gas takes place at the anode and the electrons are carried via an electrical circuit by which the work is done. Proton will move towards the cathode via the membrane or OH⁻ ion will reach the anode via the membrane. Meantime, oxygen will be reduced at the cathode to complete the overall fuel cell reaction. Usually, this ORR is very slow when compared to the anode reaction. It is essential to increase the rate of the reaction to improve the performance of the fuel cell. Hence, a catalyst is required to speed up the ORR in fuel cells. At present, Pt-based materials are found to be practical ORR catalysts. However, the cost of Pt is high and this prevents the manufacture of commercially viable fuel cells. Hence, a broad group of scientific people are concentrating seriously on the development of non-noble metal-based ORR catalysts, like non-noble metals, alloys, carbonaceous materials, organic compounds and their derivatives, non-noble transition metal-based compounds, etc. It is believed that they can lead to make fuel cells at low cost.

10.9 SUMMARY AND FUTURE PERSPECTIVES

For many decades, researchers had been looking for alternatives to Pt as an electrocatalyst to reduce oxygen in fuel cells. Numerous novel strategies and models were developed to improve ORR activity by conniving absorption capacity, doping matching and nanostructure fabrication. Innovative carbon-based nanomaterials (carbon allotropes or polymers) demonstrated efficient ORR activity in fuel cells. Doped non-noble metal-based composites are intriguing for more active sites, but a more accurate manufacturing method is needed to achieve the best catalysts. It is exciting to obtain ORR catalysts with high efficiency in the acidic media as opposed to alkaline media in this day and age. As a result, our future research should be associated with the high surface area of nanostructures, which might lead to more active sites and improved performance. Despite various challenges in this field, the growing interest in electrochemical reduction by non-noble metal catalysts indicates an exciting outlook advancement. Supplementary efforts must be made to rationally propose extremely competent electrodes and investigate reduction mechanisms. As a result, non-noble metal catalysts may have a wide range of applications in ORR. Because of the lack of metal leakage and metal ion impurity, metal-free electrocatalysts are more stable in the acidic media. Regardless of the fact that many metal-free catalysts have already been established through various approaches over the last several decades, there is still a long way to go before ORR catalysts are mass commercialized in real-world applications. Half-cell studies are an excellent strategy for analyzing and investigating the inherent activity and ORR kinetics of electrocatalysts. The architecture and strategy of electrode catalyst nanomaterials are extremely important for improving growth and progress in fuel cell technology. By facilitating mass transfer, advanced design can increase the efficiency of active sites. Furthermore, it develops sophisticated technology and mechanisms to improve electrocatalytic activity and improved productivity.

ACKNOWLEDGEMENTS

RR acknowledges Indian Institute of Technology Madras (IITM) for Institute Postdoctoral Fellowship (ID No. MM21R002) and Prof. Dr.-Ing Lakshman Neelakantan for his constant support. VM acknowledges Prof. Kothandaraman Ramanujam for his support.

REFERENCES

An, H., R. Zhang, Z. Li, L. Zhou, M. Shao and M. Wei. 2016. Highly efficient metal-free electrocatalysts toward oxygen reduction derived from carbon nanotubes@polypyrrole core-shell hybrids. J. Mater. Chem. A. 4: 18008–18014. https://doi.org/10.1039/c6ta08892a

Ania, C.O., M. Seredych, E. Rodriguez-castellon and T.J. Bandosz. 2015. New copper/GO-based material as an efficient oxygen reduction catalyst in an alkaline medium: The role of unique Cu/rGO architecture. Appl. Catal. B. Environ. 163: 424–435. https://doi.org/10.1016/j.apcatb.2014.08.022

Behret, H., H. Binder, G. Sandstede and G.G. Scherer. 1981. On the mechanism of electrocatalytic oxygen reduction at metal chelates, Part III. Metal phthalocyanines. J. Electroanal. Chem. 117: 29–42. https://doi.org/10.1016/S0022-0728(81)80448-2

Bozzini, B., P. Bocchetta and A. Gianoncelli. 2015. Coelectrodeposition of ternary Mn-oxide/polypyrrole composites for ORR electrocatalysts: a study based on micro-x-ray absorption spectroscopy and x-ray fluorescence mapping. Energies. 8: 8145–8164. https://doi.org/10.3390/en8088145

Chen, X., Y. Chen, Z. Shen, C. Song, P. Ji, N. Wang, et al., 2020. Self-crosslinkable polyaniline with coordinated stabilized CoOOH nanosheets as a high-efficiency electrocatalyst for oxygen evolution reaction. Appl. Surf. Sci. 529: 147173. https://doi.org/10.1016/j.apsusc.2020.147173

Chhetri, K., B. Dahal, T. Mukhiya, A.P. Tiwari, A. Muthurasu, T. Kim, et al., 2021. Integrated hybrid of graphitic carbon-encapsulated CuxO on multilayered mesoporous carbon from copper MOFs and polyaniline for asymmetric supercapacitor and oxygen reduction reactions. Carbon. 179: 89–99. https://doi.org/10.1016/j.carbon.2021.04.028

Choi, H.C., W.M. Chung, C.H. Kwon, H.S. Parka and I.S. Woo. 2013. B, N- and P, N-doped graphene as highly active catalysts for oxygen reduction reactions in acidic media. J. Mater. Chem. A. 1: 3694–3699. https://doi.org/10.1039/c3ta01648j.

Chung, H.T., G. Wu, Q. Li and P. Zelenay. 2014. Role of two carbon phases in oxygen reduction reaction on the Co-PPy-C catalyst. Int. J. Hydrogen Energy. 39: 15887-15893. https://doi.org/10.1016/j.ijhydene.2014.05.137

Ćirić-Marjanović, G., S. Mentus, I. Pašti, N. Gavrilov, J. Krstić, J. Travas-sejdic, et al., 2014. Synthesis, characterization, and electrochemistry of nanotubular polypyrrole and polypyrrole-derived carbon nanotubes. J. Phys. Chem. C. 118(27): 14770–14784. https://doi.org/10.1021/jp502862d

Coutanceau, C., M.J. Croissant, T. Napporn and C. Lamy. 2000. Electrocatalytic reduction of dioxygen at platinum particles dispersed in a polyaniline film. Electrochim. Acta. 46(4): 579–588. https://doi.org/10.1016/S0013-4686(00)00641-1

Cui, C.Q. and J.Y. Lee. 1994. Effect of polyaniline on oxygen reduction in buffered neutral solution. J. Electroanal. Chem. 367(1–2): 205–212. https://doi.org/10.1016/0022-0728 (93)03048-T

Feng, W., H. Li, X. Cheng, T.C. Jao, F.B. Weng, A. Su, et al., 2012. A comparative study of pyrolyzed and doped cobalt-polypyrrole eletrocatalysts for oxygen reduction reaction. Appl. Surf. Sci. 258(8): 4048–4053. https://doi.org/10.1016/j.apsusc.2011.12.098

Fu, L., S.J. You, G.Q. Zhang, F.L. Yang, X.H. Fang and Z. Gong. 2011. PB/PANI-modified electrode used as a novel oxygen reduction cathode in microbial fuel cell. Biosens. Bioelectron. 26(5): 1975–1979. https://doi.org/10.1016/j.bios.2010.08.061

Fu, X., Y. Liu, X. Cao, J. Jin, Q. Liu and J. Zhang. 2013. FeCo–Nx embedded graphene as high performance catalysts for oxygen reduction reaction. Appl. Catal. B: Environ. 130–131: 143–151. https://doi.org/10.1016/j.apcatb.2012.10.028

Gong, K., F. Du, Z. Xia, M. Durstock and L. Dai. 2009. Nitrogen-doped carbon nanotube arrays with high electrocatalytic activity for oxygen reduction. Science. 323: 760–764. https://doi.org/10.1126/science.1168049

Góral-kurbiel, M. and A. Drelinkiewicz. 2014. Palladium content effect on the electrocatalytic activity of palladium—polypyrrole nanocomposite for cathodic reduction of oxygen. Electrocatalysis. 23–40. https://doi.org/10.1007/s12678-013-0155-0

Hossen, M., K. Artyushkova, P. Atanassov and A. Serov. 2017. Synthesis and characterization of high performing Fe-N-C catalyst for oxygen reduction reaction (ORR) in alkaline exchange membrane fuel cells. J. Power Sources. 375: 214–221. https://doi.org/10.1016/j. jpowsour.2017.08.036

Huang, D., Y. Luo, S. Li, B. Zhang, Y. Shen and M. Wang. 2014. Active catalysts based on cobalt oxide @ cobalt/N–C nanocomposites for oxygen reduction reaction in alkaline solutions. Nano Res. 7: 1054–1064. https://doi.org/10.1007/s12274-014-0468-1

Jeon, I-Y., H-J. Choi, M. Choi, J-M. Seo, S-M. Jung, M-J. Kim, et al., 2013a. Facile, scalable synthesis of edge-halogenated graphene nanoplatelets as efficient metal-free eletrocatalysts for oxygen reduction reaction. Sci. Rep. 3: 1810. https://doi.org/10.1038/ srep01810.

Jeon, I-Y., S. Zhang, L. Zhang, H-J. Choi and J-M. Seo, Z. Xia, et al., 2013b. Edge-selectively sulfurized graphene nanoplatelets as effi cient metal-free electrocatalysts for oxygen reduction reaction : the electron spin effect. Adv. Mater. 25: 6138–6145. https:// doi.org/10.1002/adma.201302753

Kozuch, S. and S. Shaik. 2008. Kinetic-quantum chemical model for catalytic cycles: the haber-bosch process and the effect of reagent concentration. J. Phys. Chem. A. 112: 6032–6041. https://doi.org/10.1021/jp8004772

Kozuch, S. and S. Shaik. 2011. How to conceptualize catalytic cycles? The energetic span model. Acc. Chem. Res. 44(2): 101–110. https://doi.org/ 10.1021/ar1000956.

Lai, E.K.W., P.D. Beattie, F.P. Orfino, E. Simon and S. Holdcroft. 1999. Electrochemical oxygen reduction at composite films of Nafion®, polyaniline and Pt. Electrochim. Acta. 44(15): 2559–2569. https://doi.org/10.1016/S0013-4686(98)00389-2

Le, T.H., Y. Kim and H. Yoon. 2017. Electrical and electrochemical properties of conducting polymers. Polymers. 9(4). https://doi.org/10.3390/polym9040150

Li, L., X. Yuan, Z. Ma and Z. Ma. 2015. Properties of pyrolyzed carbon-supported cobalt-polypyrrole as electrocatalyst toward oxygen reduction reaction in alkaline media. 162(4): 359–365. https://doi.org/10.1149/2.0081504jes.

Li, H., J. Yin, Y. Meng, S. Liu and T. Jiao. 2020. Nickel/Cobalt-containing polypyrrole hydrogel-derived approach for efficient ORR electrocatalyst. Colloids Surf. A. 586: 124221. https://doi.org/10.1016/j.colsurfa.2019.124221.

Liang, Y., Y. Li, H. Wang, J. Zhou, J. Wang, T. Regier, et al., 2011. Co_3O_4 nanocrystals on graphene as a synergistic catalyst for oxygen reduction reaction. Nature Materials. 10: 780–786. https://doi.org/10.1038/nmat3087.

Lin, Z., G.H. Waller, Y. Liu and M. Liu. 2013. 3D Nitrogen-doped graphene prepared by pyrolysis of graphene oxide with polypyrrole for electrocatalysis of oxygen reduction reaction. Nano Energy. 2: 241–248. http://dx.doi.org/10.1016/j.nanoen.2012.09.002

Liu, Z., F. Peng, H. Wang, H. Yu, W. Zheng and J. Yang. 2011. Phosphorus-doped graphite layers with high electrocatalytic activity for the O_2 reduction in an alkaline medium. Angew. Chem. Int. Ed. 50: 3257–3261. https://doi.org/10.1002/anie.201006768

Liu, F.L. and J. Tao. 2017. Hysteretic two-step spin-crossover behavior in two two-dimensional hofmann-type coordination polymers. Chem. Eur. J. 23(72): 18252–18257. https://doi.org/10.1002/chem.201704276

Ma, Z., S. Dou, A. Shen, L. Tao, L. Dai and S. Wang. 2014. Sulfur-doped graphene derived from cycled lithium—sulfur batteries as a metal-free electrocatalyst for the oxygen reduction reaction. Angewandte. 54: 1888–1892. https://doi.org/10.1002/anie.201410258

Ma, R., B.Y. Xia, Y. Zhou, P. Li and Y. Chen. 2016. Ionic liquid-assisted synthesis of dual-doped graphene as efficient electrocatalysts for oxygen reduction. Carbon. 102: 58–65. https://doi.org/10.1016/j.carbon.2016.02.034

Magdalena, W., O. Magdalena, B. Magnus and E.D. Głowacki. 2020. Electrogeneration of hydrogen peroxide via oxygen reduction on polyindole films. J. Electrochem. Soc. 1–14. https://doi.org/10.1149/1945-7111/ab88bb

Manesh, K.M., P. Santhosh, A.I. Gopalan and K.P. Lee. 2006. Electrocatalytic dioxygen reduction at glassy carbon electrode modified with polyaniline grafted multiwall carbon nanotube film. Electroanalysis. 18(16): 1564–1571. https://doi.org/10.1002/elan.200603567

Meng, Y., J. Yin, T. Jiao, J. Bai, L. Zhang, J. Su, et al., 2020. Self-assembled copper/cobalt-containing polypyrrole hydrogels for highly efficient ORR electrocatalysts. J. Mol. Liq. https://doi.org/https://doi.org/10.1016/j.molliq.2019.112010

Nguyen-cong, H., V.de la G. Guadarrama, J.L. Gautier and P. Chartier. 2003. Oxygen reduction on $Ni_xCo_{3-x}O_4$ spinel particles/polypyrrole composite electrodes: hydrogen peroxide formation. Electrochim. Acta. 48: 2389–2395. https://doi.org/10.1016/S0013-4686(03)00252-4

Nguyen-thanh, D., A.I. Frenkel, J. Wang, S.O. Brien and D.L. Akins. 2011. Cobalt–polypyrrole-carbon black (Co–PPY–CB) electrocatalysts for the oxygen reduction reaction (ORR) in fuel cells: composition and kinetic activity. Appl. Catal. B. Environ. 105: 50–60. https://doi.org/10.1016/j.apcatb.2011.03.034

Nguyen, M.T., B. Mecheri, A. Iannaci, A. D'Epifanio and S. Licoccia. 2016. Iron/polyindole-based electrocatalysts to enhance oxygen reduction in microbial fuel cells. Electrochim. Acta. 190: 388–395. https://doi.org/10.1016/j.electacta.2015.12.105

Nilekar, A.U. and M. Mavrikakis. 2008. Surface Science-improved oxygen reduction reactivity of platinum monolayers on transition metal surfaces. Surf. Sci. 602: L89–L94. https://doi.org/10.1016/j.susc.2008.05.036.

Olson, T.S., S. Pylypenko, P. Atanassov, K. Asazawa, K. Yamada and H. Tanaka. 2010. Anion-exchange membrane fuel cells: dual-site mechanism of oxygen reduction reaction in alkaline media on cobalt—polypyrrole electrocatalysts. J. Phys. Chem. C. 114(11): 5049–5059. https://doi.org/10.1021/jp910572g

Osmieri, L., A.H.A.M. Videla and S. Specchia. 2016. Optimization of a Fe-N-C electrocatalyst supported on mesoporous carbon functionalized with polypyrrole for oxygen reduction reaction under both alkaline and acidic conditions. Int. J. Hydrogen Energy. 41: 19610–19628. https://doi.org/http://dx.doi.org/10.1016/j.ijhydene.2016.05.270

Qiu, J., L. Shi, R. Liang and G. Wang. 2012. Controllable deposition of a platinum nanoparticle ensemble on a polyaniline/graphene hybrid as a novel electrode material for electrochemical sensing. Chem. Eur. J. 18: 7950–7959. https://doi.org/10.1002/chem.201200258.

Ramírez-pérez, A.C., J. Quílez-bermejo, J.M. Sieben, E. Morallón and D. Cazorla-amorós. 2018. Effect of nitrogen-functional groups on the ORR activity of activated carbon fiber-polypyrrole-based electrodes. Electroanalysis. 9: 697–705. https://doi.org/https://doi.org/10.1007/s12678-018-0478-y

Ratso, A.S., I. Kruusenberg and K. Maike. 2017. Transition metal-nitrogen co-doped carbide-derived carbon catalysts for oxygen reduction reaction in alkaline direct methanol fuel cell. Appl. Catal. B. Environ. 219: 276–286. https://doi.org/10.1016/j.apcatb.2017.07.036

Ren, S., S. Ma, Y. Yang, Q. Mao and C. Hao. 2015. Hydrothermal synthesis of Fe_2O_3/polypyrrole/graphene oxide composites as highly efficient electrocatalysts for oxygen reduction reaction in alkaline electrolyte. Electrochim. Acta. 178: 179–189. https://doi.org/10.1016/j.electacta.2015.07.181

Saremi, M. and M. Salehisaki. 2014. The catalytic effect of polypyrrole/Pt-Cu on oxygen reduction reaction. Electroanalysis. 26: 1606–1611. https://doi.org/10.1002/elan.201300569

Shang, X., L. Suo, W. Li and S. Chen. 2015. Synthesis of Fe nanoparticles on polypyrrole covered graphite for oxygen reduction reaction. Adv. Eng. Res. Icectt. 562–565. https://doi.org/https://doi.org/10.2991/icectt-15.2015.107

Sheng, Z., H. Gao, W. Bao, F. Wang and X. Xia. 2012. Synthesis of boron-doped graphene for oxygen reduction reaction in fuel cells. J. Mater. Chem. 22: 390–395. https://doi.org/10.1039/c1jm14694g

Shi, Z. and J. Zhang. 2007. Density functional theory study of transitional metal macrocyclic complexes, dioxygen-binding abilities and their catalytic activities toward oxygen reduction reaction. J. Phys. Chem. C: Nanomater. Interfaces. 111: 19, 7084–7090. https://doi.org/10.1021/jp0671749.

Skúlason, E., S. Siahrostami, J.K. Nørskov, J. Rossmeisl and V. Tripkovi. 2010. The oxygen reduction reaction mechanism on Pt (111) from density functional theory calculations. Electrochim. Acta. 55: 7975–7981. https://doi.org/10.1016/j.electacta.2010.02.056

Stamenović, U., N. Gavrilov, I.A. Pašti, M. Otoničar, G. Ćirić-Marjanović, S.D. Škapin, et al., 2018. One-pot synthesis of novel silver-polyaniline-polyvinylpyrrolidone electrocatalysts for efficient oxygen reduction reaction. Electrochim. Acta. 281: 549–561. https://doi.org/10.1016/j.electacta.2018.05.202

Sun, T., W. Li, M. Yang, H. Chen, Y. Liu and H. Li. 2019. Nitrogen and iron co-doped porous carbon spheres derived from tetrazine-based polyindole as efficient catalyst for oxygen reduction reaction in acidic electrolytes. J. Power Sources. 434: 226738. https://doi.org/10.1016/j.jpowsour.2019.226738

Tsakova, V. and R. Seeber. 2016. Conducting polymers in electrochemical sensing: Factors influencing the electroanalytical signal. Anal. Bioanal. Chem. 408(26): 7231–7241. https://doi.org/10.1007/s00216-016-9774-7

Tran, T.N., M.Y. Song, K.P. Singh, D.S. Yang and J.S. Yu. 2016. Iron–polypyrrole electrocatalyst with remarkable activity and stability for ORR in both alkaline and acidic conditions: a comprehensive assessment of catalyst preparation sequence. J. Mater. Chem. A. 4: 8645–8657. https://doi.org/10.1039/C6TA01543C

Varga, T., G. Ballai, L. Vásárhelyi, H. Haspel, Á. Kukovecz and Z. Kónya. 2018. Co_4N/ nitrogen-doped graphene: a non-noble metal oxygen reduction electrocatalyst for alkaline fuel cells. Appl. Catal. B: Environ. 237: 826–834. https://doi.org/10.1016/j. apcatb.2018.06.054

Verma, C.J., A. Kumar, R.P. Ojha and R. Prakash. 2020. Au-V2O5/Polyindole composite: an approach for ORR in different electrolytes. J. Electroanal. Chem. 861: 113959. https://doi.org/10.1016/j.jelechem.2020.113959.

Wang, S., L. Zhang, Z. Xia, A. Roy, D.W. Chang, J. Baek, et al., 2012. BCN graphene as efficient metal-free electrocatalyst for the oxygen reduction reaction. Angew. Chem. Int. Ed. 51: 4209–4212. https://doi.org/10.1002/anie.201109257

Wang, X., J. Wang, D. Wang, S. Dou, Z. Ma, J. Wu, et al., 2014. One-pot synthesis of nitrogen and sulfur co-doped graphene as efficient metal-free electrocatalysts for the oxygen reduction reaction. Chem. Commun. 50: 4839–4842. https://doi.org/10.1039/ c4cc00440j

Wang, J., R. Ma, Z. Zhou, G. Liu and Q. Liu. 2015. Magnesiothermic synthesis of sulfur-doped graphene as an efficient metal-free electrocatalyst for oxygen reduction. Sci. Rep. 5: 9304. https://doi.org/10.1038/srep09304.

Wang, X., L. Zou, H. Fu, Y. Xiong, Z. Tao, J. Zheng, et al., 2016. Noble metal-free oxygen reduction reaction catalysts derived from prussian blue nanocrystals dispersed in polyaniline. ACS Appl. Mater. Interfaces. 8(13): 8436–8444. https://doi.org/10.1021/ acsami.5b12102

Wei, P. and H. Tanabe. 2011. Synergy effects between single-walled carbon nanotubes and polypyrrole on the electrocatalysis of their composites for the oxygen reduction reaction. Carbon. 49: 4877–4889. https://doi.org/10.1016/j.carbon.2011.07.010

Xiao, D., J. Ma, Chen, C., Q. Luo, L. Zheng and X. Zuo. 2018. Oxygen-doped carbonaceous polypyrrole nanotubes-supported Ag nanoparticle as electrocatalyst for oxygen reduction reaction in alkaline solution. Mater. Res. Bull. 105: 184–191. https://doi.org/https://doi. org/10. 1016/j.materresbull.2018.04.030

Xu, J., G. Dong, C. Jin, M. Huang and L. Guan. 2013. Sulfur and nitrogen co-doped, few-layered graphene oxide as a highly efficient electrocatalyst for the oxygen-reduction reaction. ChemSusChem. 6: 493–499. https://doi.org/10.1002/cssc.201200564

Xue, Y., D. Yu, L. Dai, R. Wang, D. Li, A. Roy, et al., 2013. Three-dimensional B, N-doped graphene foam as a metal-free catalyst for oxygen reduction reaction, Phys. Chem. Chem. Phys. 15: 12220–12226. https://doi.org/10.1039/c3cp51942b

Yang, L., S. Jiang, Y. Zhao, L. Zhu, S. Chen, X. Wang, et al., 2011. Boron-doped carbon nanotubes as metal-free electrocatalysts for the oxygen reduction reaction. Angew. Chem. Int. Ed. 50: 7132–7135. https://doi.org/10.1002/anie.201101287

Yang, Z., Yao, G. Li, G. Fang, H. Nie, Z. Liu, et al., 2012. Sulfur-doped graphene as an efficient metal-free cathode catalyst for oxygen reduction. ACS Nano. 6(1): 205–211. http://dx.doi.org/10.1021/nn203393d.

Yang, M., Y. Liu, H. Chen, D. Yang and H. Li. 2016. Porous N-doped carbon prepared from triazine-based polypyrrole network: an highly efficient metal-free catalyst for oxygen reduction reaction in alkaline electrolytes. ACS Appl. Mater. Interfaces. 8: 28615–28623. https://doi.org/10.1021/acsami.6b09811

Yang, Y., R. Zeng, Y. Xiong, F.J. Disalvo and J. Accepted. 2019. Rock-salt-type $MnCo_2O_3/C$ as efficient oxygen reduction electrocatalysts for alkaline fuel cells. Chem. Mater. https://doi.org/10.1021/acs.chemmater.9b02801

Yao, Z., H. Nie, Z. Yang, X. Zhou, Z. Liu and S. Huang. 2012. Catalyst-free synthesis of iodine-doped graphene via a facile thermal annealing process and its use for electrocatalytic oxygen reduction in an alkaline medium. Chem. Commun. 48: 1027–1029. https://doi.org/10.1039/c2cc16192c.

Yin, Z.S., T.H. Hu, J.L. Wang, C. Wang, Z.X. Liu and J.W. Guo. 2014. Preparation of highly active and stable polyaniline-cobalt-carbon nanotube electrocatalyst for oxygen reduction reaction in polymer electrolyte membrane fuel cell. Electrochim. Acta. 119: 144–154. https://doi.org/10.1016/j.electacta.2013.12.072.

Yuan, X., H. Sha, X. Ding and H. Kong. 2014. Comparative investigation on the properties of carbon-supported cobalt-polypyrrole pyrolyzed at various conditions as electrocatalyst towards oxygen reduction reaction. Int. J. Hydrogen Energy. 39: 1593–15947. https://doi.org/10.1016/j.ijhydene.2014.03.205.

Zamani, P., D. Higgins, F. Hassan, G. Jiang, J. Wu, S. Abureden, et al., 2014. Electrospun iron-polyaniline-polyacrylonitrile derived nanofibers as non-precious oxygen reduction reaction catalysts for PEM fuel cells. Electrochim. Acta. 139: 111–116. https://doi.org/10.1016/j.electacta.2014.07.007.

Zeng, R., Y. Yang, X. Feng, H. Li, L.M. Gibbs, F.J. Disalvo, et al., 2022. Nonprecious transition metal nitrides as efficient oxygen reduction electrocatalysts for alkaline fuel cells. Sci. Adv. 35–37. https://doi.org/10.1126/sciadv.abj1584.

Zhong, H., H. Zhang, Z. Xu, Y. Tang and J. Mao. 2012. A nitrogen-doped polyaniline carbon with high electrocatalytic activity and stability for the oxygen reduction reaction in fuel cells. ChemSusChem. 5(9): 1698–1702. https://doi.org/10.1002/cssc.201200178

Zhou, Q., C.M. Li, J. Li and J. Lu. 2008. Electrocatalysis of template-electrosynthesized cobalt-porphyrin/polyaniline nanocomposite for oxygen-reduction. Russ. J. Phys. Chem. C. 112(47): 18578–18583. https://doi.org/10.1021/jp8077375

Zhu, H., S. Zhang, Y. Huang, L. Wu and S. Sun. 2013. Monodisperse $M_xFe_{3-x}O_4$ (M = Fe, Cu, Co, Mn) nanoparticles and their electrocatalysis for oxygen reduction reaction. Nano Lett. 13(6): 2947–2951. https://doi.org/10.1021/nl401325u.

Chapter 11

Novel Contributions to the Fundamental Role of Structural Engineering in Polymeric Membranes for Alkaline Fuel Cells

Hilda M. Alfaro-López[1]*, H. Rojas-Chávez[2],
and O. Solorza Feria[3]

[1]Escuela Superior de Ingeniería Mecánica y Eléctrica-
Departamento de Ingeniería Eléctrica-Edif. 2, Instituto Politécnico Nacional,
U.P.A.L.M., Col. Lindavista, C.P. 07738, CDMX, Mexico.

[2]Tecnológico Nacional de México,
Instituto Tecnológico de Tláhuac II, Camino Real 625,
Tláhuac, CDMX, 13550, Mexico.

[3]Química-CINVESTAV. Av. IPN 2508,
Zacatenco, 07360 CDMX, Mexico.

11.1 INTRODUCTION

With an accelerating worldwide demand for energy consumption and an increasing environmental concern, results related to energy advances play a key role in development of ecological energy sources for replacement of conventional consumption of fossil-origin fuels. New novel nanomaterials for their use in

*For Correspondence: Email: hilmar105@hotmail.com

different types of fuel cells are an important factor in pushing toward the deal of this technology, regarding low-cost raw materials. The electrocatalytic conversion of renewable energy into chemical one has attracted significant attention as it may help to mitigate the worldwide energy crisis, where chemical fuels such as hydrogen can be stored and reconverted into electricity via electrochemical reaction processes in fuel cells (Ke et al., 2018; Anson and Stahl, 2020; Miller et al., 2020). The interfaces created by electrocatalysts and membranes are key components to secure the development, deployment and storage of sustainable energy sources. This is achieved through the performance of membrane-electrode assembly (MEA) to determine the reactivity of the novel produced materials, valuable in a great deal of applications of fuel cells. It is worth noting that these cells have been created to efficiently generate electricity from ecological fuels, gaining increased attention as promising electrochemical devices for energy production and diverse applications. For this reason, they are used in devices for vehicular transport, as well as stationary and cross-cutting applications, focusing on reducing cost and improving durability (Hren et al., 2021).

Alkaline fuel cells (AFC) have widely been treated in literature (Dekel, 2018; Yang et al., 2022) which are substantially related to the technical development directions for the next-generation of high-power-density for improving the membrane electrode assembly and its basic components (Jiao et al., 2021; Wu et al., 2022). In this context, such results have been used to improve the design and development of synthetic or naturally-derived anion exchange membrane (AEM), focusing on the progress of ionic conductivity (IC), ion exchange capacity (IEC), fuel crossover, durability, stability and cell performance (Masa et al., 2021). Also, the conductive bipolar plate is considered as the major component of polymeric fuel cells and a critical contributor to the fuel cell performance, which is improved by using conductive polymer composites based on graphite-filled polymer blend (Ramírez-Herrera et al., 2021; Tariq et al., 2022). Even more, the development of more effective and inexpensive Pt-based and Pt group-free (PGM-free) catalysts have also been reported (Nie et al., 2015; Cruz-Martínez et al., 2019, 2021; Peera et al., 2021; Yang et al., 2022). When AFC was devised, the relevance of this electrochemical technology took an important place in the development of spacecraft missions. The main advantages of having chosen AFC were their high efficiency and minimum cost, due to the possibility of using economical PGM-free catalysts. However, it is worth pointing out that this type of fuel cell presented some problems; therefore, the development of the cathode catalyst is receiving greater emphasis because its characteristics strongly determine the operation of the entire device. Another problem is the poisoning by carbonation (due to CO_2) of the electrolyte by the difficulty in using a liquid electrolyte as a means of ion conductor. To address the problems associated with the utilization of liquid electrolytes (e.g. NaOH and KOH), the use of solid ion exchange membrane electrolytes has received outstanding attention in recent years (Venugopal et al., 2022). So far as conductivity and physicochemical stable ion transport properties are concerned, technologies of novel membranes have been developed in sustainable energy conversion for reducing adverse environmental impacts. Consequently, those have diverse applications, including membrane-based separation process in microfiltration, ultrafiltration, nanofiltration and reverse osmosis for wastewater treatment and reuse (Sun et al., 2021). Even more, they are used as separators in plant microbial fuel cells (Ramírez-Nava et al., 2021).

Anion exchange membrane fuel cells (AEMFCs) are becoming highly attractive energy-conversion devices due to a reduced-cost technology for clean energy conversion (Reyes-Rodríguez et al., 2018; Guo et al., 2020; Mandal et al., 2020; Chen and Young, 2021; Ge et al., 2022). This fuel cell technology offers further advantages of employing economic PGM-free metals as electrocatalysts. In addition, the problem of carbonation poisoning is resolved. The increased number of studies in AEMFCs offers a great deal of advantages for this technology in catalysts; i.e. less severe corrosion of structural materials over the proton exchange membrane fuel cells (PEMFC). Although in principle, the two technologies are comparable, the application of AEMFCs creates an alkaline pH cell ambience offering several potential advantages over the mature PEMFC technology.

A single AEMFC is constituted by these components, an anode, a cathode and a solid anionic alkaline membrane, which together are the so-called membrane electrode assembly (MEA). Figure 11.1 shows a typical schematic drawing characteristic of an AEMFC where the anion polymer electrolyte membrane is responsible for the hydroxyl ion conductivity, which allows the transport of OH^- from the cathode to the anode, constituting the key component of the electrochemical device (Bagostky, 2009; Ramaswamy and Mukerjee, 2019; Hren et al., 2021). In AEMFCs working at low temperature, the hydrogen charged on the anode side originates a reaction with hydroxyl anions giving way to water and electrons. Then the electrons are conveyed via the external circuit to the cathode. After that, oxygen originates a reaction in a reduction process on the catalytic surface of the electrode with water to generate hydroxyl (OH^-) ions. The electric current released to the external circuit is used in a specific application. The basic electrochemical reactions occurring simultaneously on both sides of the electrodes of MEA of the AEMFCs are (Pu et al., 2021):

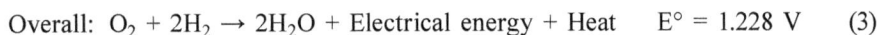

Anode: $2H_2 + 4OH^- \rightarrow 4H_2O + 4e^-$ $E° = -0.829$ V   (1)

Cathode: $O_2 + 2H_2O + 4e^- \rightarrow 4OH^-$ $E° = 0.40$ V   (2)

Overall: $O_2 + 2H_2 \rightarrow 2H_2O$ + Electrical energy + Heat $E° = 1.228$ V   (3)

Although AEMFCs are considered as a very efficient technology to transform the chemical energy contained in hydrogen into electrical one, different issues need solving for their large-scale production and applications. For example, the anodic reaction even catalyzed by the state of the Pt catalysts is at least two orders of magnitude lower than that in acidic PEM systems (Wang et al., 2021). Furthermore, the chemical stability and conductivity of the AEMs are restricted so far. In recent years great advances have also been made in the design of novel catalysts for anodic and cathodic reactions, as well as continuous enhancement in the performance of AEMs, including electrical conductivity, mechanical properties, and chemical stability, as has been demonstrated and reported in literature (Cheng et al., 2020; Firouzjaie and Mustain, 2020).

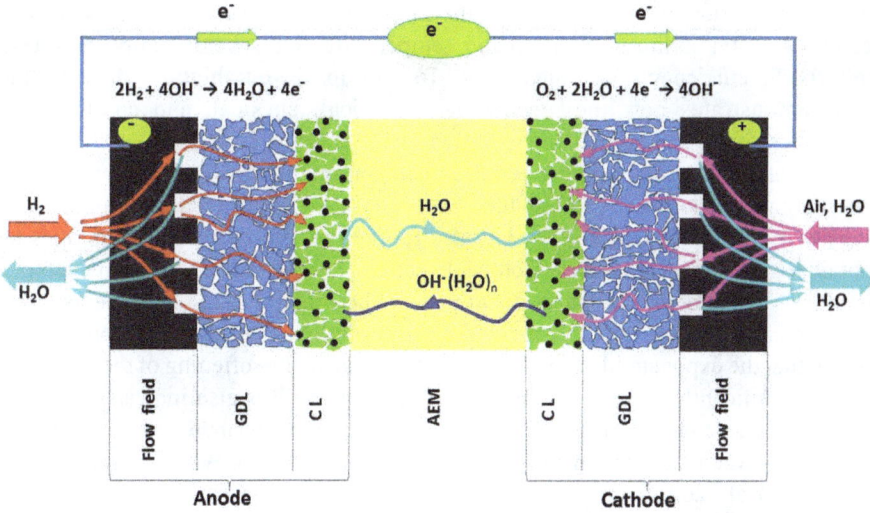

Figure 11.1 Schematic drawing of transport characteristics in an H_2/O_2–AEMFC and the main species conveyed via the cell (Reprinted with permission from Dekel, 2018).

Therefore, the aim of this review is to present a scope of recent trends of the strategies investigated to enhance the performance, conductivity, mechanical strength, chemical and electrochemical stability of the AEMFCs and the materials' performance documented in new contributions, which could help in the advancement of this line of research.

11.2 ALKALINE ANION EXCHANGE MEMBRANES (AAEM)

In 1964, the use of AFCs became outstandingly relevant during the aerospace developed project (Stone and Morrison, 2002) because of their high efficiency and low cost. Worth noting that in those days, there was the possibility of using cheaper catalysts that were not based on noble metals, such as Pt. Another advantage was the kinetic reaction of oxygen reduction in an alkaline medium that turned out to be faster (Mandal, 2020). However, a few years ago, this type of cells showed poisoning problems due to the presence of CO_2 in the electrolyte and they were difficult to be used as a liquid electrolyte for means of transportation. A decade ago, the transition from a liquid alkaline electrolyte membrane (NaOH, KOH) to a solid polymeric anion exchange membrane was achieved, which led to overcoming the problem of carbonation poisoning and the ability to use AFCs in portable devices (Hermida-Castro et al., 2013; Gottesfeld et al., 2018).

Currently, the challenge to overcome is to meet certain specific requirements such as: high IC, IEC, low electro-osmosis, chemical stability, high selectivity, mechanical integrity, high thermal stability, minimum fuel permeability degradation (or alkaline stability), and low cost. The main tests contemplated the determination of the alkaline stability where the membrane undergoes critical conditions of alkalinity, at a certain temperature, for several hours; this test reflects the efficiency of the membrane. To sum up, the durability or degradation test demonstrates how the mechanical, chemical, physical, and conductivity properties of the membrane get modified under extreme conditions.

Nonetheless, many of these requirements optimizing membrane performance are mutually exclusive; therefore, the challenge is not only to further minimize the sensitivity of the AAEM to CO_2, but also to improve and optimize membranes to obtain highly-efficient devices (Gottesfeld et al., 2018; Xu et al., 2020). An example of this is the amount of water contained in the membrane as it influences the mechanical stiffness, IEC and membrane degradation. Recent simulation studies reveal that the exponential absorption of water is due to the softening of the polymer, which significantly reduces the mechanical properties. It is also indicated that the alkyl group side chains, attached to the base structure, contribute to the softening, increasing water absorption (WU), and promoting IEC by water activity (WA) (Barnett et al., 2022).

Recently, several research groups have been interested in the synthesis and characterization of these alkaline anion exchange membranes from numerous polymeric materials applying different synthesis techniques. Conversely, there are companies in the market that develop AAEM with good results, such as AHA by Tokuyama, Morgane-ADP by Solvay, and Tosflex© by Tosoh. In the following section is presented the most outstanding and recent innovations based on these membranes.

11.2.1   Structural Conformation

Basically AEMs are made up of two parts linked directly or indirectly: a polymeric structure, which is the base chain and a positive side chain functional group. Cationic groups are attributed to the mobility of the OH^- ions from the cathode to the anode through the membrane (Hren et al., 2021). Base chains included in this review are poly (vinyl alcohol) (PVA) which is currently used to generate degradable materials; poly (ether-ether ketone) (PEEK) which is one of the most used membranes for PEM because it is an alternative to fluorinated membranes such as Nafion© that is the base structure applied to AEM; poly (aryl ether ketones) (PAEK) which are structures similar to PEEK with an aromatic structural modification; poly (ether ether ketone ketone) (PEEKK); and polyphenylene oxide (PPO); polysulfone (PS); poly (ether sulfone) (PES) polynorbornene (PNB), among others. Figure 11.2 shows all these base chemical structures.

Figure 11.3 shows common side chain groups grafted to the base chain: imidazole, quaternary ammonium, tertiary amines (e.g. trimethylamine, N-methyl-piperidine and N, N, N′, N′, N″-pentaethylguanidine), secondary amines (e.g. crown ethers) and graphene oxide (GO), among other more specific functional groups.

It is important to keep in mind that obtaining functional structures depends on the various types of syntheses that are appropriate for each structure. Table 11.1 shows the base chains related to the grafted side groups and their respective syntheses.

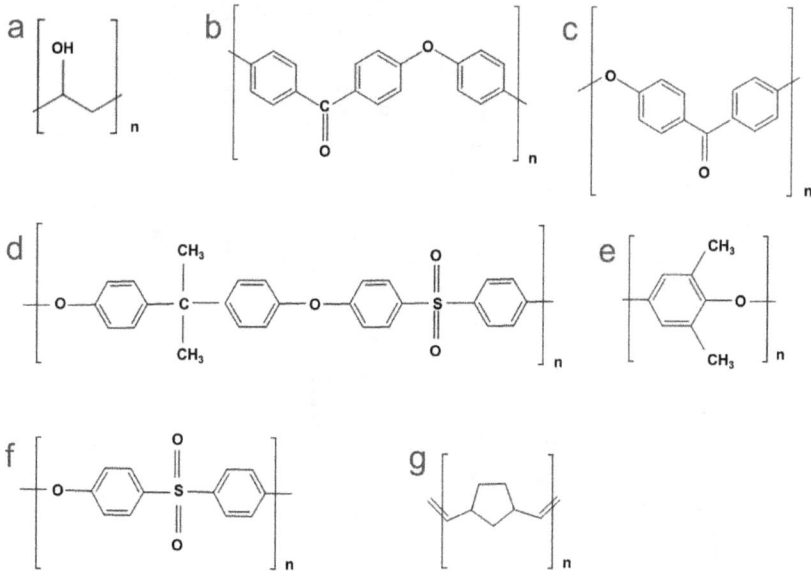

Figure 11.2 Base structures: (a) PVA, (b) PEEK, (c) PAEK, (d) PS, (e) PPO, (f) PES and (g) PNB.

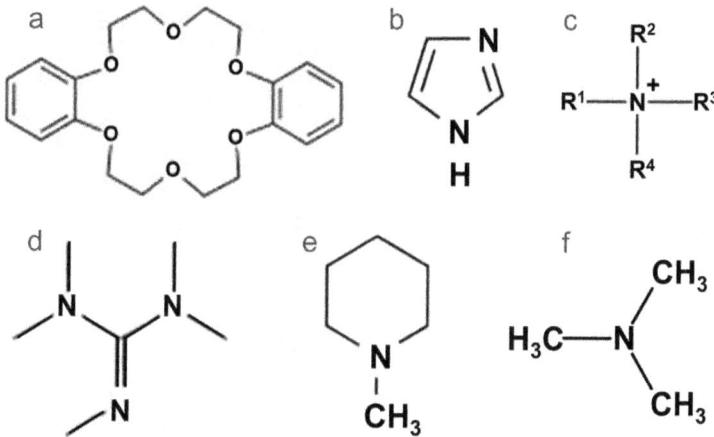

Figure 11.3 Most representative side chain structures. (a) crown ether/Dibenzo-18-crown-6, (b) imidazole, (c) quaternary ammonium, (d) N, N, N', N', N''-pentaethylguanidine, (e) N-methylpiperidine, and (f) trimethylamine.

11.2.2 Factors that Influence the Transport of OH⁻ Ions

The study of AEM ionic transport has been very limited, both in studies and reports in literature, fundamentally due to the structural diffusion mechanism of OH⁻ ions through hydrated AEM.

In most cases, the understanding of the OH⁻ ions transport mechanism is associated with structural engineering as a qualitative parameter applied in each

Table 11.1 AEM syntheses, their base and functional ionic group

Base group	Functional ionic group	Synthesis	References
PVA	Dibenzo-18-crown-6 ether	Reduction of Schiff bases by condensation	Wang et al., 1990
	Imidazolium	Acetalization and direct quaternization reaction	Albayrak and Şimşek, 2020; Xiong et al., 2008
PEEK	Imidazolium	Condensation polymerization	Son et al., 2020
PAEK	Trimethylamine, N-methyl-piperidine, N, N, N′, N′, N″-pentaethylguanidine	Polyhydroxyalkylation reaction catalyzed by superacid	Li et al., 2021
PPO	Oxyhexyltrimethylammonium	Via solution casting	Zhang and Xu, 2020
	Triazole groups	Quaternization reaction	Chu et al., 2021
	PS + TiO_2	Quaternization reaction	Msomi et al., 2020
	Imidazolium	Quaternization reaction	Sheng et al., 2020
PS	Imidazolium	Via solution casting	Rambabu et al., 2020
PES	Piperidinium	Via nucleophilic polycondensation, demethylation and Williamson reactions	Shen et al., 2020; Lin et al., 2016
	N-spirocyclic cations	Cyclo-polycondensations of tetrakis (bromomethyl) benzene and dipiperidines under mild conditions	Liu et al., 2020; Pham et al., 2017
PESN	9,9-bis(4-hydroxy-3-methylphenyl) fluorene	Nucleophilic substitution polycondensation reaction	Lai et al., 2020
PNB	Hexyl-3-methyl imidazolium chloride	Menshutkin reaction	Huang et al., 2020

research work, which is based on trial and error and on the similarity with works including base columns and side chain groups. The quantitative parameters are based on the characterization of IC, diffusion, permeability, operating temperature and relative humidity, among others. Furthermore, specific transport mechanisms for H^+ ions are reported in PEM literature, such as the Grotthus mechanism, convection, diffusion, and migration (Castañeda and Ribadeneira, 2018). Therefore, experts propose that PEM and AEM ion transport characteristics could be analogous (Feng et al., 2016). Grotthus transport theory establishes a mechanism by which an 'exceeding' proton or a defect in protonic charge (H_3O^+) is integrated into the water network and solvated by water molecules, forming a non-permanent hyper coordinated complex $H_9O_4^+$ transferred through the hydrogen bonds of water molecules or other liquids with hydrogen bonds, via the formation or breaking of covalent bonds. The solvation of these excess protons in water is represented in two complexed forms $H_9O_4^+$ (Eigen cation) and $H_5O_2^+$ (Zundel cation) (Ma and Tuckerman, 2011).

Theory simulations suggest that OH^- ions form complexed aggregates of the same nature as hydronium ions; that is, the OH^- ion is solvated and forms a hyper coordinated ion $OH^-(H_2O)_3$. In this case, the OH^- ion accepts three bonds with the hydrogen of adjacent water molecules and a hyper coordinated ion OH^- $(H_2O)_4$ is transiently formed when a hydrogen bond of a water molecule is broken (Fig. 11.4) (Tuckerman et al., 2002).

As for this mechanism, OH^- ion circulates via the hydrogen-bonded network of water molecules due to the formation and breaking of covalent bonds. The movement of solvated OH^- ion has been proposed to be driven by a hyper coordinating water molecule and another electron-donor water molecule, which causes hydrogen bonds to rearrange, reorient, and transfer H^+ ions (Merle et al., 2011).

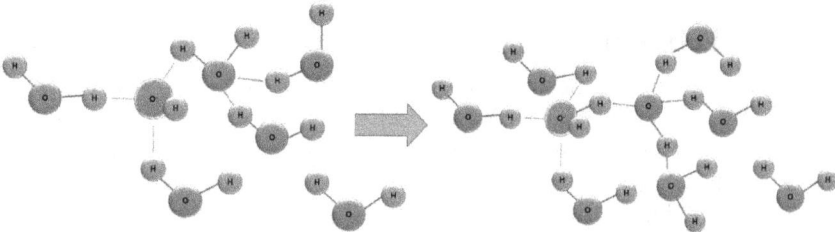

Figure 11.4   Transport of OH^- ions in water $(OH^-(H_2O)_3 \rightarrow OH^-(H_2O)_4)$.

Based on these guidelines, structural engineering should enable to find structures that allow OH^- ions to migrate through the base structure with grafted side chain support and these two factors are focused in various researches.

11.2.3   Base Chain and Side Chain Groups

11.2.3.1   Polyvinyl Alcohol (PVA)

The structure of PVA provides hydrophilic zones as a result of OH groups which generate chemical stability. Shang and co-workers (2020) reported a PVA base material with symmetrical bis-crown ether grafts (B-C30%-PVA) to prepare AEM with an application to AEMFC. The integration of symmetric bis-crown ether to the PVA main chain gives the structure negative charges due to the multiple oxygen atoms of the graft. The purpose of these negative charges is to select complex metal ions, which in turn leads to electrostatic adsorption of the OH⁻ ions. According to Shang et al. (2020), AEM showed alkaline resistance, high IEC, high IC, water uptake (WU), and swelling ratio (SR) in a test cell that was relatively higher than those reported in literature (Table 11.2). In another study, Albayrak and Şimşek (2020) reported two membranes with application in DMFC. In the first case, a membrane was prepared where PVA was functionalized with imidazolium (PVA-Im) and deposited on a porous PC resulting PVA-Im/PC; in the second one, a membrane was prepared functionalizing PVA with quaternary ammonium (PVA-QAm) and deposited on the PC matrix resulting PVA-QAm/PC. The polycarbonate matrix (PC) addition aims to significantly improve mechanical properties (10^7 Pa at 140°C) and control the membrane swelling of the membrane because of PC hydrophobicity. PVA-Im/PC showed higher IC compared to PVA-QAm/PC. However, IEC and IC showed lower values (Table 11.2).

Furthermore, PVA-Im/PC showed the lowest methanol permeation rate, the highest membrane selectivity, and a high alkaline and oxidative stability (Albayrak and Şimşek, 2020). It is important to mention that the grafts have the role of OH⁻ charge carriers in these works. However, it should be taken into account that the greater WU in the membrane, the IC, the degree of swelling, and the ionic mobility are more proportional. Consequently, the ion exchange capacity is favored in function of the ionic group solvation. Therefore, in the case of PVA-Im/PC, the imidazolium ring structure and the PC hydrophobic character limit the interaction of water, which is the reason for the reduced properties.

11.2.3.2 Poly (ether ether ketone) (PEEK)

One of the most applied membranes for PEM is PEEK, which is an alternative to fluorinated membranes, such as Nafion©. PEEK base membranes have excellent mechanical and thermal stability as well as low fuel permeability. However, they show several technical problems for AEM application. It is worth highlighting that the most serious problem is the reduced stability in anion conduction under alkaline conditions. Son et al. (2020) proposed direct synthesis of ion-conducting polymers using monomers that have functional moieties to prevent gelation and allow control of cationic functional groups. For this reason, a series of PEEKs with imidazolium groups have been synthesized by direct condensation polymerization (Table 11.1), using monomers with multiple imidazole terminations (DI-PEEK). As can be seen in Table 11.2, DI-PEEK-30 membrane showed that the IC and the mechanical stability improved.

It is worth mentioning that such values are analogous to Fumatech commercial sample (FAA-3-50). In this structure, the application of structural engineering lies in the use of a type of nucleophilic substitution, where a pair of free electrons from a nucleophile attacks an electrophilic center and binds to it, expelling another group. Thus the degree of polymer degradation through this reaction can

Table 11.2 Features of the membranes and their properties

Membrane	IC (mS·cm^{-1})	Power density$_{(max)}$ (mW·cm^{-2})	Current density (mA·cm^{-2})	IEC (meq·g^{-1})	WU (%)	SR (%)	References
B-C30%-PVA*	235(80°C)	197.0(80°C)	400(80°C)	5.03	–	–	Shang et al., 2020
PVA-Im/PC**	16(50°C)	–	–	–	0.72	0	Albayrak and Şimşek, 2020
DI-PEEK-30*	78(80°C)	98.0(60°C)	277(60°C)	0.51	14	–	Son et al., 2020
Fumatech (FAA-3-50)	150	114.0	266	1.50	25	2.0	
PPAEK*	99.8(80°C)	92.0(80°C)	154(80°C)	1.52	79.0	18.5	Li et al., 2021
Q$_6$PEKBO-2.0*	41.1(30°C)	–	–	1.73	54.0	32.9	Zhang and Xu, 2020
PPO-C-IQA*	40.6(20°C)	141.3	320	1.76	79.0	15.0	Chu et al., 2021
QPPO/PSF/TiO$_2$**	54.7(80°C)	118.0(60°C)	300(60°C)	2.86	136.0	14.0	Msomi et al. 2020
PPO-TMIm*	31.7 (25°C)	22.1(60°C)	128 (60°C)	2.06	53.2	10.1	Sheng et al., 2020
ImPS/10% ZrO$_2$*	80.2(50°C)	270.0	640	2.84	170.0	41.6	Rambabu et al., 2020
bPES-Pip-10–5*+	105.4(80°C)	136.9(80°C)	292.5(80°C)	2.03	55.8	21.2	Shen et al., 2020
PES-NS-10%*	95.5(80°C)	110.1(80°C)	300.0(80°C)	1.70	13.3	5.7	Liu et al., 2020
CFPESN-0.6*	91.5 (80°C)	91.3(80°C)	210.0(80°C)	1.71	32.5	8.8	Lai et al., 2020
rPNB-O-Im-30*	61.8(80°C)	98.4(80°C)	208.0(80°C)	1.37	31.8	8.0	Huang et al., 2020

* AAEMFC H$_2$/O$_2$ (Alkaline Anion Exchange Membrane Fuel Cell).
*+ AAEMFC H$_2$/air.
** DMFC (Direct Alcohol Fuel Cell).

be reduced by replacing the alkyl ammonium with bulky cations, which stabilize the α-C-N bond by delocalizing the charge of the cations (Son et al., 2020).

11.2.3.3   Poly (arylene ether) (PAEK)

AEMs composed of cyclic quaternary ammonium salts show improvement in WU, SR, IC and elongation at breaking in comparison to AEM containing simple quaternary ammonium salts. Unfortunately, as aforementioned, something that impedes the optimal efficiency of AEMs is the lack of control of the functionalized sites during the synthesis mechanism (Yan et al., 2014). In this context, Li et al. (2021) proposed a based poly AEM (aryl ether ketones) (PAEK) where the synthesis method is due to polyhydroxyalkylation reaction between 4, 4'-bis (phenoxy) benzophenone (BPBP) and 1-(4-bromobutyl) indoline-2,3-dione (BID), applying trifluoromethane sulfonic acid (TFSA) as a catalyst to graft bromoalkyl side chain groups onto PAEK. Tertiary amines, such as trimethylamine, N-methyl-piperidine, and N, N, N', N', N''-pentaethylguanidine were chosen to give as a result, ionomers QPAEK, PPAEK and GPAEK. According to Li et al. (2021), among the different AEMs, PPAEK showed the highest WU (79%) and high IC at 80°C (Table 11.2), which made these membranes potential AEM candidates for fuel cells. Both properties, alkaline stability end thermal one, did not increase with the change of cationic group because they depend on the nature of the polymer main chain, although they are also presented optimally.

A preponderant factor in AEM is alkaline stability; for example, there are other aromatic polymers containing pendant anion exchange groups attached to aromatic backbones, through relatively long alkyl spacers, that show superior alkaline stability to benzyl groups (Marino and Kreuer, 2015). Zhang and Xu (2020) described the comb-like structure as that of Nafion$^{©}$ tending to form nanophases, one hydrophobic, and one hydrophilic, which offer two advantages. The former improves IC through interconnections of hydrophilic channels and the latter keeps hydrophobic backbones protected from the highly alkaline hydrophilic environment. Zhang and Xu (2020) synthesized PAEK (Q6PEKBP-y/x) containing oxyhexyltrimethylammonium groups ($-O-(CH_2)_6-N^+(CH_3)_3$ PF_6^-, OHeTMA). The result of this synthesis was PAEK (Q_6PEKBP-y/x) with controlled contents of OHeTMA side chain groups. As seen in Table 11.2, Q_6PEKBP-2.0 showed high IC and an acceptable SR of 32.9% at 80°C, which could be due to both microblocked main chain and QAs tightly packed along the chain. It is worth noting that Q_6PEKBP-2.0 did not show any degradation after immersion in [1 M] NaOH at 60°C for 90 days. PAEK containing long side chain type QA are prospects as membrane materials in fields such as AFC and electrolyzers.

11.2.3.4   Poly(2,6-dimethyl-1,4-phenyleneoxide) (PPO)

Chu et al. (2021) proposed a structural design based on the base structure poly (2,6-dimethyl-1,4-phenylene oxide) (PPO) that was quaternized with several lengths (n or m) of cationic side chains; PPO-C-nQA (with a linker including triazole groups) and PPO-mQA (without triazole groups in the side chains). The PPO-C-nQA membrane showed higher conductivity of OH$^-$ ions because the triazole groups provided more sites for water/ion transport and consequently, higher WU while the PPO-mQA membrane was made up of a microphase-separated structure. It should be noted that those membranes with very long side chains do not have good conductivity while PPO-C-1QA in AEMFC H_2/O_2 with the highest conductivity showed maximum power density and acceptable current density as well as an extraordinary alkaline stability, when immersed in [10 M] NaOH at 80°C for 250 h.

In another study, Msomi et al. (2020) synthesized an AEM composed of two polymeric bases based on quaternized PPO and polysulfone (PS) mixed with TiO_2 resulting in (QPPO/PSF/TiO_2). In that study, the possibility of introducing an inorganic material that gives new properties to the AEM is explored. One benefit of inorganic nanofillers is that they perform like a Lewis acid, promoting water storage in membrane and therefore, increasing IC (Aslan and Bozkurt, 2014). This is justified by IC results at 80°C with excellent alkaline stability and good performance when assembled in an alkaline medium of a methanol fuel cell and the same occurred with QPPO/PSF/TiO_2 composite membrane at 2 wt.% TiO_2 at 60°C with the highest power density (Table 11.2) (Msomi et al., 2020).

Based on structural engineering, Sheng et al. (2020) designed a PPO from the poly (2,6-dimethyl-1,4-phenylene oxide) (PPO) backbone due to its commercial availability, viable functionalization as well as thermal, chemical, and mechanical stability. PPO was functionalized by direct quaternization with imidazolium side chains in four variants selected for their adequate WU, IC, thermal stability, mechanical properties, and alkalinity. Variants were based on the different substitutions that were detected at the places of imidazolium rings and were grafted on to the PPO backbone as a structure/property relationship study, e.g. PPO-NMIm, PPO-TMIm, PPO-BIm, and PPO-TPIMm, grafts from lowest to highest substitutions, respectively. According to Sheng et al. (2020), PPO-TMIm AEM showed the highest WU and moderate OH⁻ conductivity (Table 11.2) at room temperature in comparison to PPO-TPIm AEM. However, PPO-TPIm showed alkaline stability after 192 hours of immersion in [1 M] NaOH at 80°C due to the steric hindrance presence in the imidazole ring positions. In an H_2/O_2 fuel cell, a maximum power density at 60°C (Table 11.2) was obtained for the PPO–TMIm co-polymer as AEM, presumably due to the acceptable IC and alkaline stability.

As can be observed, triazole chains allow greater WU and therefore, greater IC (Chu et al., 2021), comparable to AEMs with inorganic nanofillers (Msomi et al., 2020), as well as surprising alkaline stability. However, in cases with ringed side chains with minimal substitutions, alkaline stability is acceptable due to the steric hindrance of the structure; it is worth noting that IC gets reduced (Sheng et al., 2020).

11.2.3.5   Polysulfones (PS)

One of the most widely used AEM-based polymers is polysulfone (PS) due to its improved IC, thermal, mechanical and chemical stability at room temperature. When it is functionalized with imidazolium, hydroxide conductivity and selective solubility of the hybrid membranes increases, but mechanical resistance decreases (Dai et al., 2016). For this reason, inorganic particles were introduced to increase the mechanical resistance of the membrane without excluding other properties provided by the structure itself. Rambabu et al. (2020) used imidazole-functionalized PS (ImPS) as a base, introducing ZrO_2 particles into the structure and giving way to an $ImPS/ZrO_2$ membrane that showed higher WU favoring IC, which increased from 21% to 47% and the power density from 35% to 39% in comparison to pure ImPS (Table 11.2).

11.2.3.6 Poly (ether sulfone) (PES)

In the same way as in PPOs, structural engineering is essential for PES. One of the most outstanding works is that of Shen et al. (2020) who synthesized a series of multiblock PES with long side chains densely finished with piperidinium (bPES-Pip-XY). In this research, hydrophilic/hydrophobic microphases are formed due to the multi-block structure and conglomeration of hydrophilic ionic groups on the side chains. The highest IC of the bPES-Pip-10-5 membrane reached significant values at 80°C and the MEA based on bPESPip-10-5 with moderate power density reached significant values at 60°C (Table 11.2). Importantly, the bPES-Pip-10-5 membrane still maintained 88% of the original conductivity after being treated in a [1 M] KOH solution at 60°C for 336 h (Shen et al., 2020). In this work, it showed that the MEA improved its conductivity and alkaline stability due to the piperidinium groups in the side chain, which cause ion agglomeration, giving way to microphase separation.

In another study, Liu et al. (2020) synthesized an AEM based on PES and dense N-spirocyclic cations (NS) as side chain resulting in MEA PES-NS-x. Given this structure, there is a microphase formation due to the agglomeration of N-spirocyclic cations. The steric hindrance of the spirocyclic structure and the electron donating effect of the alkyl spacers, substantially improved the stability of the cations under severe alkaline conditions and high hydration. PES-NS-10% reached the highest CI at 80°C with an SR of 7.69% (Table 11.2). Additionally, during alkaline stability test under [2 M] NaOH at 80°C for 864 h, the membrane showed good alkaline stability with IEC and IC retentions of 86.2% and 84.3%, respectively. Performance was tested at 60°C and a higher power density was reached for PES-NS-10 (Table 11.2) (Lui et al., 2020).

In summary, up to this point, it has been commented that studies on this type of membranes are based on two structural strategies. First, the synthesis of the defined structures separated by microphases between hydrophilic cationic structures and waterproof polymeric structures. This strategy promotes the aggregation of conductive anion groups that form ionic clusters acting as ion transport channels, which is an efficient way to transport hydroxide ions and therefore, cause an improvement in the ionic conductivity of EMAs as it is reported (Zhu et al., 2018). Second, research on Nafion[©] structure developed comb-shaped AEMs where cationic groups were grafted on to or next to long, flexible side chains, which can form nanoscale ion clusters (Shukla and Shahi, 2018). Therefore, the length of the flexible side chains can make the ion channel achieve high conductivity, while the long flexible side chains have hydrophobicity and a steric hindrance effect that can also increase the alkaline stability of AEMs. However, the optimization of some factors excludes other ones in ion transport through water molecules when the transport pathway is controlled by hydrophobic aliphatic structures (Gao et al., 2019). Given these aspects, Lai et al. (2020) proposed an AEM with a comb-shaped structure resulting in a fluorene-based poly (arylene ether sulfone nitrile) membrane (CFPESN-x) and long comb-shaped C8 alkyl side chains containing QA and N, N-dimethyloctylamine groups (DOA) as a quaternization reagent to ensure good alkaline stability. The content of nitrile groups was varied because they showed an impact on the morphology and properties of the membrane. The increase in nitrile groups in the CFPESN-x backbone showed more interconnected ion clusters that form more efficient ion transport channels, giving way to higher conductivities without excessive WU; thus CFPESN-0.6 membranes showed moderate IC at 80°C (Table 11.2). Furthermore, CFPESN-0.6 also showed good mechanical properties, thermal stability, and optimized alkaline stability (Lai et al., 2020).

11.2.3.7 Polynorbornene (PNB)

This summary shows that the main chains—also called backbones—of the AEM structure are susceptible to degradation at high pH values and the side chains allow the reduction of this disadvantage. An alternative to stable structures at high pH values is the cross-linking of the structure. However, in some studies, a notable decrease in the IC was observed since it prevents ionic transport and a very dense and hydrophobic cross-linking causes the AEM to become rigid, brittle apart from poorly conductive of OH^- ions (Gao et al., 2019). Huang and co-workers (2020) designed a series of polynorbornene membranes as a self-cross-linking ion-conducting base. Structural engineering is present in the microphase separation of the cross-linked hydrophilic side chain of synthesis, that is, simple structure and polymer backbone. Variants of the cross-linked AEM series, such as 5-norbornene-2-methylene glycidyl ether (NB-MGE) with significant self-cross-linking, improved dimensional stability, while 5-norbornene-2-alkoxy-1-hexyl chloride-3-methylimidazolium ($NB-O-Im^+Cl^-$) provided high conductivity due to the hydrophilic part. Furthermore, according to Huang et al. (2020), cross-linking significantly reduces WU and SR, providing excellent solvent resistance and better thermal and mechanical properties. Overall, these EMAs showed remarkable alkaline stability and moderate IC (Table 11.2).

11.2.4 Polymeric Structures with Grafted Graphene Oxide

GO and reduced GO (rGO) are reported in literature to be typical fillers of nanocomposite membranes (He et al., 2014). In this work, it is reported that a wide analysis due to the number of functional groups in the GO sheet provide surface modifications; this is a new alternative to the addition of inorganic compounds, such as ZrO_2 and TiO_2, among others. Initially, GO and rGO were added to PEM to improve proton conductivity and thermally stabilize the membrane. A great deal of studies of AEMs have been accomplished, analyzing the incorporation of

GO. In these studies, functionalized GO (FGO) has been combined in different ways; for instance, with polymer PVA/PDA SIPNs (Semi-interpenetrating Polymer Networks), with quaternized graphene (QG), quaternized polysulfone (QPSU), ionic liquid functionalized GO (GO-IL), and PPO functionalized with imidazolium (ImPPO). In all these membranes, IC as well as thermal and mechanical stability have been improved (Yang et al., 2018). Furthermore, AEM series have been tested through simple thermal cross-linking between rGO and azide groups of QPSU. Although the cross-linked membranes showed improved properties, such as alkaline stability and permeability to methanol, their IC decreased drastically due to the inhibition of WU, since cross-linking generates very compact structures (Hu et al., 2017). Furthermore, polysulfone-based AEMs with stable cross-linked rGO (Bai et al., 2019) were constructed. In that study, π-π interactions in polymer were designed with a non-covalent modification to rGO with a small pyrene molecule tertiary amine. Functionalized rGO (PrGO and TrGO) was prepared as a cross-linker and filler to make AEM with QPSU and (CQPSU-X-TrGO and CQPSU-X-PrGO) base. According to Bai et al. (2019), PrGO and CQPSU-2%-PrGO cross-linked membranes had a relevant IC at 80°C (Table 11.2). Therefore, the strategy was used efficiently to obtain new types of cross-linked organic-inorganic nanohybrid AEMs with superior chemical stability and high IC. It is worth noting that the cross-linked membranes can compact the internal packing structure, which optimizes alkaline resistance, IC, and oxidative stability of AEMs (Bai et al., 2019).

In 2021, Gorgieva and co-workers designed a new generation of membranes constituted by nanocomposites based on chitosan (CS), biobased material, and three variants of N-doped graphene with different composition and structure. Such structures have been obtained with different order of dimensions, (N-rGO)-doped reduced N graphene oxides, and quasi-1D nanoribbons (N-rGONR), as well as 3D graphitized polyenaminone (N-pEAO)-doped N porous particles were synthesized in a 2D way. During the application of these membranes to direct alkaline ethanol fuel cells, it was observed that doped graphene variants with concentrations between 0.01% wt. and 0.07% wt. changed the shape of the CS membrane significantly due to the interactions between the CS functional groups and doped N graphene derivatives. CS/NrGONRs (0.07%) particularly showed a power density of 3.7 mW·cm^{-2}. In general, tensile modulus, crystallinity, KOH absorption, tensile strength, and ethanol permeability are relevant in these composite membranes as as π-π interactions that allow good conductivity in these types of structures (Gorgieva et al., 2021). However, research on this type of structures with GO grafts continues due the peculiarities of this new material.

11.3   AFC PERFORMANCE

11.3.1   Single-cell Performance

AEMs have been widely studied for their applications in electrochemical devices, such as redox flux batteries, electrolyzers, electrodialysis, sensors, and fuel cells in electric vehicular transport, since the alkaline medium favors the faster kinetics of the ORR and allows avoiding the use of Pt and PGMs as electrocatalysts (Yang et al., 2021). This enables high anion-selective conductivity in membrane for high-performance neutral organic-based aqueous redox flow battery by microstructure design (Si et al., 2019; Chu et al., 2019; Kim and Kang, 2020; Arunachalam et al., 2021; Xiao et al., 2022). Although long-term alkaline stability is used as electrolyte of hydroxide conduction in electrochemical fixed and portable devices, it has been reported that the presence of antioxidants as scavengers can improve the properties and long lifetime stability of fuel cells (Ye et al., 2018; Shin et al., 2019). Figure 11.5 shows the fuel cell performance of dual antioxidants doped alkaline membranes synthesized by selected hindered phenols and organic diphosphite as dual antioxidants based on ammonium quaternized poly(4-vinylbenzyl chloride-styrene) with a hindered 2,4,6-tri-ter-butylphenol, MB, which conducted to QP(VBC-St)-MB/PB$_{0.6}$ membrane (Peng et al., 2021). The phenolic ester in the membrane preparation was selected as scavenger of free radicals produced to improve the chemical stability of the membranes via synergistic effect. The performance of the membrane used as an electrolyte, in a single AEMFC, attained an open circuit voltage of 1.02 V under the test conditions at 60°C, pointing out that the membrane has a compact structure with low gas permeability. Such results indicated that the antioxidant-doped membrane is a possible candidate to be applied as electrolyte in alkaline fuel cells (Table 11.3).

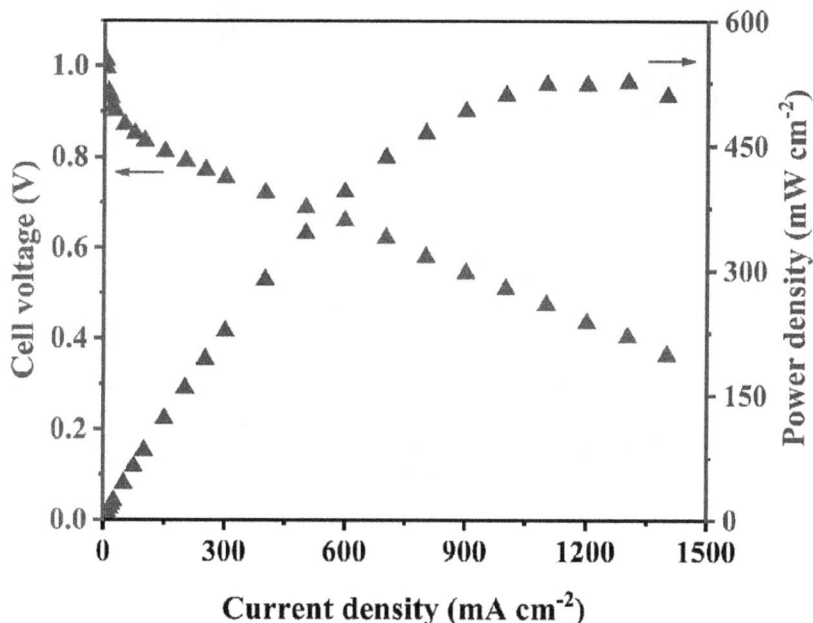

Figure 11.5 Single AMFC design considering the QP(VBC-St)-MB/PB$_{0.6}$ membrane at 60°C fueling with humidified hydrogen and oxygen (Reprinted with permission from [Peng et al., 2021]).

Besides the antioxidants' incorporation to membranes, quaternary ammonium group was used in the preparation of anionic membranes and characterization in a MEA (Yang et al., 2021). Figure 11.6 shows the performance of a prepared membrane synthesized from a sequence of semi-interpenetrating polymer network, obtained by adding flexible polyvinyl alcohol to the rigid photo-cross-linked poly (2,6-dimethyl-1,4-phenylene oxide) network (peak power density) (Table 11.3). Furthermore, due to the synergetic effect of the semi-interpenetrating polymer network, both stabilities, dimensional and alkaline, of AEMs were enhanced especially at high temperature (80°C), which is due to the effective ion channels and better dimensional stability of SIPNS membranes. It is worth noting that such results indicate that polymer network membranes could be an effective strategy to improve the properties of AEMs in fuel cell applications.

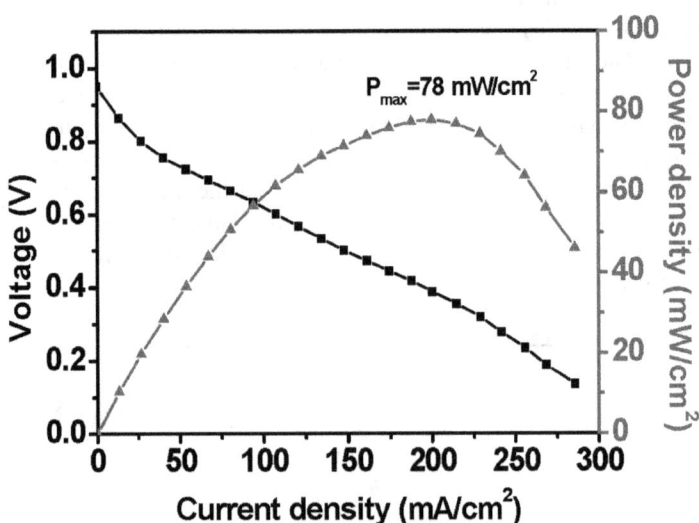

Figure 11.6 Single fuel cell performance of a semi-interpenetrating polymer network AEM under H_2/O_2 condition at 60°C (Reprinted with permission from [Yang et al., 2021]).

Table 11.3 Features of the special membranes and their properties

Membrane	Open circuit voltage (V)	Power density$_{(max)}$ (mW·cm^{-2})	Current density (mA·cm^{-2})	References
QP(VBC-St)-MB/PB$_{0.6}$	1.02$^{(60°)}$	526$^{(60°)}$	1400$^{(60°)}$	Peng et al., 2021
SiPNs	–	78$^{(80°C)}$	–	Yang et al., 2021
TQ-PDBA-70%	1.10$^{(80°C)}$	158.8$^{(80°C)}$	360$^{(80°C)}$	Gao, et al., 2020
QAPPGG	–	622$^{(2-70°C)}$	–	Zhang et al., 2021
D1M2PPO–CBT D1M2PPO	–	315$^{(60°C)}$ 150$^{(60°C)}$	–	Wu et al., 2022
HDPEBMA* HDPEBMA	0.9 0.9	2 100	7 44	Adabi et al., 2021

*AAEMFC H_2/air.

Some issues regarding the improvement of AEMs remain to be solved for diverse applications associated with poor alkaline stability and low conductivity. Liu et al. (2020) implemented novel strategies, i.e. the hydrophobic fluorine-based polymer backbone was used during the chemical synthesis, bearing special functional sites together with hydrophilic long flexible multi-cation side chain. The synthesized partial diallyl bisphenol a poly(-arylene ether) TQ-PDBA-70% polymers presented high IEC retention and maintained good mechanical properties and OH⁻ ion conductivity (Gao, X.L. et al., 2020). Figure 11.7 shows the performance of a single cell assembled with the synthesized membrane, fed with H_2/O_2 and evaluated at 80°C (Table 11.3). Results of this structure open a new strategy for developing high performance membrane with a good durability, attributed to the robust chemical stability and excellent dimensional stability of the membrane, prerequisite to attain a potential application in AEFCs as described in literature (Mandal et al., 2020; Sun et al., 2018).

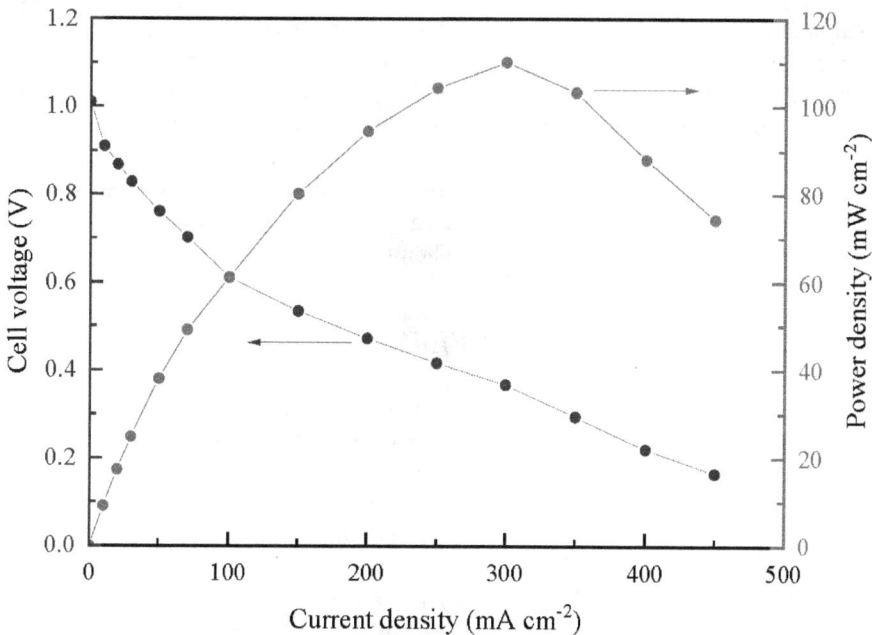

Figure 11.7 H_2/O_2 single alkaline fuel cell design. High frequency resistance test of the TQ-PDBA-70% MEA at 80°C (Reprinted with permission from [Sun et al., 2018]).

Advances in supra-molecular chemistry are reported by Zhang et al. (2021) creating inter/intra-molecular cation-dipole interactions between pendent quaternary ammoniums cations and polar polyethylene glycol grafts (QAPPGG) in an AEM. Results of this interaction produced a membrane with 3-dimensional pathways for both ion transport and water. The result is an anionic membrane which exhibited high OH⁻ conductivity (49 mS·cm⁻¹ at 30°C) and a high performance in a single H_2/O_2 alkaline fuel cell (Table 11.3). Wu et al. (2022) recently reported a covalent bonding-triggered (CBT) strategy for filling of pre-formed membrane pores with ion conductive polyelectrolyte formation of CBT-adopted to ion transport channels (CITCs) obtaining a high-performance AEM. The fabrication of a single H_2/O_2 AEMFC reported high peak power density (Table 11.3), which is 2.1 times higher than the power output using a membrane denoted as D1M2BPPO synthesized from a poly(2,6-dimethyl-1,4-phenylene oxide immersed in an homogeneous solution of poly(2-dimethylamino ethyl methacrylate-co-methylmethacrylate co-polymers).

Recently, Adabi et al. (2021) reported high performance of H_2/O_2 and H_2/air AAFCs using commercial Fe-N-C cathode in a high-density polyethylene (HDPE) membrane with benzyltrimethylammonium cationic head groups. This was in order based PGM-free catalyst in order to enhance the oxygen reduction kinetics at the cathode side electrode to either reduce or eliminate the use of Pt. Experimental results demonstrated that design and durability of AEMFCs increased considerably (Table 11.3) (above 1 W·cm^{-2} with H_2/air) with testing voltage durability for more than 150 h of continuous operation without severe loss of performance. By using a second configuration, Fe-N-C cathodes paired with a low-loading PtRu/C anodes (0.125 mg PtRu per cm^2, 0.08 mg Pt per cm^2), a specific power of 10.4 W per mg PGM (16.25 W per mg Pt) was achieved. Cathode ensembles with Fe-N-C electrocatalysts load reached a high peak power density superior with a mass-transport-tuned current density above, achieving an exceeding DOE target for PGM-free ORR cathodes (Table 11.3). These results demonstrated that Fe-N-C catalysts showed intrinsic activities very close to those of carbon-supported Pt catalysts, to become a fuel cell alternative to PEMFC as analyzed and discussed previously by Thompson et al. (2020).

11.4 CONCLUSION AND PERSPECTIVES

This review attempts to give an overview of relevant progress on alkaline membrane reported recently in literature, focusing on the design of novel structure of AEMs with outstanding performance in a single H_2/O_2 alkaline fuel cell. This is attributed to their high ionic hydroxide conductivity, mechanical properties, improved chemical stability, and long-term durability. Notorious research efforts are observed on novel innovative membranes during the development of a diversity of electrochemical energy conversion applications and materials science, in conjunction with most advanced analytical sensor works.

As a perspective, there is an ongoing effort to develop novel hydroxide membranes. However, AEMs with great alkaline stability and high IC are necessary for the promising massive production and successful applications of AAFC in their commercialization as electrochemical energy-conversion devices. The open perspective in the field of membranes is the production of a combination of cation-exchange membranes and anion-exchange membranes to form a bipolar membrane for electrochemical renewable energy applications in solar-to-fuel devices systems.

REFERENCES

Adabi, H., A. Shakouri, N.U. Hassan, J.R. Varcoe, B. Zulevi, A. Serov, et al., 2021. High-performing commercial Fe–N–C cathode electrocatalyst for anion-exchange membrane fuel cells. Nat. Energy. 6: 834–843.

Albayrak, A.G. and Ö. Şimşek. 2020. Imidazolium functionalized poly (vinyl alcohol) membranes for direct methanol alkaline fuel cell applications. Polym. Int. 69: 644–652.

Anson, C.W. and S.S. Stahl. 2020. Mediated fuel cells: soluble redox mediators and their applications to electrochemical reduction of O_2 and oxidation of H_2, alcohols, biomass, and complex fuels. Chem. Rev. 120: 3749–3786.

Arunachalam, M., A. Sinopoli, F. Aidoudi, S.E. Creager, R. Smith, B. Merzougui, et al., 2021. High performance of anion exchange blend membranes based on novel phosphonium cation polymers for all-vanadium redox flow battery applications. ACS Appl. Mater. Interfaces. 13: 45935–45943.

Aslan, A. and A. Bozkurt. 2014. Nanocomposite membranes based on sulfonated polysulfone and sulfated nano-titania/NMPA for proton exchange membrane fuel cells. Solid State Ion. 255: 89–95.

Bai, Y., Y. Yuan, Y. Yang and Ch. Lu. 2019. A facile fabrication of functionalized rGO cross-linked chemically stable polysulfone-based anion exchange membranes with enhanced performance. Int. J. Hydrog. Energy. 44: 6618–6630.

Bagostky, V.S. 2009. Fuel cells: Problems and Solutions. Chap. 6. Alkaline fuel cells. pp. 107–121. John Wiley & Sons. USA.

Barnett, A., J.J. Karnes, J. Lu, D.R. Major Jr., J.S. Oakdale, K.N. Grew, et al., 2022. Exponential water uptake in ionomer membranes results from polymer plasticization. Macromolecules. 55(15): 6762–6774.

Castañeda-Ramírez, S. and R. Ribadeneira-Paz. 2018. Hydroxide transport in anion-exchange membranes for alkaline fuel cells. pp. 51–69. *In:* S. Karakuş (ed.). New Trends in Ion Exchange Studies. IntechOpen. Cerrahpaşa. Estambul. Turquía.

Chen, N. and M. Young. 2021. Anion exchange polyelectrolytes for membranes and isonomers. Prog. Polym. Sci. 113: 101345.

Cheng, H., R. Gui, S. Liu, Y. Xie and Ch. Wu. 2020. Local structure engineering for active sites in fuel cell electrocatalysts. Sci. China Chem. 63: 1543–1556.

Chu, X., Y. Shi, L. Liu, Y. Huang and N. Li. 2019. Piperidinium-functionalized anion exchange membranes and their application in alkaline fuel cells and water electrolysis. J. Mat. Chem. A: Mat. Energy Sust. 7: 7717–7727.

Chu, X., J. Liu, S. Miao, L. Liu, Y. Huang, E. Tang, et al., 2021. Crucial role of side-chain functionality in anion exchange membranes: Properties and alkaline fuel cell performance. J. Membr. Sci. 625: 119172.

Cruz-Martínez, H., M.M. Tellez-Cruz, O.X. Guerrero-Gutiérrez, C.A. Ramírez-Herrera, M.G. Salinas-Juárez, A. Velázquez-Osorio, et al., 2019. Mexican contributions for the improvement of electrocatalytic properties for the oxygen reduction reaction in PEM fuel cells. Int. J. Hydrog. Energy. 44: 12477–12491.

Cruz-Martínez, H., H. Rojas-Chávez, P.T. Matadamas-Ortiz, J.C. Ortiz-Herrera, E. López-Chávez, O. Solorza-Feria, et al., 2021. Current progress of Pt-based ORR electrocatalysts for PEMFCs: An integrated view combining theory and experiment. Mater. Today Phys. 19: 100406.

Dai, J., G. He, X. Ruan, W. Zheng, Y. Pan and X. Yan. 2016. Constructing a rigid cross-linked structure for enhanced conductivity of imidazolium functionalized polysulfone hydroxide exchange membrane. Int. J. Hydrog. Energy. 41: 10923–10934. (In text it is 2014?)

Dekel, D.R. 2018. Review of cell performance in anion exchange membrane fuel cells. J. Power Sources. 375: 158–169.

Feng, L., X. Zhang, Ch. Wang, X. Li, Y. Zhao, X. Xie, et al., 2016. Effect of different imidazole group positions on the hydroxyl ion conductivity. Int. J. Hydrog. Energy. 41: 16135–16141.

Firouzjaie, H.A. and W.E. Mustain. 2020. Catalytic advantages, challenges, and priorities in alkaline membrane fuel cells. ACS Catal. 10: 225–234.

Gao, X.L., Q. Yang, H.Y. Wu, Q.H. Sun, Z.Y. Zhu, Q.G. Zhang, et al., 2019. Orderly branched anion exchange membranes bearing long flexible multi-cation side chain for alkaline fuel cells. Int. J. Hydrogen Energy. 45: 11148–11157.

Gao, L., C.S. Lu, S.Y. Ma, X.M. Yan, X.B. Jiang, X.M. Wu, et al., 2020. Flexibly cross-linked and post-morpholinium-functionalized poly (2,6-dimethyl-1,4-phenylene oxide) anion exchange membranes. Int. J. Hydrog. Energy. 45: 29681–29689.

Gao, X.L., L.X. Sun, H.Y. Wu, Z.Y. Zhu, N. Xiao, J.H. Chen, et al., 2020. High conductive fluorine-based anion exchange membranes with robust alkaline durability. J. Mat. Chem. A: Mat. Energy and Sustainability. 8: 13065–13076.

Ge, X., F. Zhang, L. Wu, Z. Yang and T. Xu. 2022. Current challenges and perspectives of polymer electrolyte membranes. Macromolecules. 55: 3773–3787.

Gorgieva, S., A. Osmic, S. Hribernik, M. Bozic, J. Svete, V. Hacker, et al., 2021. Efficient chitosan/nitrogen-doped reduced graphene oxide composite membranes for direct alkaline ethanol fuel cells. Int. J. Mol. Sci. 22: 1740.

Gottesfeld, S., D.R. Dekel, M. Page, Ch. Bae, Y. Yan, P. Zelenay, et al., 2018. Anion exchange membrane fuel cells: current status and remaining challenges. J. Power Sources. 375: 170–184.

Guo, Y., Z. Pan, Zhefei and L. An. 2020. Carbon-free sustainable energy technology: Direct ammonia fuel cells. J. of Power Sources. 476: 228454.

He, Y., C. Tong, L. Geng, L. Liu and C. Lu. 2014. Enhanced performance of the sulfonated polyimide proton exchange membranes by graphene oxide: size effect of graphene oxide. J. Membr. Sci. 458: 36–46.

Hermida-Castro, M.J., D. Hermida-Castro, X.M. Vilar Martínez and J.A. Orosa. 2013. Comparative study of fuel cell applications and future plant conservation applications. Int. J. Energy Sci. 3: 357–361.

Hren, M., M. Bozic, D. Fakin, K.S. Kleinschek and S. Gorgieva. 2021. Alkaline membrane fuel cells: anion exchange membranes and fuels. Sustain. Energy Fuels. 5: 604–637.

Hu, B., L. Miao, Y. Zhao and C. Lu. 2017. Azide-assisted cross-linked quaternized polysulfone with reduced graphene oxide for highly stable anion exchange membranes. J. Membr. Sci. 530: 84–94.

Huang, S., X. He, Ch. Cheng, F. Zhang, Y. Guo and D. Chen. 2020. Facile self-cross-linking to improve mechanical and durability of polynorbornene for alkaline anion exchange membranes. Int. J. Hydrog. Energy. 45: 13068–13079.

Jiao, K., J. Xuan, Q. Du, Z. Bao, B. Xie, B. Wang, et al., 2021. Designing the next generation of proton-exchange membrane fuel cells. Nature. 595: 361–369

Ke, X., J.M. Prahl, J.I.D. Alexander, J.S. Wainright, T.A. Zawodzinski and R.F. Savinell. 2018. Rechargeable redox flow batteries: flow fields, stacks and design considerations. Chem. Soc. Rev. 47: 8721–8743.

Kim, Do-H. and M.S. Kang. 2020. Pore-filled anion-exchange membranes with double cross-linking structure for fuel cells and redox flow batteries. Energies. 13: 4761.

Lai, A.N., Z. Wang, Q. Yin, R.Y. Zhu, P.Ch. Hu, J.W. Zheng, et al., 2020. Comb-shaped fluorene-based poly (arylene ether sulfone nitrile) as anion exchange membrane. Int. J. Hydrog. Energy. 45: 11148–11157.

Li, Z., R. Yu, Ch. Liu, J. Zheng, J. Guo, T.A. Sherazi, et al., 2021. Carbon-nitrogen. Preparation and characterization of side-chain poly (aryl ether ketone) anion exchange membranes by superacid-catalyzed reaction. Polymers. 222: 123639.

Lin, C.X., X.L. Huang, D. Guo, Q.G. Zhang, A.M. Zhu, M.L. Ye, et al., 2016. Side-chaintype anion exchange membranes bearing pendant quaternary ammonium groups via flexible spacers for fuel cells. J. Mater. Chem. 4: 13938–13948.

Liu, F.H., Q. Yang, X.L. Gao, H.Y. Wu, Q.G. Zhang, A.M. Zhu, et al., 2020. Anion exchange membranes with dense N-spirocyclic cations as side-chain. J. Membrane Sci. 595: 117560.

Ma, Z. and M.E. Tuckerman. 2011. On the connection between proton transport, structural diffusion, and reorientation of the hydrated hydroxide ion as a function of temperature. Chem. Phys. Lett. 511: 177–182.

Mandal, M. 2020. Recent advancement on anion exchange membranes for fuel cell and water electrolysis. Chem. Electro. Chem. 7: 1–11.

Mandal, M., G. Huang, N.U. Hassan, X. Peng, T. Gu, A.H. Brooks-Starks, et al., 2020. The importance of water transport in high conductivity and high-power alkaline fuel cells. J. Electrochem. Soc. 167: 054501.

Marino, M.G. and K.D. Kreuer. 2015. Alkaline stability of quaternary ammonium cations for alkaline fuel cell membranes and ionic liquids. Chem. Sus. Chem. 8: 513–523.

Masa, H., B. Mojca, F. Darinka, K.K. Stana and G. Selestina. 2021. Alkaline membrane fuel cells: anion exchange membranes and fuels. Sustain. Energy Fuels. 5(3): 604–637.

Merle, G., M. Wessling and K. Nijmeijer. 2011. Anion exchange membranes for alkaline fuel cells: a review. J. Membrane Sci. 377: 1–35.

Miller, E.L., S.T. Thompson, K. Randolph, Z. Hulvey, N. Rustagi and S. Satyapal. 2020. US department of energy hydrogen and fuel cell technologies perspectives. MRS Bull. 45: 57–64.

Msomi, P.F., P.T. Nonjola, P.G. Ndungu and J. Ramontja. 2020. Poly(2,6-dimethyl-1, 4-phenylene)/polysulfone anion exchange membrane blended with TiO_2 with improved water uptake for alkaline fuel cell application. Int. J. Hydrogen Energy. 45: 29465–29476.

Nie, Y., L. Li, Z. Wei, Yao Nie, L. Li and Z. Wei. 2015. Recent advancements in Pt and Pt-free catalysts for oxygen reduction reaction. Chem. Soc. Rev. 44: 2168–2201.

Pham, T.H., J.S. Olsson and P. Jannasch. 2017. N-spirocyclic quaternary ammonium ionenes for anion-exchange membranes. J. Am. Chem. Soc. 139: 2888–2891.

Peera, S.G., T. Maiyalagan, Ch. Liu, S. Ashmath, T.G. Lee, Z. Jiang, et al., 2021. A review on carbon and non-precious metal-based cathode catalysts in microbial fuel cells. Int. J. Hydrog. Energy. 46(4): 3056–3089.

Peng, X., Y. Yang, N. Ye, S. Xu, D. Zhang, R. Wan, et al., 2021. Synergy effects of hindered phenol and diphosphite antioxidants on promoting alkali resistance of quaternary ammonium functionalized poly(4-vinylbenzyl chloride-styrene) anion exchange membranes. Electrochim. Acta. 380: 138249.

Pu, Z., G. Zhang, A. Hassanpour, D. Zhen, S. Wang, S. Liao, et al., 2021. Regenerative fuel cells: recent progress, challenges, perspectives and their applications for space energy system. Appl. Energy. 283: 116376.

Rambabu, K., G. Bharath, A.F. Arangadi, S. Velu, F. Banat and P.L. Show. 2020. ZrO$_2$ incorporated polysulfone anion-exchange membranes for fuel cell applications. Int. J. Hydrog. Energy. 45: 29668–29680.

Ramaswamy, N. and S. Mukerjee. 2019. Alkaline anion-exchange membrane fuel cells: challenges in electrocatalysis and interfacial charge transfer. Chem. Rev. 119: 11945–11979.

Ramírez-Herrera, C.A., M.M. Tellez-Cruz, J. Pérez-González, O. Solorza-Feria, A. Flores-Vela and J.G. Cabañas-Moreno. 2021. Enhanced mechanical properties and corrosion behavior of polypropylene/multi-walled carbon nanotubes/carbon nanofibers nanocomposites for application in bipolar plates of proton exchange membrane fuel cells. Int. J. Hydrogen Energy. 46: 26110.

Ramírez-Nava, J., M. Martínez-Castrejón, R.L. García-Mesino, J.A. López-Díaz, O. Talavera-Mendoza, A. Sarmiento-Villagrana, et al., 2021. The implications of membranes used as separators in microbial fuel cells. Membranes. 11: 738.

Reyes-Rodríguez, J.L., H. Cruz-Martínez, M.M. Tellez-Cruz, A. Velázquez-Osorio and O. Solorza-Feria. 2018. pp. 229–244. *In*: E. Rincón-Mejía, and A. de las Heras (eds). Recent Contributions in the development of fuel cell technologies in sustainable energy technologies. CRC Press. Taylor & Francis Group.

Shang, Ch., Z. Wang, L. Wang and J. Wang. 2020. Preparation and characterization of a polyvinyl alcohol grafted bis-crown ether anion-exchange membrane with high conductivity and strong alkali stability. Int. J. Hydrog. Energy. 45: 16738–16750.

Shen, B., B. Sana and H. Pu. 2020. Multi-block poly (ether sulfone) for anion exchange membranes with long side chains densely terminated by piperidinium. J. Membr. Sci. 615: 118537.

Sheng, W., X. Zhou, L. Wu, Y. Shen, Y. Huang, L. Liu, et al., 2020. Quaternized poly (2,6-dimethyl-1,4-phenylene oxide) anion exchange membranes with pendant sterically protected imidazoliums for alkaline fuel cells. J. Membr. Sci. 601: 117881.

Shin, S.H., A. Kodir, D. Shin, S.H. Park and B. Bae. 2019. Per-fluorinated composite membranes with organic antioxidants for chemically durable fuel cells. Electrochim. Acta. 298: 901–909.

Shukla, G. and V.K. Shahi. 2018. Poly (arylene ether ketone) co-polymergrafted with amine groups containing a long alkyl chain by chloroacetylation for improved alkaline stability and conductivity of anion exchange membrane. ACS Appl. Energy Mater. 1: 1175–1182.

Si, J., Y. Lv, S. Lu and Y. Xiang. 2019. Microscopic phase-segregated quaternary ammonia polysulfone membrane for vanadium redox flow batteries. J. Power Sources. 428: 88–92.

Son, T.Y., D.J. Kim, V. Vijayakumar, K. Kim, D.S. Kim and S.Y. Nam. 2020. Anion exchange membrane using poly (ether ether ketone) containing imidazolium for anion-exchange membrane fuel cell (AEMFC). J. Ind. Eng. Chem. 89: 175–182.

Stone, Ch. and A.E. Morrison. 2002. From curiosity to "power to change the world®". Solid State Ionics. 152–153: 1–13.

Sun, Z., B. Lin and F. Yan. 2018. Anion-exchange membranes for alkaline fuel-cell applications: the effects of cations. Chem. Sus. Chem. 11: 58–70.

Sun, M., X. Wang, L.R. Winter, Y. Zhao, W. Ma, T. Hedtke, et al., 2021. Electrified membranes for water treatment applications. ACS EST Eng. 1: 725–752.

Tariq, M., U.N.A. Syed, A.H. Behravesh, R. Pop-Iliev and G. Rizvi. 2022. Synergistic enrichment of electrically conductive polypropylene-graphite composites for fuel cell bipolar plates. Int. J. Energy Res. 46: 10955.

Thompson, S.T., D. Peterson, D. Ho and D. Papageorgopoulos. 2020. Perspective—the next decade of AEMFCs: near-term targets to accelerate applied R&D. J. Electrochem. Soc. 167: 084514.

Tuckerman, M.E., D. Marx and M. Parrinello. 2002. The nature and transport mechanism of hydrated hydroxide ions in aqueous solution. Nature. 417: 925–929.

Venugopal, A., L.H.T. Egberts, J. Meeprasert, E.A. Pidko, B. Dam, T. Burdyny, et al., 2022. Polymer modification of surface electronic properties of electrocatalysts. ACS Energy Lett. 7: 1586–1593.

Wang, D.F., D.J. Wang and H.W. Hu. 1990. Studies on double crown ethers (VII)- synthesis of Schiff base and secondary amines dibenzo-18 crown 6. Chem. J. Chin. Univ. 11: 266–270.

Wang, Y.H., X.T. Wang, H. Ze, X.G. Zhang, P.M. Radjenovic, Y.J. Zhang, et al., 2021. Spectroscopic verification of adsorbed hydroxy intermediates in the bifunctional mechanism of the hydrogen oxidation reaction. Angew. Chem. Int. Ed. 60: 5708–5711.

Wu, H., S. Lu and B. Yang. 2022. Carbon-dot-enhanced electrocatalytic hydrogen evolution. Acc. Mat. Res. 3: 319–330.

Xiao, Y., L. Hu, L. Gao, M. Di, X. Sun, J. Liu, et al., 2022. Enabling high anion-selective conductivity in membrane for high-performance neutral organic based aqueous redox flow battery by microstructure design. Chem. Eng. J. 432: 134268.

Xiong, Y., J. Fang, Q.H. Zeng and Q.L. Liu. 2008. Preparation and characterization of cross-linked quaternized polyvinyl alcohol) membranes for anion exchange membrane fuel cells. J. Membr. Sci. 311: 319–325.

Xu, F., Y. Su and B. Lin. 2020. Progress of alkaline anion exchange membranes for fuel cells: the effects of micro-phase separation. Front. Mater. 7: 4.

Yan, X., S. Gu, G. He, X. Wu, W. Zheng and X. Ruan. 2014. Quaternary phosphonium functionalized poly (ether ether ketone) as highly conductive and alkali-stable hydroxide exchange membrane for fuel cells. J. Membr. Sci. 466: 220–228.

Yan, Z. and E. Thomas. 2021. Bipolar membranes for ion management in (Photo) electrochemical energy conversion. Acc. Mat. Res. 2: 1156–1166.

Yang, Q., C.X. Lin, F.H. Liu, L. Li, Q.G. Zhang, A.M. Zhu, et al., 2018. Poly (2,6-dimethyl-1,4-phenylene oxide)/ionic liquid functionalized graphene oxide anion exchange membranes for fuel cells. J. Membr. Sci. 552: 367–376.

Yang, W., P. Xu, X. Li, Y. Xie, Y. Liu, B. Zhang, et al., 2021. Mechanically robust semi-interpenetrating polymer network (sIPN)) via thiol-ene chemistry with enhanced conductivity for anion exchange membranes. Int. J. Hydrog. Energy. 46: 10377–10388.

Yang, Y, C.R. Peltier, R. Zeng, R. Schimmenti, Q. Li, X. Huang, et al., 2022. Electrocatalysis in alkaline media and alkaline membrane-based energy technologies. Chem. Rev. 122(6): 6117–6321.

Ye, N., Y. Xu, D. Zhang, J. Yang and R. He. 2018. Inhibition mechanism of the radical inhibitors to alkaline degradation of anion exchange membranes. Polym. Degrad. Stabil. 153: 298–306.

Zhang, Z. and T. Xu. 2020. Facile synthesis of poly (arylene ether ketone) with pendent oxyhexyltrimethylammonium groups for robust anion exchange membranes. Polymer. 210: 123035.

Zhang, J., Y. He, K. Zhang, X. Liang, R. Bance-Soualhi, Y. Zhu, et al., 2021. Cation-dipole interaction that creates ordered ion channels in an anion exchange membrane for fast OH⁻ conduction. AIChE J. 67: e17133.

Zhu, Y., L. Ding, X. Liang, M.A. Shehzad, L. Wang, X. Ge, et al., 2018. Beneficial use of rotatable-spacer side-chains in alkaline anion exchange membranes for fuel cells. Energy Environ. Sci. 11: 3472–3479.

Chapter 12

An Overview of Metal-Air Battery and Applications

Dr. D. Sridharan

Senior Research Scientist, Vioma Motors Pvt. Ltd.,
Mumbai–400028, India.
Email: sridharanchan@gmail.com

12.1 INTRODUCTION

Climate change has been a global issue in recent years due to overutilization of fossil fuels and other human activities. The overpopulation and the scarcity of non-renewable fuel sources are facing big challenges on the earth. In such a case, battery technologies play an important role in the present and future perspectives. It is an emerging area in various applications and solves challenges against the present situation. The batteries have been used essentially from electronic gadgets to heavy automobile industries. This chapter discusses an oxygen reduction reaction (ORR) and oxygen evolution reaction (OER) based on nanoelectrocatalyst in the metal-air battery (MAB) and its types, working principles, the role of various metal electrodes, electrolytes, and separators. MAB is an inexpensive battery compared with lithium-ion batteries (LIBs); on the other hand, it has many challenges, such as corrosion of metal electrodes, various components of MAB batteries, and behavior of reversibility.

Metal-air batteries are an attractive technology in the electrochemical energy storage (EES) industry. They are commonly backed as a way out for posterity in EES for use like grid energy storage and electric vehicles (EVs) because they are safer and have theoretical energy density that is substantially greater when compared with LIBs. The environmental air contains oxygen as a cathode which helps in lowering the cost and the weight considerably. The utilization of cheap metals as an anode further supports in lowering the cost of battery.

Currently, the global perception of low-carbon emission and the concept of sustainability greatly support the growth of chemical power sources. Recent battery chemistries, especially rechargeable lithium batteries, have achieved marketable success in powering electric vehicles (EVs) and portable consumer devices. Emerging applications of batteries in electrified transportation, such as electric and plug-in hybrid electric vehicles and energy storage for the integration of sustainable energy, help to operate smart grid technology. It demands high energy density, is inexpensive, offers increased security, and is respectful of the environment. Therefore, innovation in battery technology is necessary to build better sources of energy for our current lifestyle requirements.

Metal-air batteries are a century-old technology when the first MAB, Zn-air battery was invented (Smee, 1840). Maiche demonstrated functional primary Zn-air batteries which fused a porous platinized carbon cathode (Maiche, 1878). Commercial products of primary Zn-air batteries were first introduced in the market in 1932 (Heise, 1933). They are frequently utilized in medical applications, like hearing aids, due to their high energy density and low power yield. NantEnergy Inc. first made the rechargeable Zn-air batteries available for sale in 2012, although it was noted that their energy density was quite low. This is due to the fact that the oxygen evolution reaction (OER), which is also possible at the cathode, consists of multiple electron transfers and is particularly slow in nature, causing low current density and high electrode polarization. Most accessible zinc electrodes have poor cyclability due to deterioration and dendrite growing of zinc metal during recharge. The Al-air batteries offer many an advantage, like higher range, energy density, safety, longer life-cycle, among others. Combining advancement of science, nanotechnology, algorithms, and more, our multidisciplinary approach allows us to achieve multiple technological breakthroughs in metal-air technology, and create efficient, stable, and durable metal-air systems that overcome the major obstacles of the past.

The existing battery technology is lithium-ion battery having a promising technology; on the other hand, highly expensive, and sometimes explodes under operating conditions. To overcome these problems, growing attention is being given to MABs research in battery technology. Finally, MABs are mainly demanded in the automotive production industry, their merits and their various applications are discussed in this chapter.

12.2   WORKING PRINCIPLE OF METAL-AIR BATTERY

Metal-air batteries generate electricity through an oxidation and a reduction (redox) reaction between metal and oxygen in air. Figure 12.1 shows the schematic configuration of metal-air battery. It is presented with the open cell structure, which declares the supply of cathode material incessantly passing oxygen and is substantially from air.

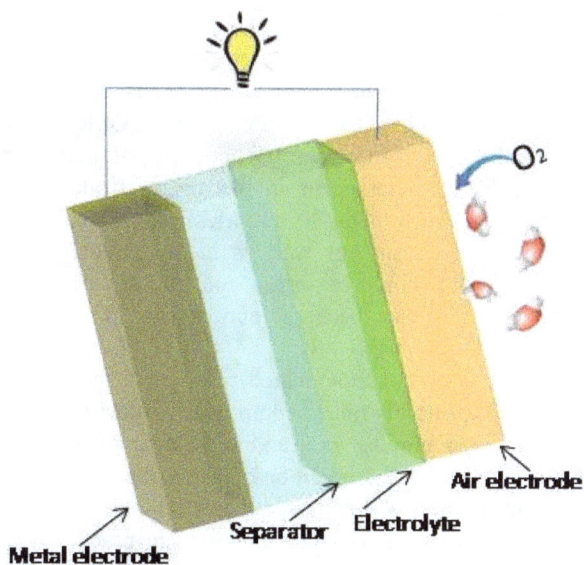

Figure 12.1 Schematic representation of a metal-air battery.

The following reaction mechanisms (Eq. 1–6) generally occur in the metal-air batteries during discharge.

The anode oxidation (half-reaction),

$$M \rightarrow M^{n+} + ne^- \ (M - metal) \tag{1}$$

The air cathode reduction (half-reaction),

$$O_2 + 2H_2O + 4e^- \rightarrow 4OH^- \tag{2}$$

The oxygen reduction reaction (ORR) is the critical cathodic reaction in metal-air batteries. ORR proceeds through an electrochemical reaction due to multi-electron transfer that can happen at the cathode through dual pathways, namely the transfer of $2e^-$ to form hydrogen peroxide and the direct transfer of $4e^-$ to make H_2O, as shown by the following equations:

Direct $4e^-$ pathway:

$$O_2 + 2H_2O + 4e^- \rightarrow 4OH^- \tag{3}$$

$2e^-$ pathway:

$$O_2 + 2H^+ + 2e^- \rightarrow H_2O_2 \tag{4}$$

$$H_2O_2 + 2H^+ + 2e^- \rightarrow 2H_2O \tag{5}$$

As the $2e^-$ path generates peroxide, which is harmful to air-cathode, the $4e^-$ pathway is preferred in this situation.

Overall reaction,

$$M + O_2 + 2H_2O \rightarrow M^{n+} + 4OH^- \tag{6}$$

The amalgamation of a metal anode with high energy density and an air cathode open structure to draw cathode active materials exposed as oxygen from air is the most noticeable feature of MAB.

12.3   AIR CATHODES

Dissolution and deposition on the metal electrode as an anode, and oxygen reduction reaction (ORR) and oxygen evolution reaction (OER) on the air electrode as a cathode, constitute the fundamental chemistry of metal-air batteries. It is primarily ORR and OER that occur in the air electrode, posing significant technological challenges for all metal-air batteries. The critical goal is to achieve high capacity and power density, as well as high round-trip efficiency and a long cycling life. Catalysts should be introduced into air electrodes to facilitate electrochemical reactions during discharge and charge, in addition to improving transport kinetics.

Because air electrodes can be contaminated, resulting in unresolvable by-products and decrease in cycle life, stability must be improved further. Furthermore, enabling operation in ambient air, removing CO_2 and H_2O contamination, and preventing liquid electrolyte evaporation are difficult tasks. Thus, the assembly of air cathodes in different types of metal-air batteries needs to be designed and optimized accordingly. Subsequently, oxygen electrochemistry in practical MABs is a very complicated process. When using an electrocatalyst, lowering the over-potential in discharge and charge is of primary importance to prevent corrosion and decrease electrolytic oxidation. Nobel metals, usually as catalysts, have the properties of abundant activities and better stability; however, the drawbacks include high cost and inadequacy. As a result, Pt substitutes have been thoroughly investigated and entail use of further transition metals to either completely or partly exchange Pt. The ORR and OER responses can be influenced by the catalyst, electrolytes, electrode materials, oxygen pressure, and even electrochemical stressing (Sheng et al., 2010; Xin-hui and Xia, 2010; Lau et al., 2011).

12.3.1   Role of Nanoelectrocatalyst

Current challenges in nanoelectrocatalysts have drawn considerable attention in numerous applications due to the distinct properties of electrochemistry. One of the electrocatalyst principles is to lower the over-potential of precise electrochemical properties of oxygen. Subsequently the electrocatalyst may efficiently ease the electron transfer among the electrode and electrolyte interface considering that the size and form of the electrocatalyst are very important in electrochemical reactions. Nanotechnology is a new area of study that has recently gained traction in the fuel cell and energy application industries. In order to obtain a greater surface area, encouraging spatial confinement, strong mechanical strength, and higher electrocatalytic performance, a number of nanoelectrocatalysts have been produced with various advantages. Many applications, including MABs, were successfully utilized using the nanomaterial-based catalysts.

Fuel cells, metal-air batteries, and water-splitting processes are just a few of the sustainable energy-related devices and applications that have benefited from the development of nanoscale-based electrocatalysts in recent years. Improving metal-air batteries is currently of great interest since nanocatalysts can have a major impact on their characteristics. All metal-air batteries require improved catalysts that can lower over-potentials and improve ORR since the reaction kinetics of ORR in the non-catalyst cathode is frequently slow due to over-potentials. Based on this, numerous studies have been conducted to explore viable ORR catalysts, including precious metals and alloys, metal macrocyclic compounds, carbonaceous materials, transition metal oxides, and chalcogenides (Yu et al., 2021), etc. The catalysts for the cathode oxygen reduction/evolution are selectively measured from materials chemistry to electrode properties and its role in battery applications: Pt, Au nanoparticles and Pt-based alloys, carbonaceous materials like graphene nanosheets, transition-metal oxides, such as perovskites and Mn-based spinels, and organic-inorganic composites. The development of greatly effective nanoelectrocatalysts for green energy storage technologies has both opportunities and challenges in the future.

In order to realize the practical applications of MABs, highly efficient and inexpensive non-noble metal-based oxygen electrocatalysts are urgently required. Graphene is one of the versatile materials for various applications due to its atypical properties, such as specific surface area, intrinsic mobility, thermal conductivity, optical transmittance, and good electrical conductivity (Jamesh et al., 2021). Several studies have been stated to improve the electrochemical performance of catalysts, and the synergistic relationship of catalysts with graphene-based materials is a favourable method to produce more active sites, which can increase the electrical conductivity and chemical stability. The structure of graphene has the robust in-plane C–C bonds, whereas the out-of-plane p-bonds contribute to the delocalized electronic network responsible for electronic conduction in graphene, resulting in weaker interactions between graphene layers or between graphene and the catalyst.

Because of their high activity and long-term stability, many heteroatom-doped graphene-materials containing B, N, P, S, and halogens have been concentrated on oxygen electrocatalysts in metal-air battery applications. Study is being done on various catalysts, like metal oxides, non-noble metals, noble metals, graphene with non-metals, perovskites, carbides, sulphides, nitrides, and other carbon composites, in order to improve the activity of heteroatom-doped with graphene. This can result in an improvement in the bifunctional activity of MABs due to graphene containing heteroatoms, and other carbon composites have also been reported in the aspects of high ORR/OER activity. Moreover, the current advancement on the ZABs comprise the approaches used to progress the high cycling performance, capacity, power density, and energy density. In order to create potential ZABs, the ORR/OER activities of ZABs and the performance of several air catalyst types based on graphene have been identified. The different types of materials have proven very promising in terms of catalytic activity and stability.

12.3.2 Oxygen Reduction Reaction (ORR) and Oxygen Evolution Reaction (OER)

Consequently, the prepared nanoelectrocatalyst exhibits higher catalytic activity and durability in the oxygen reduction (ORR) and oxygen evolution (OER) reactions than the commercial Pt catalyst. The nanocatalyst is a promising candidate for rechargeable batteries and it can be acted as bifunctional activities of those reactions. MAB's cathode reaction is typically an electrochemical process, and the performance of the battery is greatly influenced by the electrocatalyst used in the air electrode. Fuel cell technologies, which generally entail a cathode oxygen reduction, have been the focus of considerable investigations on ORR catalysts for many years. In theory, the majority of catalytic components used in fuel cells can also be used in MABs, as might the methods and procedures used to increase cathode productivity.

Future studies will focus on the creation of bifunctional catalysts to develop both ORR and OER processes at once. Figure 12.2 shows the mechanism of electrocatalyst in MEAs. Although there has been minimal research on Li-O$_2$ cells, bifunctional catalysts have been researched for the use of rechargeable ZABs in aqueous conditions. In alkaline solution, mixed-metal oxides with the spinel, pyrochlore and perovskite structures have received major attention in research on bifunctional oxygen catalysts. Li-air batteries urgently need non-precious bifunctional electrocatalysts to encourage the creation and degradation of Li$_2$O$_2$ and to control the development of further lithium complexes. Utilizing various synthesis techniques, it is possible to tune the intrinsic catalytic activity and increase the density of the active sites by changing the physicochemical characteristics of the catalysts, such as composition, conductivity, phase, structure, defects, morphology, size, surface area, and valence state. On the other hand, doped or metal-nitrogen complexed nanostructured carbonaceous materials hold promise as plentiful, affordable oxygen catalytic catalysts and are deserving of further study and development. For the catalysts activity and the enhancement of energy conversion efficiency, the electrode structure's logical designs are also crucial. Finally, the role of bifunctional catalyst supports ORR and OER. This performance of cathodic reaction helps reversibility of the MABs. Subsequently, the battery has a higher discharge capacity and can sustain better cycling stability.

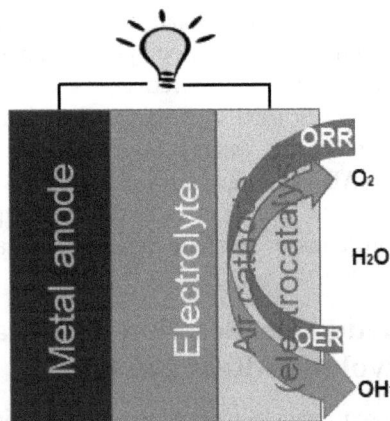

Figure 12.2   Mechanism of Electrocatalyst in MEAs.

12.4 SELECTION OF METAL ANODES

The use of MABs, which generally encompass those that use oxygen as a fuel, has rekindled attention. They are put together using an appropriate electrolyte, a cathode and a metal anode. At this point, the metal anode can be alkali metals, such as Li, Na, and K, alkaline earth metals, such as Mg, and first-row transition metals, such as Fe and Zn, all of which have favorable electrochemical equivalence; the electrolyte can be either aqueous or non-aqueous, subject to the type of anode used; and the oxygen-conscious cathode frequently has an open porous construction that is enabling. The design elements of both conventional batteries and fuel cells are combined in metal-air batteries.

12.5 TYPES OF ELECTROLYTES

According to the kind of electrolytes, MABs fall into two major types. The first type is established on protic, which is unaffected by water or moisture, while the second type is established on either non-aqueous or aprotic electrolyte that can be influenced by water or moisture. In contrast to aqueous MABs, which are still in their early stages, non-aqueous electrolyte is frequently employed by metals that are extremely reactive in aqueous electrolyte. Metals like Ca, Al, Fe, and Zn are good for the aquatic system with open-air friendliness, but Li, Na, K, and Mg are generally suitable for the non-aqueous system, where their stability is constrained by atmospheric conditions. Based on the design and compactness of MABs, recently, researchers have also investigated the solid-state electrolyte and ionic liquid electrolyte in MABs.

12.5.1 Aqueous Electrolyte

The first primary zinc-air battery was designed (Maiche, 1878), and its commercial products began entry into the market in 1932. In the 1960s, aqueous batteries, such as iron-air, aluminium-air and magnesium-air were developed. In aqueous media, metals like Al, Fe, Mg, and Zn, are thermodynamically unstable; but, under specific conditions, matching oxides or hydroxides can passivate their surfaces, and render them well-suited with aqueous electrolytes. For use in aqueous MABs, these metals have been investigated in recent years. Metal takes place oxidation at the anode during discharge, and O_2 from the environment is reduced on a catalyst sustained at the gas-diffusion cathode. In previous studies on MABs, scientists were interested in aqueous ZABs for their affordability and usability.

12.5.2 Non-aqueous Electrolyte

About two decades ago, non-aqueous-based metal-air batteries first appeared for LABs, and presently for Na-air and K-air batteries. Although being relatively new, issues with metal anodes, air catalysts, and electrolytes have had a negative impact on the development of metal-air batteries. It is also unknown if they will be able to replace lithium-ion batteries in electric vehicle (EV) applications in future. Metals with high reactivity in aqueous solutions include Li, Na, and K. On the other hand, aprotic electrolytes are currently used in Li, K, and Na metal-air batteries. Notably, ORR in non-aqueous electrolyte that works by a completely different mechanism than it does in watery electrolytes.

As its benefits in energy density and rechargeability over protic ZABs today, with the world's energy needs constantly increasing, non-aqueous Li-air batteries have gained interest on a global scale. Interestingly, despite the distinct reaction processes, oxygen catalysts for aprotic systems also exhibit encouraging capabilities. Since both schemes are comparable, there are many possibilities for developing non-aqueous Li-air batteries, and significant advancement has been made in this area. The non-aqueous rechargeable lithium-air battery (Abraham et al., 1996) consists of a lithium metal anode, a porous cathode, and a non-aqueous electrolyte and has theoretically high specific energy density of 3,505 Wh kg^{-1} (3,436 Wh L^{-1}) (Bruce et al., 2012). However, non-aqueous LABs are still in their early stages and there are a number of difficulties to be overcome in addition to selecting the appropriate electrolyte, cathode, and anode for oxygen electrocatalysts. With the increasing interest in non-aqueous oxygen catalysis/electrode, research emphasis will be on MABs.

12.5.3 Solid-state Electrolyte (SSE)

SSEs differ from protic electrolytes in two ways: wettability and ion conduction. MABs have an exceptional wetting property in aqueous electrolytes and can come in full contact with the cathode at three-phase boundaries. For a solid-based electrolyte, the three-phase interface reaction can be restricted by the poor wetting property of immobilized electrolyte, thereby interfacial transporting resistance of OH$^-$ may be remarkably higher than that of an aqueous system (Wang et al., 2013). Alkaline gel electrolytes, consisting of low molecular weight polymer and alkaline solutions, have been developed to mitigate these issues in primary lithium-air batteries (Li et al., 2013).

Al-air solid-state batteries are perfect due to their durability, thermal stability, and potential for preventing electrolyte leakage. Based on their properties, polymer-based electrolytes have been studied for their use in Al-air batteries (Zhang et al., 2014b; Mokhtar et al., 2015). The acrylamide-based polymer gels (Tan et al., 2018) and a solid electrolyte composed of acryl resin (Ma et al., 2019) used inorganic materials including Sn, Zn and In in an Al-air battery combined with an air-cathode material composed of iron carbide. When xanthan and κ-carrageenan (Di Palma et al., 2017) were used as hydrogel additives, it was found that these additives were conductive toward Al ions in which the performances of these additives were compared and their conductivities were in the following order: the 1 M KOH liquid electrolyte solvents. Despite these findings, solid-state electrolytes have the potential to raise cell resistances and lower battery capacity when compared to those made by sing liquid-form electrolytes.

12.5.4 Ionic Liquid Electrolyte

Ionic liquids are also a type of non-aqueous liquid electrolytes, including two types of cations: large organic cations with organic/inorganic anions and alkali metal ions in an organic solvent, such as organic carbonates, ethers, and esters (Xu et al., 2014a; Mokhtar et al., 2015). Lithium salts, such as $LiPF_6$, $LiAsF_6$, $LiN(SO_2CF_3)_2$, and $LiSO_3CF_3$ are commonly used in LABs (Daming et al., 2013). PYR14TFSI–TEGDME–$LiCF_3SO_3$ are also employed in LABs (Cecchetto et al., 2012), consisting of $LiCF_3SO_3$ in tetraethylene glycol dimethyl ether (TEGDME) and pure PYR14TFSI (1-butyl-1-methylpyrrolidinium bis(trifluoromethanesulfonyl) imide). The carbonates, which ingest the electrolyte and block the electrode surface pores, also pose problems for ionic liquid electrolytes. The reaction of oxygen in ion fluid is poorly understood. This prevents ionic fluid electrolytes from being used effectively. MABs rarely use neutralised salt solutions, acidic solutions, or hybrid electrolytes, whereas alkaline solution electrolytes and ionic liquid electrolytes are common.

12.6 ROLE OF SEPARATORS

The separators are important in the MABs. The separator is a porous membrane placed between electrodes of opposite polarity, penetrable to ionic flow preventing any electric contact between them. Generally, separators play a key role in the batteries, keeping the cathode and anode apart and preventing electrical short circuits, simultaneously allowing for rapid transport of ionic charge carriers, which is necessary to complete the circuit during the passage of electrical current in a cell. They should be very good electric insulators as well as possess the capability of conducting ions via either their intrinsic ionic conductivity or soaking electrolyte, thus minimizing any processes that adversely affect the electrochemical energy efficiency of the battery (Besenhard, 1999; Linden et al., 2002; Berndt, 2003).

12.7 TYPES OF MABS

The different metals that can be used as anode are first-group alkali metals, such as lithium, sodium, potassium, etc. and in second group, alkaline earth metals like magnesium, calcium, etc. and in third group the post transition metal like aluminium and a few transition metals from eighth group, like Fe and second-group Zn. Metal-air batteries are actually made up of a combination of the design with metal anode and air cathode in working of traditional and fuel cell batteries.

12.7.1 Lithium-Air Battery (LAB)

The first secondary LAB was invented by using conductive organic polymer electrolyte (Abraham and Jiang, 1996). It had an air cathode with carbon incorporated at the anode, which had Li+ in the form of a membrane that was electrically conducting. Maximum Li-air batteries, which theoretically have an energy density of 11,140 Wh kg^{-1} across all MABs range and are quite advanced compared to Li-ion batteries, are a good alternative for energy storage systems (EES). Since the oxygen electrode's creation in 1996, extensive research has been done to increase its electrochemical reversibility. Figure 12.3 presents a graphical representation of theoretical specific energies of different MABs.

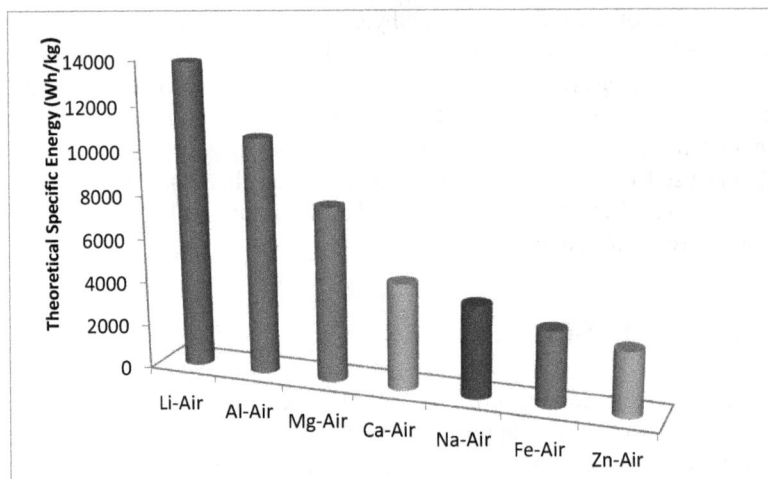

Figure 12.3 Graphical representation of theoretical specific energies of different metal-air batteries.

On the other hand is the development of their closed systems in non-aqueous metal-air batteries. Instead of the gaseous $O_2 \leftrightarrow$ solid superoxide/peroxide phase transformation that involves large structural reorganization and hence huge energy barrier, batteries based on the more rapid solid superoxide $\leftrightarrow$ peroxide $\leftrightarrow$ oxide phase transformation have recently emerged (Zhi et al., 2016). Because of the oxygen/water/side products crossover in non-aqueous LABs, the reactions at the interfaces of the electrolytes and lithium anode are far more difficult. It is important to select an aqueous electrolyte with a high energy density and high stability towards the solid state electrolyte membrane. Better anode protection coatings and a way to remove CO_2 before it enters the cell are also necessary for aqueous electrolyte-based LABs. The difficulty with the non-aqueous lithium-air concept is in preventing water intrusion. Significant future works focusing on these challenges are essential to realize a safe, fully rechargeable lithium-air battery with high capacity, high energy density, and long cycling life for future EV industries (Tan et al., 2017).

12.7.2 Aluminium-Air Battery (AAB)

The aluminium air battery (AAB) is highly suitable for the primary battery (Zaromb, 1962) and also secondary battery for EVs as an energy source. It has a remarkable theoretical energy density of 8140 Wh kg^{-1} which is far greater than LIBs. A new AAB is reported with an innovative organic non-aqueous electrolyte, which consists of tetra-butyl-ammonium fluoride trihydrate salt dissolved in non-aqueous solvent, like PC/TEG-DME, CAN, etc. (Natasha et al., 2020). Another type of aluminium air battery has the conventional 1-ethyl-3-methylimidazolium chloride and AlCl$_3$ electrolyte, as also Al foil and Pt/C as gas diffusion electrode (Bogolowski and Drillet, 2017; Yisi et al., 2017).

Al anode with high corrosion rate in aqueous alkaline solution is of major concern in terms of Al usage and safety. This battery utilizes 1-ethyl-3-methylimidazolium oligo-fluoro-hydrogenate at room-temperature ionic liquid (RTIL). The ionic liquid type Al-air battery is, in terms of formulation and conductivity, feasible of reversible oxygen reduction/oxidation reactions, as well as reversible and efficient (Gelman et al., 2014). The solid electrolyte made up of AlCl$_3$, CMC, glycerine, and urea has been used for solid-state rechargeable AAB. The battery exhibited stable electrochemical reactions as confirmed by the charge-discharge characterisation (Ryohei, 2019). Al-air batteries have a lot of potential for practical use because of their high energy density. The idea of developing rechargeable Al-air batteries has been explored. The prevention of self-corrosion and the generation of by-products is the main goal in the improvement of aluminium alloy anodes and pure aluminium. The electrolytes in Al-air batteries play a crucial role in defining their capacity for rechargeability. Electrolyte additives are a significant concern because they can enhance electrochemical properties by reducing corrosion and hydrogen evolution. Due to its slower reaction kinetics than electrochemical reactions at the anode, this process is essentially a rate-determining step in the AABs. Therefore, air-cathode materials are important elements that have a big impact on the entire battery's attributes.

12.7.3 Magnesium-Air Battery (MAB)

Magnesium-air battery is a combination of Mg as the anode and air is reduced at cathode (Carson and Kent, 1966). Activated carbon, a catalyst, a thin coating of an aquaphobic polymer substance, and a metal sheet are typically employed as the cathode and are based on the electrode position of the electrolyte material. Finding the ideal combination of electrolytes has been one of the key hurdles in the development of secondary magnesium batteries. A biocompatible ionic liquid embedded Mg-air with a polymeric electrolyte material was demonstrated to explain their electrochemical behaviour (Zhang et al., 2014a). The discharge behavior and characterization of Mg air battery with tri-hexyl (tetradecyl) phosphonium chloride ionic liquid as an electrolyte and their characterization was explained by FTIR, XPS and mass spectroscopy techniques (Timothy et al., 2013).

12.7.4 Calcium-Air Battery (CAB)

More calcium than sodium and magnesium can be found in the crust of the earth. One of the most effective, safe, and well-known metals in MABs is calcium which is demonstrated to have a wide range of uses when combined with an aqueous electrolyte. A calcium air battery with solid electrolyte of binary molten salt of $CaCl_2$ and Cao were fabricated (Nirupama et al., 1988). Calcium as metal, for metal air batteries, has the potential to achieve a high electrical density, low cost, and high safety (Reinsberg et al., 2016).

12.7.5 Potassium Air Battery (PAB)

In 2013, potassium air batteries were invented in Ohio State University (Xiaodi and Yiying, 2013). The batteries might be more capable when compared with current LIBs and capable to store twofold as much energy. Another potassium air battery is designed with KPF_6 dissolved in ether as an electrolyte.

12.7.6 Sodium-Air Battery (SAB)

The SAB is a type of MAB with a high specific energy of 1683 Wh kg^{-1}. Table 12.1 presents different metal-air batteries with individual electrochemical properties. Due to the abundance of sodium, low cost and eco-friendly nature, SAB is finding application in EVs (Xiaolong et al., 2019). A sodium air battery which comprised of carbon-fiber-based gas diffusion layer (GDL) and sodium triflate salt in diethylene glycol dimethyl ether as electrolyte was reported and its electrochemical properties were studied by EDS, XRD, SEM and Raman spectroscopy, etc. (Hartmann et al., 2012).

Table 12.1 Different metal-air batteries with individual electrochemical properties

Metal-air battery	Theoretical specific energy Wh kg^{-1} (including oxygen)	Theoretical specific energy Wh kg^{-1} (excluding oxygen)	Open circuit voltage (V)
Calcium-air	2990	4180	3.12
Iron-air	1431	2044	1.30
Aluminium-air	4300	8140	1.20
Lithium-air	5210	11140	2.91
Sodium-air	1683	2260	2.30
Magnesium-air	2789	6462	2.93
Potassium-air	935	1700	2.48
Zinc-air	1090	1353	1.65

12.7.7 Iron-Air Battery (IAB)

Electrically rechargeable iron-air batteries (Ojefors and Carlsson, 1978), having a capacity of 30 kWh and reaching a specific energy of 80 Wh kg^{-1}, have been reported. However, iron-air systems are limited with respect to specific power and coulombic efficiency due to the kinetics and the high amount of parasitic hydrogen evolution caused by the iron electrode.

12.7.8 Zinc-Air Battery (ZAB)

The first ZAB which was developed in 1878 is a sort of MABs. The zinc-air batteries are most appropriate for small-current applications, such as hearing aids (Zhang et al., 2019). They are the only ones in this category that have been successful and commercialized as primary cells. Although their shelf-life and capacity for recharging are constrained, zinc-oxygen systems present perhaps the quickest and most reliable route to a workable secondary metal-air battery. The multiphase electrolytes are intended to present the Zn deposition and the OER using a polymer electrolyte-assisted and dendrite-resistant ZAB. The variety of methods for processing and using zinc materials will make it possible to research and develop zinc for unconventional electrochemical power sources and energy storage systems in future.

12.8 RECHARGING OF MABS

12.8.1 Electrical Recharging

Given the superior performance of primary zinc-air batteries, significant growth has been made over 60 years in the progress of rechargeable zinc-air batteries. These zinc-air batteries, dual energy-storage systems, which combine a high-power source for acceleration and regenerative braking with a high-energy-density source for base-load power, have also been proposed for EVs by Tesla (Stewart et al., 2012). The longevity and reversibility of the electrodes are the key obstacles for rechargeable zinc-air batteries. The rechargeable zinc-air battery classifies the interferences into dendritic growth at the anode, lack of higher performance bifunctional catalyst loaded in air electrode systems and electrolyte-related problems (Peng et al., 2017). Recently, many researchers have studied various approaches used in rechargeable zinc-air batteries to overcome these barriers (Park et al., 2016; Thippani et al., 2016). A bifunctional air electrode has been used in this instance, and the reactions are reversible. The electrode has been used with a variety of catalysts, including nickel, MnO_x, CoO_x, RuO_2, TiO_2, IrO_2, and RhO_3, to accelerate oxygen evolution. Nickel is the most practical and commonly used catalyst. To improve the electrode's stability and cyclability, several different designs have been created. To prevent overvoltage during the charging process, a third electrode was included separately in some battery designs. The electrode surface on bi-layer structure that works on both ORR and OER has also been utilised because it is difficult to find a single catalyst that can support both oxygen reduction and evolution. The structure is sustained by a low-impedance current collector, is hydrophobic and permeable to oxygen. A high-pores separator with low electrochemical resistance and the capacity to hold the alkali electrolyte separates the flat zinc negative from the air electrode positive. Controlling the air and wetness drive into and outside of the cell is essential for an electrically-rechargeable ZAB to sustain the composition of electrolyte needed for long-term reversibility. One of the difficulties is that a significant amount of water will evaporate from the cell, drying it out and reduced oxygen diffusion will reduce the cell's ability to discharge. Accordingly, it is preferable to cover the outside of zinc-air batteries with an oxygen/water vapour selective membrane.

AER Energy Resources, Inc. developed a membrane with limited oxygen/ water vapour selectivity (Alan and Thomas, 2005). The membrane is made up of a layer of a polymeric perfluoro substance on top of a gas-permeable substrate film. It has been observed for the selectivity between oxygen and water vapour. Due to the significant difference in discharge and charge voltages, these electrically rechargeable zinc-air cells have an energy efficiency of roughly 50%. A number of serious issues plague electrically rechargeable zinc-air batteries, such as zinc dendrite growth and change of shape, as well as the variability of carbon-based cathodes through charging while oxygen is produced. Considerable studies have been done for rechargeable MABs. The challenges of rechargeable MABs should be addressed at the earliest to help in solving the scarcity of energy in a sustainable manner in future.

12.8.2 Mechanical Recharging

The viability of mechanically recyclable MAB has been investigated in the light of these issues and shorter cycle life of electrically-reversible MABs, like zinc-air and aluminium-air batteries. A number of techniques for mechanically recharging an aluminium and zinc anode have had their configurations examined. The wasted electrode is sent to recyclers and oxide consumers, using mechanically rechargeable processes. Second, a mechanically rechargeable technique involves recharging the anode outside the cell in a separate apparatus. Thirdly, a mechanically rechargeable technique regenerates the fresh anode in a specialized facility at the service station. It can be seen that the cost of aluminium as an anode can be as low as US\$ 1.1/kg as long as the reaction product is recycled (Shaohua and Knickle, 2002). Great power sources used in motor power applications have given rise to mechanically rechargeable systems. Although this system needs servicing and upkeep after every discharge, some applications, such as stationary backup power supply, might not be feasible.

12.8.3 Hydraulic Recharging

Another method for addressing the issue of recharging is hydraulically rechargeable MABs. In this technique, the electrodes are continuously provided with renewed metal by flowing electrolyte, as opposed to the cell being recharged electrically or the expended active materials being mechanically replaced with new metal. Individual cells receive electrolyte-fed zinc fuel in the form of particulates. According to the theory, the issues with dendrite formation and shape change as well as the challenges of using a bifunctional air electrode are eliminated by a zinc-air system with hydraulic refuelling. The wasted fuel is kept inside the similar system and is treated to create zinc metal particles in the regenerative-hydraulically-feeding fuel cell. The used zinc oxide and electrolyte mixture is collected and moved from the system with *ex-situ* regeneration to another system, where new zinc fuel is made. While the system based on reprocessing is basic in manufacturing, the system with *in-situ* regeneration is complex in terms of system design and engineering. Zinc residue contained in moving electrolyte is pushed from a tank during discharge via a set of tubular cells before being pumped into the tank. The zinc recharge device is made up of a reservoir, a circulating pump, and a zinc electrolyzer. The electrolyte has two functions: it first keeps the dissolution of zinc in solution and then moves the zinc particles into the anode compartment. In another design developed by the Lawrence Livermore National Laboratory in the 1990s, the refuelling ports are located on the top of the cell stack (Cooper et al., 1994). Electrolyte slurry containing zinc particles overflows the discharged cell plates.

Metallic Power Inc. developed a regenerative zinc-air technology that embodied the functions and facilities of conventional fuel cells and fuel reservoir were physically separate. By adjusting the size of the electrolyte reservoir, this method enabled the separate scaling of the system's energy (Metallic Power Inc., 2002). These technologies are probably not suitable for automobile applications due to the high energy systems requirement for huge amounts of electrolyte. The ZAB has potential applicant for stationary systems, like backup in energy storage, where inexpensive and dependability are more crucial and when zinc serves as an energy carrier in zinc-air batteries that can be hydraulically recharged.

12.9 APPLICATIONS OF THE MEA'S BATTERIES

12.9.1 Electric Vehicle Applications

The growth of electrical vehicles industry, including hybrid electric vehicles, plug-in hybrid electric vehicles, and stationary energy storage (Xu, 2014b; Tripathi et al., 2010) has pushed the boundary of high-energy storage systems constantly. Although rechargeable Li-ion batteries have been used in these applications for more than two decades due to their relatively high energy density and long lifecycle, they still cannot satisfy the increasing demands for long-range electric vehicles that require an energy storage system with an even higher energy density (Li et al., 2016). Among the alternative energy storage systems, the Li-air battery has attracted global interest over the last two decades due to their much higher theoretical specific energy than the conventional Li-ion batteries with a typical energy density of about 200 Wh kg^{-1} (Peng et al., 2012). The Al-air battery, electrochemical energy storage system can be used as a battery on board vehicles, electric or hybrid motor vehicles or two wheelers (Gonzalez and Renaud, 2015). The Al-air battery has proved to be very attractive as an efficient and sustainable technology for energy storage and conversion (Yisi et al., 2017). Figure 12.4 depicts different applications of metal-air batteries.

Figure 12.4 Applications of metal-air batteries.

12.9.2 Military Applications

Norwegian Defence Research Centre uses 120 W saline Al-air batteries for military communications. Al-air batteries are used to power military communications systems and applications for unmanned aerial and undersea vehicles. Magnesium-air cells and batteries are also frequently used as so-called reserve batteries in marine applications and as a power source for emergency lighting in marine lifejacket lights, and buoys, etc. (Linden and Reddy, 2002).

12.9.3 Smart Grid Applications

Zinc-air batteries could really excel at reducing the cost of storing renewable energy. They have several advantages over the lithium-ion batteries. Using low-cost zinc metal and oxygen to produce electricity, the battery is inherently affordable, can store more energy and is safer than flammable lithium-ion batteries. With a cheaper catalyst, zinc-air batteries become a promising alternative to the lithium-ion batteries that currently use large-scale storage grids (Bev Betkowski, 2022).

12.9.4 Biomedical Applications

Zinc-air button disposable batteries are employed in many applications. Zinc-air batteries can be inactive until the factory-sealed sticker is taken off since they are air-activated. When the sticker covering the battery's back is removed, oxygen will react with the zinc and cause the battery to turn on. Wait for about a minute after removing the sticker for the zinc-air battery to fully activate before inserting it into the hearing aid for the best results. Once the sticker is taken off, the battery will continue to function until power gets exhausted because replacing the sticker won't turn it off. When kept dry and at room temperature, zinc-air batteries can be stored for up to three years without losing their stability. On the other hand, magnesium-air batteries drive low-power implantable biomedical devices (IMDs), such as cardiac pacemakers or bio-monitoring systems (Xiaoteng et al., 2014). Moreover, recently developed all solid-state zinc-air batteries hold promise for portable and flexible applications, such as human-like electronic skins and intelligent bracelets (Peng et al., 2017). Therefore, collaborations between researchers with different backgrounds are anticipated to overcome critical scientific and technical problems of metal-air batteries and further promote their practical applications.

12.10   SUMMARY

The high specific energies of MABs have the ability to encounter the rising demands of electrical energy storage for numerous developing applications, like electric vehicles and smart grids, etc. MABs are recommended as the next-generation of battery technology. Finding greatly robust nanocatalysts with improved stability for ORR and OER in MABs is crucial if we are to fully realise the potential. In both aqueous Zn-air and non-aqueous Li-air batteries, metal oxides are the non-precious catalysts that have received much attention in research. In addition, catalysts of metal oxides comprising metals, perovskites and nitrogen as well as heteroatom-doped carbonaceous catalytic components, have also revealed potential candidates. Besides, research outcomes have revealed that non-oxide materials, like carbon-based quantum dots, carbides, metal nitrides, MOFs and single-atom catalysts can also be applied as next-generation catalytic components with great ORR activity. Although they have received less research in conductive polymers, they are useful complements. For aqueous to non-aqueous systems, there are a number of obstacles that must be overcome in order to fully benefit from research efforts and advancements.

ZAB, AAB, and LABs are considered three of the most promising candidates for real-world applications among the several metal-air batteries explored to date. A possible option for high-energy-density devices is zinc-air batteries. The creation of high-drain batteries with substantially bigger capabilities than those already in the market would be made possible by a high-power zinc anode. Better control of the zinc-electrode morphology in secondary zinc batteries should lengthen their lifecycle and enable them to be used in commonplace commercial applications. Although primary lithium-air batteries have made great advancements, more research is needed before recyclable lithium-air batteries can be used in real-world applications. It is necessary to look into and comprehend the basic reaction mechanisms involved in the charge and discharge of a non-aqueous lithium-air cell in more detail. In the present scenario, the long-term objectives of mobile and static applications, LAB technology is one of the few competitors that can outperform lithium-ion technology in terms of energy density. A better bifunctional catalyst must also be created in order to increase power rate and decrease voltage hysteresis while enhancing battery reversibility.

In general, the production of recyclable AABs is critical and can be used in smart grid energy systems. These recyclable AABs can also be used in EVs to endorse an ecologically friendly future. AABs are considered promising for next-generation energy storage applications because of their high theoretical energy density, which is considerably higher than that of existing lithium-ion batteries. Despite these difficulties, the fact remains that modern society needs energy storage technologies with exceptionally high energy densities more than ever. Finally, metal-air batteries offer enormous promise for use in more effective, high-performance, and affordable energy storage systems.

REFERENCES

Abraham, K.M. and Z. Jiang. 1996. A polymer electrolyte based rechargeable lithium oxygen battery. J. Electrochem. Soc. 143.

Alan and Thomas. 2005. AER Energy Resources Inc. JP 3689432 B2 2005.8.31

Berndt, D. 2003. Maintenance-free Batteries, 3rd ed. Research Studies Press Ltd., Taunton, Somerset, England.

Besenhard, J.O. (ed.). 1999. Handbook of Battery Materials, Wiley-VCH: Weinheim, Germany.

Bev Betkowski. 2022, Researcher working to build a better, cheaper battery for power grids. Folio. https://www.ualberta.ca/folio/2022/05/researcher-working-to-build-a-better-cheaper-battery-for-power-grids.html.

Bogolowski, N. and J.F. Drillet. 2017. An electrically rechargeable Al-air battery with aprotic ionic liquid electrolyte. ECS Transactions. 7522: 85–92.

Bruce, P.G., S.A. Freunberger, L.J. Hardwick and J.-M. Tarascon. 2012. Li-O_2 and Li-S batteries with high energy storage. Nat. Mater. 11: 19e29. https://doi.org/10.1038/nmat3191.

Carson, W.N. and C.E. Kent. 1966. The magnesium-air cell. pp. 119-131. *In*: D.H. Collins (ed.). Power Sources. Pergamon Press: Oxford, U.K.

Cecchetto, L., M. Salomon, B. Scrosati and F. Croce. 2012. Study of a Li-air battery having an electrolyte solution formed by a mixture of an ether-based aprotic solvent and an ionic liquid. J. Power Sources. 213: 233–238.

Cooper, J.F., D. Fleming, L. Keene, A. Maimoni, K. Peterman and R. Koopman. 1994. Demonstration of a zinc/air fuel battery to enhance the range and mission of fleet electric vehicles. The 29th Intersociety Energy Conversion Engineering Conference. Lawrence Livermore National Laboratory.

Daming, G., W. Yu, G. Shuo and C. Zhang. 2013. Research progress and optimization of non-aqueous electrolyte for lithium-air batteries. Acta Chim. Sin. 71: 1354–1364.

Di Palma, T.M., F. Migliardini and D. Caputo. 2017. Xanthan and κ-carrageenan based alkaline hydrogels as electrolytes for Al-air batteries. Carbohydr. Polym. 157: 122–127.

Gelman, D., B. Shvartsev and Y. Ein-Eli. 2014. Aluminium–air battery based on an ionic liquid electrolyte. J. Mater. Chem. A. 2: 20237–20242.

Gonzalez, S. and R. Renaud. 2015. Aluminium-air battery and accumulator system. US 2015/0093659 A1.

Hartmann, P., L. Conrad, M. Bender Vracar, D.A. Katharina, A. Garsuch, U. Janek, et al., 2012. A rechargeable room-temperature sodium superoxide NaO_2 battery. Nat. Mater. 12(3): 228–232.

Heise, G.W. 1933. Air depolarized primary battery. US Pat. 1899615.

Jamesh, M.-I., P. Moni, A.S. Prakash and M. Harb. 2021. ORR/OER activity and zinc-air battery performance of various kinds of graphene-based air catalysts. Mater. Sci. Energy Technol. 4: 1–22.

Lau, K.H., L.A. Curtiss and J. Greeley. 2011. Density functional investigation of the thermodynamic stability of lithium oxide bulk crystalline structures as a function of oxygen pressure. J. Phys. Chem. C. 115(47): 23625–23633.

Li, F., H. Kitaura and H. Zhou. 2013. The pursuit of rechargeable solid-state Li-air batteries. Energy Environ. Sci. 6: 2302–2311.

Li, Y., X. Wang and S. Dong. 2016. Recent advances in non-aqueous electrolyte for rechargeable $Li-O_2$ batteries. Adv. Energy Mater. 6(18): 1600751.

Linden, D. and T.B. Reddy. 2002. Handbook of Batteries. 3rd ed. McGraw-Hill, New York.

Maiche, L. 1878. French Patent 127069.

Ma, Y., A. Sumboja and W. Zang. 2019. Flexible and wearable all-solid-state Al-air battery based on iron carbide encapsulated in electrospun porous carbon nanofibers. ACS Appl. Mater. Interfaces. 11: 1988–1995.

Mokhtar, M., M.Z. MeorTalib, E.H. Majlan, S.M. Tasirin, W.M.F.W. Ramli, W.R.W. Daud, et al., 2015. Recent developments in materials for aluminium–air batteries: a review. J. Ind. Eng. Chem. 32: 1–20.

Mokhtar, M., M. Zainal and E.H. Majlan. 2015. Recent developments in materials for aluminium–air batteries: a review. J. Ind. Eng. Chem. 32: 1–20.

Natasha, R.L., L. Shira, E. Yohanan and Y.E. Eli. 2020. Hybrid ionic liquid propylene carbonate based electrolytes for aluminium-air batteries. ACS Applied Energy Mater. 3(3): 2585–2592.

Nirupama, P.U., W.S. Krystyna and A.F. Sammells. 1988. A calcium-oxygen secondary battery. J. ECS. 135: 260.

Ojefors, L. and L. Carlsson. 1978. An iron-air vehicle battery. J. Power Sources. 2: 287–296.

Park, M.G., D.U. Lee, M.H. Seo, Z.P. Cano and Z. Chen. 2016. 3D ordered mesoporous bifunctional oxygen catalyst for electrically rechargeable zinc-air batteries. Small. 12: 2707.

Peng, Z., S.A. Freunberger, Y. Chen and P.G. Bruce. 2012. A reversible and higher-rate $Li-O_2$ battery. Science. 337(6094): 563–566.

Peng, G., M. Zheng, Q. Zhao, X. Xiao, H. Xue and H. Pang. 2017. Rechargeable zinc–air batteries: a promising way to green energy. J. Mater. Chem. A. 5: 7651–7666.

Reinsberg, P., J.B. Christoph and H. Baltruschat. 2016. Calcium-oxygen batteries as a promising alternative to sodium oxygen batteries. J. Phys. Chem. 120(39): 22179–22185.

Ryohei, M. 2019. All solid state rechargeable aluminium-air battery with deep eutectic solvent based electrolyte and suppression of by-products formation. RSC Adv. 9: 22220–22226.

Shaohua, Y. and H. Knickle. 2002. Design and analysis of aluminium/air battery system for electric vehicles. J. Power Sources. 112: 162–173.

Sheng, S., D.F. Zhang and R. Jeffrey. 2010. Discharge characteristic of a non-aqueous electrolyte Li-O_2 battery. J. Power Sources. 195: 1235–1240.

Smee, A. 1840. On the galvanic properties of the metallic elementary bodies, with a description of a new chemico-mechanical battery. Dublin Philos. Mag. J. Sci. 16(103): 315–321.

Stewart, S.G., S.I. Kohn, K.R. Kelty and J.B. Straubel (Tesla Motors, Inc.). 2012. Electric Vehicle Extended Range Hybrid Battery Pack System. US 20120041624 A1.

System of and Method for Power Management. 2002. WO200210877. Metallic Power. USA and Teck Metals, Canada.

Tan, P., H.R. Jiang, X.B. Zhu, L. An, C.Y. Jung, M.C. Wu, et al., 2017. Advances and challenges in lithium-air batteries. Appl. Energy. 204: 780–806.

Tan, M.J., B. Li and P. Chee. 2018. Acrylamide-derived freestanding polymer gel electrolyte for flexible metal-air batteries. J. Power Sources. 400: 566–571.

Timothy, K., A. Somers, A.J. Angel, D.R. Torriero, M. Farlane, H. Patrick, et al., 2013. Discharge behavior and interfacial properties of a magnesium battery incorporating trihexyl (tetradecyl) phosphonium-based ionic liquid electrolytes. Electrochim. Acta. 87: 701–708.

Tripathi, R., T.N. Ramesh, B.L. Ellis and L.F. Nazar. 2010. Scalable synthesis of tavorite $LiFeSO_4F$ and $NaFeSO_4F$ cathode materials. Angew. Chem. Int. Ed.. 49(46): 8738–8742.

Thippani, T., S. Mandal, G. Wang, V.K. Ramani and R. Kothandaraman. 2016. Probing oxygen reduction and oxygen evolution reactions on bifunctional non-precious metal catalysts for metal–air batteries. RSC Adv. 6: 71122–71133.

Wang, Y.J., J. Qiao, R. Baker and J. Zhang. 2013. Alkaline polymer electrolyte membranes for fuel cell applications. Chem. Soc. Rev. 42: 5768–5787.

Xiaodi, R. and W. Yiying. 2013. A low-over potential potassium oxygen battery based on potassium superoxide. J. Am. Chem. Soc. 135: 2923–2926.

Xiaoteng, J., Y. Yang, C. Wang, C. Zhao and R. Vijayaraghavan. 2014. Biocompatible Ionic liquid—biopolymer electrolyte enable thin and compact mg air batteries. ACS Applied materials and Interfaces. 6(23): 21110–21117.

Xiaolong, X., S.H. Kwan, A.D. Duc, N.H. Kwun and H. Wang. 2019. Recent advances in hybrid sodium–air batteries. Mater. Horiz. 6: 1306–1335.

Xin-hui, Y. and Y.-Y. Xia. 2010. The effect of oxygen pressures on the electrochemical profile of lithium/oxygen battery. J. Solid State Electrochem. 14: 109–114.

Xu, K. 2014a. Electrolytes and interphases in Li-ion batteries and beyond. Chem. Rev. 114(23): 11503–11618.

Xu, K. 2014b. Non-aqueous liquid electrolytes for lithium-based rechargeable batteries. Chem. Rev. 104: 4303–4418.

Yisi, L., Q. Sun, W. Li, K.R. Adair, J. Li and X. Sun. 2017. A comprehensive review on recent progress in aluminium-air batteries. Green Energy Environ. 2: 246–277.

Yu, L., F. Wu, J. Qian, M. Zhang, Y. Yuan, Y. Bai, et al., 2021. Metal chalcogenides with heterostructures for high-performance rechargeable batteries. Small Sci. 1: 2100012.

Zaromb, S. 1962. The use and behavior of aluminium anodes in alkaline primary batteries. J. Electrochem. Soc. 109: 1125–1130.

Zhang, T., T. Zhanliang and J. Chen. 2014a. Magnesium-air batteries: from principle to application. Mater. Horiz. 1: 196–206.

Zhang, Z., C. Zuo and Z. Liu. 2014b. All-solid-state Al-air batteries with polymer alkaline gel electrolyte. J. Power Sources. 251: 470–475.

Zhang, J., Q. Zhou, Y. Tang, L. Zhang and Y. Li. 2019. Zinc-air batteries: are they ready for prime time? Chem. Sci. 10: 8924–8929.

Zhi, Z., A. Kushima, Z. Yin, L. Qi, K. Amine, J. Lu, et al., 2016. Anion-redox nanolithia cathodes for Li-ion batteries. Nat. Energy. 1–7.

Chapter 13

Carbon Materials and their Performance as Support for Catalytic Nanoparticles

Elvia Teran-Salgado[1], Jose Luis Reyes-Rodriguez[2],
Adrián Velázquez-Osorio[3] and Daniel Bahena-Uribe[4*]

[1]Centro de Investigación en Ingeniería y Ciencias Aplicadas-(IICBA),
Universidad Autónoma Del Estado de Morelos,
Av. Universidad 1001, C.P, 62209, Cuernavaca, Morelos, México.

[2]Escuela Superior de Ingeniería Química e Industrias Extractivas (ESIQIE),
Instituto Politécnico Nacional (IPN). Av. Luis Enrique Erro S/N,
Unidad Profesional Adolfo López Mateos, Zacatenco,
Alcaldía Gustavo A. Madero, C.P. 07738, Ciudad de México, México.
Tel. +52 1 55 3084 9664. Email: jlreyes@ipn.mx.

[3]Departamento de Química,
Centro de Investigación y de Estudios Avanzados (Cinvestav IPN),
Av. IPN 2508, Col. San Pedro Zacatenco,
Alcaldía Gustavo A. Madero, C.P. 07360, Ciudad de México, México.
Tel. +52 (55) 5747 3715. Email: aovelazquez@gmail.com

[4*]Laboratorio Avanzado de Nanoscopía Electrónica (LANE),
Centro de Investigación y de Estudios Avanzados (Cinvestav IPN),
Av. IPN 2508, Col. San Pedro Zacatenco, Alcaldía Gustavo A. Madero,
C.P. 07360, Ciudad de México, México.
Tel.+52 (55) 5747 3800 Exts. 1740, 1742. Email: dbahenau@cinvestav.mx.

13.1 INTRODUCTION

Carbon is one the most abundant elements on earth and has been a source of energy for thousands of years. Carbon materials play a significant role in the

*For Correspondence: Email: dbahenau@cinvestav.mx (D. Bahena-Uribe).

development of clean and sustainable energy. Some of its applications include integration into electrodes for supercapacitors, lithium–ion batteries, optically transparent and electronically conductive films for solar cells, among others. In the case of fuel cells, carbon materials are commonly used as nanoparticles supports for electrocatalysts (Candelaria et al., 2012; Xin et al., 2012; Ibrahim et al., 2020; Chen et al., 2021). Various nanostructured materials have been developed for catalytic processes, such as, activated carbon, carbon black, mesoporous carbon, graphite, and graphitized materials (Sharma and Pollet 2012; Figueiredo 2018; Chen et al., 2021). Additionally, there are many types of carbon blacks, such as, acetylene black, Ketjen Black, Black Pearl, or Vulcan carbon XC-72 (VC XC-72). The latter is extensively used as a support material in the anode and cathode electrodes of Polymer Electrolyte Membrane Fuel Cells (PEMFC), Direct Methanol Fuel Cells (DMFC), Alkaline Fuel Cells (AFC), Microbial Fuel Cells (MFC), etc., (Zeng et al., 2006; Bayrakçeken et al., 2009; Fan et al., 2021). VC XC-72 provides excellent conductivity, versatile chemical and physical properties, and presents few impurities or organic residues (Bayrakçeken et al., 2009). Despite their excellent properties, carbon materials are susceptible to corrosion and may develop ohmic resistance and mass transfer issues that affect performance when used in electrochemical devices (Fan et al., 2021).

Platinum (Pt) continues to be the choice catalyst for high-activity electrochemical reactions carried out in fuel cells (Yu and Ye 2007; Seselj et al., 2015). Its catalytic activity has been extensively studied experimentally and computationally. Worldwide Pt reserves are scarce and the continued need for high amounts of this material have prevented early commercialization of fuel cell technology. For this reason, it is necessary to decrease the amount of Pt used in fuel cell catalysts without negatively affecting the catalytic activity. Currently there are three ways to achieve this:

1. Tuning the morphology of the nanoparticles (cubes, octahedrons, etc.). —nanoparticles with a high faceting index exhibit higher catalytic activity (Lim et al., 2008; Xia et al., 2009; Nosheen et al., 2013).

2. Modifying the electronic properties of Pt—by blending Pt with some other transition metal (lower cost) to make bimetallic compounds (Pt-Pd, Pt-Ni, etc.) or trimetallic compounds (Pt-Cr-Cu, etc.), it is possible to improve catalytic activity while simultaneously decreasing the amount of Pt necessary (Marković et al., 2001; Seselj et al., 2015; Hwang et al., 2016; Tinoco-Muñoz et al., 2016; Stacy et al., 2017).

3. Improving Pt/support carbon interaction—this interaction plays an important role over the catalytic activity, providing a higher performance of the cell (Calvillo et al., 2011).

In terms of Pt/C nanocatalysts synthesis, efforts have been targeted to maximize their surface area and decrease the total amount of metal employed. Important approaches include:

1. Reduction of metal salts with sodium boron hydride ($NaBH_4$), ascorbic acid ($C_6H_8O_6$), ethylene glycol, and others where the chemical reductions of carbon and Pt precursors occur simultaneously. Hexachloroplatinic

acid (H_2PtCl_6), potassium tetrachloroplatinate (K_2PtCl_4), or sodium hexachloroplatinate (Na_2PtCl_6) are some precursors for Pt nanoparticle formation. H_2PtCl_6 is the Pt precursor most commonly used in aqueous systems due its higher solubility in polar solvents, and a low reduction temperature that favors the formation of small particles (Yu and Ye 2007; Şanli et al., 2016).

2. Polyol method (Pullamsetty et al., 2015)
3. High energy mechano-chemical milling (Cheng et al., 2022).
4. Sonochemistry-assisted synthesis (Zou et al., 2012).

At its essence, nanoparticle synthesis consists of six stages: dissolution, reduction, nucleation, growth, agglomeration, and stabilization (Xia et al., 2009; Jia and Schüth 2011). Synthetic routes exist to ensure adequate control of dispersity, size, and shape of the metal nanoparticles in the support. However, the simultaneous control over morphology and distribution of the catalyst continues to be a challenge.

The Pt–carbon interaction has been shown to enhance the catalytic properties and stability of the electrocatalyst (Kumar et al., 2011; Dong et al., 2014; Reyes-Rodríguez et al., 2015; Prithi et al., 2021). The mechanism of this electronic interaction depends on the metal-support interface. The surface chemistry of support materials may be modified by oxidation treatments of the carbon surface; these treatments increase hydrophilicity and improve Pt dispersion. The oxidation of the carbon surface is achieved by using different oxidants: HNO_3, H_2SO_4, H_3PO_4, etc. to carry out the introduction of functional groups. The most common functional groups are carboxylic (R–(C=O)–OH), hydroxyl (R–OH), carbonyl groups (R–(C=O)–R′), etc. (Yu and Ye 2007; Figueiredo and Pereira 2010; Calvillo et al., 2011). Thus, functional groups act as anchoring sites for the strong adsorption of platinum nanoparticles on carbon, regulating their growth, structure, and dispersion on the support (Wu et al., 2015; Xin et al., 2016).

The stability and catalytic activity of a catalyst is also influenced by the degree of graphitization or amorphicity of the carbonaceous support, its functional surface chemistry, active surface area, porosity, among other factors. One of the most stable carbonaceous materials is graphite. Its laminar structure of stacked carbon atoms gives it a high degree of hydrophobicity and no chemical reactivity for catalytic purposes. The latter can be solved if functionalization of the graphitic material is carried out by incorporating functional groups during an oxidative chemical or electrochemical process. In the first case, by adding oxidizing agents during the synthesis, or in the latter case, by exfoliating the material in the presence of an acid and applying electrical potential. In either case, graphitic materials ranging from graphene oxide to electrochemically-oxidized graphite are obtained.

This chapter will address in greater detail the use of an oxidative electrochemical method to obtain exfoliated graphite as a potential support for Pt nanoparticles with catalytic activity towards ORR. Physical characterization of synthesized samples is done by XRD, XPS, SEM and TEM. Additionally, a discussion is presented on the electrochemical evaluation towards the ORR comparisons made with respect to reference Pt/Vulcan carbon catalysts.

13.2   CORROSION OF CARBON AND Pt NANOPARTICLES

Carbon corrosion, and Pt dissolution or aggregation, are the main reasons for electrochemical activity loss under PEM fuel cell operating conditions. The PEMFC has a cell voltage around 1.0 V during the no-load stage or 1.4 V during start-stop operation. Pt oxides are formed at potentials higher than 0.6 $V_{RHE,}$ leading to Pt dissolution in the electrodes of the cell (Schonvogel et al., 2017). On the other hand, carbon corrosion is one of the main reasons for catalyst degradation (Shao et al., 2009; Linse et al., 2011; Zhao et al., 2021). During carbon corrosion, carbon monoxide (CO) or carbon dioxide (CO_2) may be formed, causing a performance decrease due to the accelerated loss of active surface area. This means that fewer sites are available to accept Pt nanoparticles, causing an irreversible nanoparticle agglomeration effect. Furthermore, a thickening of the supporting structure results in a drop in performance and an increase in the ohmic resistance (Borup et al., 2006; Shao et al., 2009; Fan et al., 2021). Siroma et al., found that carbon corrosion takes place even when the cell is kept at the open circuit potential (OCP) (Siroma et al., 2007).

At the electronic level, carbon corrosion/oxidation starts at edges and corners of basal planes in the carbon structure due to their unsaturated valences and free electrons available. Simultaneously, the oxygen functional groups are susceptible to be decomposed due their accelerated oxidation in presence of humidity and high temperature conditions (Maass et al., 2008; Zhao et al., 2021; Sim et al., 2022).

It has been noted that graphitization slows down the kinetics of the carbon oxidation reaction, and in the case of the popular Vulcan carbon support, even slower oxidation rates may be achieved than with other types of carbon (Lee et al., 2022).

13.3   NANOSTRUCTURED CARBON MATERIALS AS NANOPARTICLE SUPPORT

Various carbon materials have been widely investigated as catalyst supports for fuel cells, for example: mesoporous carbon, with pore sizes of 2–50 nm that provide high surface area and conductivity. Manufacturing of this material is a complex process that requires care to avoid damaging the carbon's orderly structure (Lee et al., 2022,). Carbon aerogel (CA) is another amorphous carbon material that exhibits high stability and high specific surface area (600–1100 m^2 g^{-1}). However, it has poor electrochemical activity, low corrosion resistance, and low graphitization (Singh et al., 2017). The goal of creating nanostructured supports with fast electron transfer and high electrocatalytic activity has led to use graphitized supports such as carbon nanotubes (CNTs), carbon nanofibers (CNF), graphene, etc., due to their high stability under fuel cell operation conditions. CNTs are 2D nanostructures, typically composed of nanotubes formed by rolled up single sheets of hexagonally-arranged carbon atoms. CNTs may be single walled or multi-walled (SWCNT/MWCNT). Various methods like impregnation, ultrasound, polyol and microwave-assisted polyol, electrochemical deposition, etc. have been widely explored for the deposition of the electrocatalyst nanoparticles on CNTs (Sahoo et al., 2015; Ortiz-Herrera et al., 2022; Yu et al., 2022).

Functionalization to introduce surface oxygen groups (using strong acids like HNO_3 and H_2SO_4) is used to make the surface more hydrophilic and improve the catalyst support interactions. Pristine CNTs are chemically inert, marking their interaction with metal nanoparticles difficult. Recently, Ruiz-Camacho et al., investigated the application of carbon nanotubes as an alternative support for ORR cathodes in an alkaline medium. This group synthesized Pt and Pt–Ag nanomaterials supported on the CNT by a sonochemical method. The monometallic Pt samples supported in CNT and in Vulcan carbon showed higher electrochemical stability than carbon supported Pt–Ag bimetallic samples. Furthermore, their EIS results showed that Pt/CNT exhibited a lower resistance to electron transfer than conventional Pt/C and Pt-Ag/CNT (Ruiz-Camacho et al., 2022). On the other hand, Yu et al., studied an Ag–Ni electrocatalyst supported on multiwalled carbon nanotubes (MWCNTs) as a new type of electrocatalyst cathode for direct borohydride–hydrogen peroxide fuel cells (DBHPFCs). $Ag_{41}Ni_{59}$/MWCNTs exhibited the highest maximum power density, improved catalytic activity and stable performance compared to monometallic electrocatalysts (Au/MWCNTs, Ag/MWCNTs, and Ni/MWCNTs) (Yu et al., 2022). Although CNT are much more electrochemically stable due to their higher corrosion resistance, their application in fuel cells is limited by large-scale production issues and cost limitations.

Another carbon material used as support is carbon nanofibers (CNF). CNFs were first synthesized by the decomposition of hydrocarbons over metal particles. Since then, CNFs have been extensively researched as fuel cell supports (Sharma and Pollet 2012). Carol A. Bessel et al. demonstrated that 5 wt. % Pt on graphite nanofibers (GNF) exhibits 400% greater oxidation activity than Vulcan carbon (XC-72) in methanol oxidation studies. Their result shows a notable increase in catalyst stability thanks to the optimization of the carbon support composition, crystallinity, and morphology of the nanoparticles (Bessel et al., 2001). Secondly, Steigerwalt et al., prepared Pt-Ru nanocrystals dispersed on herringbone graphitic carbon nanofibers (GCNF). The nanostructures resulted in appreciably higher activity than Pt/Vulcan carbon nanocomposites and showed less susceptibility to CO poisoning than Pt/Vulcan carbon electrocatalysts. The structure of the GCNF support influences Pt nanocrystal morphology and is responsible for the enhanced performance of the Pt-Ru/GCNF anode catalysts. However, more comparative tests are required to conclusively support the data (Steigerwalt et al., 2021). Zheng et al., synthetized platinum/carbon nanofiber (Pt/CNF) nanocomposites with a platinum loading of 15 wt. % prepared by a modified electrophoretic deposition (EPD) method. The composite has a low metal loading and its Tafel plot suggests strong ORR activity. This is attributed to the small particles formed by EPD and the metal-support interaction. For comparison, they also synthesized a Pt/CNF composite prepared by chemical reduction and a metal loading of 40 wt. %. The resulting platinum nanoparticle size was about 3–5 nm, and this electrocatalyst was less active towards electrochemical capacity, demonstrating that the Pt/CNF composites synthesized by EPD are more effective for ORR (Zheng et al., 2008).

13.4 GRAPHENE AND GRAPHENE OXIDE SUPPORT

Graphene is a planar monolayer of carbon atoms arranged into a two-dimensional (2D) honey-comb lattice with a carbon–carbon bond length of 0.142 nm. Graphene possesses a variety of properties, such as, high electron mobility at room temperature (250,000 cm^2/Vs), exceptional thermal conductivity (5000 $Wm^{-1} K^{-1}$), and superior mechanical properties with a measured Young modulus of 1 TPa (Geim and Novoselov 2007; Ye et al., 2018). Graphene applications include sensors (biosensors, flexible pressure sensor, photoelectrochemical sensors), transparent electrodes, polymer compounds, nanocomposites, photodetectors, solar cells, and energy storage devices, including supercapacitors, lithium- ion batteries, and fuel cells (Jiping et al., 2014; Yang et al., 2016). These properties and applications of graphene have attracted the interest of researchers. A pure graphene sheet is chemically inert due to the lack of functional groups. In order to increase hydrophilicity and to improve an adequate interaction of metals with graphene, it is vitally important to promote the formation of chemical bonds in the graphene to facilitate interfacial charge transfer sites. The Pt-carbon interactions can lead the formation of such bonds as functional groups, defects on the structure, doping or impurities addition, and their inclusion causes enhanced catalytic activity of the graphitic materials (Eftekhari and Garcia 2017).

Graphene oxide (GO) consists of individual graphene sheets with attached oxygen functional groups on both the basal planes and edges. These oxygen functional groups are hydroxyl and epoxy groups present on the basal plane, and carboxy, carbonyl, and phenol groups at the edge of GO (Tian et al., 2021). Chemical oxidation is the most widely used process to produce graphene oxide, some chemical oxidation approaches include the Brodie method, Staudenmaier method, and the most commonly used Hummers or modified Hummers method (Singh et al., 2016). These methods consist of using acids and strong oxidizing agents for the oxidation of graphite.

Historically, Brodie used $KClO_4$ to oxidize a suspension of graphite in HNO_3 (Brodie 1859). Sometime later, Staudenmaier improved Brodie's preparation by adding concentrated H_2SO_4 to increase the acidity of the mixture (Staudenmaier 1898). Afterward, Hummers and Offeman developed an alternative to the oxidation method; their approach consisted in reacting graphite in a mixture of potassium permanganate and concentrated sulfuric acid, reaching oxidation levels like those of Brodie and Staudenmaier's GO (Hummers and Offeman 1958). These methods are very popular due to their scalability and high performance. The main purpose of introducing oxygenated functional groups on the carbon surface is to decrease the size of the Pt particles to the nanoscale level (2–3 nm); to ensure adequate dispersion of these nanoparticles in the carbon support so as to maximize the surface area (by preventing the aggregation of nanoparticles), and to increase the electrocatalytic activity (Xin et al., 2016).

Ghosh et al., synthesized functionalized graphene (FG) by chemical oxidation, a 20 wt.% Pt/FG catalysts was prepared using a precipitation method. The electrochemical surface area of the catalyst (Pt/FG) was found to be more than 45% greater in comparison to a commercial carbon-supported platinum catalyst.

The stability of the synthesized catalyst was also significantly higher than the commercial Pt/C reference. The maximum power densities of the fuel cell were found to be 314, 426, and 455 mW cm^{-2} using Pt/C, Pt/graphene, and Pt/FG, respectively (Ghosh et al., 2013). In another case, Pushkareva et al., synthesized catalysts with a relatively high Pt content to form thinner catalytic layers using ethylene glycol and formaldehyde as reductants. The simultaneous reduction of Pt and graphene oxide precursors during synthesis provided a significantly better PEMFC performance. The maximum power densities measured for Pt/C and Pt/RGO(s) were ca. 0.358 and 0.421 Wcm^{-2}, respectively. This difference in power density is attributed to the higher electrochemically surface area (ECSA), higher Pt utilization, and less agglomerated support structure, which provided effective mass transport for the ORR. The use of this catalyst on the cathode of a PEMFC led to an increase in its maximum power density of up to 17%, and significantly enhanced its performance at high current densities (Pushkareva et al., 2021). Other investigations have agreed that graphene shows promising results as a support for nanoparticles (Seselj et al., 2015; Higgins et al., 2016; Saravanan and Subramanian 2016).

As mentioned previously, the presence of functional groups such as carboxyl, hydroxyl, and carbonyl on the support is responsible for anchoring metal ions during the synthesis of the supported metal catalyst. For this reason, these functional groups result in better Pt precursor dispersion and a higher long-term stability (Ghosh et al., 2013). However, some researchers have found that surface functionalization of the carbon support affects electrocatalytic performance because different oxygen-containing groups exhibit distinct decomposition behaviors. For example, carboxyl and lactone groups generally begin to decompose at 250°C, while hydroxyl shows a higher decomposition temperature close to 400°C (Chen et al., 2020). Additionally, according to Yu and Ye 2007 and Zhao et al., 2021, oxygen-containing functional groups, generated on the carbon surface, weaken the interaction between the carbon and the Pt nanoparticles, resulting in their shedding from support. There is evidence that a better platinum dispersion and an improved resistance to sintering are obtained when the pre-graphitized carbon black support is free of oxygen surface groups (Yu and Ye 2007). A Pt/rGO catalyst with the lowest concentration of oxygen-containing functional groups exhibited the highest catalytic activity and durability compared with a highly oxidized graphene support (Zhao et al., 2021).

For this reason, some synthesis alternatives are needed to chemically oxidize graphite without considerably damaging the structure of the support material promoting oxygenated functional groups necessary for the anchoring and distribution of Pt nanoparticles.

13.5 ELECTROCHEMICALLY-OXIDIZED GRAPHITE AS A NANOPARTICLE SUPPORT

Electrochemical oxidation consists of applying a potential to a graphite electrode immersed in an electrolytic solution. The experimental setup contains a working electrode (graphite rods, sheets, HOPG, etc.), a counter electrode (graphite rods or platinum mesh), electrolyte (aqueous or organic), and a power supply (Figure 13.1a). The electrochemical oxidation of graphite is a three-stage reaction (Beck et al., 1995; Cao et al., 2017; Gurzęda et al., 2016; Tian et al., 2017).

1. The first stage is the formation of interlaminar or graphite intercalation compounds (GIC). The distance between graphite layers (0.33 nm) allows the insertion molecules or ions between them. Applying a potential (7 V) causes electrons to be released from the anode, generating a positive charge (+). This charge attracts anionic intercalants ($NO_3^-, SO_4^-, PO_4^{3-}$) and causes an increase of the layer separation, prompting a transition from graphite to GIC:

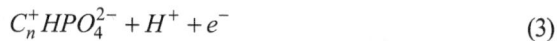

$$C_n^+ NO_3^- + H^+ + e^- \tag{1}$$

$$C_n^+ HSO_4^- + H^+ + e^- \tag{2}$$

$$C_n^+ HPO_4^{2-} + H^+ + e^- \tag{3}$$

   These compounds are of the donor type, and they transfer an e⁻ to the graphite layer. GICs are a reaction intermediate, and they are important for the formation of graphite oxide that enables the diffusion of oxidizing agents to replace existing intercalators.

2. In the second stage, the oxidation of the graphite surface is carried out from the co-intercalation of the H_2O molecules between the graphite layers. This co-intercalation occurs due to the nucleophilic attack of H_2O towards the (+) charge of the carbon lattice:

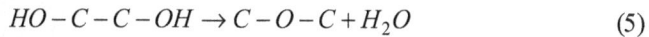

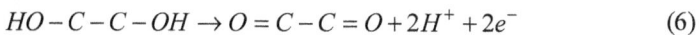

$$C = C + 2H_2O \rightarrow HO-C-C-OH + 2H^+ + 2e^- \tag{4}$$

$$HO-C-C-OH \rightarrow C-O-C + H_2O \tag{5}$$

$$HO-C-C-OH \rightarrow O=C-C=O + 2H^+ + 2e^- \tag{6}$$

3. Finally, the decomposition of the GICs takes place and electrochemically oxidized graphite (EOG) is formed. Both GIC decomposition and EOG formation are competing reactions. The intercalated H_2O molecules releases oxygen and causes an expansion of the layers to promote exfoliation. O_2 evolution may be dominant due to electrolysis of H_2O:

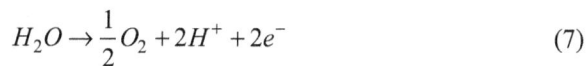

$$H_2O \rightarrow \frac{1}{2}O_2 + 2H^+ + 2e^- \tag{7}$$

The electrochemical oxidation/exfoliation of graphite is strongly related to the structure of the precursor and its purity. Among the family of carbon materials used as electrodes are highly oriented pyrolytic graphite-HOPG (the most crystalline of carbonaceous sources), flakes, graphite paper, graphite rods, etc. Liu et al., also experimented using graphite rods extracted from pencils as anode and cathode (Liu et al., 2013). Botas et al. mention that oxides obtained from a low-crystallinity graphite will contain a considerable amount of hydroxyl and carboxyl groups; in contrast, the oxides obtained from a crystalline type of graphite will mostly result in epoxide groups (Botas et al., 2012). In other words, the type of carbon precursor plays an important factor in defining the characteristics of the oxygenated functional groups present in the graphite structure. Another important oxidation parameter is the electrolyte used. Some commonly used electrolytes include ionic liquids, aqueous acids (HBr, HCl, H_2SO_4, $HClO_4$, HNO_3) or aqueous inorganic salts ((NH_4)$_2SO_4$, $NaSO_4$, K_2SO_4) (Lu et al., 2009; Kakaei 2013; Coroş et al., 2016; Munuera et al., 2016). H_2SO_4 and HNO_3 electrolytes are considered to be strong electrolytes, and in aqueous solution form, they tend to dissociate completely. Electrolyte concentration is also important. A higher acid concentration generates large graphite particles, but if the acid concentration is low, the exfoliation efficiency improves due to a decrease in the number of ions available to position themselves in-between the graphite layers. Therefore, it is essential to strike a balance between the type and concentration of electrolyte (Parvez et al., 2014). The acid provides the ionic molecules that introduce themselves between the graphite layers while the H_2O produces hydroxyl radicals that aid in intercalation, oxidation, and subsequent exfoliation. In this way, the level of graphite oxidation depends on both the acid concentration and the amount of H_2O used.

To illustrate the effect of the acid electrolyte on the formation of electrochemically oxidized graphite, some experimental results, derived from the study previously reported by Teran-Salgado et al., 2019, will be discussed.

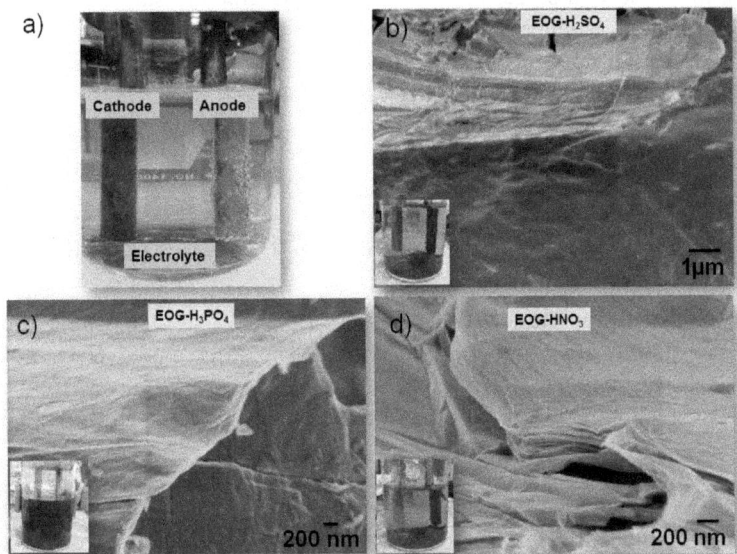

Figure 13.1 a) Electrolytic cell. The bubbling at the anode is evidence that an electrolysis process is taking place. SEM image of the EOG sheet after the oxidation and exfoliation process in b) H_2SO_4, c) H_3PO_4 and d) HNO_3. The experimental setup inset in figure shows the changes in the working electrode by the electrolysis process.

Figure 13.1a shows the electrochemical oxidation of two graphite electrodes used as anode and cathode, respectively, during a chronoamperometry test in a two-electrode cell configuration at a fixed potential of 7 V. Scanning electron microscopy (SEM) micrographs reveal the formation of large graphite oxide sheets when using H_2SO_4 y H_3PO_4 electrolytes (Figs. 13.1b-d). In contrast, the oxidation by HNO_3 causes the formation of overlapping EOG sheets. An alternative to promote exfoliation of overlapping sheets is to sonicate with a solvent other than H_2O such as dimethylformamide (DMF) because it is a compound with a larger-size molecules. X-ray diffraction (XRD) patterns in left side of Figure 13.2 reveal the appearance varying levels of oxidation depending on the acid electrolyte used. The broadness of the main peak around 26° obeys the following tendency, $HNO_3 >$ $H_3PO_4 > H_2SO_4$ and corresponds to the (002) crystallographic plane. The appearance of a peak between $10° - 15°$ is characteristic of graphene oxide (GO) (Li et al., 2007). A GO peak decrease due to lower functionalization degree can be noticed in Figure 13.2, and it obeys the following tendency, $HNO_3 > H_2SO_4 > H_3PO_4$.

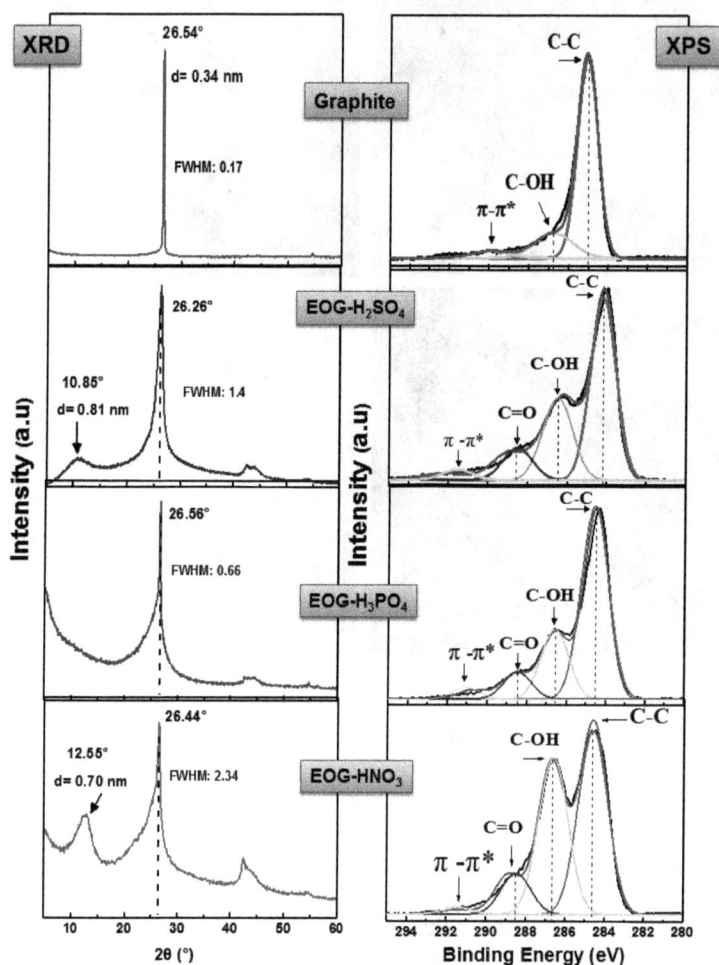

Figure 13.2 Left side, XRD pattern. Right side, deconvoluted XPS spectra C1s for pristine graphite and synthesized EOG samples, respectively.

XPS shows the formation of functional groups (C–C/C=C, C–OH, C=O) as result of various electrochemical oxidation treatments [Figure 13.2 (right side)]. A limitation of this EOG method is the inefficient distribution of electrical current to all graphite layers, especially after the initial expansion. This prevents intercalation and electrochemical oxidation from being completed and represents a challenge in the production of highly reproducible electrochemically oxidized graphite (Cao et al., 2017). Despite this, electrochemical oxidation is an easy technique to implement that enables control over the level of oxidation from the selection of the type and concentration of an appropriate electrolyte. For this reason, it has become an ideal synthesis method for many applications.

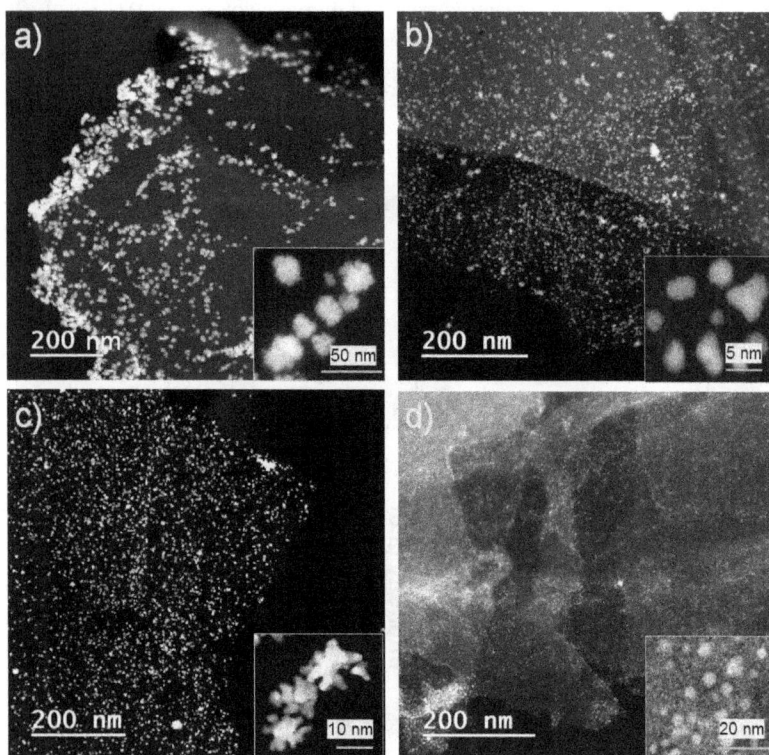

Figure 13.3 STEM micrographs showing the distribution of nanoparticles a) Pt/graphite, b) Pt/EOG-H_2SO_4, c) Pt/EOG-H_3PO_4, and d) Pt/EOG-HNO_3. The inset in figure shows size and shape of Pt nanoparticles.

Figure 13.3 shows graphite STEM micrographs for Pt nanoparticles supported in electrochemically-oxidized graphite (EOG) using H_2SO_4, HNO_3, and H_3PO_4 as electrolyte media. Visible differences in size and dispersion of nanoparticles is appreciated. The synthesis of the nanoparticles was carried out from the chemical reduction of metal salts using K_2PtCl_6 as precursor and $NaBH_4$ as reducing agent. Pt particles supported on pure graphite are concentrated on the graphite edges. The average particle size is 16.78 nm. The Pt/EOG-H_2SO_4 catalyst presents a better distribution, as observed from the edges and the basal plane. The nanoparticles are polymorphous with an average size of 4.56 nm, indicating that the functional groups adhered to the support influenced both the dispersion and the size. For the Pt/EOG-H$_3$PO$_4$ catalyst, the nanoparticles appear well distributed, with an average particle size of 5.97 nm. Finally, for the Pt/EOG-HNO$_3$ catalyst, Pt nanoparticles show a more defined morphology with an average particle size of 4.05 nm. However, a larger number of Pt agglomeration zones can be observed and are attributed to the overlap of graphite sheets and their poor degree of exfoliation.

Compared to pure graphite, these results are an indication that the functional groups acted as anchor sites for the growth, distribution and smaller size of Pt nanoparticles (Lakshmi et al., 2006; Dong et al., 2014; Xin et al., 2016). The Pt/EOG-H$_2$SO$_4$ catalyst presents better results for size and distribution of the Pt nanoparticles in the carbon. For this reason, the catalytic activity of Pt/EOG-H$_2$SO$_4$ was also compared to that of the high-performance Pt/Vulcan carbon catalyst. A Vulcan carbon support was oxidized using a 20 v/v % HNO$_3$ solution with the purpose of introducing functional groups within its structure to modify its surface chemical composition. The catalyst compositions were 20/80 wt. % of Pt-NPs in the respective support material.

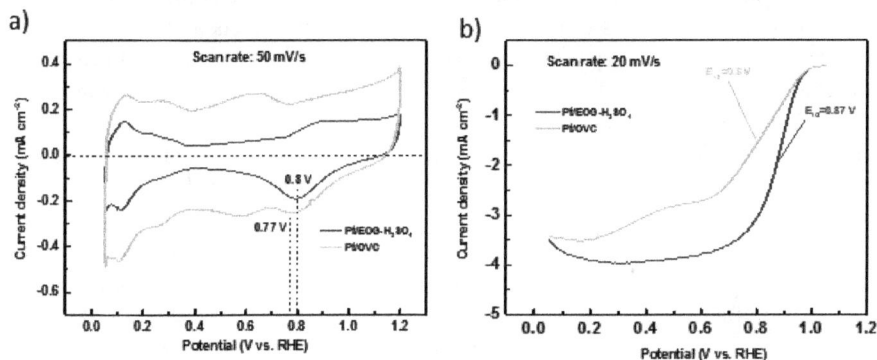

Figure 13.4 a) Cyclic voltammetries for Pt/EOG and Pt/OVC catalysts, b) ORR steady-state polarization curves for catalysts.

The electrocatalytic activity was evaluated by a rotating disk electrode (RDE) technique, described in the method reported by Garsany et al., (Garsany et al., 2010). Figure 13.4 shows characteristic cyclic voltammograms (CV) for Pt bulk electrodes and polycrystalline Pt/Carbon nanomaterials in acid media (Paulus et al., 2001; Mayrhofer et al., 2008; Gasteiger et al., 2005). The high magnitude of the oxygen reduction potential indicates a rapid reduction of oxidized species for the formation of water and this points to higher electrocatalytic activity towards ORR (Reyes-Rodríguez et al., 2013; Taylor et al., 2016). The cathodic peaks around 0.8 V/RHE for Pt/EOG and 0.77 V/RHE for Pt/OVC are associated to the Pt-oxide reduction process (Arenz et al., 2005; Reyes-Rodríguez et al., 2013; Taylor et al., 2016). Therefore, the positive shift of the oxygen cathodic peak toward higher potentials (i.e., lower overpotential to ORR close to the thermodynamic potential for the ORR ca.1.23 V$_{RHE}$) constitutes a first qualitative feature to highlight better catalytic activity. The Electrochemical Surface Area (ECSA) represents the number of electrochemically active sites available per mass of noble metal (Pt) in m^2 g^{-1} for the catalytic reaction towards ORR (Shinozaki et al., 2015; Chen et al., 2016). ECSA value for Pt/EOG catalyst was 14.0 m^2 g^{-1} and 18.6 m^2 g^{-1} for Pt/OVC (Table 13.1).

Table 13.1 Electrochemical parameters for the ORR
at 0.90 V for Pt/EOG, Pt/OVC catalysts.

Material	ECSA (m^2/g)	Specific activity (mA cm^{-2} real)	Mass activity (A mg^{-1})
Pt/EOG	14.0	0.665	0.09
Pt/OVC	18.6	0.237	0.044

ECSA differences may be directly related to the variation in Pt particle size in the catalyst, as seen in the STEM micrographs. The electrocatalytic activity of the synthesized catalysts towards the ORR was evaluated from steady-state polarization curves at different rotation rates between 400 and 2500 rpm, under O_2 saturation, at a scan rate of 20 mV s^{-1}, in a potential window of 0.05–1.05 V_{RHE}, as per to the procedure described by Teran-Salgado et al., 2019. The anodic sweep polarization curves were corrected for capacitive effects and normalized by geometrical area at 1600 rpm and scan rate of 20 mV s^{-1}. Polarization curves show the characteristic profile for Pt-based catalysts (Garsany et al., 2010; Mayrhofer et al., 2008; Paulus et al., 2001). A comparison of polarization curves reveals that the Pt/EOG catalysts presents better diffusion control when compared to the synthesized Pt/OVC. In the case of the latter, the capacitive effects (product of the functionalization of OVC) make it difficult to maintain constant limit current. According to the half-wave potential ($E_{1/2}$) determined for catalysts, Pt/EOG ($E_{1/2}$ = 0.87 V/RHE) presents the least overpotential, indicating improved electrocatalytic activity towards ORR as compared to Pt/OVC ($E_{1/2}$ = 0.78 V). A practical and conventionally accepted representation of the electrocatalytic activity is the Specific Activity (SA) and the Mass Activity (MA). The former tracks the current density per real Pt surface area in mA cm^{-2} Pt while the latter measures the current density per milligram of Pt in A mg^{-1}. Both parameters are determined at 0.9 V_{RHE} after the capacitive current subtraction and the mass-transport corrections were made (Shinozaki et al., 2015). MA is the most representative parameter for catalytic activity because it tracks the current per mass of Pt used; this is closely related to the cost/benefit of the catalyst material in low-temperature fuel cells. Table 13.1 shows that the Pt/EOG material has a higher SA (0.665 mA cm^{-2} Pt) and MA (0.092 A mg^{-1}) with respect to Vulcan carbon (0.24 mA cm^{-2} Pt and 0.04 A mg^{-1}). Therefore the Pt/rGO material displays a better performance towards ORR.

Although it is necessary to further improve the SA and MA performance of the Pt/EOG catalyst, the data discussed shows that Pt supported on the electrochemically-oxidized graphite has better catalytic efficiency than the Pt/OVC. Thus, it can be concluded that electrochemical oxidation is a good synthesis method to oxidize graphite and to functionalize it for the purpose of supporting Pt nanoparticles. It should also be highlighted that this method does not require special synthesis conditions or strong acids that could damage the structure of the graphite. Therefore, this method is a promising approach for optimizing the performance of carbon-supported Pt materials for fuel cells.

REFERENCES

Arenz, M., K. Mayrhofer, V. Stamenkovic, B. Blizanac, T. Tomoyuki, P. Ross, et al., 2005. The effect of the particle size on the kinetics of CO electrooxidation on high surface area Pt catalysts. J. Am. Chem. Soc. 127: 6819–6829.

Bayrakçeken, A., A. Smirnova, U. Kitkamthorn, M. Aindow, L. Türker, C. Erkey, et al., 2009. Vulcan-supported Pt electrocatalysts for PEMFCs prepared using supercritical carbon dioxide deposition. Chem. Eng. Commun. 196: 194–203.

Beck, F., J. Jiang and H. Krohn. 1995. Potential oscillations during galvanostatic overoxidation of graphite in aqueous sulphuric acids. J. Electroanal. Chem. 389: 161–165.

Bessel, C.A., K. Laubernds, N.M. Rodriguez and R.K. Baker. 2001. Graphite nanofibers as an electrode for fuel cell applications. J. Phys. Chem. B. 105: 1115–1118.

Borup, R.L., J.R. Davey, F.H. Garzon, D. Wood and M.A. Inbody. 2006. PEM fuel cell electrocatalyst durability measurements. J. Power Sources. 1: 76–81.

Botas, C., P. Álvarez, C. Blanco, R. Santamaría, M. Granda, P. Ares, et al., 2012. The effect of the parent graphite on the structure of graphene oxide. Carbon. 50: 275–282.

Brodie, B.C. 1859. On the atomic weight of graphite. Philos. Trans. R. Soc. London. 149: 249–259.

Calvillo, L., V. Celorrio, R. Moliner and M. Lázaro. 2011. Influence of the support on the physicochemical properties of Pt electrocatalysts: Comparison of catalysts supported on different carbon materials. Mater. Chem. Phys. 127(1-2): 335–341.

Candelaria, S.L., Y. Shao, W. Zhou, X. Li, J. Xiao, J.-G. Zhang, et al., 2012. Nanostructured carbon for energy storage and conversion. Nano Energy. 1(2): 194–220.

Cao, J., P. He, M.A. Mohammed, X. Zhao, R.J. Young, B. Derby, et al., 2017. Two-step electrochemical intercalation and oxidation of graphite for the mass production of graphene oxide. J. Am. Chem. Soc. 139(48): 17446–17456.

Chen, G., M. Li, K.A. Kuttiyiel, K. Sasaki, F. Kong, C. Du, et al., 2016. Evaluation of oxygen reduction activity by the thin-film rotating disk electrode methodology: the effects of potentiodynamic parameters. Electrocatalysis. 7: 305–316.

Chen, W., S. Chen, G. Qian, L. Song, D. Chen, X. Zhou, et al., 2020. On the nature of Pt-carbon interactions for enhanced hydrogen generation. J. Catal. 389: 492–501.

Chen, J., Z. Ou, H. Chen, S. Song, K. Wang and Y. Wang. 2021. Recent developments of nanocarbon based supports for PEMFCs electrocatalysts. Chinese J. Catal. 42(8): 1297–1326.

Cheng, Y., X. Gong, S. Tao, L. Hu, W. Zhu, M. Wang, et al., 2022. Mechano-thermal milling synthesis of atomically dispersed platinum with spin polarization induced by cobalt atoms towards enhanced oxygen reduction reaction. Nano Energy. 98: 107341.

Coroş, M., F. Pogăcean, M.-C. Roşu, G. Borodi, L. Mageruşan, A.R. Biriş, et al., 2016. Simple and cost-effective synthesis of graphene by electrochemical exfoliation of graphite rods. RSC Advances. 6(4): 2651–2661.

Dong, L., J. Zang, J. Su, Y. Jia, Y. Wang, J. Lu, et al., 2014. Oxidized carbon/nano-SiC supported platinum nanoparticles as highly stable electrocatalyst for oxygen reduction reaction. Int. J. Hydrog. Energy. 39(29): 16310–16317.

Eftekhari, A. and H. Garcia. 2017. The necessity of structural irregularities for the chemical applications of graphene. Mater. Today Chem. 4: 1–16.

Fan, L., J. Zhao, X. Luo and Z. Tu. 2021. Comparison of the performance and degradation mechanism of PEMFC with Pt/C and Pt black catalyst. Int. J. Hydrogen Energy. 47(8): 5418–5428.

Figueiredo, J.L. and M.F.R. Pereira. 2010. The role of surface chemistry in catalysis with carbons. Catal. Today. 150(1–2): 2–7.

Figueiredo, J.L. 2018. Nanostructured porous carbons for electrochemical energy conversion and storage. Surf. Coat. Technol. 350: 307–312.

Garsany Naval, Y., O.A. Baturina and S.S. Kocha. 2010. Experimental methods for quantifying the activity of platinum electrocatalysts for the oxygen reduction reaction. Anal. Chem. 82(15): 6321–6328.

Gasteiger, H., S. Kocha, B. Sompalli and F. Wagner. 2005. Activity benchmarks and requirements for Pt, Pt-alloy, and non-Pt oxygen reduction catalysts for PEMFCs. Appl. Catal. B Environ. 56(1–2): 9–35.

Geim, A. and K. Novoselov. 2007. The rise of graphene. Nature Mater. 6: 183–191.

Ghosh, A., S. Basu and A. Verma. 2013. Graphene and functionalized graphene supported platinumcatalyst for PEMFC. Fuel Cells. 13(3): 355–363.

Gurzęda, B., P. Florczak, M. Kempiński, B. Peplińska, P. Krawczyk and S. Jurga. 2016. Synthesis of graphite oxide by electrochemical oxidation in aqueous perchloric acid. Carbon. 100: 540–545.

Higgins, D., P. Zamani, A. Yu and Z. Chen. 2016. The application of graphene and its composites in oxygen reduction electrocatalysis: a perspective and review of recent progress. Energy Environ. Sci. 9: 357–390.

Hummers, W.S. and R.E. Offeman. 1958. Preparation of graphitic oxide. J. Am. Chem. Soc. 80(6): 1339.

Hwang, S.-K., A.T. Vilian, C.H. Kwak, S.Y. Oh, C.-Y. Kim, G.-w. Lee, et al., 2016. Pt-Au bimetallic nanoparticles decorated on reduced graphene oxide as an excellent electrocatalysts for methanol oxidation. Synthetic Metals. 219: 52–59.

Ibrahim, I.D., E.R. Sadiku, T. Jamiru, Y. Hamam, Y. Alayli and A.A. Eze. 2020. Prospects of nanostructured composite materials for energy harvesting and storage. J. King Saud Univ. Sci. 32(1): 758–764.

Jia, C.-J. and F. Schüth. 2011. Colloidal metal nanoparticles as a component of designed catalyst. Phys. Chem. Chem. Phys. 13(7): 2457-2487.

Jiping, Z., R. Duan, Z. Sheng, J. Nan, Y. Zhang and J. Zhu. 2014. The application of graphene in lithium ion battery electrode materials. SpringerPlus. 3: 585.

Kakaei, K. 2013. One-pot electrochemical synthesis of graphene by the exfoliation of graphite powder in sodium dodecyl sulfate and its decoration with platinum nanoparticles for methanol oxidation. Carbon. 51: 195-201.

Kumar, S.M.S., N. Hidyatai, J.S. Herrero, S. Irusta and K. Scott. 2011. Efficient tuning of the Pt nano-particle mono-dispersion on Vulcan XC-72R by selective pre-treatment and electrochemical evaluation of hydrogen oxidation and oxygen reduction reactions. Int. J. Hydrogen Energy. 36(9): 5453–5465. https://doi.org/10.1016/j.ijhydene.2011.01.124

Lakshmi, N., N. Rajalakshmi and K.S. Dhathathreyan. 2006. Functionalization of various carbons for proton exchange membrane fuel cell electrodes: Analysis and characterization. J. Phys. D: Appl. Phys. 39(13): 2785–2790.

Lee, F., M. Ismail, K. Hughes, D. Ingham, K. Hughes, L. Ma, et al., 2022. Alternative architectures and materials for PEMFC gas diffusion layers: A review and outlook. Renewable Sustainable Energy Rev. 166: 112640.

Li, Z.Q., C.J. Lu, Z.P. Xia, Y. Zhou and Z. Luo. 2007. X-ray diffraction patterns of graphite and turbostratic carbon. Carbon. 8: 1686–1695.

Lim, B., X. Lu, M. Jiang, P.H. Camargo, E.C. Cho, E.P. Lee, et al., 2008. Facile synthesis of highly faceted multioctahedral pt nanocrystals through controlled overgrowth. Nano Lett. 8(11): 4043–4047.

Linse, N., L. Gubler, G.G. Scherer and A. Wokaun. 2011. The effect of platinum on carbon corrosion behavior in polymer electrolyte fuel cells. Electrochim. Acta. 56(22): 7541–7549.

Liu, J., H. Yang, S. Zhen, C. Poh, A. Chaurasia, J. Luo, et al., 2013. A green approach to the synthesis of high-quality graphene oxide flakes via electrochemical exfoliation of pencil core. RSC Advances. 3(29): 11745–11750.

Lu, J., J.-X. Yang, J. Wang, A. Lim, S. Wang and K. Loh. 2009. One-pot synthesis of fluorescent carbon graphene by the exfoliation of graphite in ionic liquids. ACS Nano. 3(8): 2367–2375.

Maass, S., F. Finsterwalder, G. Frank, R. Hartmann and C. Merten. 2008. Carbon support oxidation in PEM fuel cell cathodes. J. Power Sources. 176(2): 444–451.

Markovic, N.M., T.J. Schmidt, V. Stamenkovic and P.N. Ross. 2001. Oxygen reduction reaction on Pt and Pt bimetallic surfaces: a selective review. Fuel Cells. 1(2): 105–116.

Mayrhofer, K.J.J., D. Strmcnik, B.B. Blizanac, V. Stamenkovic, M. Arenz and N. Markovic. 2008. Measurement of oxygen reduction activities via the rotating disc electrode method: from Pt model surfaces to carbon-supported high surface area catalysts. Electrochim. Acta. 53(7): 3181–3188.

Munuera, J.M., J.I. Paredes, S. Villar-Rodil, M. Ayán-Varela, A. Martínez-Alonso and J. Tascón. 2016. Electrolytic exfoliation of graphite in water with multifunctional electrolytes: en route towards high quality, oxide-free graphene flakes. Nanoscale. 8(5): 2982–2998.

Nosheen, F., Z.-C. Zhang, J. Zhuang and X. Wang. 2013. One-pot fabrication of single-crystalline octahedral Pt–Cu nanoframes and their enhanced electrocatalytic activity. Nanoscale. 5(9): 3660–3663.

Ortiz-Herrera, J.C., H. Cruz-Martínez, O. Solorza-Feria and D. Medina. 2022. Recent progress in carbon nanotubes support materials for Pt-based cathode catalysts in PEM fuel cells. Int. J. Hydrogen Energy. 47(70): 30213–30224.

Parvez, K., Z.-S. Wu, R. Li, X. Liu, R. Graf, X. Feng, et al., 2014. Exfoliation of graphite into graphene in aqueous solutions of inorganic salts. J. Am. Chem. Soc. 136(16): 6083–6091.

Paulus, U., T. Schmidt, H. Gasteiger and R. Behm. 2001. Oxygen reduction on a high-surface area Pt/Vulcan carbon catalyst: a thin-film rotating ring-disk electrode study. J. Electroanal. Chem. 495(2): 134–145.

Prithi, J.A., R. Vedarajan, G. Ranga Rao and N. Rajalakshmi. 2021. Functionalization of carbons for Pt electrocatalyst in PEMFC. Int. J. Hydrogen Energy. 46(34): 17871–17885.

Pullamsetty, A., M. Subbiah and R. Sundara. 2015. Platinum on boron doped graphene as cathode electrocatalyst for proton exchange membrane fuel cells. Int. J. Hydrogen Energy. 40(32): 10251–10261.

Pushkareva, I.V., A.S. Pushkarev, V.N. Kalinichenko, R.G. Chumakov, M.A. Soloviev, Y. Liang, et al., 2021. Reduced graphene oxide-supported Pt-based catalysts for PEM fuel cells with enhanced activity and stability. Catalysts. 11(2): 256.

Reyes-Rodríguez, J., F. Godínez-Salomón, M. Leyva and O. Solorza-Feria. 2013. RRDE study on Co@Pt/C core–shell nanocatalysts for the oxygen reduction reaction. Int. J. Hydrogen Energy. 38(28): 12634–12639.

Reyes-Rodríguez, J.L., K. Sathish-Kumar and O. Solorza-Feria. 2015. Synthesis and functionalization of green carbon as a Pt catalyst support for the oxygen reduction reaction. Int. J. Hydrogen Energy. 40(48): 17253–17263.

Ruiz-Camacho, B., A. Medina-Ramíreza, R. Fuentes-Ramírez, R. Navarro, C. Martínez Goméz and A. Pérez-Larios. 2022. Pt and Pt–Ag nanoparticles supported on carbon nanotubes (CNT) for oxygen reduction reaction in alkaline medium. Int. J. Hydrogen Energy. 47(70): 30147–30159.

Sahoo, M., K. Scott and S. Ramaprabhu. 2015. Platinum decorated on partially exfoliated multiwalled carbon nanotubes as high perfor- mance cathode catalyst for PEMFC. Int. J. Hydrogen Energy. 40(30): 9435–9443.

Şanli, L.I., V. Bayram, B. Yarar, S. Ghobadi and S.A. Gursel. 2016. Development of graphene supported platinum nanoparticles for polymer electrolyte membrane fuel cells: Effect of support type and impregnationereduction methods. Int. J. Hydrogen Energy. 41(5): 3414–3427.

Saravanan, G. and M. Subramanian. 2016. Pt nanoparticles embedded on reduced graphite oxide with excellent electrocatalytic properties. Appl. Surf. Sci. 386: 96–102.

Schonvogel, D., J. Hülstede, P. Wagner, I. Kruusenberg, K. Tammeveski, A. Dyck, et al., 2017. Stability of Pt nanoparticles on alternative carbon supports for oxygen reduction reaction. J. Electrochem. Soc. 164: F995–F1004.

Seselj, N., C. Engelbrekt and J. Zhang. 2015. Graphene-supported platinum catalysts for fuel cells. Sci. Bull. 60(9): 864–876.

Shao, Y., J. Wang, R. Kou, M. Engelhard, J. Liu, Y. Wang, et al., 2009. The corrosion of PEM fuel cell catalyst supports and its implications for developing durable catalysts. Electrochim. Acta. 54(11): 3109–3114.

Sharma, S. and B.G. Pollet. 2012. Support materials for PEMFC and DMFC electrocatalysts —A review. J. Power Sources. 208: 96–119.

Shinozaki, K., J.W. Zack, R.M. Richards, B.S. Pivovar and S.S. Kocha. 2015. Oxygen reduction reaction measurements on platinum electrocatalysts utilizing rotating disk electrode technique. J. Electrochem. Soc. 162(10): F1144–F1158.

Sim, J., M. Kang, K. Min, E. Lee and J.-Y. Jyoung. 2022. Effects of carbon corrosion on proton exchange membrane fuel cell performance using two durability evaluation methods. Renew. Energ. 190: 959–970.

Singh, R., R. Kumar and D. Singh. 2016. Graphene oxide: strategies for synthesis, reduction and frontier applications. RSC Advances. 6(69): 64993–65011.

Singh, R., M. Singh, S. Bhartiya, A. Singh, D. Kohli, P. Ghosh, et al., 2017. Facile synthesis of highly conducting and mesoporous carbon aerogel as platinum support for PEM fuel cells. Int. J. Hydrogen Energy. 42(16): 11110–11117.

Siroma, Z., M. Tanaka, K. Yasuda, K. Tanimoto, M. Inaba and A. Tasaka. 2007. Electrochemical corrosion of carbon materials in an aqueous acid solution. Electrochemistry. 75(2): 258–260.

Stacy, J., Y.N. Regmi, B. Leonard and M. Fan. 2017. The recent progress and future of oxygen reduction reaction catalysis: A review. Renewable Sustainable Energy Rev. 69: 401–414.

Staudenmaier, L. 1898. Verfahren zur Darstellung der Graphitsäure. Eur. J. Inorg. Chem. 31(2): 1481–1487.

Steigerwalt, E.S., G.A. Deluga and D.E. Clif. 2021. A Pt–Ru/Graphitic carbon nanofiber nanocomposite exhibiting high relative performance as a direct-methanol fuel cell anode catalyst. J. Phys. Chem. B. 105(34): 8097–8101.

Taylor, S., E. Fabbri, P. Levecque, T. Schmidt and O. Conrad. 2016. The effect of platinum loading and surface morphology on oxygen reduction activity. Electrocatalysis. 7: 287–296.

Teran-Salgado, E., D. Bahena-Uribe, P.A. Marquez-Aguilar, J.L. Reyes-Rodriguez, R. Cruz-Silva and O. Solorza-Feria. 2019. Platinum nanoparticles supported on electrochemically oxidized and exfoliated graphite for the oxygen reduction reaction. Electrochim. Acta. 298: 172–185.

Tian, Z., P. Yu, S.E. Lowe, A.G. Pandolfo, T.R. Gengenbach, K.M. Nairn, et al., 2017. Facile electrochemical approach for the production of graphite oxide with tunable chemistry. Carbon. 112: 185–191.

Tian, Y., Z. Yu, L. Cao, X. Li, C. Sun and D.-W. Wang. 2021. Graphene oxide: an emerging electromaterial for energy storage and conversion. J. Energy Chem. 55: 323–344.

Tinoco-Muñoz, C.V., J.L. Reyes-Rodríguez, D. Bahena-Uribe, M. Leyva, J.G. Cabañas-Moreno and O. Solorza-Feria. 2016. Preparation, characterization and electrochemical evaluation of Ni-Pd and Ni-Pd-Pt nanoparticles for the oxygen reduction reaction. Int. J. Hydrog. Energy. 41(48): 23272–23280.

Wu, S., J. Liu, Z. Tian, Y. Cai, Y. Ye, Q. Yuan, et al., 2015. Highly dispersed ultrafine Pt nanoparticles on reduced graphene oxide nanosheets: in situ sacrificial template synthesis and superior electrocatalytic performance for methanol oxidation. ACS Appl. Mater. Interfaces. 7(41): 22935–22940.

Xia, Y., Y. Xiong, B. Lim and S.E. Skrabalak. 2009. Shape-controlled synthesis of metal nanocrystals: simple chemistry meets complex physics? Angew. Chem. Int. Ed. 48: 60–103.

Xin, S., Y.-G. Guo and L.-J. Wan. 2012. Nanocarbon networks for advanced. Acc. Chem. Res. 45(10): 1759–1769.

Xin, L., F. Yang, S. Rasouli, Y. Qiu, Z.-F. Li, A. Uzunoglu, et al., 2016. Understanding Pt Nanoparticle anchoring on graphene supports through surface functionalization. ACS Catalysis. 6(4): 2642–2653.

Yang, Y., C. Han, B. Jiang, J. Iocozzia, C. He, D. Shi, et al., 2016. Graphene-based materials with tailored nanostructures for energy conversion and storage. Mater. Sci. Eng.: R: Rep. 102: 1–72.

Ye, M., Z. Zhang, Y. Zhao and L. Qu. 2018. Graphene platforms for smart energy generation and storage. Joule. 2(2): 245–268.

Yu, X. and S. Ye. 2007. Recent advances in activity and durability enhancement of Pt/C catalytic cathode in PEMFC. J. Power Sources. 172(1): 145–154.

Yu, S., T. Lee and T. Oh. 2022. Ag–Ni nanoparticles supported on multiwalled carbon nanotubes as a cathode electrocatalyst for direct borohydride–hydrogen peroxide fuel cells. Fuel. 123151.

Zeng, J., J.Y. Lee and W. Zhou. 2006. Activities of Pt/C catalysts prepared by low temperature chemical reduction methods. Appl. Catal. A: Gen. 308: 99–104.

Zhao J., X. Huang, H. Chang, S.H. Chan and Z. Tu. 2021. Effects of operating temperature on the carbon corrosion in a proton exchange membrane fuel cell under high current density. Energy Convers. Manage.: X. 10: 100087.

Zheng, J.-S., M.-X. Wang, X.-S. Zhang, Y.-X. Wu, P. Li, X.-G. Zhou, et al., 2008. Platinum/carbon nanofiber nanocomposite synthesized by electrophoretic deposition as electrocatalyst for oxygen reduction. J. Power Sources. 175: 211–216.

Zou, Z., K. Lin, L. Chen and J. Chang. 2012. Ultrafast synthesis and characterization of carbonated hydroxyapatite nanopowders via sonochemistry-assisted microwave process. Ultrason. Sonochem. 19(6): 1174–1179.

Index

For Product Safety Concerns and Information please contact our EU
representative GPSR@taylorandfrancis.com
Taylor & Francis Verlag GmbH, Kaufingerstraße 24, 80331 München, Germany

www.ingramcontent.com/pod-product-compliance
Lightning Source LLC
Chambersburg PA
CBHW060806220326
41598CB00022B/2550